# FIAS Interdisciplinary Science Series



For further volumes:
http://www.springer.com/series/10781

Walter Greiner
Editor

# Exciting Interdisciplinary Physics

Quarks and Gluons
Atomic Nuclei
Relativity and Cosmology
Biological Systems

Springer

*Editor*
Prof. Dr. Dr. h.c. mult. Walter Greiner
Frankfurt Institute for Advanced Studies
Goethe Universität Frankfurt
Frankfurt
Germany

ISBN 978-3-319-00046-6 ISBN 978-3-319-00047-3 (eBook)
DOI 10.1007/978-3-319-00047-3
Springer Cham Heidelberg New York Dordrecht London

Library of Congress Control Number: 2013932771



Printed on acid-free paper

Springer is part of Springer Science+Business Media (www.springer.com)

# Preface

Nuclear physics is an exciting, broadly faceted field. It spans a wide range of topics, reaching from nuclear structure physics to high-energy physics, astrophysics, and medical physics (heavy ion tumor therapy). The *Symposium on Exciting Physics* held from 13 to 20 November 2011 at Makutsi Safari Farm, South Africa has aimed to bring together specialists from many of these fields for a constructive dialogue. New developments were presented and the status of research was reviewed.

A major focus was put on nuclear structure physics, dealing with superheavy elements and with various forms of exotic nuclei: strange nuclei, very neutron rich nuclei, nuclei of antimatter. Also quantum electrodynamics of strong fields was addressed, which is linked to the occurrence of giant nuclear systems in, e.g., U+U collisions. At high energies nuclear physics joins with elementary particle physics. Various speakers addressed the theory of elementary matter at high densities and temperature, in particular the quark gluon plasma which is predicted by quantum chromodynamics (QCD) to occur in high-energy heavy ion collisions. In the field of nuclear astrophysics, the properties of neutron stars and quark stars were discussed. A topic which transcends nuclear physics and which is close to my heart was discussed in two talks: The proposed pseudocomplex extension of Einstein's General Relativity leads to the prediction that there are no black holes and that big bang cosmology has to be revised. Finally, the interdisciplinary nature of the conference was further accentuated by talks on protein folding and on magnetoreception in birds and many other animals.

Much of the research discussed at the symposium is also conducted at the Frankfurt Institute for Advanced Studies (FIAS) which is dedicated to interdisciplinary theoretical research in the natural sciences. The establishment of FIAS goes back to the initiative of the former president of Frankfurt University, Prof. Rudolf Steinberg, who invited me and the neuroscientist Prof. Wolf Singer (Max Planck Institute for Brain Research) in 2004 to lay the foundation for a new independent broad-based research institute. To make this possible we had to search for support from funding institutions and also from private sponsors. Very important in this respect was my contact with Prof. Carlo Giersch and his wife

Senator h.c. Karin Giersch who sponsored the FIAS-building (see the picture below), to my mind one of the nicest scientific institutes worldwide. More than 150 scientists (physicists, biologists, chemists, ...) are presently working at FIAS as fellows, guest scientists, doctoral students, etc.; quite a number of them are attending and contributing to this conference.

I suggested to my South-African colleague and friend Prof. Zeblon Vilakazi to hold this conference at Makutsi, which I have known from former visits. Makutsi gives us the feeling of "real Africa", not like the big towns like Johannesburg and Cape Town which represent "Europe in Africa".

The Makutsi Safari Farm was founded in the late 1970s by the German Dr. Gerhard Weber. A mathematician by training, Dr. Weber came to South Africa on a professional assignment, fell in love with the country and decided to settle there, founding the Makutsi Farm at a wonderful location, combining a beautiful landscape and fascinating wildlife. Dr. Weber organized our lecturing hall in the neighboring camp and the daily trips back and forth through the bushveld: fantastic! Our sincere thanks go to him, his family and his staff for having arranged all this so excellently.

My special thanks go also to my Capetown colleagues Dr. Richard Newman and Prof. Zeblon Vilakazi for their cordial help, and in particular to Mrs. Laura Quist, my secretary, and to Mrs. Daniela Radulescu, our 'chief financial officer' at the Frankfurt Institute für Theoretische Physik. Dr. Joachim Reinhardt in his attentive and reliable way has coordinated and prepared the presentations for the printing. They all helped to make this symposium a success!

The proceedings of the Makutsi symposium are published as the first volume of the "FIAS Interdisciplinary Science Series". This new series of books published

The Frankfurt Institute for Advanced Studies

by Springer-Verlag will comprise monographs, multi-author volumes and conference proceedings dedicated to all areas of research pursued at FIAS. I am grateful to Springer-Verlag for making this possible and in particular to Dr. Thorsten Schneider for constructive cooperation and continuous support.

Frankfurt, November 2011 Walter Greiner

# Contents

## Part II Nuclear Structure and Reactions



## Part III High-Energy Nuclear Physics

# Part I
# Superheavy Elements

# Exciting Physics: Superheavy, Superneutronic, Superstrange Nuclear Clusters

**Walter Greiner**

**Abstract** The extension of the periodic system into various new areas is investigated. Experiments for the synthesis of superheavy elements and the predictions of magic numbers are reviewed. Investigations on hypernuclei and the possible production of antimatter-clusters in heavy-ion collisions are reported. Various versions of the meson field theory serve as effective field theories at the basis of modern nuclear structure and suggest structure in the vacuum which might be important for the production of hyper- and antimatter.

## 1 Introduction

When we discussed various topics of Exciting Physics at one of the Program Advisory Meetings at Dubna the idea of a symposium on these topics arose and Prof. Zeblon Vilakazi suggested to hold it in South Africa. I supported this idea and suggested Makutsi as an exciting place for Exciting Physics. We will see and hear more about this wonderful place from Dr. Gerhard Weber and his family as our symposium progresses. Dr. Richard Newman from Itemba will be helpful on our side.

Exciting Physics is a broad field, covering not only areas of nuclear and elementary matter physics, but also topics in field theory, biology (magneto-reception of birds, fishes, animals in general), chemistry, nano-clusters and their structure, channeling of charged particles in bent crystals as new undulators leading perhaps to lasers in the MeV range and beyond. We are not able to discuss this all in detail; some of the topics can only be touched, some have to be left for future meetings of this kind.

Let me begin with the progress in exotic nuclear matter clusters. There are fundamental questions in science, like e.g. "how did life emerge" or "how does our brain

---

W. Greiner (✉)
Frankfurt Institute for Advanced Studies, J. W. Goethe–Universität,
60438 Frankfurt am Main, Germany
e-mail: greiner@fias.uni-frankfurt.de

W. Greiner (ed.), *Exciting Interdisciplinary Physics*,
FIAS Interdisciplinary Science Series, DOI: 10.1007/978-3-319-00047-3_1,

work" and others. However, the most fundamental of those questions is "how did the world originate?". The material world has to exist before life and thinking can develop. Of particular importance are the substances themselves, i.e. the particles the elements are made of (baryons, mesons, quarks, gluons), i.e. elementary matter. The vacuum and its structure is closely related to that. We want to report on these questions, beginning with the discussion of modern issues in nuclear physics.

The elements existing in nature are ordered according to their atomic (chemical) properties in the *periodic system*, which was developed by Dmitry Mendeleev and Lothar Meyer. The heaviest element of natural origin is uranium. Its nucleus is composed of $Z = 92$ protons and a certain number of neutrons ($N = 128–150$). They are called the different uranium isotopes. The transuranium elements reach from neptunium ($Z = 93$) via californium ($Z = 98$) and fermium ($Z = 100$) up to lawrencium ($Z = 103$). The heavier the elements are, the larger are their radii and their number of protons. Thus, the Coulomb repulsion in their interior increases, and they undergo spontaneous fission. In other words: the transuranium elements become more unstable as they get bigger. In the late sixties, the dream of the superheavy elements arose. Theoretical nuclear physicists around S. G. Nilsson (Lund) and from the Frankfurt school [1–4] predicted that so-called closed proton and neutron shells should counteract the repelling Coulomb forces. Atomic nuclei with these special *"magic" proton and neutron numbers* and their neighbours could again be rather stable. These magic proton ($Z$) and neutron ($N$) numbers were thought to be $Z = 114$ and $N = 184$ or 196. Typical predictions of their life-times varied between seconds and many thousand years. Figure 1 summarizes the expectations at the early time. One can see the islands of superheavy elements around $Z = 114$, $N = 184$ and 196, respectively, and the one around $Z = 164$, $N = 318$.

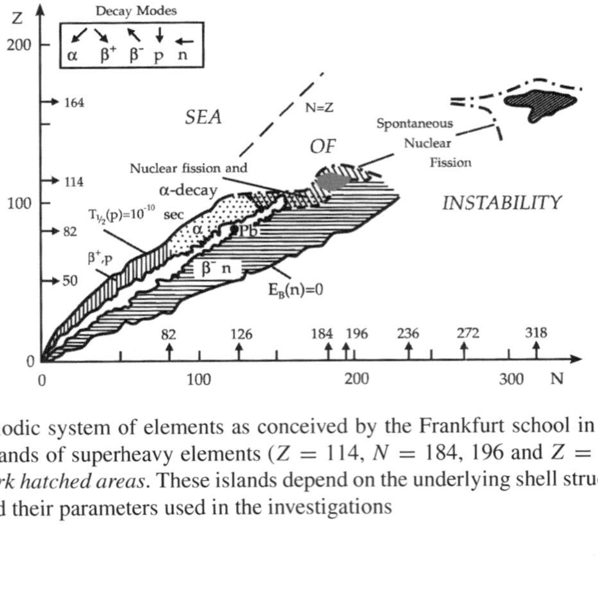


**Fig. 1** The periodic system of elements as conceived by the Frankfurt school in the late sixties. The possible islands of superheavy elements ($Z = 114$, $N = 184$, 196 and $Z = 164$, $N = 318$) are shown as *dark hatched areas*. These islands depend on the underlying shell structure, i.e. on the shell models and their parameters used in the investigations

## 2 Cold Valleys in the Potential

The important question was, and still is, how to produce these superheavy nuclei. There were many attempts, but only little progress was made. It was not until the middle of the seventies that the Frankfurt school of theoretical physics together with foreign guests (R. K. Gupta (India), A. Sandulescu (Romania)) [5, 6] theoretically understood and substantiated the concept of bombarding of double magic lead nuclei with suitable projectiles, which had been proposed intuitively by the Russian nuclear physicist Y. Oganessian [7]. The two-center shell model, which is essential for the description of fission, fusion and nuclear molecules, was developed in 1969–1972 by W. Greiner and his students U. Mosel [1, 2] and J. Maruhn [8, 9]. It showed that the shell structure of the two final fragments was visible far beyond the barrier into the fusioning nucleus. The collective potential energy surfaces of heavy nuclei, which were calculated utilizing the Strutinsky renormalization in the framework of the two-center shell model, exhibit pronounced valleys.

These valleys provide promising doorways to the fusion of superheavy nuclei for certain projectile-target combinations (Fig. 2). If projectile and target approach each other through those "*cold*" *valleys* [5, 6, 10], they get only minimally excited and the barrier, which has to be overcome (fusion barrier) is lowest (as compared to the neighbouring projectile-target combinations). In this way, the correct projectile- and target-combinations for fusion were predicted. Indeed, Sigurd Hofmann and Gottfried Münzenberg and their group at GSI [11] have followed this approach. With the help of the SHIP mass-separator (which had been proposed by H. Ewald–G. Münzenberg was his Ph.D. student at Giessen University) and the position sensitive detectors, which were especially developed by S. Hofmann (upon suggestion of his PhD-advisor E. Kankeleit at the Technical University Darmstadt), they produced the pre-superheavy elements $Z = 106, 107, \ldots 112$, each of them with the theoretically predicted projectile-target combinations, and only with these. Everything else failed. This is an impressive success, which crowned the laborious construction

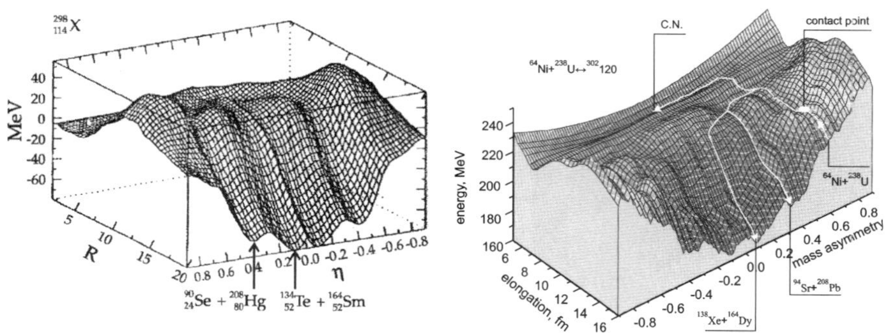


**Fig. 2** *Left* The collective potential energy surface of $^{184}114$ calculated within the two center shell model by J. Maruhn et al. shows clearly the cold valleys, which reach up to the barrier and beyond. *Right* Collective potential energy surface of the element $^{302}120$

work of many years. The last but one example of this success, was the discovery of element 112 and its long $\alpha$-decay chain. The Dubna group produced the six isotopes of $Z = 113 \ldots 118$ by bombarding $^{244}$Pu–$^{248}$Cf with $^{48}$Ca [12]. These are also nearly cold valley reactions (in this case due to the combination of a spherical and a deformed nucleus), as predicted by Gupta, Sandulescu and Greiner in 1977 [5, 6]. The production of the element $Z = 117$ by the $^{48}$Ca+$^{249}$Bk reaction came about through the help of the Vanderbilt (J. Hamilton)—Oak Ridge group providing the short-living and expensive Berkelium-target. There exist also cold valleys for which both fragments are deformed [10], or have non-axial orientations [13], but these have not yet been verified experimentally. The cold valleys also play an important role in nuclear fission giving rise to asymmetric and superasymmetric [14, 15] fission and to cluster radioactivity [16].

## 3 Shell Structure in the Superheavy Region

Decay properties and stability of the heaviest nuclei with $Z \leq 132$ were recently studied within the macro-microscopical approach for nuclear ground-state masses and phenomenological relations for the half-lives with respect to $\alpha$-decay, $\beta$-decay and spontaneous fission [17]. It was found (see Fig. 3) that the $\beta$-stable isotopes $^{291}$Cn and $^{293}$Cn with a half-life of about 100 years are the longest-living superheavy nuclei located at the first island of stability. Remember that such lifetime estimates depend also on the collective masses. This is a challenging task in itself! Because of their short half-lives he search for superheavy nuclei in nature may be performed only in cosmic rays. Under terrestrial conditions a measurable amount of superheavies is unlikely to exist. Note, that fusion reactions lead to the proton-rich nuclei along the proton drip line. The heaviest synthesized nuclei with $Z = 118$ are situated already quite close to the border of 1 μs half-life. It means that the synthesis and detection of nuclei with $Z > 120$ produced in fusion reactions may be difficult at existing experimental facilities due to their short half-lives (shorter than 1 μs). This prediction should be taken into account when planning new experiments and experimental setups. One may see as well that the nearest neutron-rich isotopes of superheavy elements with $111 \leq Z \leq 115$ to those synthesized recently in Dubna with $^{48}$Ca-induced fusion reactions are found to be $\beta^+$-decaying. This fact may significantly complicate their experimental identification. However, the existence of this area of $\beta^+$-decaying nuclei gives us a possibility to reach the center of the island of stability. One way to produce $^{291}$Cn is the triple $\beta^+$ (or EC) decay of $^{291}$115 which in turn could be, for example, synthesized after $\alpha$-decay of $^{295}$117 in the reaction $^{48}$Ca+$^{249}$Bk $\rightarrow ^{295}$ 117+2n. The proposed method of reaching the island of stability hopefully may be realized in future with the progress of experimental techniques. We found as well the second area of stability of superheavy nuclei (still with shorter half-lives) situated in the region of $Z \sim 124$ and $N \sim 198$. It is separated from the "continent" by the "gulf" of short-living nuclei with half-lives shorted than 1 μs.

Studies of the shell structure of superheavy elements in the framework of the meson field theory and the Skyrme-Hartree-Fock approach have recently shown that the magic shells in the superheavy region are very isotope dependent [18–28]. Additionally, there is a strong dependence on the parameter set and the model. Some forces hardly show any shell structure, while others predict the magic numbers $Z =$ 114, 120 and 126. Using the heaviest known even-even nucleus Hassium $^{264}_{156}108$ as a criterium to find the best parameter sets in each model, it turns out that PL-40 and SkI4 produce best its binding energy. However, these two forces yield conflicting predictions for the magic numbers in the superheavy region: SkI4 predicts $Z = 114$, 120 and PL-40 predicts $Z = 120$. Most interesting, $Z = 120$ *as magic proton number seems to be as probable as* $Z = 114$. Calculations of deformed systems within the two models [18–20] reveal again different predictions: Though both parametrizations predict $N = 162$ as the deformed neutron-shell closure, the deformed proton-shell closures are $Z = 108$ (SkI4) and $Z = 104$ (PL-40) (see Fig. 4). Calculations of the potential energy surfaces [18–20] show single humped barriers; their heights and widths strongly depend on the predicted magic number. Furthermore, recent investigations in a chirally symmetric mean-field theory (see also below) result also in the prediction of these two magic numbers [21–28]. The corresponding magic neutron numbers are predicted to be $N = 172$ and to a lesser extent $N = 184$. Thus, this region provides an open field of research. The charge distribution of the $Z = 120$, $N = 184$ nucleus indicates a hollow inside. This may suggest that it might be essentially a fullerene consisting of 60 $\alpha$-particles and one binding neutron per alpha. The cold valleys in the collective potential energy surface are basic for understanding this exciting area of nuclear physics! It is a master example for understanding the *structure of elementary matter*, which is so important for other fields, especially astrophysics, but even more so for enriching our "Weltbild", i.e. the status of our understanding of the world around us.

The investigation of the neutron drip line by extended Hartree-Fock+BCS calculations led to a great surprise: extremely neutron rich nuclei along the magic neutron numbers become stable against one-and two-neutron separation [29, 30], see Fig. 5. The standard production of superheavy nuclei by fusing two smaller stable nuclei leads automatically to neutron poor isotopes near the proton drip line (therefore the lifetime of the produced superheavies is so small). Only a few superheavy atoms are produced this way. This leads us directly to the question of how superheavies with larger neutron numbers (and therefore having larger lifetimes: up to thousands of years) can be produced. One also wants to produce such long–living superheavies in macroscopic quantities (milligrams, grams,. . .) so that they can eventually be used medically and technically. This can be done either by double (or multiple) underground atomic bomb explosions or by pulsed reactors with very high neutron flux ($\simeq 10^{21}$ neutrons/ sec cm$^2$) see Figs. 3 and 6 [31, 32].

The idea to take advantage of the shell effects for the production of SH nuclei in the multi-nucleon transfer processes of low-energy heavy ion collisions was proposed in [33]. The shell effects are known to play an important role in fusion of heavy ions with actinide targets driving the nuclear system via the quasi-fission channels into the deep lead and tin valleys and, thus, decreasing the fusion probability.

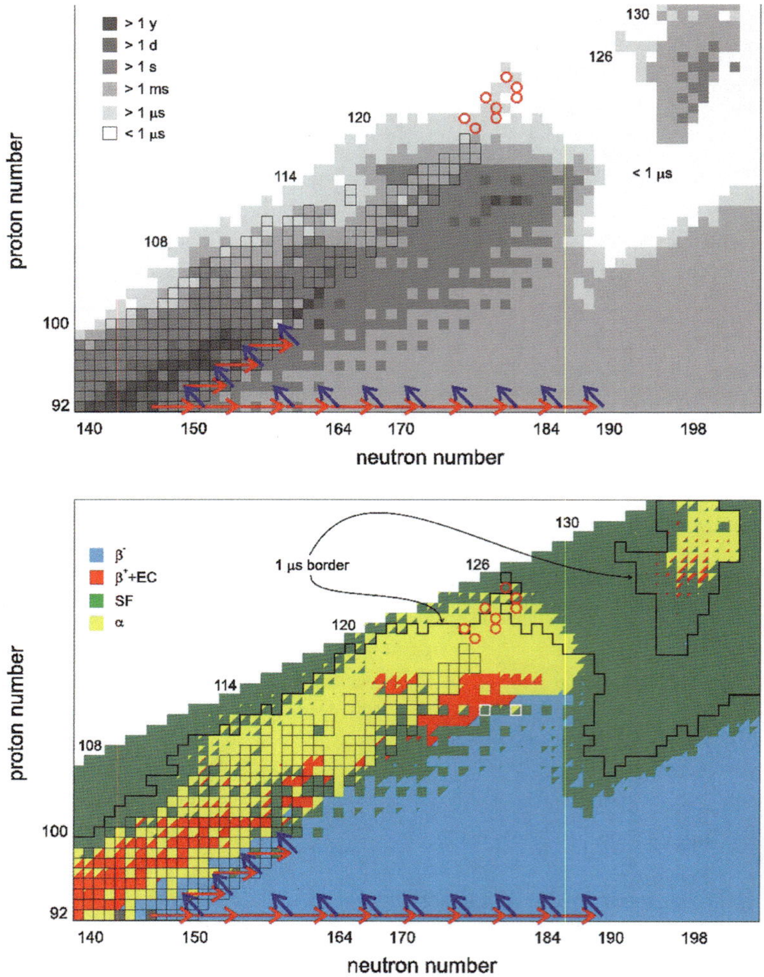


**Fig. 3** Half-lives (*top*) and decay modes (*bottom*) of nuclei in the *upper part* of the nuclear map. The *circles* show the nuclei with $Z = 119 - 124$, which may be synthesized in the $3n$ channel of fusion reactions $^{50}$Ti $+^{249}$Bk, $^{249}$Cf and $^{54}$Cr, $^{58}$Fr $+^{248}$Cm, $^{249}$Bk, $^{249}$Cf. The bounded cells correspond to the experimentally known nuclei. The bounded nuclei with the *white color* border are the most stable Copernicium isotopes $^{291}$Cn and $^{293}$Cn. Schematic view of slow (terminated at the short-lived fission Fermium isotopes) and fast neutron capture processes with subsequent $\beta^-$ decays are shown by the *arrows*

On the contrary, in the transfer reactions the same effects may lead to enhanced yield of SH nuclei. It may occur if one of the heavy colliding nuclei, say $^{238}$U, gives away nucleons approaching to double magic $^{208}$Pb nucleus, whereas another one, say $^{248}$Cm, accepts these nucleons becoming superheavy in the exit channel. This is the so-called "inverse" (anti-symmetrizing) quasi-fission process. The potential energy surface of the giant nuclear system formed in the collision of $^{238}$U and $^{248}$Cm nuclei

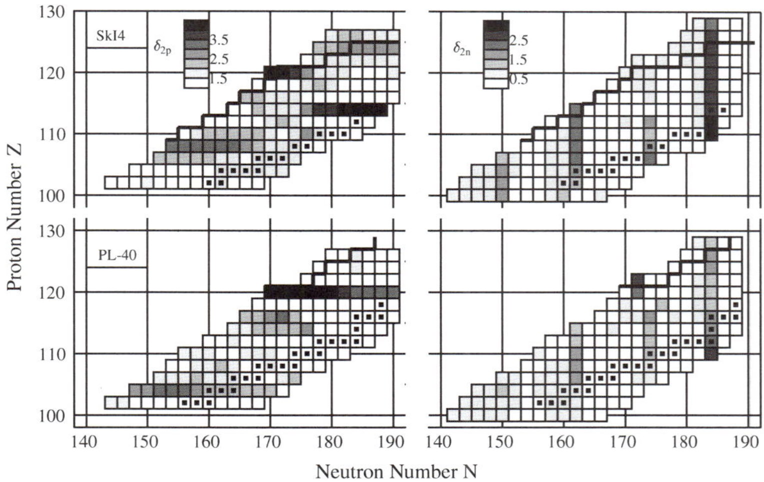


**Fig. 4** *Grey scale* plots of proton gaps (*left column*) and neutron gaps (*right column*) in the $N - Z$ plane for deformed calculations with the forces SkI4 and PL- 40. Besides the spherical shell closures one can see the deformed shell closures for protons at $Z = 104$ (PL-40) and $Z = 108$ (SkI4) and the ones for neutrons at $N = 162$ for both forces

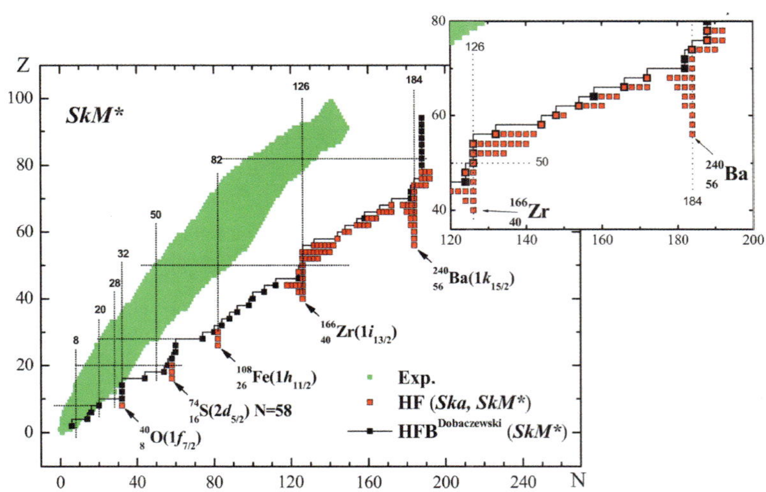


**Fig. 5** Fragment of the neutron drip line and elements (*red squares*) that are stable against one neutron emission [29, 30]. One recognizes the formation of stability peninsulas along neutron magic numbers

is shown in Fig. 7. In low-energy damped collisions of heavy ions just the potential energy surface regulates to a great extent the evolution of the nuclear system driving it to the minimal values of potential energy in the multidimensional space of collective variables. In the course of nucleon exchange the most probable path of the nuclear system formed by $^{238}$U and $^{248}$Cm lies along the line of stability with formation

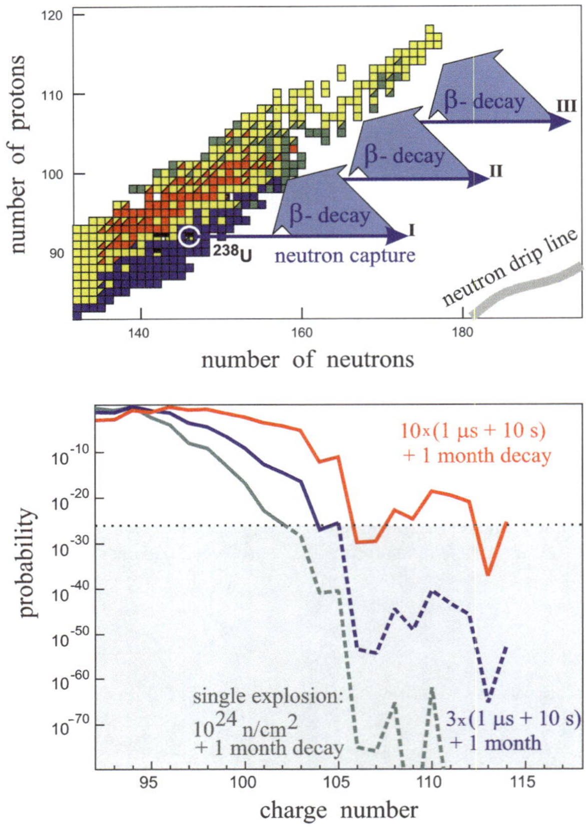


**Fig. 6** Schematic picture for multiple neutron irradiation of initial $^{238}$U material (*top*) and probability for formation of heavy nuclei (*bottom*) in bomb explosion processes (one, three and ten subsequent explosions). The *dotted line* denotes the level of few atoms

of SH nuclei which have many more neutrons as compared with those produced in the "cold" and "hot" fusion reactions. Due to fluctuations even more neutron rich isotopes of SH nuclei may be formed in such transfer reactions. The calculated cross sections for formation of primary fragments in low-energy collisions of $^{238}$U with $^{248}$Cm target are shown in Fig. 7 by the contour lines in logarithmic scale. As can be seen, the superheavy nuclei located very close to the center of the island of stability may be produced in this reaction with rather high cross section of one microbarn. This region of the nuclear map cannot be reached in any fusion reaction with stable projectiles and long-living targets. Of course, the question arises whether these excited superheavy primary fragments may survive. The calculated cross sections for formation of neutron-rich SH nuclei in low-energy collisions of $^{238}$U with $^{248}$Cm target are shown in Fig. 8 for final surviving fragments. These SH nuclei are located very close to the center of the island of stability and cannot be produced in any fusion reactions with stable projectiles and long-living targets. These are the shell effects

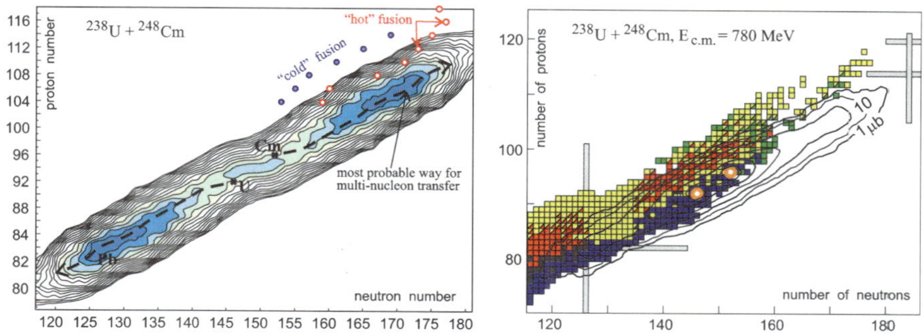


**Fig. 7** Landscape of potential energy surface (*on the left*) and the cross sections (microbarns, logarithmic scale) for production of primary fragments (*on the right*) in collision of $^{238}$U with $^{248}$Cm

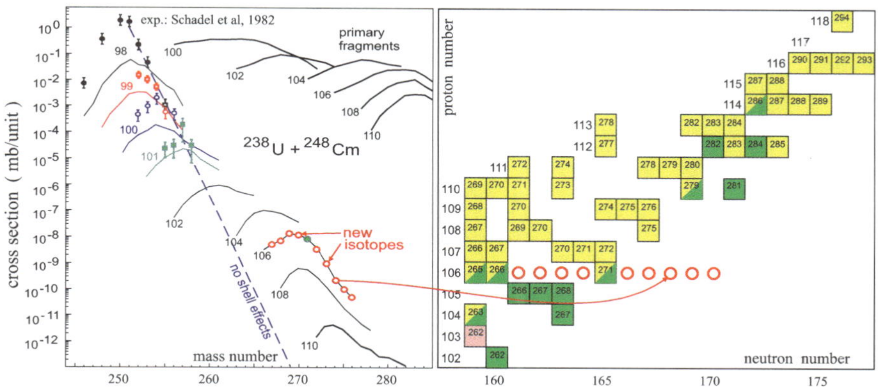


**Fig. 8** Yield of primary and surviving isotopes of SH nuclei produced in collisions of $^{238}$U with $^{248}$Cm at 800 MeV center-of-mass energy. The *dashed line* shows the expected locus of transfer reaction cross sections without the shell effects

which give us a significant gain as compared to a monotonous exponential decrease of the cross sections with increasing number of transferred nucleons.

We found that the nuclear system consisting of two very heavy nuclei may hold in contact rather long in some cases [34]. During this time the giant nuclear system moves over the multidimensional potential energy surface with almost zero kinetic energy (result of large nuclear viscosity). The reaction time distribution is shown in Fig. 9 for the $^{238}$U+$^{248}$Cm collision. With increase of the energy loss and mass transfer the reaction time becomes longer and its distribution becomes more narrow. The lifetime of a giant composite system of more than $10^{-20}$ s is quite enough to expect the positron line structure emerging on top of the dynamical positron spectrum due to spontaneous $e^+e^-$ production from the supercritical electric fields as a fundamental QED process ("decay of the vacuum", Fig. 10) [35]. Formation of the background positrons in these reactions forces one to find some additional trigger

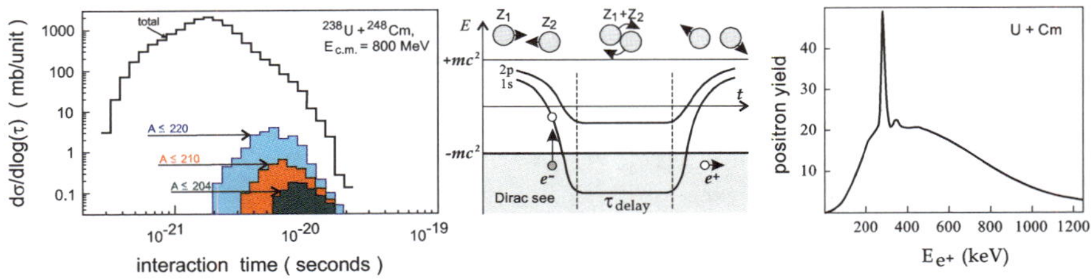


**Fig. 9** Reaction time distribution and spontaneous positron formation in supercritical electric field of long-lived giant quasi-atom formed in the collision of U+Cm. The *sharp positrons* line shown in the *right* figure would only appear if the giant nuclear system lives infinitely long. In reality one deals with time distributions as shown in the first figure. In this latter case the positron spectrum looks like in Fig. 10 [36]

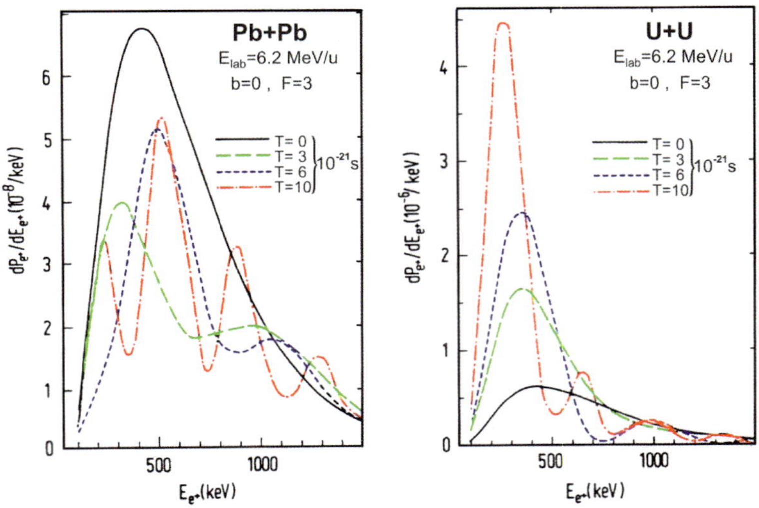


**Fig. 10** Positron spectra in central Pb+Pb and U+U collisions at $E_{\text{lab}}/A = 6.2\,\text{MeV}$ assuming various nuclear delay times. The subcritical system displays destructive interference while in the supercritical system spontaneous positron production leads to the build-up of a peak in the spectrum [35]

for the longest events. For the considered case of $^{238}$U+$^{248}$Cm collision at 800 MeV center-of-mass energy, the detection of the surviving nuclei in the lead region at the laboratory angles of about 25° and at the low-energy border of their spectrum (around 1000 MeV for Pb) could be a real trigger for longest reaction time.

It was recently found that low-energy collision of actinides may lead to quite an exotic process of three-body clusterization, the so-called true ternary quasifission, leading to formation of two lead-like fragments and some heavy third particle in between [36]. This type of processes is quite possible because the shell effects significantly reduce the potential energy of the three-cluster configurations with two

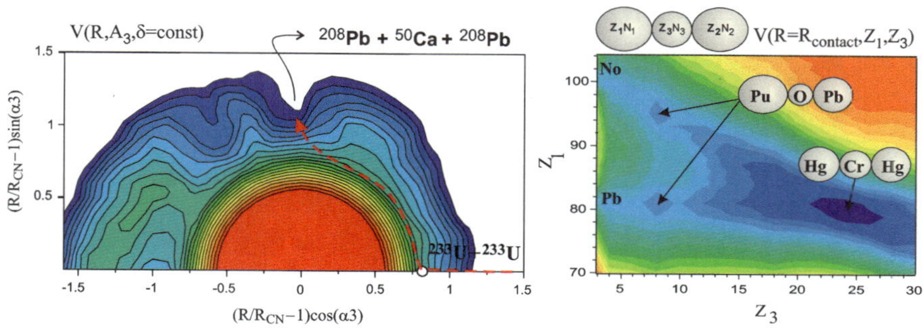


**Fig. 11** (*Left panel*) Potential energy (macroscopic plus shell corrections) for ternary quasifission of the giant nuclear system formed in $^{233}$U+$^{233}$U collision, depending on elongation and mass of third fragment ($\alpha_3 = \pi \cdot A_3/100$, where $A_3$ is the mass number of the third fragment). The probable path for the triple fission is indicated. (*Right panel*) Landscape of potential energy of three-body contact configurations of giant nuclear system formed in the collision of $^{238}$U+$^{238}$U

strongly bound lead-like fragments. In Fig. 11 the landscape of the potential energy surface is shown for a three-body clusterization of the nuclear system formed in the collision of U+U. It is seen (left panel) that the shell correction at contact configurations yields a very deep minimum for the "lead-calcium-lead" ($A_3 = 50$) clusterization. In the right panel the potential energy is shown as a function of three variables, $Z_1$, $Z_3$ (charges of the first and third fragments) and system elongation $R$ (minimized over the neutron numbers) at fixed (equal) deformations of the fragments being in contact. As can be seen, the giant nuclear system, consisting of two touching uranium nuclei, may split into the two-body exit channel with formation of a lead-like fragment and a complementary superheavy nucleus (the so-called anti-symmetrizing quasifission process which may lead to an enhanced yield of SH nuclei in multi-nucleon transfer reactions [33]). Beside the two-body Pb-No clusterization and the shallow local three-body minimum with formation of a light intermediate oxygen-like cluster, the potential energy has the very deep minimum corresponding to the Pb-Ca-Pb-like configuration (or Hg-Cr-Hg) caused by the $N = 126$ and $Z = 82$ nuclear shells. The extreme clustering process of formation of two lead-like doubly magic fragments in collisions of actinide nuclei is a very interesting subject for experimental study. Such measurements, in our opinion, are not too difficult. It is sufficient to detect two coincident lead-like ejectiles (or one lead-like and one calcium-like fragments) in U+U collisions to conclude unambiguously about the ternary fission of the giant nuclear system.

Calculations of half-lives of superheavy nuclei (SH) show an unexpected result: for some of them heavy particle radioactivity (HPR) dominates over alpha decay—the main decay mode of the majority of recently discovered SHs. The result is important for theory and future experiments producing heavier SHs with a substantial amount of funding. The standard identification technique by alpha decay chains will be impossible for these cases. HPR had been predicted in 1980 by Sandulescu, Poenaru and W. Greiner [37] four years before the first experiment. The daughter is mainly the doubly magic $^{208}$Pb. Poenaru, Gherghescu and Greiner changed the concept of HPR to allow emitted particles with $Z_e > 28$ from parents with $Z > 110$ (daughter around $^{208}$Pb) and found a trend toward shorter half-lives and larger branching ratios relative to alpha decay for heavier SHs.

## 4 Extension of the Periodic System into the Field of Hyper- and Antimatter

Nuclei that are found in nature consist of nucleons (protons and neutrons) which themselves are made of u (up) and d (down) quarks. However, there also exist s (strange) quarks and even heavier flavours, called charm, bottom, top. The latter has been discovered in 1995. Let us stick to the s quarks. They are found in the "strange" relatives of the nucleons, the so-called hyperons ($\Lambda$, $\Sigma$, $\Xi$, $\Omega^-$). The $\Lambda$-particle, e.g., consists of one u, d, and s quark, the $\Xi$-particle even of one u or d and two s quarks, while the $\Omega^-$ (sss) contains strange quarks only.

If such a hyperon is taken up by a nucleus, a *hyper-nucleus* is created. Hypernuclei with one hyperon have been known for 20 years [38]. Several years ago, Carsten Greiner, Jürgen Schaffner and Horst Stöcker [39] theoretically investigated nuclei with many hyperons, *hypermatter*, and found that the binding energy per baryon of strange matter is in many cases even higher than that of ordinary matter (composed only of u and d quarks). This leads to the idea of extending the periodic system of elements in the direction of strangeness.

One can also ask for the possibility of building atomic nuclei out of *antimatter*, that means searching, e.g., for anti-helium [40], anti-carbon, anti-oxygen. Figure 12 depicts this idea. Due to charge conjugatiton symmetry, antinuclei should have the same magic numbers and the same spectra as ordinary nuclei. However, as soon as they get in touch with ordinary matter, they annihilate with it and the system explodes. Now the important question arises, how these strange matter and antimatter clusters can be produced. First, one thinks of collisions of heavy nuclei, e.g., lead on lead, at high energies (energy per nucleon $\geq$200 GeV). Calculations with the URQMD-model of the Frankfurt school show that through *nuclear shock waves* [41] nuclear matter gets compressed to 5 ... 10 times of its usual value, $\rho_0 \simeq 0.17\,\mathrm{fm}^{-3}$, and heated up to temperatures of $kT \simeq 200\,\mathrm{MeV}$. As a consequence, about 10,000 pions, 100 $\Lambda$'s, 40 $\Sigma$'s and $\Theta$'s and about as many antiprotons and many other particles are created in a single collision. It seems conceivable that it is possible in such a scenario for some $\Lambda$'s to get captured in a nuclear cluster. This happens indeed rather frequently for one or two $\Lambda$-particles; however, more of them get built into nuclei with rapidly decreasing probability only. This is due to the low probability for finding the right conditions for such a capture in the phase space of the particles: the numerous particles travel with all possible momenta (velocities) in all directions. The chances for hyperons and antibaryons to meet get rapidly worse with increasing

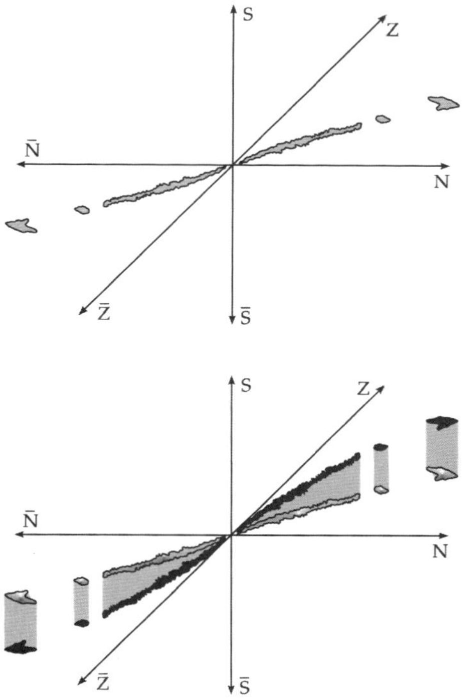


**Fig. 12** The extension of the periodic system into the sectors of strangeness $S, \overline{S}$ and antimatter $\overline{ZN}$. The stable valley winds out of the known proton (Z) and neutron (N) plane into the $S$ and $\overline{S}$ sector, respectively. The same can be observed for the antimatter sector. In the *upper part* of the figure only the stable valley in the usual proton (Z) and neutron (N) plane is plotted, however, extended into the sector of antiprotons and antineutrons. In the *second part* of the figure it has been indicated, how the stable valley winds out of the Z-N-plane into the strangeness sector

number. In order to produce multi-$\Lambda$-nuclei and antimatter nuclei, one has to look for a different source.

In the framework of the meson field theory, the energy spectrum of baryons has a peculiar structure, depicted in upper part of Fig. 13. It consists of an upper and a lower continuum, as it is known for electrons (see, e.g., Ref. [42, 43]). Of special interest in the case of the baryon spectrum is the potential well, built of the scalar and the vector potential, which rises from the lower continuum. Naftali Auerbach and collaborators noticed this first [44]. It is known since P. A. M. Dirac (1930) that the negative energy states of the lower continuum have to be occupied by particles (electrons or, in our case, baryons). Otherwise our world would be unstable, because the "ordinary" particles are found in the upper states can decay through the emission of photons into lower lying states. However, if the "underworld" is occupied, the Pauli-principle will prevent this decay. Holes in the occupied "underworld" (Dirac sea) are antiparticles.

The occupied states of this underworld, including up to 40,000 occupied bound states of the lower potential well, represent the *vacuum*. The peculiarity of this strongly correlated vacuum structure in the region of atomic nuclei is that—depending on the size of the nucleus—more than 20,000 up to 40,000 (occupied) bound nucleon states contribute to this polarization effect. Obviously, we are dealing here with a *highly correlated vacuum*. Pronounced shell structure can be recognized

[45, 46]. Holes in these states have to be interpreted as bound antinucleons (antiprotons, antineutrons). If the primary nuclear density rises due to compression, the lower well increases while the upper decreases and is soon converted into a repulsive barrier. This compression of nuclear matter can only be carried out in relativistic nucleus-nucleus collision with the help of shock waves, which have been proposed by the Frankfurt school (see W. Scheid et al., Ref. [47]) and which have since then been confirmed extensively (see, e.g., Ref. [48]). These *nuclear shock waves* are accompanied by heating of the nuclear matter. Indeed, density and temperature are intimately coupled in terms of the hydrodynamic Rankine–Hugoniot equations. Heating as well as the violent dynamics cause the creation of many holes in the very deep (measured from $-M_B c^2$) vacuum well. These numerous bound holes resemble antimatter clusters which are bound in the medium; their wave functions have large overlap with antimatter clusters. When the primary matter density decreases during the expansion stage of the heavy-ion collision, the potential wells, in particular the lower one, disappear.

The bound antinucleons are then pulled down into the (lower) continuum. In this way antimatter clusters may be set free. Of course, a large part of the antimatter will annihilate on ordinary matter present in the course of the expansion. However, it is important that *this mechanism for the production of antimatter clusters out of the highly correlated vacuum does not proceed via the phase space*. The required coalescence of many antiparticles in phase space suppresses the production of antimatter-clusters, while it is favored by the direct production out of the highly correlated vacuum. In a certain sense, the highly correlated vacuum is a kind of cluster vacuum (vacuum with cluster structure). The shell structure of the vacuum levels (see Fig. 13) supports this interpretation. Figure 14 illustrates this idea. Recently the STAR Collaboration at RHIC observed Anti-$^4$He with production rate in excess of coalescent nucleosynthesis production [40]. My colleagues Reinhard Stock and Thorsten Kollegger played a leading role in this pioneering experiment. Prof. Stock will report about it at this symposium. Figures 15 and 16 show their extremely exciting results.

The mechanism is similar for the production of multi-hyper nuclei ($\Lambda$, $\Sigma$, $\Theta$, $\Omega^-$). Meson field theory predicts also for the $\Lambda$ energy spectrum at finite primary nucleon density the existence of upper and lower wells. The lower well belongs to the vacuum and is fully occupied by $\Lambda$'s.

Dynamics and temperature then induce transitions ($\Lambda\overline{\Lambda}$ creation) and deposit many $\Lambda$'s in the upper well. These numerous bound $\Lambda$'s are sitting close to the primary baryons: in a certain sense a giant multi-$\Lambda$ hypernucleus has been created. When the system disintegrates (expansion stage) the $\Lambda$'s distribute over the nucleon clusters (which are most abundant in peripheral collisions). In this way multi-$\Lambda$ hypernuclei can be formed. Of course this vision has to be worked out and probably refined in many respects. This requires much more thorough investigations in future. It is particularly important to gain more experimental information on the properties of the lower well by $(e, p)$ or $(e, \overline{p})$ and also $(\overline{p}_c p_b, p_c \overline{p}_b)$ reactions at high energy ($\overline{p}_c$ denotes an incident antiproton from the continuum, $p_b$ is a proton in a bound state; for the reaction products the situation is just the opposite). Also the reaction (p, p' d), (p, p' $^3$He), (p, p' $^4$He) and others of similar type need to be investigated in

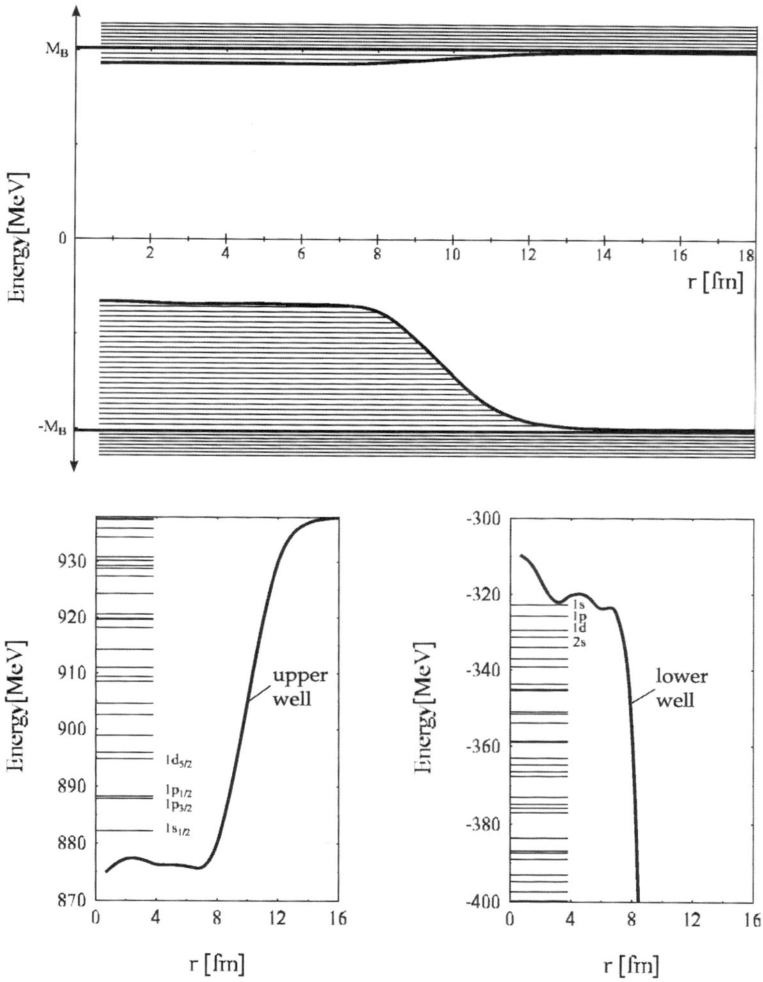


**Fig. 13** Baryon spectrum in a nucleus. Below the positive energy continuum exists the potential well of real nucleons. It has a depth of 50–60 MeV and shows the correct shell structure. The shell model of nuclei is realized here. However, from the negative continuum another potential well arises, in which about 40,000 bound particles are found, belonging to the vacuum. A part of the shell structure of the *upper well* and the *lower* (vacuum) well is depicted in the *lower figures*

this context. The systematic studies of antiproton scattering on nuclei can contribute to clarify these questions. Various effective theories, e.g. of the Walecka-type on the one side and theories with chiral invariance on the other side, have been constructed to describe strongly interacting dense matter. It is important to note that they seem to give different strengths of the potential wells and also different dependence on the baryon density.

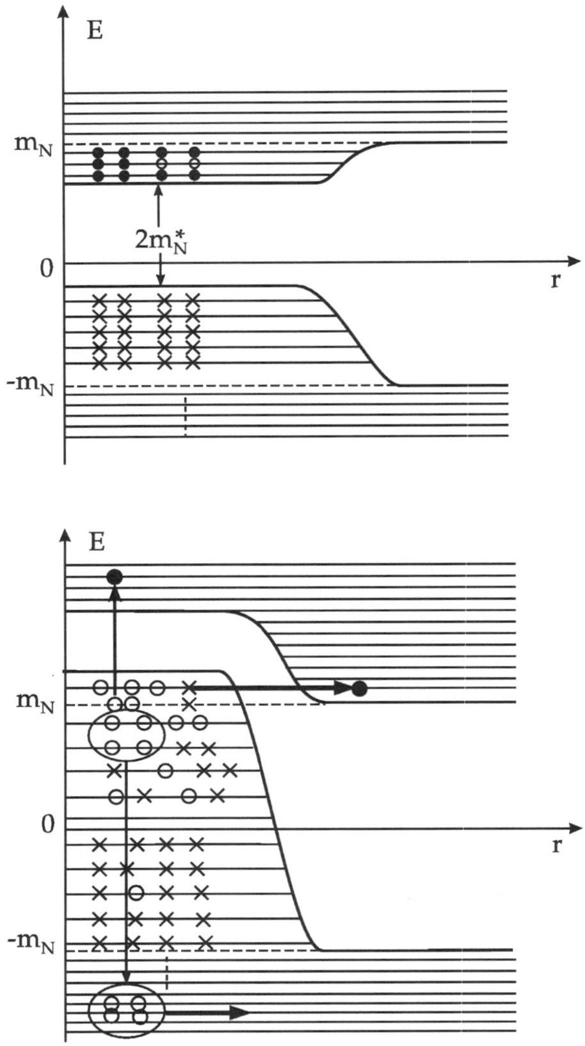


**Fig. 14** Due to the high temperature and the violent dynamics, many bound holes (antinucleon clusters) are created in the highly correlated vacuum, which can be set free during the expansion stage into the lower continuum. In this way, antimatter clusters can be produced directly from the vacuum. The *horizontal arrow* in the *lower part* of the figure denotes the spontaneous creation of baryon-antibaryon pairs, while the antibaryons occupy bound states in the lower potential well. Such a situation, where the lower potential well reaches into the upper continuum, is called supercritical. Four of the bound holes states (bound antinucleons) are encircled to illustrate a "quasi-antihelium" formed. It may be set free (driven into the lower continuum) by the violent nuclear dynamics

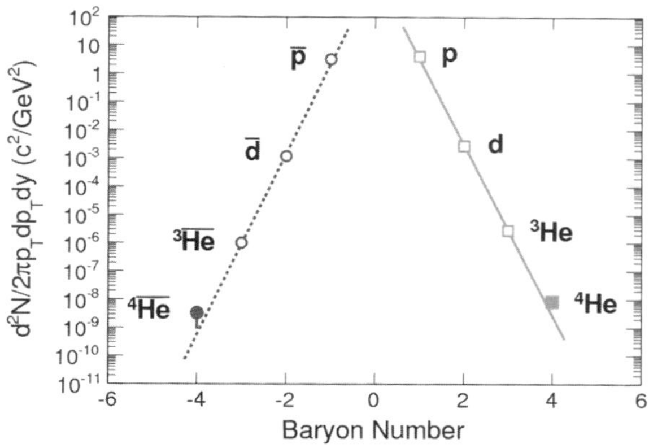


**Fig. 15** Light nucleus/antinucleus production in central Au+Au collisions at $\sqrt{s} = 200\,\text{GeV}$ in STAR [40]

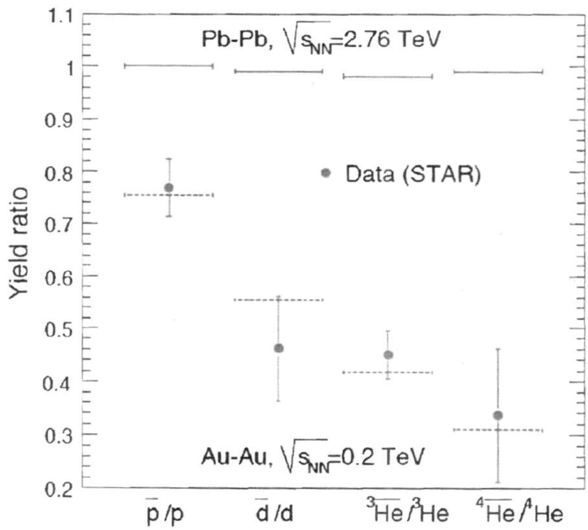


**Fig. 16** Antinucleus to nucleus ratios at top RHIC energy. We refer to [40] for further description of this encouraging experiment

According to chirally symmetric meson field theories, the antimatter-cluster- production and multi-hypermatter-cluster production out of the highly correlated vacuum takes place at approximately the same heavy-ion energies as compared to the predictions of the Dürr-Teller-Walecka type meson field theories. This in itself is a most interesting, quasi-fundamental question to be clarified. In the future, the question of the nucleonic substructure (form factors, quarks, gluons) and its influence on the highly correlated vacuum structure has to be studied. The nucleons are possibly strongly modified in the correlated vacuum: the $\Delta$ resonance correlations are probably important. Is this highly correlated vacuum state, especially during the compression, a preliminary stage to the quark-gluon cluster plasma? To which extent is it similar or perhaps even identical with it? This still has to be clarified!

## References


1. U. Mosel, W. Greiner, Z. Phys. 217, 256 (1968), see also Memorandum zur Errichtung eines gemeinsamen Ausbildungszentrums für Kernphysik Hessischer Hochschulen, this was the initiative for building GSI
2. U. Mosel, W. Greiner, Z. Phys. 222, 261 (1969)
3. S.G. Nilsson, J.R. Nix, A. Sobiczewski, Z. Szymanski, S. Wycech, C. Gustafson, P. Möller, Nucl. Phys. A **115**, 545 (1968)
4. S.G. Nilsson, S.G. Thompson, C.F. Tsang, Phys. Lett. B **28**(7), 458 (1969)
5. A. Săndulescu, R.K. Gupta, W. Scheid, W. Greiner, Phys. Lett. B **60**(3), 225 (1976)
6. R.K. Gupta, A. Săndulescu, W. Greiner, Phys. Lett. B **64**, 257 (1976)
7. Yu. Ts, Oganessian, A.G. Demin, A.S. Iljinov, S.P. Tretyakova, A.A. Pleve, Yu.E. Penionzhkevich, M.P. Ivanov, Yu.P. Tretyakov. Nucl. Phys. A **239**, 157 (1975)
8. J.A. Maruhn, W. Greiner, Phys. Rev. Lett. **32**, 548 (1974)
9. J.A. Maruhn, Ph.D. thesis, J. W. Goethe Universität, Frankfurt an Main, 1973
10. R.K. Gupta, G. Münzenberg, W. Greiner, J. Phys. G **23**, L13 (1997)
11. S. Hofmann, G. Münzenberg, Rev. Mod. Phys. **72**, 733 (2000)
12. Yu. Ts, Oganesian, V.K. Ityonkov, K.J. Moody. Sci. Am. **282**, 4 (2000)
13. Ş. Mişicu, W. Greiner, Phys. Rev. C **66**, 044606 (2002)
14. M.G. Itkis, S.M. Luk'yanov, V.N. Okolovich, YuE Penionzkevich, A. Ya. Rusanov, V.S. Salamatin, G.N. Smirenkin, G.G. Chubaryan. Yad. Fiz. **52**, 23 (1990). [Sov. J. Nucl. Phys. 52, 15 (1990)]
15. F. Gönnenwein, A. Gagarski, G. Petrov, I. Guseva, T. Zavarukhina, M. Mutterer, J. von Kalben, Yu. Kopatch, G. Tiourine, W. Trzaska, M. Sillanpäa, T. Soldner, V. Nesvizhevsky, AIP Conf. Proc. **1224**, 338 (2010)
16. D.N. Poenaru, W. Greiner, Cluster radioactivity in Clusters in Nuclei 1, ed. by C. Beck. Lecture notes in physics, vol. 818 (Springer, 2010), pp. 1–56
17. A.V. Karpov, V.I. Zagrebaev, Y. Martinez Palenzuela, L. Felipe Ruiz, and Walter Greiner. Int. J. Mod. Phys. E **21**, 1250013 (2012)
18. K. Rutz, M. Bender, T. Bürvenich, T. Schilling, P.-G. Reinhard, J.A. Maruhn, W. Greiner, Phys. Rev. C **56**, 238 (1997)
19. T. Bürvenich, K. Rutz, M. Bender, P.-G. Reinhard, J.A. Maruhn, W. Greiner, Eur. Phys. J. A **3**, 139 (1998)
20. M. Bender, K. Rutz, P.-G. Reinhard, J.A. Maruhn, W. Greiner, Phys. Rev. C **58**, 2126 (1998)
21. B.D. Serot, J.D. Walecka, *Advances in Nuclear Physics*, vol. 16 (Plenum Press, New York, 1986)
22. J. Theis, G. Graebner, G. Buchwald, J. Maruhn, W. Greiner, H. Stöcker, J. Polonyi, Phys. Rev. D **28**, 2286 (1983)
23. S. Klimt, M. Lutz, W. Weise, Phys. Lett. B **249**, 386 (1990)
24. I.N. Mishustin, *in Proceedings of International Conference on Structure of Vacuum and Elementary Matter, South Africa, Wilderness* (World Scientific, Singapore, 1996), p. 499
25. P. Papazoglou, D. Zschiesche, S. Schramm, J. Schaffner-Bielich, H. Stöcker, W. Greiner, Phys. Rev. C **59**, 411 (1999)
26. I.N. Mishustin, L.M. Satarov, H. Stöcker, W. Greiner, Phys. Rev. C **62**, 034901 (2000)
27. P. Papazoglou, Ph.D. thesis, University of Frankfurt, 1998
28. Ch. Beckmann, P. Papazoglou, D. Zschiesche, S. Schramm, H. Stöcker, W. Greiner, Phys. Rev. C **65**, 024301 (2002)

29. K.A. Gridnev, V.N. Tarasov, D.V. Tarasov, D.K. Gridnev, V.V. Pilipenko, W. Greiner, Int. J. Mod. Phys. E **19**, 449 (2010)
30. V.N. Tarasov, K.A. Gridnev, D.K. Gridnev, D.V. Tarasov, S. Schramm, and W. Greiner, Stability Peninsulas near neutron drip line, arXiv:1106:5910 [nucl-th] (2011)
31. A. Botvina, I. Mishustin, V. Zagrebaev, W. Greiner, Int. J. Mod. Phys. E **19**, 2063 (2010)
32. V.I. Zagrebaev, A.V. Karpov, I.N. Mishustin, W. Greiner, Superheavies: theoretical incitements and predictions, in Exploring Fundamental Issues in Nuclear Physics, ed. by D. Bandyopadhyay (World Scientific, Singapore, 2012) p. 32
33. V. Zagrebaev, Yu. Oganessian, M. Itkis, W. Greiner, Phys. Rev. C **73**, 031602 (2006)
34. V. Zagrebaev, W. Greiner, J. Phys. G **34**, 1 (2007)
35. J. Reinhardt, U. Müller, B. Müller, W. Greiner, Z. Phys, A **303**, 173 (1981)
36. V.I. Zagrebaev, A.V. Karpov, W. Greiner, Phys. Rev. C **81**, 044608 (2010)
37. A. Sandulescu, D.N. Poenaru, W. Greiner, Sov. J. Part. Nucl. Phys. **11**, 528 (1980)
38. B. Povh, Rep. Prog. Phys. **39**, 823 (1976)
39. J. Schaffner, C. Greiner, H. Stöcker, Phys. Rev. C **46**, 322 (1992)
40. The STAR Collaboration, Nature 473, 353 (2011)
41. H. Stöcker, W. Greiner, W. Scheid, Z. Phys, A **286**, 121 (1978)
42. W. Greiner, B. Müller, J. Rafelski, *QED of Strong Fields* (Springer Verlag, Heidelberg, 1985)
43. For a more recent review see W. Greiner, J. Reinhardt, Supercritical fields in heavy-ion physics, in Proceedings of the 15th Advanced ICFA Beam Dynamics Workshop on Quantum Aspects of Beam Physics, (World Scientific, Singapore, 1998)
44. N. Auerbach, A.S. Goldhaber, M.B. Johnson, L.D. Miller, A. Picklesimer, Phys. Lett. B **182**, 221 (1986)
45. I. Mishustin, L.M. Satarov, J. Schaffner, H. Stöcker, W. Greiner, J. Phys. G **19**, 1303 (1993)
46. P.K. Panda, S.K. Patra, J. Reinhardt, J. Maruhn, H. Stöcker, W. Greiner, Int. J. Mod. Phys. E **6**, 307 (1997)
47. W. Scheid, J. Hofmann, W. Greiner, *in Proceedings of Symposium Physics with Relativistic Heavy Ions* (Lawrence Berkeley Lab, Berkeley, California, 1974)
48. H. Stöcker, W. Greiner, Phys. Rep. **137**, 279 (1986)

# Overview and Perspectives of SHE Research at GSI SHIP

**Sigurd Hofmann**

**Abstract** Theoretical studies present a quite detailed view of the stability of nuclei in the region of the heaviest elements. Three regions of increased stability exist. Two for deformed nuclei at proton and neutron numbers $Z = 100$ and $N = 152$ and at $Z = 108$ and $N = 162$. The third region is located at $Z = 114$, 120 or 126 and $N = 184$ for spherical nuclei due to closed shells or subshells. Experimentally, the existence of these regions of increased stability is established by synthesis of nuclei in cold fusion reactions using lead or bismuth targets and in hot fusion reactions based on actinide targets. Present experiments are trying to consolidate existing data and to explore the extension of the island of spherical nuclei into the direction of still heavier elements. The present status of experiments at the GSI SHIP is given as well as an outlook on investigations planned for the near future.

## 1 Introduction and Status of Experiments

For the synthesis of heavy and superheavy nuclei (SHN) fusion-evaporation reactions are used. Two approaches have been successfully employed. Firstly, reactions of a medium mass ion beam impinging on targets of stable lead and bismuth isotopes (cold fusion). These reactions have been successfully applied for producing elements up to $Z = 112$ at the GSI SHIP [1] and to confirm the results of these experiments at RIKEN [2] and LBNL [3]. Recently, a number of neutron deficient odd element isotopes were produced in a combination with $^{208}$Pb targets and odd element projectiles at LBNL [4, 5]. Using a $^{209}$Bi target the isotope $^{278}113$ was synthesized at RIKEN [6].

S. Hofmann (✉)
GSI Helmholtzzentrum für Schwerionenforschung, 64291 Darmstadt, Germany
e-mail: S.Hofmann@gsi.de

S. Hofmann
Fachbereich Physik, Goethe-Universität Frankfurt, 60438 Frankfurt, Germany

W. Greiner (ed.), *Exciting Interdisciplinary Physics*,
FIAS Interdisciplinary Science Series, DOI: 10.1007/978-3-319-00047-3_2,

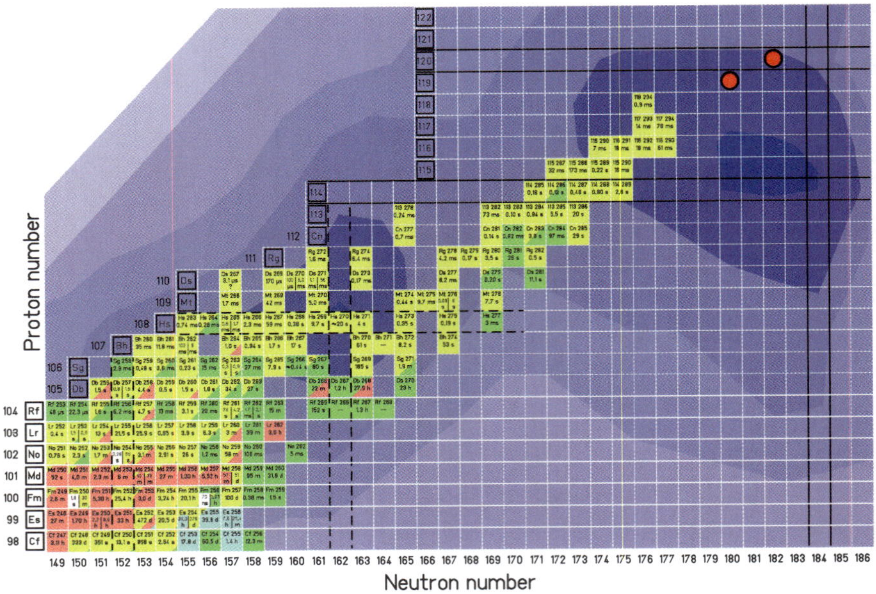


**Fig. 1** *Upper end* of the chart of nuclei showing the presently (2012) known isotopes. For each known nucleus the element name, mass number, and half-life are given. The magic numbers for the protons at element 114 and 120 and for the neutrons at $N = 184$ are emphasized. The *bold dashed lines* mark proton number 108 and neutron numbers 152 and 162. Nuclei with that number of protons or neutrons have increased stability. However, they are deformed contrary to the spherical superheavy nuclei. In the region of the crossing between *bold* and *dashed lines* at $Z = 114$ and $N = 162$ it is uncertain, whether nuclei there are deformed or spherical. The *red dots* mark compound nuclei which can be formed in the reactions $^{50}$Ti +$^{249}$Bk or $^{51}$V +$^{248}$Cm ($^{299}$119*) and $^{54}$Cr +$^{248}$Cm ($^{302}$120*). The background structure shows the calculated shell correction energy according to the macroscopic-microscopic model [16, 17]

Heavier isotopes of the element copernicium and new elements up to $Z = 118$ were produced in reactions with beams of $^{48}$Ca and radioactive actinide targets (hot fusion) at FLNR [7, 8]. The results of four of these reactions, $^{48}$Ca+$^{242}$Pu [9–11], $^{48}$Ca+$^{238}$U [12], $^{48}$Ca+$^{244}$Pu [13], and $^{48}$Ca+$^{248}$Cm [14] were confirmed in independent experiments. A new isotope of element 114, $^{285}$114 and its decay daughters, was synthesized by evaporation of five neutrons in the reaction $^{48}$Ca+$^{242}$Pu at LBNL [15]. Figure 1 summarizes the data as they are presently known.

Element 112, 'copernicium', is presently the last element in the Periodic Table, which has received a name. Agreement between element-112 data of the GSI-SHIP work and the confirmation experiments was stated in a IUPAC Technical Report in 2009 [18]. Similarly, IUPAC has assigned priority of the discovery of elements 114 and 116 to the Dubna-Livermore group [19]. The names proposed by this group are 'flerovium' and 'livermorium', respectively [20].

Besides the insight that nuclei with such a high number of protons and resulting extremely high repulsive Coulomb forces are existing, two more important observations emerged. Firstly, the expectation that half-lives of the new isotopes should lengthen with increasing neutron number as one approaches the island of stability seems to be fulfilled. Secondly, the measured cross-sections for the relevant nu-clear fusion processes reach values of up to 10 pb, which is surprisingly high. Fur-thermore, the cross-sections seem to be correlated with the variation of shell-correction energies as predicted by macroscopic-microscopic calculations [16, 17, 21].

## 2 Continuation of SHN Experiments Using $^{248}$Cm Targets

A comparison of various theoretical studies reveals that the location of the next closed proton shell beyond $Z = 82$ is uncertain. The question is still open whether $Z = 120$ is a closed proton shell or if strong shell closures exist at $Z = 114$ or 126. In addition, the possibility has to be considered that the island of superheavy nuclei is relatively flat and extents between sub-shells at 114, 120 and 126. Concerning the closed neutron shell, most theories agree with $N = 184$ as a strong shell. Experimental data—longer half-lives and decreasing negative shell-correction energies with increasing neutron number—as known so far, are in agreement with this finding, too.

As an important part of our work on the synthesis and properties of SHN we proposed to study also hot fusion reactions based on actinide targets, in addition to our cold-fusion program. Together with several technical improvements this proposal was made in a medium range plan already at the end 1998 [22]. However, at the beginning of 1999 our report was rejected and the proposed program was no longer pursued.

Now, times have changed, and in 2009 we suggested to start a program for studying superheavy nuclei using reactions based on $^{248}$Cm targets. This target material has special properties which makes it favorable for the synthesis of heavy nuclei. It is one of the heaviest ($Z = 96$) and most neutron rich available targets. Increased shell effects at its neutron number $N = 152$ result in a relatively long half-life of $3.4 \times 10^5$ years and, thus, low specific activity. In combination with strongly bound projectile nuclei like $^{48}$Ca or the neutron rich isotopes of the heavier elements up to nickel, relatively low excitation energies of the compound nuclei result, which are approximately 30–40 MeV at the fusion barrier. This advantageous property increases the probability for neutron emission instead of fission and thus results in relatively high fusion-evaporation cross-sections.

In the following, we list a number of general arguments which have to be considered selecting the best reaction with respect to cross-sections for production of new elements beyond 118, in particular the new elements 119 and 120.

Production cross-sections are strongly determined by fission barriers which again are built by shell effects in the region of SHN. The rising up of cross-sections to several picobarns for elements 114 and 116 is due to increasing shell effects when $N = 184$ is approached. This behaviour suggests using the most neutron rich projectile and target nuclei available for synthesis of the elements 119 and 120.

Shell effects and thus fission barriers are considerably reduced with increasing excitation energy of the compound nucleus. Therefore, selection of a reaction re-sulting in a minimum of excitation energy is mandatory.

Cross-sections are further strongly influenced by Coulomb re-separation in the entrance channel of the reaction due to quasi-elastic and quasi-fission processes. In order to reduce this most unwanted effect, reaction partners have to be used resulting in lowest repulsive Coulomb forces. This can be achieved using reaction partners of high asymmetry in the number of protons.

Other phenomena which also influence the cross-sections, but are difficult to predict quantitatively and in detail, originate from isotopic effects. The number of neutrons determines the nuclear radius, relatively more for the smaller projectiles, which influences the compactness of the system at the contact configuration. In the case of deformed nuclei of the actinides, nuclear orientation is another property, which strongly determines cross-section and beam energy. However, it is not possible to align the target nuclei in order to obtain an orientation which results in highest fusion probability.

Finally, the reaction must be technically possible, i.e. projectiles and targets have to be available. Heaviest isotopes which could be used as targets are $^{254}$Es ($T_{1/2}$ = 276 d) and $^{257}$Fm (100 d). However, the production of these isotopes is complex and only amounts of micrograms and nanograms, respectively, can be produced at high costs.

The next lighter isotopes available in principle are $^{252}$Cf (2.6 y), $^{249}$Cf (351 y), and $^{249}$Bk (320 d). The isotope $^{252}$Cf can be handled only with special radiation protection because of the high neutron flux being emitted from this fissioning material. The isotope $^{249}$Bk has a relatively short half-life. It must be produced on demand and it is not available regularly. Due to the relatively short half-life, also the isotope $^{249}$Cf has a high specific activity. In addition, the compound nucleus $^{299}$120, which can be made in reactions with a $^{50}$Ti beam, has three neutrons less then the compound nucleus $^{302}$120, which can be produced with a $^{248}$Cm target and a $^{54}$Cr beam, see Fig. 1. Considering all pros and cons we conclude that the reaction $^{54}\mathrm{Cr} + ^{248}\mathrm{Cm} \rightarrow ^{302}120^*$ is presently the most promising one being technically feasible to search for element 120.

Two reactions are available for the synthesis of element 119, $^{50}\mathrm{Ti} + ^{249}\mathrm{Bk}$ and $^{51}\mathrm{V} + ^{248}\mathrm{Cm}$, both forming the compound nucleus $^{299}$119, see Fig. 1. In this special case one can expect that the cross-section is independent whether the odd proton is located in the target or in the projectile. Already existing data on the synthesis of odd element isotopes using $^{208}$Pb or $^{209}$Bi targets and the corresponding odd or even element projectiles revealed similar cross-sections when the same compound nucleus is formed. A comparison was performed in [23] using experimental data measured at SHIP and at the LBNL gas-filled separator BGS.

Finally, the target $^{249}$Bk opens an interesting aspect from the reaction point of view, apart from the difficulty with the production of the material and its short half-life. Using a $^{51}$V beam the compound nucleus $^{300}$120 can be formed. In this case the odd proton is expected to be transferred to the target in an early stage of the reaction, so that the cross-section could be similar as in the case of the reaction

$^{50}$Ti $+^{250}$Cf, also resulting in the compound nucleus $^{300}$120. The homologous cold fusion reactions were not yet studied. The pairs of target nuclei $^{205}$Tl, $^{206}$Pb or $^{203}$Tl, $^{204}$Pb and beams of $^{51}$V and $^{50}$Ti, respectively, could be used supplementary to the reactions $^{208}$Pb $+^{52}$Cr and $^{209}$Bi $+^{51}$V.

Cross-sections calculated by various authors for the synthesis of elements 119 and 120 using $^{50}$Ti or $^{54}$Cr beams cover a range from 0.1 to 800 fb [24–28]. The reason for this large uncertainty is the extremely sensitive dependence of cross-sections from fusion barriers and resulting excitation energies at the barrier, from Coulomb re-separation and from fission barriers as outlined before. The model by Siwek-Wilczynska et al. [24] assumes a lower fusion barrier which results in an increase of the 3n cross-section. In the model by Nasirov et al. [25, 26] the quasi-fission processes result in strong reduction of cross-sections with increasing symmetry, whereas this effect changes the cross-section only within a factor of ten in the model by Zagrebaev and Greiner [27]. Finally, in the paper by Adamian et al. [28] various mass formula and various damping parameters of the fission barrier at increasing excitation energy were compared. Two of the results predict cross-sections differing by two and three orders of magnitude.

Experimental limits were obtained for reactions with $^{58}$Fe and $^{64}$Ni beams and targets of $^{244}$Pu and $^{238}$U at FLNR [29] and at SHIP [30], respectively. Although these limits are still high, they allow to reject unusually high fission barriers at element 120. Using the rule of thumb that a 1 MeV change of the fission barrier changes the cross-section by one order of magnitude at least [24], we obtain experimental limits for the fission barrier of element 120 isotopes of $<8.9$ and $<8.3$ MeV, respectively. As a starting point in this estimate we used the calculation of Zagrebaev and Greiner [27], who determined their cross-sections with a fission barrier of 7 MeV. At a fission barrier of 8.3 MeV their cross-section estimates would be a factor of 20 higher. This simple consideration and the very different predictions show that a sufficiently accurate estimate for the beam time necessary to produce the elements 119 and 120 cannot be made on the basis of the presently existing data and calculations.

Half-life and decay mode are nuclear properties which could hamper the identification, although isotopes of element 120 could be produced with high enough cross-sections. Theoretical calculations show that the heaviest elements decay by $\alpha$ emission. This result is proved by experimental data on elements up to $^{294}$118, which has a measured $Q_\alpha$ value of 11.81 MeV and decays with a half-life of 890 μs [7]. In the region of interest, $\beta$ decay and spontaneous fission are predicted to have significantly longer half-lives. This result is in agreement with the measured $\alpha$-decay chains which end by spontaneous fission only at copernicium or below.

Whereas fission barriers and deduced fission half-lives are difficult to calculate, the access to $Q_\alpha$ values as difference of masses of neighbouring nuclei and deduced partial $\alpha$ half-lives is easier. In the following we compare experimental $Q_\alpha$ values of an established decay chain with few but representative theoretical predictions. In Fig. 2, calculated $Q_\alpha$ values are shown over a wide range from element 104 to 122 for the chain passing $^{292}$116. Showing this figure, we are also aiming to obtain a sense for the uncertainties related to predictions on the stability of isotopes of the so far

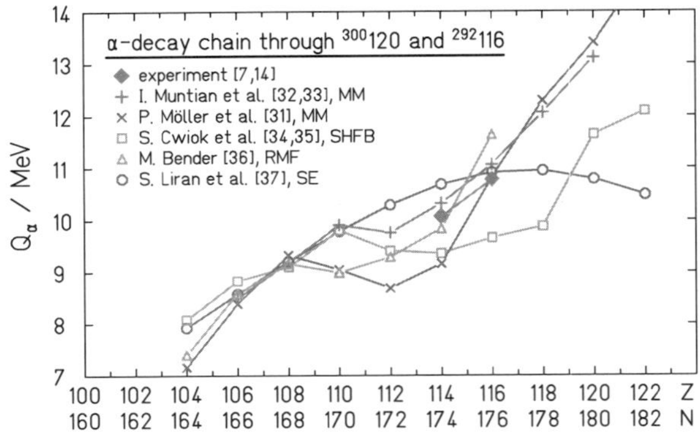


**Fig. 2** Comparison of measured and calculated $Q_\alpha$ values of the $\alpha$-decay chain passing the isotope $^{292}$116. Nuclei of this decay chain belong to the most neutron rich nuclei which can be produced in the laboratory. They are of special interest with respect to a future synthesis of so far unknown elements beyond $Z = 118$

unknown elements 119 and 120, their synthesis is presently the aim at the research centres JINR, RIKEN, and GSI.

Two of the theoretical data shown are based on the macroscopic-microscopic (MM) model [31–33], one on the self-consistent mean field model using the Skyrme-Hartree-Fock-Bogoliubov (SHFB) method [34, 35], one on the relativistic mean-field (RMF) model [36], and one on a semiempirical (SE) shell-model mass equation having $Z = 126$ and $N = 184$ as spherical proton and neutron shells after the double magic $^{208}$Pb [37].

Obviously, the considered range of $Q_\alpha$ values can be subdivided in three parts concerning the variations of the predictions. One for elements below darmstad-tium, one for elements between darmstadtium and $Z = 116$, and a third one for elements up to 122.

The three regions are also related to different physical properties of the nuclei. Firstly, the region of well deformed nuclei below darmstadtium and $N < 170$; in this region the shape of the nuclei is determined by stronger binding energy at large deformation due to the compression of single particle levels below the ener-gy gaps at $Z = 108$ and $N = 162$ at quadrupole deformation parameters $\beta_2 = 0.25$. The second region up to element 116 for neutron numbers of the measured $\alpha$-decay chain considered here is a transitional region of decreasing deformation into the direction of the third region extending up to element 122 and beyond, which is governed by shell effects of spherical closed shells or subshells.

Although the experimental $Q_\alpha$ values are scarce we notice that the gradient of the experimental data between elements 114 and 116 is less than in the results of the MM model [31] and the RMF model [36]. From this experimental observation we conclude that at neutron numbers 174 to 176 the proton shell strength at $Z = 114$ is less pronounced than predicted in [31, 36].

Concerning heavier elements beyond $Z = 118$, the experimental data is just at the limit which could settle the question if proton shells exist at $Z = 120$ or 126. Increasing $Q_\alpha$ values as predicted by the MM models would rule out shell closures at 120 and 126. As a consequence, the lifetimes of elements beyond 120 would fall below 1 μs which is the limit of present detection methods. The elements 119 and 120 would be the last ones which could be detected in the near future. At $Z = 120$ the 1 μs limit is reached at $Q_\alpha = 13.3$ MeV and at $Z = 126$ at 14.0 MeV.

A subshell closure at $Z = 120$ would result in relatively long $\alpha$ half-lives of element 120. At a $Q_\alpha$ value of about 11.7 MeV calculated for $^{300}120$ in [35], see Fig. 2, we obtain a half-life of 2.2 ms. In addition, also the $\alpha$ half-life of element 122 would be longer relative to the predictions of the MM models. The stronger trend to lower $Q_\alpha$ values of the semiempirical model would result in $\alpha$ half-lives of 350 ms and 43 s at $Q_\alpha = 10.8$ and 10.7 MeV [37] for isotopes of element 120 and 126 with mass numbers 300 and 310, respectively.

In the region of SHEs, fission barriers are mainly determined by ground-state shell effects. Because $Q_\alpha$ values are determined by the difference of binding energies between parent and daughter nucleus, the gradient of a $Q_\alpha$ systematics reflects the trend of increasing or decreasing fission barriers. The rapidly increasing $Q_\alpha$ values of the MM models for elements above 114 is related to increasing negative ground-state shell-correction energies and thus decreasing fission barriers. The opposite trend is valid for the semiempirical model.

The experimental $Q_\alpha$ values reveal differences to the theoretical data of up to 1 MeV, see Fig. 2. Similar differences must be expected for the ground-state shell-correction energies and the fission barriers. Fission barriers are an essential part in the calculations of cross-sections. As already reminded, a rough estimate shows that a 1 MeV increase of the fission barrier increases the cross-section by one to two orders of magnitude [24]. Uncertainties of this order of magnitude, which were revealed by the comparison of experimental and theoretical $Q_\alpha$ values, have to be considered in the discussions on the preparation of experiments aiming at searching for new elements. In other words, sufficiently long beam times have to be provided in order to perform experiments with the perspectives of being successful.

In conclusion we realize that half-lives of the isotopes of interest are predicted to be in the range from 1 to 30 μs, but could be significantly longer if the proton shell is at $Z = 126$ or 120 and not at 114. In any case special technical preparations are needed for detection of isotopes of element 120 in order to be prepared for short half-lives [43].

At SHIP, a minimum lifetime of 2 μs is needed so that the residues can pass through the separator, otherwise they will decay inside the separator. In this case, also the daughter nucleus after $\alpha$ decay will be lost with high probability due to the recoil momentum from the emitted $\alpha$ particle, an effect which reduces the transmission by a factor of ten.

The decay chains expected in the case of three and four neutron evaporation will populate isotopes of element 116, $^{290}116$ and $^{291}116$, which were measured previously at FLNR and which were confirmed indirectly by identification of the daughter nuclei $^{286}114$ and $^{287}114$ at LBNL [11]. In this case a well founded identification

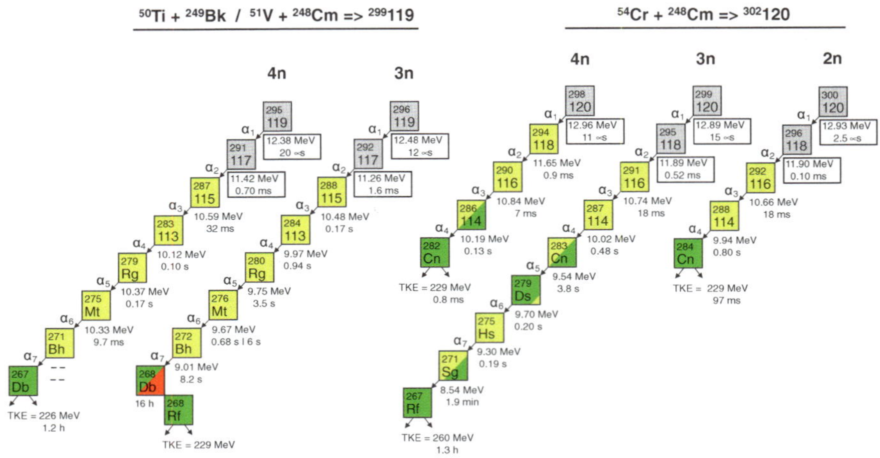


**Fig. 3** Expected decay chains of element 119 populated in the reactions $^{50}$Ti +$^{249}$Bk and $^{51}$V +$^{248}$Cm, both forming the compound nucleus $^{299}$119* and of element 120 formed in the reaction $^{54}$Cr +$^{248}$Cm $\rightarrow^{302}$ 120*. Predicted decay data of so far unknown isotopes are given in frames. Values for $E_\alpha$ of $^{295,296}$119, $^{291,292}$117, $^{300}$120, and $^{296}$118 were taken from [38]. The half-lives of these isotopes were calculated using the WKB method [39]. Energies and half-lives of $^{298,299}$120 and $^{295}$118 were taken from [40]. Concerning the uncertainty of the calculated $E_\alpha$ values see also Fig. 2. Experimental data of the known isotopes were taken from publications of the Dubna-Livermore group, namely for the decay chain of $^{287}$115 from [41], $^{288}$115 from [42], $^{294}$118, $^{291}$116, and $^{292}$116 from [7]. TKE values of the spontaneously fissioning isotopes terminating the chains were taken from Fig. 29 in [7]. Note the significant difference of the lengths of the decay chains for even-even and odd or odd-odd nuclei

of element 120 by genetic correlation to known nuclei is given. The expected decay chains are shown in Fig. 3. In the case of an also possible two neutron channel, we would observe as a granddaughter the isotope $^{292}$116 and its daughter decays, which we observed in a confirmation experiment at SHIP using the reaction $^{48}$Ca +$^{248}$Cm $\rightarrow^{292}$ 116 + 4n [14]. Similarly, the decay chains of $^{295}$119 and $^{296}$119 produced in 4n and 3n evaporation channels, respectively, will populate known decay chains after the first two $\alpha$ decays, see Fig. 3, left side.

A new search for element 120 using the reaction $^{54}$Cr +$^{248}$Cm $\rightarrow^{302}$120* has been started at SHIP in spring 2011. Main aim of this first part of 33 days was to study the performance of the targets during irradiation with a chromium beam and to condition a second wheel for further irradiation in the future. Therefore, the beam current was limited to about 400 particle nA. Nevertheless, a cross-section limit of 560 fb was reached, which is, however, still far from calculated cross-sections being mainly in the range from 30 to 100 fb. An estimate based on higher beam intensities results in an additional measuring time of about three months in order to reach such low cross-section limits. From the performance of the accelerator, the targets, the separator, the detectors, and the data acquisition system in the first part of the experiment we conclude that our experiment is well prepared for this important next step in the study of superheavy nuclei.

## 3 Conclusion and Outlook

The experimental work of the last three decades has shown that cross-sections for the synthesis of the heaviest elements do not decrease continuously as it was measured up to the production of element 113 using cold fusion reactions. Recent data on the synthesis of elements 112 to 118 in Dubna using hot fusion show that this trend is broken when the region of spherical SHN is reached. Some of the results originally obtained in Dubna were confirmed in independent experiments and with different methods, including the use of chemical specific properties of the elements. We conclude that the region of the predicted spherical SHN has finally been reached and the exploration of the 'island' has started and can be performed even on a relatively high cross-section level.

An opportunity for the continuation of experiments in the region of SHN at low cross-sections afford, among others, further accelerator developments. High current beams and radioactive beams are options for the future. A wide range of half-lives encourages the application of a wide variety of experimental methods in the investigation of SHN, from the safe identification of short lived isotopes by recoil-separation techniques to atomic physics experiments on trapped ions, and to the investigation of chemical properties of SHN using long-lived isotopes.

**Acknowledgments** The recent experiment at SHIP on confirmation of data obtained in Dubna in the reaction $^{48}$Ca $+^{248}$Cm and the experiment started to search for element 120 in the reaction $^{54}$Cr $+^{248}$Cm were performed in collaboration with the following laboratories: GSI, Darmstadt, Germany; Goethe-Universität, Frankfurt, Germany; HIM, Mainz, Germany; Comenius University, Bratislava, Slovakia; Johannes Gutenberg-Universität, Mainz, Germany; LLNL, Livermore, USA; University, Jyväskylä, Finland; JAEA, Tokai, Japan; JINR-FLNR, Dubna, Russia. The following people participated in the experiments: S. Heinz, R. Mann, J. Maurer, J. Khuyagbaatar, D. Ackermann, S. Antalic, W. Barth, M. Block, H.G. Burkhard, V.F. Comas, L. Dahl, K. Eberhardt, J. Gostic, R.K. Grzywacz, R.A. Henderson, J.A. Heredia, F.P. Heßberger, J.M. Kenneally, B. Kindler, I. Kojouharov, J.V. Kratz, R. Lang, M. Leino, B. Lommel, K. Miernik, D. Miller, K.J. Moody, G. Münzenberg, S.L. Nelson, K. Nishio, A.G. Popeko, J.B. Roberto, J. Runke, K.P. Rykaczewski, S. Saro, D.A. Shaughnessy, M.A. Stoyer, P. Thörle-Pospiech, K. Tinschert, N. Trautmann, J. Uusitalo, P.A. Wilk, and A.V. Yeremin.

## References

1. S. Hofmann, G. Münzenberg, Rev. Mod. Phys. **72**, 733 (2000)
2. K. Morita et al., J. Phys. Soc. Jpn. **76**, 043201 (2007)
3. T.N. Ginter et al., Phys. Rev. C **67**, 064609 (2003)
4. C.M. Folden III et al., Phys. Rev. Lett. **93**, 212702 (2004)
5. C.M. Folden III et al., Phys. Rev. C **73**, 014611 (2006)
6. K. Morita et al., J. Phys. Soc. Jpn. **73**, 2593 (2004)
7. Y.T. Oganessian, J. Phys. G: Nucl. Part. Phys. **34**, R165 (2007)
8. Y.T. Oganessian et al., Phys. Rev. Lett. **104**, 142502 (2010)
9. R. Eichler et al., Nucl. Phys. A **787**, 373c (2007)
10. R. Eichler et al., Nature **447**, 72 (2007)
11. L. Stavsetra et al., Phys. Rev. Lett. **103**, 132502 (2009)

12. S. Hofmann et al., Eur. Phys. J. A **32**, 251 (2007)
13. C.E. Düllmann et al., Phys. Rev. Lett. **104**, 252701 (2010)
14. S. Hofmann et al., The reaction $^{48}Ca + ^{248}Cm \rightarrow ^{296}116^{*}$ studied at the GSI SHIP, GSI Report 2011–1, vol. 197 (2011) and submitted for publication in Eur. Phys. J. (Jan 2012)
15. P.A. Ellison et al., Phys. Rev. Lett. **105**, 182701 (2010)
16. R. Smolanczuk, A. Sobiczewski, in *Proceedings of the XV Nuclear Physics Divisional Conference on Low Energy Nuclear Dynamics*, ed. by Y.T. Oganessian et al. (World Scientific, Singapore, 1995), p. 313
17. R. Smolanczuk et al., Phys. Rev. C **52**, 1871 (1995)
18. R.C. Barber et al., Pure Appl. Chem. **81**, 1331 (2009)
19. R.C. Barber et al., Pure Appl. Chem. **83**, 1485 (2011)
20. IUPAC (2011), http://www.iupac.org/web/nt/2011-12-01_name_element_114_116
21. P. Möller et al., At. Data Nucl. Data Tables **59**, 185 (1995)
22. S. Hofmann, SHIP-2000—a proposal for the study of superheavy elements, GSI-Report 99–02 pp. 1–16 (1999), indexed (1999)
23. S. Hofmann, Radiochim. Acta **99**, 405 (2011)
24. K. Siwek-Wilczynska et al., Int. J. Mod. Phys. E **19**, 500 (2010)
25. A.K. Nasirov et al., Phys. Rev. C **79**, 024606 (2009)
26. A.K. Nasirov et al., Acta Phys. Pol. B 43 (2012).
27. V. Zagrebaev, W. Greiner, Phys. Rev. C **78**, 034610 (2008)
28. G.G. Adamian et al., Eur. Phys. J. A **41**, 235 (2009)
29. Y.T. Oganessian et al., Phys. Rev. C **79**, 024603 (2009)
30. S. Hofmann et al., GSI Scientific Report 2008. GSI Report 2009–1, vol. 131 (2009)
31. P. Möller et al., At. Data Nucl. Data Tables **66**, 131 (1997)
32. I. Muntian et al., Acta Phys. Pol. B **34**, 2073 (2003)
33. I. Muntian et al., Phys. At. Nucl. **66**, 1015 (2003)
34. S. Cwiok et al., Phys. Rev. Lett. **83**, 1108 (1999)
35. S. Cwiok et al., Nature **433**, 705 (2005)
36. M. Bender, Phys. Rev. C **61**, 031302(R) (2000)
37. S. Liran et al., Phys. Rev. C **63**, 017302 (2000), arXiv:nucl-th/0102055
38. I. Muntian et al., Phys. At. Nucl. **66**, 1015 (2003)
39. J.O. Rasmussen, Phys. Rev. **113**, 1593 (1959)
40. A. Sobiczewski, Acta Phys. Pol. B **42**, 1871 (2011)
41. Y.T. Oganessian et al., Phys. Rev. C **72**, 034611 (2005)
42. Y.T. Oganessian et al., Phys. Rev. Lett. **108**, 022502 (2012)
43. R. Grzywacz et al., Nucl. Instrum. Methods B **261**, 1103 (2007)

# Nuclear Reaction Mechanisms Induced by Heavy Ions

**M. G. Itkis, I. M. Itkis, G. N. Knyazheva and E. M. Kozulin**

**Abstract** Total Kinetic Energy—Mass distributions of fission-like fragments for the reactions of $^{22}$Ne, $^{26}$Mg, $^{36}$S, $^{48}$Ca, $^{58}$Fe and $^{64}$Ni ions with actinides leading to the formation of superheavy compound systems with $Z = 108$–$120$ at energies near the Coulomb barrier have been investigated. It was found that the relative contribution of QF to the capture cross section mainly depends on the reaction entrance channel properties, but the features of asymmetric QF are determined essentially by the driving potential of composite system. A possible alternative pathway in the production of new superheavy elements near the "island of stability" is represented by the "inverse" quasifission or deep-inelastic reactions in the collision of $^{136}$Xe with actinide targets.

The collision of two massive nuclei takes a special place in nuclear reactions studies due to the large number of interacting nucleons. In this type of reactions a drastic change of the reaction partners may occur that leads to different reaction mechanisms. In reactions with heavy ions at energies close to the Coulomb barrier complete fusion, quasifission (QF) and deep-inelastic collision are competing pro-cesses [1–3]. The balance between these processes strongly depends on the en-trance channel properties, such as mass asymmetry, deformation of interacting nuclei, collision energy, and the Coulomb factor $Z_1 Z_2$.

Previously in the experimental investigations the symmetric fragment region with mass $A_{CN} \pm 20$ u was often attributed to compound nucleus fission (CNF). However, a realistic description of the mass, energy and angular distributions of the reaction fragments formed in deep inelastic scattering, QF and CN-fission pro-cesses in low energy heavy ion collisions shows [4] that the potential energy sur-face for these systems is strongly modulated by shell effects and leads to the ap-pearance of deep valleys corresponding to the formation of well bound magic nuclei. In accordance with these calculations, at least three paths leading to the formation of fission-like

M. G. Itkis (✉) · I. M. Itkis · G. N. Knyazheva · E. M. Kozulin
FLNR, Joint Institute for Nuclear Research, Moscow region, Dubna141980 , Russia
e-mail: itkis@jinr.ru

W. Greiner (ed.), *Exciting Interdisciplinary Physics*,
FIAS Interdisciplinary Science Series, DOI: 10.1007/978-3-319-00047-3_3,

fragments can be distinguished: asymmetric QF caused by the influence of proton shells with $Z = 28, 82$ and neutron shells with $N = 50, 126$; symmetric QF determined by the shells with $Z = 50$ and $N = 82$; CN-fission leading to the formation of symmetric fragments.

It is known that in superheavy composite systems QF mainly leads to the formation of asymmetric fragments with mass asymmetry ~0.4 [1]. This type of QF process, so-called asymmetric quasifission (QFasym), is characterized by asymmetric angular distributions in the center-of-mass system and thus fast reaction times (~$10^{-21}$ s) [5, 6]. The total kinetic energy (TKE) for these fragments is ob-served to be higher than that for CNF [1, 5] and hence this process is colder than CNF. Due to this reason shell effects in QF are more pronounced [7].

Figure 1 shows the mass-energy distributions of binary fragments obtained in the reactions of $^{16}$S, $^{48}$Ca, $^{64}$Ni ions with an uranium target. The Coulomb factors are 1472, 1840 and 2576 for the $^{16}$S+$^{238}$U, $^{48}$Ca+$^{238}$U and $^{64}$Ni+$^{238}$U, respectively. Some noteworthy features of the QFasym component of fragment mass distributions for the studied reactions can be highlighted at this point. Generally, in heavy-ion-induced reactions the formation of QFasym fragments is connected with the strong influence of the nuclear shell at $Z = 82$ and $N = 126$ (doubly magic lead). In fact, as was shown in Ref. [8], for the $^{48}$Ca + $^{238}$U reaction the maximum yield corresponds to fragments with masses 208 u. However, in reactions with lighter projectiles on a uranium target, the asymmetric QF peak shifts toward more symmetric masses [9]. By contrast, for the heavier projectile $^{64}$Ni, the maximum yield of QFasym fragments corresponds to the heavy mass 215 u [8]. This trend is illustrated in Table 1 , where the positions of heavy QF fragments for these reactions are presented. But,

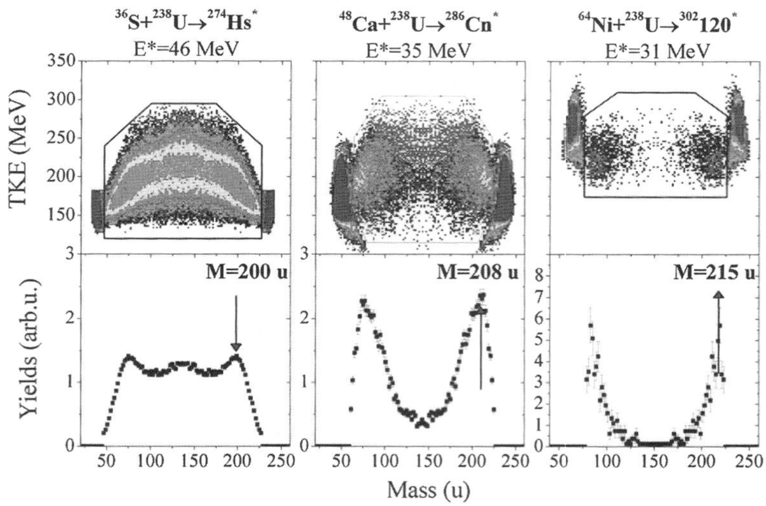


**Fig. 1** Mass-energy distributions for the reactions $^{36}$S, $^{48}$Ca, $^{64}$Ni +$^{238}$U at energies close to the Coulomb barrier

**Table 1** Positions of heavy peaks in the primary mass distributions of QFasym fragments in reactions with heavy ions

| Reaction | $Z_1Z_2$ | $M_H$ | $M_H^{\mathrm{Shell}}$ | Exchanged nucleons | Reference |
|---|---|---|---|---|---|
| $^{26}Mg + ^{248}Cm$ | 1152 | 185±5 | 202.5 | 68 | I. Itkis et al. [10] |
| $^{30}Si + ^{238}U$ | 1288 | 178 | 199.3 | 60 | K. Nishio et al. [9] |
| $^{36}S + ^{238}U$ | 1472 | 200±3 | 202.5 | 38 | I. Itkis et al. [10], K.Nishio et al. [9] |
| $^{40}Ar + ^{238}U$ | 1656 | 204 | 204.5 | 34 | K. Nishio et al. [9] |
| $^{48}Ca + ^{238}U$ | 1840 | 208 ±2 | 208.5 | 30 | E. Kozulin et al. [8] |
| $^{64}Ni + ^{238}U$ | 2576 | 215±3 | 216.5 | 23 | E. Kozulin et al. [8] |

in the formation of the asymmetric QF component, also the closed shell in the light fragment at $Z = 28$ and $N = 50$ could be effective, together with the shells $Z = 82$ and $N = 126$, and could lead to the shift of the asymmetric QF peak. Based on the simple assumption of an $N/Z$ equilibration, the masses of the light and heavy fragments corresponding to these closed shells were calculated. In Table 1 $M_H^{\mathrm{Shell}}$ is a heavy fragment mass averaged over all these shells. The obtained values of $M_H^{\mathrm{Shell}}$ are in good agreement with the experimental ones, except for the more asymmetric $^{26}Mg + ^{248}Cm$ and $^{30}Si + ^{238}U$ reactions. For these reactions the Coulomb repulsion is expected to be smaller. This may lead to longer reaction times before separation for asymmetric QF and thus allow for larger numbers nucleons to be exchanged. For the other more symmetric reactions with heavier projectiles, the major part of the asymmetric QF peak fits into the region of the $Z = 82$, $N = 126$ and $Z = 28$, $N = 50$ shells. The maximum yield of the asymmetric QF component is a mixing between all these shells.

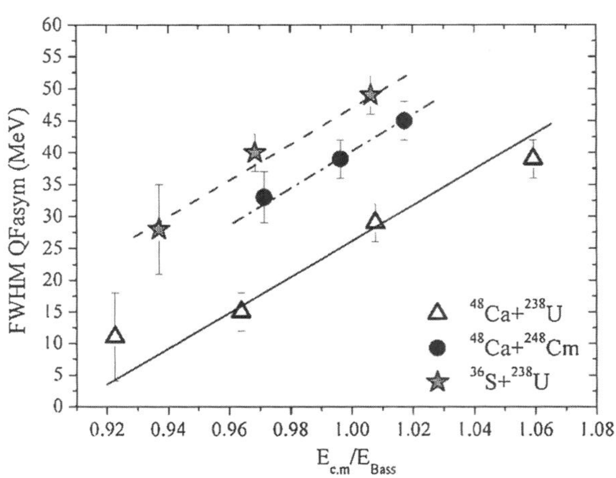


**Fig. 2** The widths of QFasym mass distributions as a function of the energy above the Bass barrier for the reactions $^{36}S+^{238}U$, $^{48}Ca+^{238}U$ and $^{48}Ca+^{248}Cm$

Besides the position of peaks in the mass distributions of QFasym fragments, also the widths of these peaks vary for different "ion-target" combinations even in the case of the formation of the same composite systems. Figure 2 presents the width of QFasym mass distributions as a function of the energy above the Bass barrier for the reactions $^{36}S+^{238}U$, $^{48}Ca+^{238}U$ and $^{48}Ca+^{248}Cm$. Notice, that for all studied reactions the width increases with a similar slope with increasing collision energy. Nevertheless, the absolute values are different: at energy of the Bass barrier the width is about 47 u for the $^{36}S+^{238}U$, about 26 u for the $^{48}Ca+^{238}U$ and about 40 u for the $^{48}Ca+^{248}Cm$. So, it is not a function of mass or charge number of composite system. For these reactions the driving potentials as a function of mass asymmetry and distance between mass centers have been calculated in the diabatic approximation using the proximity model with the help of Nuclear Reaction Vision Project (NRV) [11]. These potentials are shown in Fig. 3 for the case of distance between mass centers of about 13 fm (scission point). It is clearly seen that the deepest narrow minimum at mass close to 208 u corresponds to the reaction $^{48}Ca+^{238}U$ that is the case of the narrowest QFasym mass distribution. Not only the deepnesses of the minima around the mass of 208 u, but also their positions are different for these reactions. The calculated position of the minimum of the diving potential agrees with the position of peaks in the experimental QFasym mass distributions. Thereby, the driving potential calculated in diabatic approximation qualitatively reproduces the main features of QFasym mass distributions. While the relative contribution of QF to the capture cross section mainly depends on the reaction entrance channel properties, the features of asymmetric QF are determined essentially by the driving potential of composite system.

Coming back to the $^{26}Mg+^{248}Cm$ and $^{36}S+^{238}U$ reactions leading to the formation of the same composite system, the difference between the peak positions of the

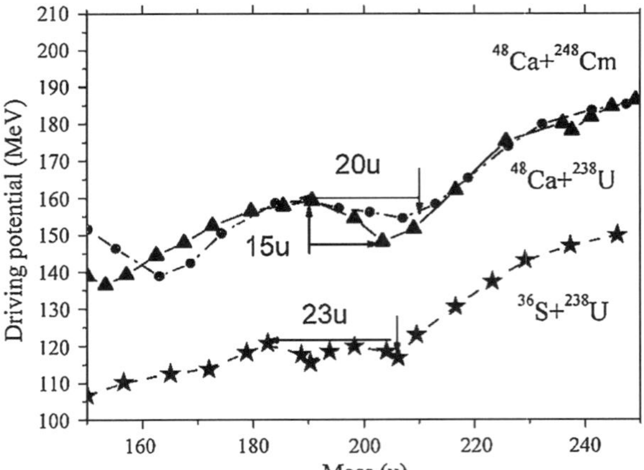


**Fig. 3** The driving potentials as a function of reaction fragment mass at scission point for the reactions $^{36}S+^{238}U$, $^{48}Ca+^{238}U$ and $^{48}Ca+^{248}Cm$

QFasym fragment mass distributions may be connected also with the shape of the driving potential. The driving potential for this composite system has two wide small local minima at fragment mass of about 190 and 205 u (see Fig. 3). Due to the small value of $Z_1 Z_2 = 1152$ for the $^{26}$Mg+$^{248}$Cm reaction the composite system is closed to the CN shape than in the case of the $^{36}$S+$^{238}$U ($Z_1 Z_2 = 1472$). Thus, due to the difference in the entrance channels for these reactions the QFasym process may come by different ways and lead to the formation of various fragments.

For the reactions $^{58}$Fe+$^{244}$Pu and $^{64}$Ni+$^{238}$U leading to the formation of the same composite system of $^{302}$120, the positions of the QFasym peaks are approximately the same, but the width of these peaks are different: 22 u for the former reaction and 11 u for the latter one. The left panel of Fig. 4 presents the driving potential for the composite system $^{302}$120 at scission point. The mass-energy distributions for the reactions $^{58}$Fe+$^{244}$Pu and $^{64}$Ni+$^{238}$U are shown in the right panel of Fig. 4. As in the case of $^{26}$Mg+$^{248}$Cm and $^{36}$S+$^{238}$U the entrance channels are different: $Z_1 Z_2 = 2444$ for the $^{58}$Fe+$^{244}$Pu reaction and 2576 for the $^{64}$Ni+$^{238}$U reaction. The larger Coulomb repulsion in the entrance channel may lead to the faster interaction time and, consequently, transfer of less number of nucleons from one nucleus to another. Nevertheless, the structure of the driving potential is clearly seen in the mass distributions of fission-like fragments formed in the both reactions.

The experimental mass distribution of the fission-like fragments formed in the reaction $^{48}$Ca+$^{248}$Cm at energy close to the Coulomb barrier is presented in the top panel of Fig. 5. The symmetric fragment mass distribution (the open circles in Fig. 5) has been extracted from the experimental mass distribution of the all fission-like fragments using the Gaussian fitting procedure of asymmetric peaks. The driving potential for this composite system at scission point calculated in the frame of the proximity model is also shown in the same figure. As it was mentioned above the

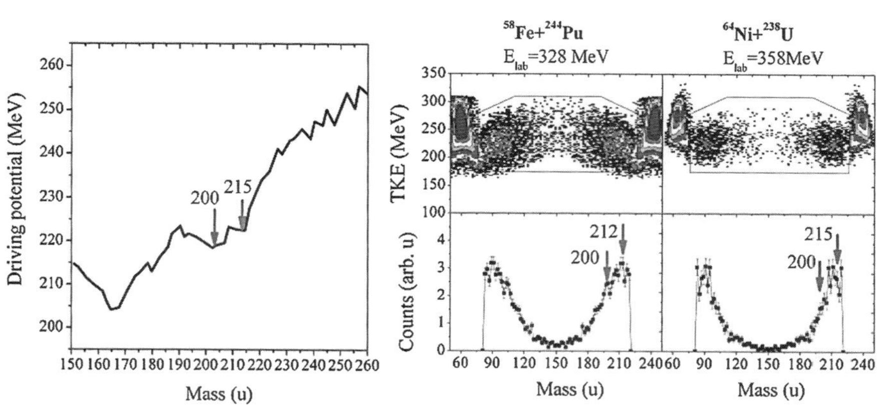


**Fig. 4** The driving potential as a function of reaction fragment mass at scission point for the superheavy composite system $^{302}$120 (*left* panel). The mass-energy distributions of binary fragments (*top*) and mass distribution for fission-like fragments (*bottom*) for the reactions $^{58}$Fe+$^{244}$Pu and $^{64}$Ni+$^{238}$U at energies of about 1.06 MeV above the Coulomb barrier

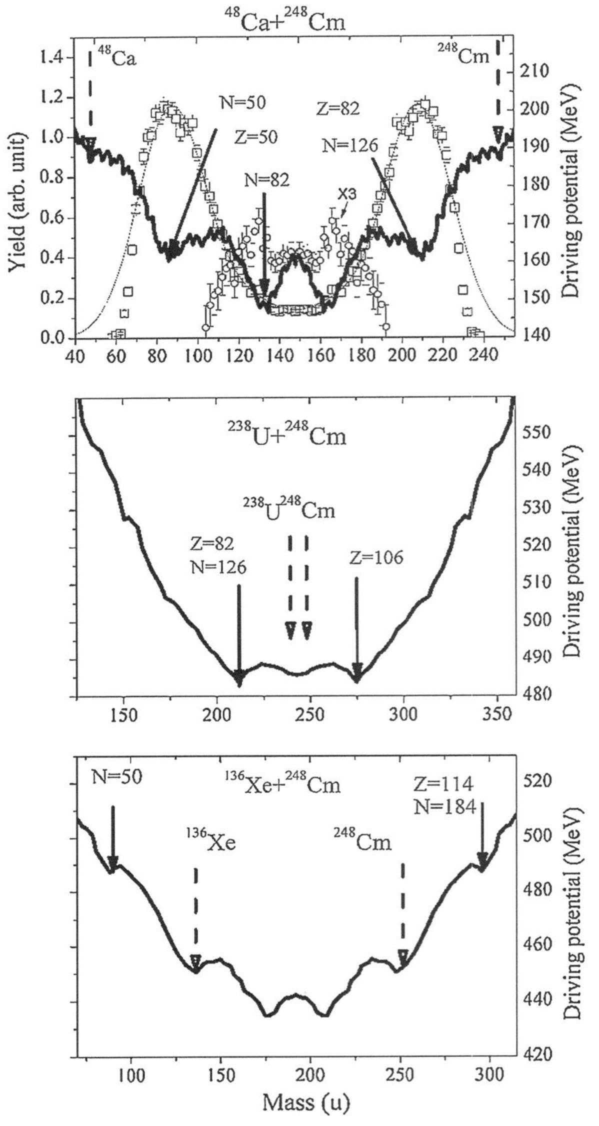


**Fig. 5** The driving potentials as a function of reaction fragment mass at scission point for the reactions $^{48}$Ca+$^{248}$Cm, $^{238}$U+$^{248}$Cm and $^{136}$Xe+$^{248}$Cm

maximum fragment yields correspond to the positions of the local minima of the driving potential as in the case of formation of asymmetric fragments, as well as symmetric ones.

In the middle and bottom panels of Fig. 5 the driving potentials for the composite systems formed in the reactions $^{238}$U+$^{248}$Cm and $^{136}$Xe+$^{248}$Cm calculated in the same approximation are presented, respectively. The dashed arrows indicate the position of the entrance channels, while the solid arrows show the position of the proton and neutron closed shells. It is clearly seen that the local minima in the driving potential exist for all reactions, though in the latest two reactions these minima are located from the outside the entrance channel. Thus, we may expect an increase of the fragment yields in the mass region around these minima. W. Greiner and V. Zagrebaev proposed to call this process "inverse" quasifission in [12]. Notice, that in the case of the reaction $^{238}$U+$^{248}$Cm one of the minima corresponds to doubly magic lead valley and the complementary fragment is a superheavy nucleus around $Z = 106$. In the reaction $^{136}$Xe+$^{248}$Cm both fragments have closed shells: the light fragment is near $N = 50$, the heavy one is close to $Z = 114$ and $N = 184$ (predicted "island of the stability").

The idea of the production of superheavy nuclei in the multi-nucleon transfer reactions in the collision of U+U nuclei (or similar reactions) was already proposed in [13]. In this work it was found that at an incident energy of 7.42 MeV/u (about 22 % above the Coulomb barrier) a direct search for $\alpha$-decay or fission of superheavy nuclei being produced in a deep inelastic collision resulted in an upper cross section limit of 2 nb. Although the stronger penetration of nuclei leads to enhanced mass transfer, the higher excitation energies involved drastically reduce the survival probability of the nuclei produced. The decrease in collision energy to the Coulomb barrier energy leads to the lower total excitation and consequently to larger cross section of survived superheavy nuclei. According to the calculation of the cross section of survived superheavy nuclei formed in the reaction $^{232}$Th+$^{250}$Cf at 800 MeV center-of-mass energy (near the Coulomb barrier) from [12] there is a real chance for production of the long-lived neutron-rich superheavy nuclei in such reactions.

Figure 6 presents the chart of nuclides in the region of superheavy elements. A large success have been achieved in the synthesis of superheavy elements with $Z = 108$–118 in the reaction of "cold" and "warm" fusion. Even though the "warm" fusion reaction leads to the formation of more neutron-rich nuclei than in the case of "cold" fusion even after the deexcitation process, the isotopes of superheavy elements formed in these $^{48}$Ca induced reactions cannot reach the neutron closed shells with $N = 184$ due to the lack of 7–9 neutrons. Moreover, nuclei with $Z > 118$ cannot be synthesized in $^{48}$Ca induced reactions since $^{249}$Cf is the heaviest target material available for these purposes. From the investigation of the mass-energy distributions of binary reaction fragments obtained in the reactions $^{48}$Ca+$^{238}$U, $^{58}$Fe+$^{244}$Pu and $^{64}$Ni+$^{238}$U [8] it was found that the cross section drops three order of magnitude for the formation of the compound nucleus with $Z = 120$ obtained in the reaction $^{64}$Ni+$^{238}$U compared to the formation of the compound nucleus with $Z = 112$ obtained in the reaction $^{48}$Ca+$^{238}$U at an excitation energy of the compound nucleus of about 45 MeV. This is unfortunately a limiting factor. Furthermore, the relative contribution of the CN-fission from $^{64}$Ni+$^{238}$U is much lower than in the case of $^{58}$Fe+$^{244}$Pu, leading to the formation of the same composite system.

Recently the experiments aimed at the synthesis of isotopes of element Z=120 have been performed using the $^{244}$Pu ($^{58}$Fe, xn)$^{302-x}$120 reaction [14] and $^{238}$U($^{64}$Ni, xn)$^{302-x}$120 reaction [15]. A cross section limit of 0.4 pb at $E* = 44.7$ MeV for the

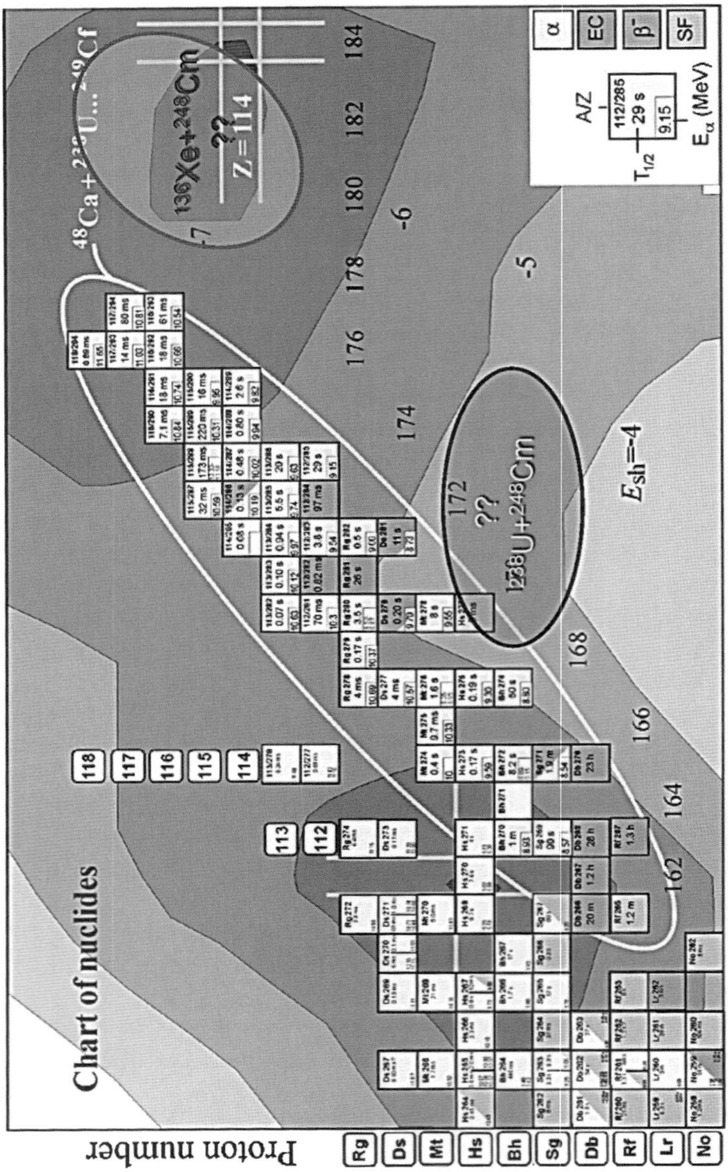


**Fig. 6** The chart of nuclides in the region of superheavy elements

former reaction and 0.09 pb at $E^* = 36.4$ MeV for the latter reaction were obtained. In the case of $^{48}$Ca induced reactions the evaporation residue cross section for 3n, 4n channels is about a few picobarns even for the heaviest nucleus with $Z = 118$.

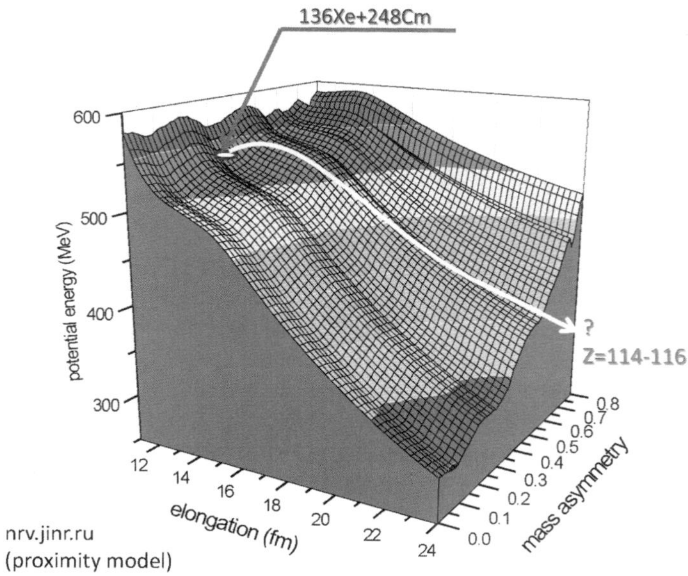


**Fig. 7** Driving potential for the nuclear system formed in the $^{136}$Xe+$^{248}$Cm collision. The *solid line* with *arrow* shows schematically the probable trajectory for the formation of fragments with $Z = 114$–$116$

A possible alternative pathway is represented by the "inverse" quasifission or deep-inelastic reactions in the collision of $^{136}$Xe, $^{232}$Th and $^{238}$U with actinide targets. According to the theoretical expectations the cross section for the survived nuclei formed in such processes is higher than in the reaction of complete fusion. The reaction $^{136}$Xe+$^{248}$Cm can be more interesting due to the fact that the heavy valley of the driving potential for this system corresponds to the nuclei close to the "island of stability", while in the case of the collision of Th and U with actinide targets the heavy valley lies outside the stability line (see Fig. 6). Driving potential as a function of elongation and mass asymmetry for the nuclear system formed in the $^{136}$Xe+$^{248}$Cm collision is shown in Fig. 7. The solid line with arrow shows schematically the possible trajectory for the formation of fragments with $Z = 114$–$116$. To estimate the formation probabilities of superheavy elements in these reactions the additional investigations are needed.

# References

1. M.G. Itkis et al., Nucl. Phys. A **787**, 150c (2007)
2. V.V. Volkov, Phys. Rep. **44**, 93 (1978)
3. R. Bock et al., Nucl. Phys. A **388**, 334 (1982)
4. V.I. Zagrebaev, W. Greiner, J. Phys. G **31**, 825 (2005)
5. G.N. Knyazheva et al., Phys. Part. Nucl. Lett. **5**, 21 (2008)

6. G.N. Knyazheva et al., Phys. Rev. C **75**, 064602 (2007)
7. M.G. Itkis et al., Nucl. Phys. A **734**, 136 (2004)
8. E.M. Kozulin et al., Phys. Lett. B **686**, 227 (2010)
9. K. Nishio et al., in Proceedings of International Symposium on Exotic Nuclei (EXON 2009), ed. by Y.E. Penionzhkevich, S.M. Lukyanov, AIP Conf. Proc. No. 1224 (AIP, Melville, NY, 2010), p. 301
10. I.M. Itkis et al., Phys. Rev. C **83**, 064613 (2011)
11. V.I. Zagrebaev et al. http://nrv.jinr.ru/nrv
12. W. Greiner, V.I. Zagrebaev, in Proceedings of International Symposium on Exotic Nuclei (EXON 2006), ed by Y.E. Penionzhkevich, E.A. Cherepanov, AIP Conf. Proc. No. 912 (AIP, Melville, NY, 2007), p. 221
13. H. Freiesleben et al., Z. Phys. A **292**, 171 (1979)
14. Yu. Ts. Oganessian et al., Phys. Rev. C **79**, 024603 (2009)
15. S. Hofmann et al., GSI Scientific Reports, **131** (2008)

# Search for Superheavy Elements in Nature (Experimental Approach)

A. G. Popeko

**Abstract** Attempts to find superheavy elements in Nature started in the middle of sixties when predictions on the possible longevity of nuclei stabilized by the closed $Z = 114$-proton and $N = 184$-neutron shells have been made. During the ensuing years hundreds of geological samples, their processing products and probes of meteorites were studied. In all experiments only upper limits of the superheavy elements concentration in the studied samples have been determined.

## 1 Background

The term "superheavy nuclei" was introduced by Wheeler in 1955 [1]. After extrapolation of mass formulae and analysis of the data on half-lives of heavy nuclei the author concluded, that one can "expect the existence of nuclei twice as heavy as known nuclear species" (up to $Z = 147$, $N = 500$), and that "the limits to the stability of *superheavy nuclei* are set primarily by neutron escape and by spontaneous fission.

The conclusion, that stability of the heaviest nuclei will be determined by the spontaneous fission, has been earlier drawn by Seaborg [2]. Half-lives of even-even nuclide (at the beginning of 50th only 10 of them were known) relative to spontaneous fission could be described by the Bohr's model [3].

The possibility to exist in Nature of superheavy elements (SHE)—elements heavier than $^{238}$U depends on two determinatives:

- it is necessary, that one of superheavy nuclides would have a lifetime long enough to cross "half" the Galaxy (of about $2 \times 10^4$ y) to be found in cosmic rays, or that, comparable with the age of the Earth (of about $4–5 \times 10^9$ y), to be found in meteorites or terrestrial samples;

A. G. Popeko (✉)
Joint Institute for Nuclear Research, Dubna, Russia
e-mail: popeko@jinr.ru

W. Greiner (ed.), *Exciting Interdisciplinary Physics*,
FIAS Interdisciplinary Science Series, DOI: 10.1007/978-3-319-00047-3_4,

- in the Universe should exist a mechanism leading to the formation of superheavy elements.

## 2 Prediction of Existence of SHE

According to the liquid drop model, there should be the linear dependency between the logarithm of a spontaneous fission half-life and the fissility parameter - $Z^2/A$. Figure 1 shows the dependency of a $\log T_{1/2}(s)$ on the fissility parameter $Z^2/A$ (the dada were taken from [2, 4]).

From the Fig. 1 one can see the giant progress in studying heavy nuclei during past 60 years. The extrapolation of the general tendency of exponential decreasing of spontaneous fission half-lives toward the region of instantaneous fission with $T_{1/2} < 10^{-20}$ s has given a value $(Z^2/A)_{\rm crit} \approx 47$ [2].

It had become obvious, that departures of $\log T_{1/2}$ by several powers of 10 from the linear dependence on the fissility parameter are principal. These deviations have been explained [5] by the suggestion of existence in the transuranium region of a "subshell" corresponding to $N = 152$ and being of a fundamentally different nature than the major closed shells.

The experimental facts indicating the influence of closed shells of $Z = 50$, 82 and of $N = 50$, 82, and 126 ("magic" numbers): numbers of isotopes and isotones, isotopic abundances, energies of radioactive decay, neutron capture cross-sections, were listed by Mayer [6].

For the first time "realistic" predictions of "magic" numbers beyond $^{208}$Pb have been done by Sobiczewski, Gareev and Kalinkin [7] after analyzing single particle nucleon levels in a Woods-Saxon potential well. The possible magic numbers were found to be 114 and $N = 184$.

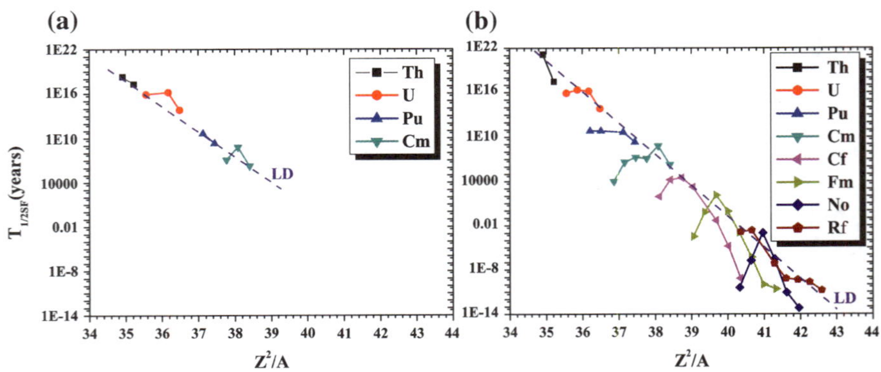


**Fig. 1** Half-lives with respect to spontaneous fission as a function of the fissility parameter $Z^2/A$, **a** as it was in 1952, **b** recent data, 2012. The *dashed lines* represent a linear fit $\log T_{1/2SF}$(y) $= 140.3 - 3.5 \cdot Z^2/A$

Theoretical predictions on the next closed shells numbers vary strongly depending on the model. Following the well-known proton and neutron shells with $Z = 82$ and $N = 126$ ($^{208}$Pb), the shell correction amplitude has a maximum for the superheavy nucleus $^{298}114$ at $N = 184$ in macro-microscopic models. After calculations performed using the Hartree–Fock–Bogoliubov (HFB)-model or self-consistent relativistic mean-field models, the closed spherical proton shells are predicted at $Z = 120$ or 126 [8].

Modern ideas of the influence of nuclear shells on stability of nuclei and results of calculations allow to conclude that the existence of superheavy nuclei with $Z \sim 114$ and $N \sim 184$ is reliably proved theoretically. Estimations of half-lives of such nuclei cover a wide range from fractions of a second till $10^{10}$ years and more.

## 3 Spontaneous Fission Properties of SHE

According to all theoretical predictions, superheavy nuclei self, or of their $\alpha$- or $\beta$-decays will undergo spontaneous fission.

Probabilities of triple fission of nuclei with $Z \sim 114$ should be $10^3$–$10^4$ times higher, than at fission of known nuclei [9]. According to [10, 11] the TKE of fission fragments of superheavy isotopes will make ~300 MeV at fission into two, and ~400 MeV at fission into three fragments.

Average number of prompt neutrons—$\overline{\nu}$ at fission of $^{298}114$ should be ~11 [10]. Authors [11] estimated $\overline{\nu} = 8 - 10$. Kolb [9] concluded, that at a double fission $\overline{\nu} = 8$, however, at a triple fission $\overline{\nu} = 2 - 4$. Hoffman [12] analyzing correlations of mass distribution modes and average numbers of fission neutrons concluded, that at symmetric fission $\overline{\nu} \sim 6$, whereas at asymmetric one $\overline{\nu} \sim 12$.

Uncertainties in predictions should be taken into account at design of detectors for SHE searches.

## 4 Possible Mechanisms of Nucleosynthesis of SHE

In an explosion of supernova the prompt capture of multiple neutrons—r-process becomes possible.

Schramm and Fowler [13], Tonder [14] and Klapdor [15] have shown that at supernova explosions formation of nuclides with $Z \sim 110$ and $A \sim 300$ is possible. The relative abundance of superheavy elements in supernova products can reach $SHE/^{238}U = 0.1 \div 1$ [13].

To the opposite conclusion had been drawn by Howard, Nix [16] and Mathews, Viola [17]. According to [16, 17] nuclide with masses $A > 275$ cannot be formed in astrophysical r-process.

More recent calculations [18] revealed that the abundance of superheavy elements in the r-process can be comparable with that of uranium, but the yield of SHE depends strongly on forecasts of properties of $\beta-$delayed, neutron-induced, and spontaneous fission.

Modern ideas about mechanisms of nucleosynthesis do not forbid formation of superheavy elements. Due to explosions or "star-quakes" interstellar clouds could be enriched by heavy and superheavy elements. Probably such nuclei are a part of nuclear component of cosmic rays. Superheavy elements, could enter in processes of evolution of planetary bodies and be captured in some rocks or minerals.

## 5 Choice of Objects to Search for SHE

The choice of objects perspective for discovery of superheavy elements is based on results of predictions of their chemical and physical properties. Most predictions are based on extrapolations of properties within considered group of elements along with calculations of electronic configurations including relativistic effects [19–23].

Elements $113 \geq Z \geq 118$ should be, correspondingly, analogues of elements from thallium to radon, and a regular filling of the 7p-shell should occurs. The next noble gas, or to be more exact—a noble liquid with the boiling temperature of +15 C should be the element with $Z = 118$ [21]. Elements 119 and 120 should behave like alkaline metals. Starting with the element $Z = 121$ there should happen the filling of $5g$- and $6f$-electron shells, and elements $121 \geq Z \geq 154$ should form a series of superactinides [21].

Unfortunately, due to uncertainties of predicted position of the center of stability "island" ($110 \geq Z \geq 130$) one can consider as light chemical analogue of SHE almost any element from the Mendeleev's periodic system of the elements.

## 6 Experimental Search for SHE in Nature

The search for heavy elements in cosmic rays, terrestrial samples, and meteorites was in the 70's and 80's one of the extensive experimental investigations (see reviews [24–27]).

### 6.1 *Search for SHE Nuclei in Cosmic Rays*

Efforts of experimentalists were appreciably forced by Fowler's studies [28] of nuclear component of cosmic rays. In photoemulsion layers exposed at an upper atmosphere, two tracks have been found out, which were attributed to stops of nuclei with $Z = 90 \pm 4$. After recalibration of emulsions the authors [28] concluded, that the observed tracks corresponded to nuclei with $Z \approx 110$.

More perspective method of studying heavy component of cosmic rays has been proved by Flerov, Otgonsuren and Perelygin [29]. The method is based on revealing of traces from stops of nuclei in minerals from some meteorites, which could accumulate tracks during many millions years.

In olivine crystals from the Marjalahti meteorite [30] two tracks, attributed to nuclei with $Z > 97$, were observed. In a more recent study of Marjalahti meteorite [31] among 853 tracks with $Z > 50$, detected in 27 olivine crystals, 4 tracks, which could be attributed to the $Th - U$ group, were found. No tracks of heavier elements were observed.

The charge and energy of ultra-heavy nuclei will be measured using silicon detectors, aerogel and acrylic Cherenkov counters, and a scintillating optical fiber hodoscope in the frame of the ENTICE—the Energetic Trans-Iron Cosmic-ray Experiment) [32].

## 6.2 Study of Isotopic Anomalies

Accumulation of fission fragments of superheavy nuclei should lead to deviations in isotope abundances of known elements. In the seventies very popular was the hypothesis of Anders, Heyman [33], and Rao [34] about the origin of excess of heavy xenon isotopes in some meteorites—due to the spontaneous fission of superheavy elements. However, this hypothesis, despite its attraction, encountered a number of serious contradictions.

## 6.3 Study of Radiation Damages in Crystals

Decay of superheavy nuclei can result as well in radiation damages in crystalline media. It is known that at $\alpha$-decay of isotopes contained in mineral inclusions, the so-called radioactive halos—spherical zones around a central mineral grain, will be formed. "Giant" halos having unusually large diameters corresponding to $\approx 14\,\text{MeV}$, an energy predicted for nuclides with $Z \sim 126$, were observed [35] in biotite from Madagascar. In order to identify elements around $Z = 126$ by characteristic $X$-rays, the inclusions—monazite crystals were irradiated by protons. In the first experiments [36] $L_{\alpha 1}$ $X$-rays of elements 116, 124 and 126 were observed. Later, a more careful analysis and advanced technique for the excitation of $X$-rays [37] disproved the early results.

## 6.4 Mass-Separation of SHE

Several authors, e.g. [38, 39] tried to separate elements with masses $A = 250 - 350$ using mass-separation technique. Because of a background of molecular complexes of uranium and thorium the sensitivity of a "simple" separation method, as a rule, does not exceed $10^{-10}$–$10^{-11}$ g/g. Application of double mass-separation [40] allows to reduce considerably a background from molecular complexes, but reduces as well the overall sensitivity of this method.

Stephens [41] applied for studying of isotopic structure of platinum ore in experiments on searches for eka-$Pt$ a more sophisticated technique—a tandem accelerator with a special beam line for ion analysis. This method is known as accelerator mass spectrometry (AMS).

In the latest experiments on the search for long-living roentgenium (Rg, $Z = 111$) isotopes in gold samples [42] using the VERA set-up, the sensitivity of $10^{-16}$ g/g was reported.

# 7 Direct Registration of SHE Decays

The most sensitive method of search for superheavy elements is connected with direct registration of their radioactive decays. At interpretation of results of such measurements one usually specify the limiting detectable dimensionless concentration $C$ g/g of a searched element, assuming that their half-life makes $T_{1/2} = 10^9$ years. The searching methods based on detection of $\beta$-particles or $\gamma$-quanta can be ruled out due to extremely high background.

## *7.1 Study of Natural α-Activities*

Experiments on studying weak natural $\alpha$-activities were carried out for many years. A history of observations of 4.4-MeV $\alpha$-activity over the period 1924–1979 in radiogenic haloes, zinc ores, monazite, thorite, huttonite, ultrabasic and other abyssal rocks, osmiridium, uranium ores, and raffinates of uranium is given in [43] (for more recent results see [42]). To search for superheavy elements in nature this method do not seems having prospects due to presence of numerous natural and man-caused $\alpha$-emitters. The use of more advanced technique of detecting coinciding $\alpha$-particles and characteristic $X$-rays [44] can give definitive results, but the overall sensitivity is insufficient for the searching experiments.

## *7.2 Study of Spontaneously Fissioning Activities*

Essentially higher sensitivity of SHE searches provide methods based on detecting of spontaneous fission. This method appears as a universal one, because superheavy nuclides or the nearest products of their $\alpha$-decay should undergo spontaneous fission. The whole accumulated experimental material allows to assert, that in nature exists only one spontaneously fissioning nuclide—uranium-238. Thus the problem of search of new elements by detecting their spontaneous fission can be reduced to reliable detecting of fission events and clearing up their possible origin due to uranium, contained in the sample.

A serious problem cause interactions of active cosmic radiation components—protons, neutrons and muons with heavy nuclei in probes or with detector parts. Whereas the nucleon component can be removed with a shielding of $\approx$10 mwe (meter of water equivalent, $\approx$4 m of concrete), for suppression of muons hundreds of meters shielding are needed. The fluxes of muons were measured in [45] up to 5200 mwe in connection with searches for rare processes.

### 7.2.1 Detecting of Fission Fragments

The first searches for SHE's spontaneous fission events were performed using solid state track detectors: Pb-glasses and glasses containing Bi, Tl, and W, which could accumulate tracks during $5 \div 200$ years [46]. As it was found out later [47, 48] with the use of sandwiched plastic track detectors, the additional background has been produced by cosmic rays.

Even higher sensitivity, but unfortunately also uncertainty, can be reached by etching internal fission tracks in suitable minerals, where they were accumulated over millions of years [49, 50].

Especially for the searches of rare spontaneous fission events of SHE, big proportional counters were developed [51]. The cathodes with an area of $\approx$1.6 $m^2$ were covered with a powdered samples layer, having the surface density $\approx$3 $mg \cdot cm^{-2}$. To reduce the "cosmic" background, the counters were operated in a room with a shield of 5 mwe.

A tricky detector - spinner [52] has been used for the observations of spontaneous fission of superheavy elements in chemical compounds, minerals, and in targets which were irradiated with 24-GeV protons [53]. The principle of operation of spinner is similar to that of bubble-chamber, but the negative pressure is produced in a working liquid by centrifugal forces. The main problem consist in the preparation of samples, which must be solvable in organic substances like ethanol.

## *7.3 Detecting of Prompt Fission Neutrons*

The method, providing the highest sensitivity at the searches for SHE, is the detection of prompt neutrons accompanying fission. Practically all samples are transparent for neutrons, thus probes of up to 100 kg can be inspected. Another advantage is that there is no need to destroy the surveyed probe.

The advanced method of detecting spontaneous fission is based on detection of multiple neutron emission [54]. The detection of 2 and more neutrons unambiguously indicates the occurrence of spontaneous fission. Several kinds of neutron multiplicity detectors based on $^3$He-filled proportional counters were designed at Flerov laboratory [55, 56].

This method provide for detecting spontaneous fission events the efficiency from 15–30 % ($\overline{\nu} \approx 2$) to about 50 % ($\overline{\nu} \approx 4$). The expected $\overline{\nu}$-values for superheavy elements were discussed earlier.

The major part of measurements was performed in the salt mine at a depth of 1100 mwe (passive shielding) with additional suppression of cosmic muons by active shielding based on Geiger-Müller counters. Long term measurements with high-purity samples (metallic lead, ferric oxide, melted quartz) showed that the "cosmic" background was less than 1 event per year. Thus, the only one source of background which should be considered is the spontaneous fission of uranium.

## 7.4 Search for SHE in Terrestrial Samples

At the first stage of searches several dozens of various ores and their processing products, products of metallurgy, ferro-manganese nodules, chemical preparations rich in platinum metals and gold, volatile metals such as Hg, Tl, Pb, Bi, rare earth elements were studied. The limits for the abundance of SHE were found to be less than $10^{-12}$ g/g [57, 58].

## 7.5 Search for SHE in Cheleken Peninsula Geothermal Waters

The heavy volatile metals content (e.g. thallium, lead, etc.) of the Cheleken Peninsula (the south-eastern coast of the Caspian Sea) water [59] is nearly 100 times that of oceanic water. These metals could escape from the Earth's crust, together with other volatile components. The extraction of heavy elements was carried out [59] using a vinyl-pyridine anion-exchange resin. Some 2,000 m$^3$ of the water were passed through a column containing 850 kg of the resin.

During a 88-day exposure of 9 kg of the saturated resin a total of 42 spontaneous fission events (0.5 events per day) have been recorded at neutron multiplicity detectors [60]. The analysis of the uranium content of the saturated resin, carried out using different methods ($(2-3) \times 10^{-8}$ g/g), had shown that the background due to spontaneous fission of uranium did not exceed 1 event from 42. Thus it was supposed that the detected spontaneously fissioning nuclide belongs to the region of superheavy elements.

Experiments were carried out aiming at the concentration of the detected activity by extraction of various chemical elements from 170 kg of the saturated resin. The fission counting rate of the obtained hydroxides was 5 counts per day, which corresponded to about 50 % of the initial spontaneous fission activity of the resin. It should be mentioned, that 1 atom of $^{252}$Cf per 5 g of saturated resin could explain the observations. Further attempts [61] to concentrate the detected activity failed.

**Table 1** Search for spontaneous fission in meteorites

| Sample | $^{238}$U g/g | Weight kg | ε% | Time days | N of events n = 2 | n = 3 | n = 4 |
|---|---|---|---|---|---|---|---|
| Saratov | $3.0 \cdot 10^{-9}$ | 5.2 | 22 | 94 | 4 | 1 | 0 |
| Allende | $1.6 \cdot 10^{-8}$ | 3.9 | 22 | 40 | 3 | 0 | 0 |
| Allende | $1.6 \cdot 10^{-8}$ | 22.5 | 12 | 55 | 10 | 1 | 0 |
| Efremovka | $4.0 \cdot 10^{-8}$ | 11.7 | 12 | 105 | 14 | 1 | 0 |
| Pb | $5.0 \cdot 10^{-9}$ | 100 | 22 | 15 | 0 | 0 | 0 |
| Empty | – | – | 22 | 200 | 0 | 0 | 0 |
| Allende | $1.6 \cdot 10^{-8}$ | 10.5 | 30 | 45 | 5 | 2 | 1 |
| $SiO_2$+$MnO_2$ | $< 10^{-9}$ | 10.0 | 30 | 70 | 0 | 0 | 0 |
| Pb | $5.0 \cdot 10^{-9}$ | 150 | 30 | 5 | 0 | 0 | 0 |
| Empty | – | – | 30 | 50 | 0 | 0 | 0 |

## 7.6 Search for SHE in Meteorites

Samples of meteorites "Saratov", "Efremovka" and "Allende" have been selected from the meteoric collection of Academy of Science of USSR. These meteorites belong to the class of chondrites and represent the less differentiated substance of the Solar system. The low uranium concentration in chondrites $3 \cdot 10^{-9}$–$4 \cdot 10^{-8}$ g/g allowed one to realize the maximum sensitivity.

Our investigations of meteorites were performed in 1972–1976 [62–64]. The results of measurements of spontaneous fission activity of meteorites along with background measurements with artificial samples analogous to these meteorites are presented in Table 1.

The observed counting rate was, on the average, 1 event per 5 days for 10 kg of a sample, what was $10 \div 30$ times higher than could originate from uranium contents and other background sources. Thus it was assumed that a long-lived spontaneously fissioning nuclide, possibly belonging to the region of SHE, is present in the studied samples of meteorites.

The isolation of the observed unknown nuclide (together with other volatile elements) has been performed by Zvara [65] from several kilograms of the Allende meteorite in hydrogen and then oxygen flows. Authors obtained an indication of its volatility in the elemental or oxidation state, but further attempts to increase the concentration of observed activity failed.

## 7.7 Discussion of Early Results

Figure 2 shows the correlation between the spontaneous fission activity observed in several samples, with the uranium concentration.

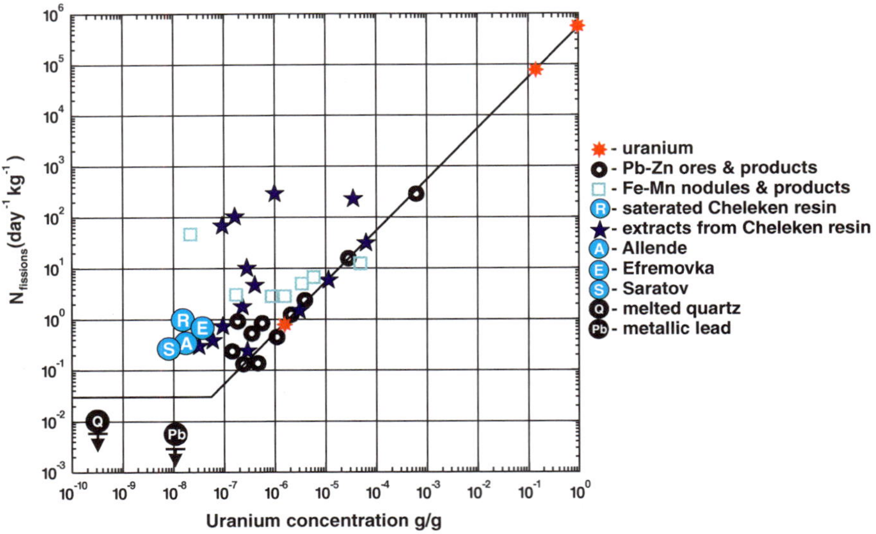


**Fig. 2** Correlation between the uranium concentration and the spontaneous fission activity detected in the samples, recalculated to $events \cdot day^{-1} \cdot \mathrm{kg}^{-1}$

As one can see the counting rate for the majority of survived samples corresponds to the spontaneous fission activity of uranium. The activity in meteorites and products of processing of Cheleken geothermal waters exceeds that from uranium. But, also it did not show any clear correlation to known elements, and it did not follow them in attempts of further concentration.

In all experiments only upper limits of the superheavy element concentration in the studied samples have been determined. All this resulted in a pessimistic view on the possible existence of SHE in nature, and the searching experiments were practically stopped in the mid of 80-th.

# 8 New Samples for SHE Searches

The experimental data accumulated, and development of modern microscopic models during passed 30 years, simulated new approach to the search for perspective objects [66]. A noticeable increase in $T_{1/2}(\alpha)$ and $T_{1/2}(SF)$ may be expected for nuclei with $Z < 110$, which have not been yet looked for.

In accordance with the calculations [67], lifetimes of the isotopes $^{290-292}$Hs and $^{290-293}$Ds fall within the range $10^{10} \div 10^{14}$ s, and according to [68], these isotopes are $\beta$-stable (or have long lifetimes).

Considering different nuclei as objects for studies, it turns out that for element 108—Hs, the chemical homologue of Os, the chances to be found in terrestrial samples could be favorable. The search for rare decays may be undertaken with a metallic sample of raw Os.

Perspectives for SHE search, can follow from the discovered high volatility of Cn ($Z = 112$) and element $Z = 114$. These elements can be gases (noble) at normal conditions—the boiling temperature of Cn is $(360 \pm 100)$ K. Thus, e.g., one can look for SHE in heavy fractions of Xe production.

These experiments are running now with an especially designed neutron multiplicity detector [56] in the underground laboratory in Modane (France).

## 9 Conclusion

The problem of the existence of superheavy elements in nature belongs to the most fundamental, because it affects nuclear and atomic physics, quantum chemistry, astrophysics, cosmology, and undoubtedly the efforts to solve it will be continued.

## References


1. J.A. Wheeler, in *Proceedings of the International Conference on the Peaceful Uses of Atomic Energy*, Geneva, August 1955, vol. 2 (United Nations, New York, 1956), p.155
2. G.T. Seaborg, Phys. Rev. **85**, 157 (1952)
3. N. Bohr, J.A. Wheeler, Phys. Rev. **56**, 426 (1939)
4. http://www.oecd-nea.org/janis, database ENDF/B-VII.0
5. A. Ghiorso et al., Phys. Rev. **95**, 293 (1954)
6. M.G. Mayer, Phys. Rev. **74**, 235 (1948)
7. A. Sobiczewski, F.A. Gareev, B.N. Kalinkin, Phys. Lett. **22**, 500 (1966)
8. A. Sobiczewski, K. Pomorski, Prog. Part. Nucl. Phys. **58**, 292 (2007)
9. D. Kolb, Phys. Lett. B **65**, 319 (1976)
10. J.R. Nix, Phys. Lett. B **30**, 1 (1969)
11. H. Schmitt, U. Mosel, Nucl. Phys. A **186**, 1 (1972)
12. D.C. Hoffman, in *Proceedings of the International Symposium on Superheavy Elments, Lubboc, Texas*, 1978, (Pergamon Press Inc., 1978), p. 89
13. D.N. Schramm, W.A. Fowler, Nature **231**, 103 (1971)
14. F. Tondeur, Z. Phys, A **297**, 61 (1980)
15. H.V. Klapdor et al., Z. Phys. **299**, 213 (1981)
16. W.M. Howard, J.R. Nix, Nature **247**, 17 (1974)
17. G.J. Mathews, V.E. Viola, Nature **261**, 382 (1976)
18. I.V. Panov, IYu. Korneev, F.-K. Thielemann, Phys. At. Nucl. **72**, 1026 (2009)
19. B. Fricke, W. Greiner, J.T. Waber, Theoret. Chim. Acta **21**, 235 (1971)
20. G.T. Seaborg, Phys. Scripta **10A**, 5 (1974)
21. B. Fricke, Struct. Bond **21**, 89 (1975)
22. P. Schwerdtfeger, M. Seth, J. Nucl. Radiochem. Sci. **3**, 133 (2002)
23. A.V. Zaitsevskii, E.A. Rykova, A.V. Titov, Rus. Chem. Rev. **77**, 205 (2008)
24. G. Herrmann, Nature **280**, 543 (1979)
25. G.N. Flerov, G.M. Ter-Akopian, Atomki Koezlemenyk **21**, 93 (1979)
26. G.N. Flerov, G.M. Ter-Akopian, Rep. Prog. Phys. **46**, 817 (1983)

27. G. Herrmann, in *The Chemistry of Superheavy Elements*, ed. by M. Schädel, (Kluwer Academic Publishers, 2003), p. 291
28. P.H. Fowler et al., Proc. Roy. Soc. A **301**, 39 (1967)
29. G.N. Flerov, O. Otgonsuren, V.P. Perelygin, Izvestia AN SSSR, ser. fiz., **39**, 388 (1975)
30. V.P. Perelygin, S.G. Stetsenko, Soviet. Lett. J.E.T.Ph. **32**, 622 (1980)
31. A.B. Aleksandrov et al., Bull. Lebedev Phys. Inst. **35**, 276 (2008)
32. W.R. Binns et al., in *Proceedings of 31st International Cosmic Ray Conference*, Łódź, 2009
33. E. Anders, D. Heyman, Science **164**, 821 (1969)
34. M.N. Rao, Nucl. Phys. A **170**, 69 (1970)
35. R.V. Gentry, Science **169**, 670 (1970)
36. R.V. Gentry et al., Phys. Rev. Lett. **37**, 11 (1976)
37. C.J. Sparks et al., Phys. Rev. Lett. **40**, 507 (1978)
38. S.G. Nilsson, S.G. Thompson, S.F. Tsang, Phys. Lett. B **28**, 458 (1969)
39. J. McMinn et al., Nucl. Instr. Meth. **139**, 175 (1976)
40. C. Stephan et al., J. Phys. **36**, 105 (1975)
41. W. Stephens et al., Phys. Rev. C **21**, 1664 (1980)
42. F. Dellinger et al., Phys. Rev. C **83**, 015801 (2011)
43. R.J. Dougan, J.D. Illige, E.K. Hulet, Preprint UCRL-52886 (University of California, Livermore, California, 1979)
44. V. Kusch et al., Atomnaya Energiya **31**, 159 (1971)
45. A.S. Malgin, O.G. Ryazhskaya, Phys. At. Nucl. **71**, 1769 (2008)
46. G.N. Flerov, V.P. Perelygin, Sov. J. Atom. Energy **26**, 603 (1969)
47. G.N. Flerov, V.P. Perelygin, O. Otgonsuren, Sov. J. Atom. Energy **33**, 1144 (1972)
48. F.H. Geisler, P.R. Phillips, R.M. Walker, Nature **244**, 428 (1973)
49. P.B. Price, R.L. Fleischer, R.T. Woods, Phys. Rev. C **1**, 1819 (1970)
50. L.L. Kahskarov et al., Microsymposium **38**, MS042 (2003)
51. G.N. Flerov et al., Yad. Fiz. **20**, 472 (1974)
52. B. Hahn, Il Nuovo Cimento **22**, 650 (1961)
53. K. Behringer et al., Phys. Rev. C **9**, 48 (1974)
54. G.M. Ter-Akopian, A.G. Popeko et al., Preprint JINR R13–5394, Dubna, 1970. Soviet Patent No 450508
55. G.M. Ter-Akopian, A.G. Popeko et al., Nucl. Instr. Meth. **190**, 119 (1981)
56. A. Svirikhin et al., AIP Conf. Proc. **1175**, 297 (2009)
57. G.N. Flerov et al., Yad. Fiz. **20**, 639 (1974)
58. G.N. Flerov et al., Yad. Fiz. **21**, 9 (1975)
59. YuT Chuburkov et al., Radiochimija **16**, 827 (1974)
60. G.N. Flerov et al., Z. Phys. A **292**, 43 (1979)
61. YuT Chuburkov, A.G. Popeko, N.K. Skobelev, Sov. Radiochem. **30**, 108 (1988)
62. A.G. Popeko et al., Phys. Lett. B **52**, 417 (1974)
63. A.G. Popeko et al., Yad. Fiz. **21**, 1220 (1975)
64. G.N. Flerov et al., Yad. Fiz. **26**, 449 (1977)
65. I. Zvara et al., Yad. Fiz. **26**, 455 (1977)
66. Yu. Ts, Oganessian. J. Phys. G **34**, R165 (2007)
67. R. Smolanczuk, Phys. Rev. C **56**, 812 (1997)
68. P. Möller, J.R. Nix, K.-L. Kratz, At. Data Nucl. Data Tables **66**, 131 (1997)

# Superheavies: Short-Term Experiments and Far-Reaching Designs

**V. I. Zagrebaev, A. V. Karpov, I. N. Mishustin and Walter Greiner**

**Abstract** Low values of the fusion cross sections and very short half-lives of nuclei with $Z > 120$ put obstacles in synthesis of new elements. However the fusion reactions of medium mass projectiles with different actinide targets still can be used for the production of the not-yet-synthesized SH nuclei. The gap of unknown SH nuclei, located between the isotopes which were produced earlier in the cold and hot fusion reactions, could be filled in fusion reactions of $^{48}$Ca with available lighter isotopes of Pu, Am, and Cm. Cross sections for the production of these nuclei are predicted to be rather large, and the corresponding experiments can be easily performed at existing facilities. The use of heavier actinide targets give us a chance to produce more neutron enriched SH isotopes. Moreover, for the first time, a narrow pathway is found to the middle of the island of stability owing to possible $\beta^+$ decay of SH isotopes which can be formed in ordinary fusion reactions of stable nuclei. Multi-nucleon transfer processes at near barrier collisions of heavy (and very heavy, U-like) ions seem to be quite realistic reaction mechanism allowing us to produce new neutron enriched heavy nuclei located in the unexplored upper part of the nuclear map. Neutron capture reactions can be also used for the production of the long-living neutron rich SH nuclei. Strong neutron fluxes might be provided by pulsed nuclear reactors and by nuclear explosions in laboratory conditions and by supernova explosions in nature. All these possibilities are discussed in the chapter.

---

V. I. Zagrebaev (✉) · A. V. Karpov
Joint Institute for Nuclear Research, Dubna, Russia
e-mail: zagrebaev@jinr.ru

A. V. Karpov
e-mail: karpov@jinr.ru

I. N. Mishustin · W. Greiner
Frankfurt Institute for Advanced Studies, Frankfurt, Germany
e-mail: mishustin@fias.uni-frankfurt.de

W. Greiner
e-mail: greiner@fias.uni-frankfurt.de

W. Greiner (ed.), *Exciting Interdisciplinary Physics*,
FIAS Interdisciplinary Science Series, DOI: 10.1007/978-3-319-00047-3_5,

# 1 Motivation

It is well known that the last element whose lifetime is comparable to the age of earth and that occurs in macroscopic quantities in nature is uranium. All the other elements with $Z > 92$ have been produced in laboratory experiments (see historical review [1]). The progress in this field is quite impressive—26 handmade new heavy elements have been synthesized within 60 years. Some transuranium elements (up to californium) are produced in considerable quantity (by neutron capture process accompanied with $\beta^-$-decay in nuclear reactors) sufficient to prepare a target which can be used for synthesis of the next superheavy (SH) elements in fusion reactions.

However the transuranium elements become more and more unstable as they get bigger. In the late sixties, the dream of the rather stable SH elements arose. Theoretical nuclear physicists around S. G. Nilsson (Lund) [2, 3] and from the Frankfurt school [4, 5] predicted that so-called closed proton and neutron shells should counteract the repelling Coulomb forces. Atomic nuclei with special "magic" proton and neutron numbers and their neighbours could again be rather stable.

Many attempts to find more or less stable SH elements in nature were not succeeded yet [6]. The "cold" fusion reactions based on the closed shell target nuclei of lead and bismuth (which looked initially very promising) lead to the production of proton rich isotopes of SH elements with very short half-lives located far from the beta-stability line [7, 8]. Many years ago it was proposed to produce the most neutron rich isotopes of SH elements in fusion of $^{48}$Ca with available actinide targets, $^{244}$Pu, $^{248}$Cm and others [9]. Such a possibility has been realized only recently. The isotope of element 112, $^{285}$Cn, observed in the decay chains of SH nuclei $^{289}$114 and $^{293}$116 produced in the 3n evaporation channels of the $^{48}$Ca+$^{244}$Pu [10] and $^{48}$Ca+$^{248}$Cm [11] fusion reactions, reveals very long half-life of about 30 s. This is five orders of magnitude longer as compared with the half-life of more neutron deficient isotope $^{277}$Cn produced in the "cold" fusion reaction [7]. This fact evidently confirms an existence of the island of stability! However one needs to add 6–8 neutrons more to reach the most stable SH nuclei of this island, which is impossible in any fusion reactions of stable beams with available targets.

Anyhow, a ten years epoch of $^{48}$Ca irradiation of actinide targets for the synthesis of SH elements is over. The heaviest available target of californium ($Z = 98$) had been used to produce the element 118 [12]. Note, that earlier predicted more or less constant value (of a few picobarns) of the cross sections for the production of SH elements with $Z = 112 \div 118$ in $^{48}$Ca induced fusion reactions [13, 14] (caused by the gradual increase of the fission barriers of the compound nuclei formed in these reactions) have been fully confirmed by the experiments performed in Dubna and later in Berkeley [15] and GSI [16, 17].

To get SH elements with $Z > 118$ in fusion reactions, one should proceed to heavier than $^{48}$Ca projectiles. The strong dependence of the calculated evaporation residue (EvR) cross sections for the production of element 120 on the mass asymmetry in the entrance channel makes the nearest to $^{48}$Ca projectile, $^{50}$Ti, most promising for further synthesis of SH nuclei [18]. The use of the titanium beam instead of $^{48}$Ca decreases the yield of SH nuclei (by factor 20 on average) mainly due to a worse fusion probability. The estimated EvR cross sections for the 119 and 120 SH elements synthesized in the $^{50}$Ti induced fusion reactions [18] ($\sim$ 0.05 pb) are quite reachable at available experimental setups, though one needs much longer time of irradiation than for the $^{48}$Ca fusion reactions. The yield of SH nuclei (number of events per day) depends not only on the cross section but also on the beam intensity and target thickness. In this connection other projectile–target combinations should be also considered. Most neutron-rich isotopes of element 120 may be synthesized in the $^{54}$Cr+$^{248}$Cm fusion reactions.

For the moment $^{249}_{98}$Cf ($T_{1/2}$ = 351 yr) is the heaviest available target that can be used in experiments. The half-life of the einsteinium isotope, $^{254}_{99}$Es, is 276 days, sufficient to be used as target material. This isotope might be produced in nuclear reactors, but it is rather difficult to accumulate the required amount of this matter (several milligrams) to prepare a target. Still the estimated cross section for the production of element 119 in the hypothetical $^{48}$Ca+$^{254}$Es fusion reaction is about 0.3 pb, which is more promising as compared with the $^{50}$Ti+$^{249}$Bk fusion reaction. The calculated excitation functions for the synthesis of SH elements 119 and 120 in the fusion reactions of $^{48}$Ca, $^{50}$Ti and $^{54}$Cr with actinide targets are shown in Fig. 1. The experiments with titanium and chromium beams aimed on the production of SH elements 119 and 120 are currently in progress at GSI.

The synthesis of these nuclei may encounter also another important problem. The proton rich isotopes of SH elements produced in these reactions are rather short-lived owing to large values of $Q_\alpha$. Their half-lives are very close to the critical value of 1 $\mu$s needed for the CN to pass through the separator up to the focal plane detector. The next elements (with $Z > 120$) being synthesized in such a way might be already beyond this natural time limit for their detection.

Thus, future studies of SH elements are obviously connected with the production of neutron enriched and longer living isotopes of SH nuclei. The possibilities of using radioactive beams, multi-nucleon transfer reactions and neutron capture processes for this purpose are discussed in Refs. [18–20]. At the same time an important area of SH isotopes located between those produced in the cold and hot fusion reactions remains

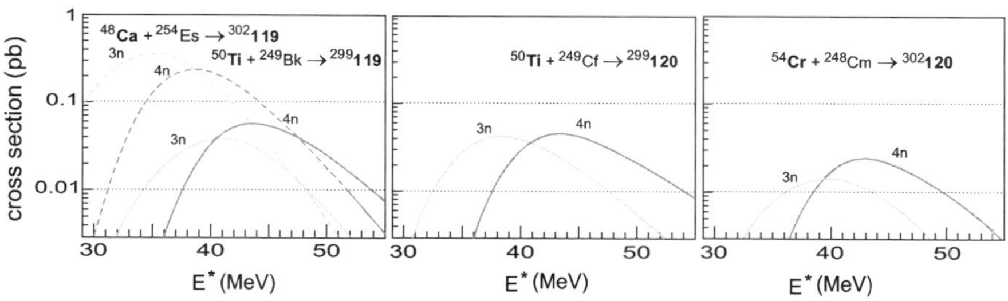


**Fig. 1** Excitation functions for the production of SH elements 119 and 120 in the 3n and 4n evaporation channels of the $^{48}$Ca+$^{254}$Es (*dashed curves*), $^{50}$Ti+$^{249}$Bk, $^{50}$Ti+$^{249}$Cf and $^{54}$Cr+$^{248}$Cm (*solid curves*) fusion reactions

unstudied yet. Approaching the island of stability (see above) testifies about strong shell effects in this area of the nuclear map. Understanding these effects as well as other properties of SH nuclei is impeded significantly by the absence or fragmentary character of experimental data on decay properties of the not-yet-synthesized isotopes of already known SH elements.

## 2 How can the Gap in Superheavy Mass Area be Filled?

As can be seen from Fig. 2, there is an important area of SH isotopes located between those produced in the cold and hot fusion reactions remains unstudied yet. This is explained by extremely low values of the corresponding production cross sections. However, recently, the synthesis of SH elements at the level of 1 pb became more or less a routine matter at several laboratories. The corresponding experiments require about 2-week irradiation time to detect several events (decay chains) of SH element formation. This means that many more unknown isotopes of SH elements could be synthesized now, and the gap between nuclei produced in the cold and hot fusion reactions could be closed at last.

Note that it can be done with the use of ordinary fusion reactions and, thus, with the use of existing recoil separators, in contrast with the mass-transfer reactions (see below) for which separators of a new kind are needed. For this purpose several (rather cheap and available) isotopes of actinide elements can be used as the targets (for example, $^{233,235}U$, $^{239,240}Pu$, $^{241}Am$, $^{243}Cm$ and so on). Besides $^{48}Ca$, the beams of $^{36}S$, $^{44}Ca$ and $^{40}Ar$ are also of interest. We found that it is more convenient (and easier) to fill the gap "from above" by synthesis of new isotopes of SH elements with larger values of $Z$, their subsequent $\alpha$ decay chains just fill the gap [21]. This unexpected finding is simply explained by greater values of survival probabilities

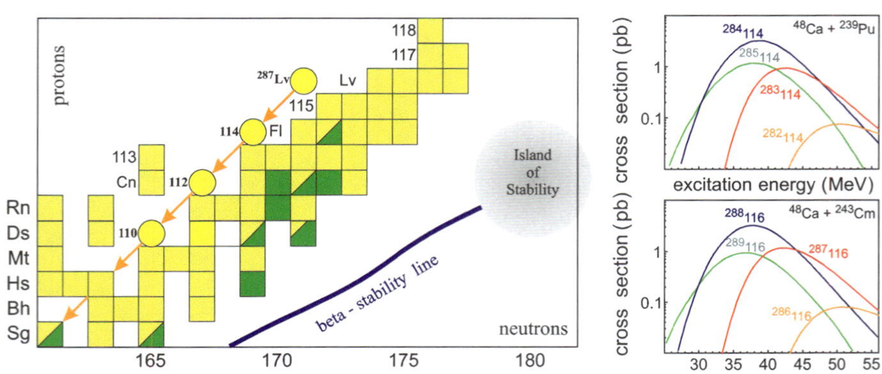


**Fig. 2** Filling the gap in SH mass area. Production cross sections for the new isotopes of elements 114 (Fl) and 116 (Lv) in the $^{48}Ca+^{239}Pu$ and $^{48}Ca+^{243}Cm$ fusion reactions. Decay chain of the isotope $^{287}Lv$ (4n channel of the $^{48}Ca+^{243}Cm$ fusion reaction) is shown on the *left panel*

of the corresponding nuclei with $Z = 115, 116$ as compared to those with $Z = 111, 112$. The values of $B_f - B_n$ are much higher for compound nuclei with $Z \sim 116$ as compared with compound nuclei of 112 element formed in fusion reactions of $^{48}$Ca with neutron deficient isotope of uranium. As a result, the corresponding survival probability of lighter CN is smaller by more than one order of magnitude.

In Fig. 2 the calculated EvR cross sections are shown for the production of new isotopes of elements 114 and 116 in the fusion reactions of $^{48}$Ca with $^{239}$Pu and $^{243}$Cm targets. More than ten new isotopes of even elements from $Z = 104$ to 116 could be produced in the $^{48}$Ca+$^{239}$Pu and/or $^{48}$Ca+$^{243}$Cm fusion reactions which just fill the gap in the superheavy mass area. The production cross sections are high enough to perform such experiments at available facilities. All the decay chains reach finally known nuclei. This fact significantly facilitates the identification of the new SH isotopes. Note, that high intensive beam of $^{40}$Ar could be also used. This material is much cheaper than $^{48}$Ca. However we found that the use of an $^{40}$Ar beam is less favorable as compared with $^{48}$Ca [21]. This is attributable to the much "hotter" character of the $^{40}$Ar+$^{251}$Cf fusion reaction. As a result, the corresponding excitation functions for this reaction are shifted to higher energies, and the cross sections are lower by one order of magnitude.

The $^{48}$Ca+$^{241}$Am fusion reaction is the best for the production of the new isotopes of odd SH elements filling the gap. The production cross sections for the new isotopes $^{284-286}$115 in this reaction are about 0.1 pb, 2 pb and 4 pb, respectively, i.e. high enough to be measured. The more neutron deficient isotopes of element 115 could be produced in the $^{44}$Ca+$^{243}$Am fusion reaction ($^{44}$Ca is a more abundant and available material as compared to $^{48}$Ca). However in this reaction the excitation energy of the formed CN is 10 MeV higher than in the $^{48}$Ca+$^{241}$Am fusion reaction. As a result, the corresponding excitation functions are shifted to higher energies at which the survival probability of the CN is much lower. Thus, the $^{48}$Ca beam is preferable also for the production of neutron deficient SH nuclei in fusion reactions with lighter isotopes of actinide targets as compared to the use of $^{42-44}$Ca beams and heavier actinide targets.

## 3 The Narrow Pathway to the Island of Stability

It is well known that there are no combinations of available projectiles and targets, fusion of which may lead to SH nuclei located at the island of stability. Only the proton rich isotopes of SH elements have been produced so far in fusion reactions (see Fig. 2). Radioactive ion beams hardly may solve this problem. Fusion cross sections for relatively light radioactive projectiles (like $^{22}$O, for example) are rather high and beam intensity of about $10^8$ pps is sufficient for synthesis of SH nuclei [18]. However the nuclei, being synthesized in such a way, would be also neutron deficient. For example, in the $^{22}$O+$^{248}$Cm fusion reaction one may produce only already known neutron deficient isotopes of rutherfordium, $^{265-267}$Rf. In fusion reactions with heavier radioactive projectiles (like $^{44}$S, for example) new neutron enriched isotopes of SH elements could be really produced, but in this case one needs to have a beam intensity of about $10^{12}$ pps to reach in experiment a 1 pb level of the corresponding EvR cross section [18], which is not realistic for the nearest future.

Still several more neutron rich actinide targets ($^{250}$Cm, $^{251}$Cf, $^{254}$Es) could be used, in principle, for production of SH nuclei shifted by one or two neutrons to the right side from those already synthesized in $^{48}$Ca induced fusion reactions (though they will be far from the beta-stability line, see Fig. 2). New neutron rich isotopes of elements 116 ($^{294,\,295}$116) and 118 ($^{295,\,296}$118) may be synthesized in 3n and 4n evaporation channels of the $^{48}$Ca+$^{250}$Cm and $^{48}$Ca+$^{251}$Cf fusion reactions with cross sections of about 1 pb [21].

More interesting feature of the fusion reactions $^{48}$Ca+$^{250}$Cm and $^{48}$Ca+$^{254}$Es (as well as the 2n evaporation channel of the reaction $^{48}$Ca+$^{249}$Bk) is an unexpected possibility to reach the middle of the island of stability just in fusion processes of "stable" nuclei. In these reactions relatively neutron rich isotopes of SH elements 114 and 115 are formed as $\alpha$ decay products of evaporation residues of the corresponding CN. These isotopes should have rather long half-lives and, thus, they could be located already in the "red" area of the nuclear map, i.e., they may be $\beta^+$-decaying nuclei [22]. In Fig. 3 the EvR cross sections are shown for the synthesis of elements 116, 117 and 119 formed in fusion reactions of $^{48}$Ca with $^{250}$Cm, $^{249}$Bk and $^{254}$Es targets.

In accordance with our calculations of decay properties of SH nuclei [22], the isotopes $^{291}$115 and $^{291}$114 may experience not only $\alpha$ decay but also electron capture with half-life of several seconds. If it is correct, the narrow pathway to the middle of the island of stability is surprisingly opened by the production of these isotopes in subsequent $\alpha$-decays of elements 116, 117 and/or 119 produced in the $^{48}$Ca+$^{250}$Cm, $^{48}$Ca+$^{249}$Bk and $^{48}$Ca+$^{254}$Es fusion reactions, see Fig. 3. The corresponding cross sections of these reactions are rather low, they are about 0.8 pb for the 3n evaporation channel of the $^{48}$Ca+$^{250}$Cm fusion reaction and 0.3 pb for the two last reactions. However, for the moment, this is the only method which is proposed for the production of SH nuclei located just in the middle of the island of stability.

## 4 Production of SH Nuclei in Multi-Nucleon Transfer Reactions

Due to the bending of the stability line forwards the neutron axis, in all fusion reactions only *proton rich* SH nuclei with a short half-life can be produced located far from the island of stability (see Fig. 2). The multi-nucleon transfer processes in low-energy collisions of heavy (and very heavy, U-like) nuclei could be quite practicable for the production of new neutron rich isotopes of SH elements [23]. Additional enhancement of the corresponding cross sections may originates here due to the shell effect. We called it "inverse quasi-fission" process [23]. In this process one of the heavy colliding partners, say $^{238}$U, transforms to lighter doubly magic nucleus $^{208}$Pb while the other one, say $^{248}$Cm, transform to the complementary SH nucleus. The role of these shell effects in damped collisions of heavy nuclei is still

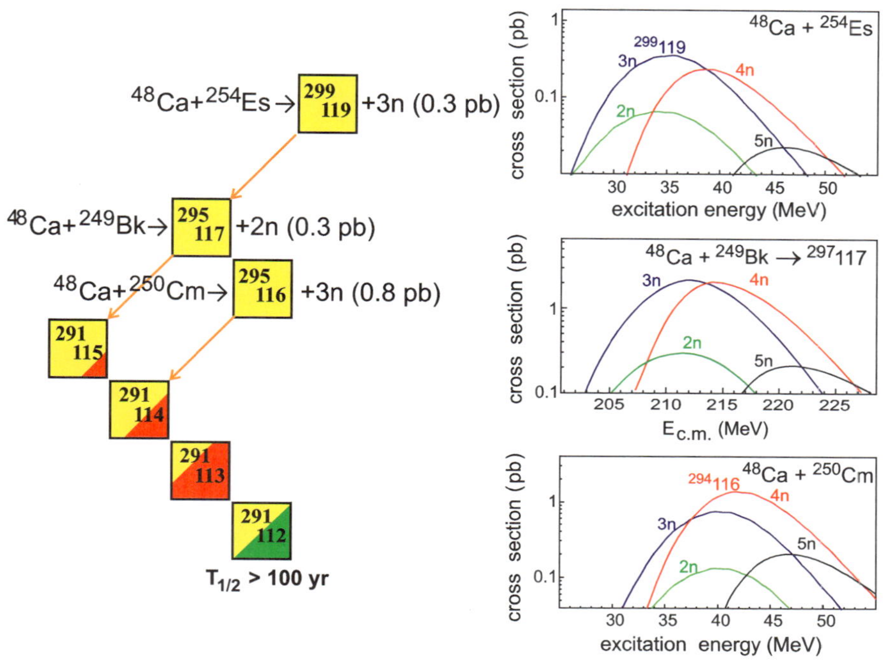


**Fig. 3** Production cross sections of elements 116, 117 and 119 in the fusion reactions of $^{48}$Ca with $^{250}$Cm, $^{249}$Bk and $^{254}$Es targets. The numbers near the *curves* indicate the corresponding neutron evaporation channels. The possible pathway to the middle of the island of stability via a $\beta^+$ decay of the isotopes $^{291}$115 and $^{291}$114 is shown

not absolutely clear and was not carefully studied experimentally. However very optimistic experimental results were obtained recently [24] confirming such effects in the $^{160}$Gd+$^{186}$W reaction, for which the similar "inverse quasi-fission" process ($^{160}$Gd→$^{138}$Ba while $^{186}$W→$^{208}$Pb) has been also predicted [25].

In multi-nucleon transfer reactions the yields of SH elements with masses heavier than masses of colliding nuclei strongly depend on the reaction combination. For example, the cross sections for the production of Fermium isotopes in the U+Cm combination are two orders of magnitude larger as compared with the U+U combination [26]. We found that the cross sections for the production of *neutron rich* transfermium isotopes in reactions with $^{248}$Cm target change sharply if one changes from medium mass (even neutron rich) projectiles to the uranium beam. In Fig. 4 the charge and mass distributions of heavy primary reaction fragments are shown for near barrier collisions of $^{238}$U, $^{136}$Xe and $^{48}$Ca with curium target. The "lead shoulder" manifests itself in all these reactions. However, for $^{136}$Xe+$^{248}$Cm and $^{48}$Ca+$^{248}$Cm collisions it corresponds to the usual (symmetrizing) quasi-fission process in which nucleons are transferred mainly from the heavy target (here it is $^{248}$Cm) to the lighter projectile. This is a well studied process both experimentally [27] and theoretically [28]. It is caused just by the shell effects leading to the deep lead valley on the multi-

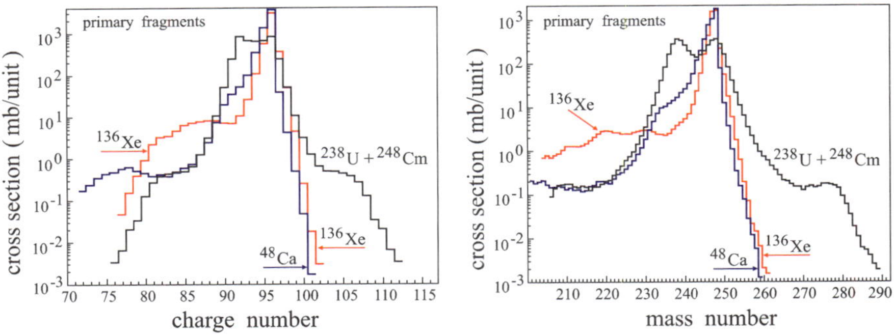


**Fig. 4** Charge and mass distributions of heavy primary reaction fragments formed in collisions of $^{238}$U, $^{136}$Xe and $^{48}$Ca with $^{248}$Cm target at $E_{\text{c.m.}} = 750$, 500 and 220 MeV, correspondingly

dimensional potential energy surface which regulates the dynamics of the heavy nuclear system at low excitation energies.

Contrary to this ordinary quasi-fission phenomena, for the $^{238}$U+$^{248}$Cm collisions we may expect an inverse process in which nucleons are predominantly transferred from the lighter partner (here is uranium) to heavy one (i.e. U transforms to Pb and Cm to 106 element). In this case, besides the lead shoulder in the mass and charge distributions of the reaction fragments, there is also a pronounced shoulder in the region of SH nuclei (see Fig. 4).

Of course, the yield of survived SH elements produced in the low-energy collisions of actinide nuclei is rather low, though the shell effects give us a definite gain as compared to a monotonous exponential decrease of the cross sections with increasing number of transferred nucleons. In Fig. 5 the calculated EvR cross sections for the

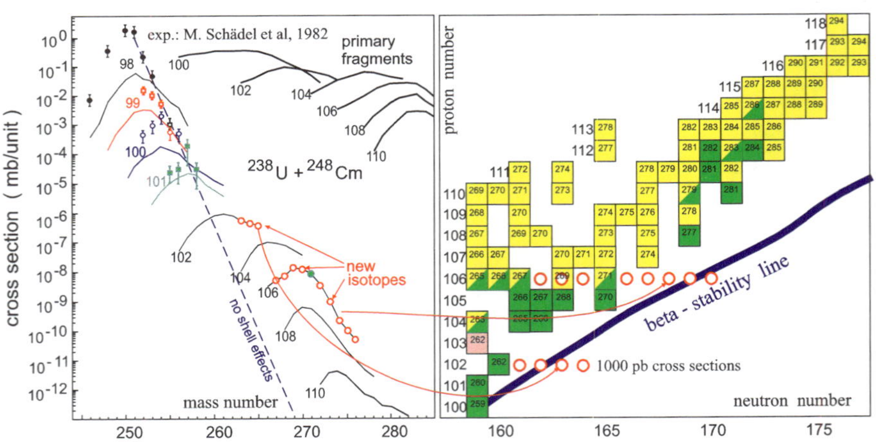


**Fig. 5** Yield of survived isotopes of SH nuclei produced in collisions of $^{238}$U with $^{248}$Cm target at $E_{\text{c.m.}} = 750$ MeV

production of SH nuclei in damped collisions of $^{238}$U with $^{248}$Cm at 750 MeV center-of-mass energy are shown along with available experimental data. As can be seen, really many new neutron-rich isotopes of SH nuclei with $Z > 100$ might be produced in such reactions.

The choice of collision energy is very important for the production of desired neutron-rich SH nuclei. With increasing beam energy the yield of primary fragments increases. However the excitation energy of these fragments also increases and thus decreases their survival probabilities. We found that the optimal beam energy for the production of neutron-rich isotopes of SH elements in multi-nucleon transfer reactions with heavy actinide nuclei (like U+Cm) is very close to the energy needed for these nuclei to reach the contact configuration (there is no ordinary barrier: the potential energy of these nuclei is everywhere repulsive). For $^{238}$U+$^{248}$Cm it is about 750 MeV center-of-mass collision energy.

## 5 Nucleosynthesis by Neutron Capture

The neutron capture process is an alternative (oldest and natural) method for the production of new heavy elements. Strong neutron fluxes might be provided by nuclear reactors and nuclear explosions under laboratory conditions and by supernova explosions in nature. It is well known that the "Fermium gap", consisting of the short-living fermium isotopes $^{258-260}$Fm located at the beta stability line and having very short half-lives for spontaneous fission, impedes the formation of nuclei with $Z > 100$ by the weak neutron fluxes realized in existing nuclear reactors. In nuclear and supernova explosions (fast neutron capture) this gap may be bypassed, if the total neutron fluence is high enough. Theoretical models predict also another region of short-living nuclei located at $Z = 106 \div 108$ and $A \sim 270$.

The synthesis of heavier nuclei in the reaction of neutron capture with subsequent beta-minus decay is a well studied process. Relative yields of the isotopes formed in such a process may be found as a solution of the following set of differential equations (somewhat simplified here)

$$\frac{dN_{Z,A}}{dt} = N_{Z,A-1} n_0 \sigma_{n\gamma}^{Z,A-1} - N_{Z,A} n_0 \sigma_{n\gamma}^{Z,A} - N_{Z,A}[\lambda_{Z,A}^{\beta-} + \lambda_{Z,A}^{fis} + \lambda_{Z,A}^{\alpha}] + N_{Z-1,A}\lambda_{Z-1,A}^{\beta-} + N_{Z+2,A+4}\lambda_{Z+2,A+4}^{\alpha}, \quad (1)$$

where $n_0$ is the neutron flux (number of neutrons per square centimeter per second) and $\lambda_{Z,A}^{i} = \ln 2/T_{1/2}^{i}$ is the decay rate of the nucleus $(Z, A)$ into the channel $i$ (i.e., beta-minus, alpha decays and fission). Neutrons generated by fission in nuclear reactors and in explosions are rather fast (far from the resonance region). In the interval of 0.1–1 MeV the neutron capture cross section is a smooth function of energy with the value of about 1 barn, which is used below for numerical estimations.

To solve Eq. (1) numerically one needs to know the decay properties of neutron rich nuclei which are not studied yet experimentally. This is the key problem, which significantly complicates any analysis of the multiple neutron capture processes. The details of calculations of the decay properties of heavy and SH nuclei can be found in [22].

In Fig. 6 the experimental data on the yield of transuranium nuclei in the test thermonuclear explosion "Mike" [29] (left panel) are compared with those calculated by Eq. (1) assuming 1 $\mu$s neutron exposure of $1.3 \times 10^{24}$ neutrons/cm$^2$ with subsequent one-month decay time. Note that elements 99 and 100 (einsteinium and fermium) were first discovered just in debris of the "Mike" explosion. As can be seen, in this case the Fermium gap does not influence the yields of nuclei with $Z > 100$.

The resulting charge number of the synthesized nuclei might be significantly increased by sequential neutron flux exposure if two or several nuclear explosions would be generated in close proximity of each other. This natural idea was already discussed many years ago [30]. At that time the experts (such as Edward Teller) concluded that technically it could be realized. Such a process is illustrated in the left panel of Fig. 6. In the right panel of this figure the probabilities of heavy element formation are shown for one, three and ten subsequent short-time (1 $\mu$s) neutron exposures of $10^{24}$ n/cm$^2$ each following one after another within a time interval of 10 s with final one month waiting time (needed to reduce the strong radioactivity of the produced material and to perform some experimental measurements).

We found that the result depends both on the neutron fluence $n = n_0\tau$ ($\tau$ is the duration of explosive neutron irradiation) and on the time interval between two exposures. The neutron fluence should be high enough to shift the produced neutron rich isotopes to the right from the second gap of unstable fissile nuclei located at $Z = 106 \div 108$ and $A \sim 270$. Dependence on the time interval between two exposures is not so crucial. The result does not almost depend on this parameter if it is longer than several milliseconds (to avoid approaching the neutron drip line after several

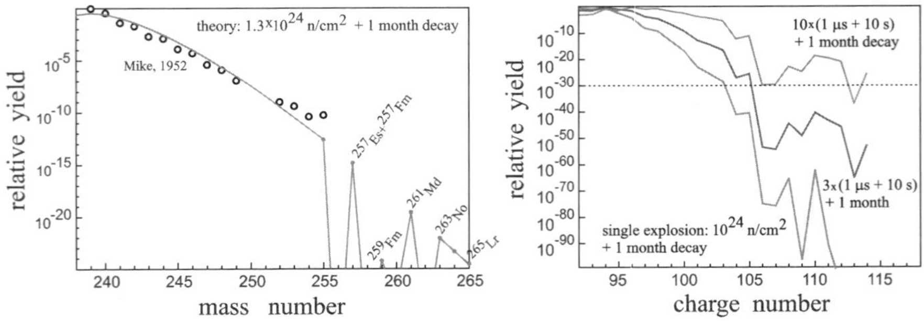


**Fig. 6** Experimental (*open dotes*) and calculated relative yields of heavy nuclei in the test nuclear explosion "Mike" [29] (*left panel*). Probability for formation of heavy nuclei in multiple neutron irradiation of initial $^{238}$U material (one, three and ten subsequent explosions). The *dotted line* denotes the level of few atoms

exposures) and shorter than a few minutes to avoid $\beta^-$-decay of the produced nuclei into the area of fission instability (Z = 106 ÷ 108 and A∼270).

Our results demonstrate for the first time that multiple rather "soft" nuclear explosions could be really used for the production of a noticeable (macroscopic) amount of neutron rich long-lived SH nuclei. Leaving aside any discussions on the possibility of such processes and associated technical problems, we want to emphasize a sharp increase of the probability for formation of heavy elements with $Z \geq 110$ in the multiple neutron irradiations: enhancement by several tens of orders of magnitude (see Fig. 6). This probability is high enough for some SH elements to perform their experimental identification.

It is interesting also to study the same process of multiple neutron exposures realized in pulsed nuclear reactors. Here the pulse duration can be much longer than in nuclear explosions (up to few milliseconds). However, the neutron fluence usually does not exceed $10^{16}$ n/cm$^2$ in existing nuclear reactors. Multi-pulse irradiation here corresponds, in fact, to the "slow" neutron capture process, in which new elements with larger charge numbers are situated close to the line of stability and finally reach the "Fermium gap" where the process stops.

The situation may change if one could be able to increase somehow the intensity of the pulsed reactor. The neutron fluence in one pulse and frequency of pulses should be high enough to bypass both gaps of short-lived nuclei on the way to the island of stability. Thus, the specification of the high-intensity pulsed reactors of next generation depends strongly on properties of heavy neutron rich nuclei located to the right of these gaps. These nuclei are not discovered yet, and undoubtedly certain experimental efforts should be made to resolve this problem. We have found that increase of the neutron fluence in the individual pulse by about three orders of magnitude as compared with existing pulsed reactors, i.e. up to $10^{20}$ neutrons/cm$^2$, could be quite sufficient to bypass both gaps.

## 6 Formation of SH Nuclei in Astrophysical r-Processes

The astrophysical r-process of nucleosynthesis is usually discussed to explain the observed abundance of heavy elements in the universe. In such a process some amount of SH elements of the island of stability might be also produced if the fast neutron flux is sufficient to bypass the two gaps of fission instability mentioned above. Strong neutron fluxes are expected to be generated by neutrino-driven proto-neutron star winds which follow core-collapse supernova explosions [31] or by the mergers of neutron stars [32]. Estimation of relative yields of SH elements is a difficult problem which depends both on the features of neutron fluxes and on the experimentally unknown decay properties of heavy neutron rich nuclei.

We made a very simple estimation of the possibility for formation of SH nuclei during the astrophysical r-process of neutron capture. This estimation is based on the following assumptions. (1) SH nuclei are relatively short-living. They are absent in stars initially, while the distribution of other elements is rather close to their abun-

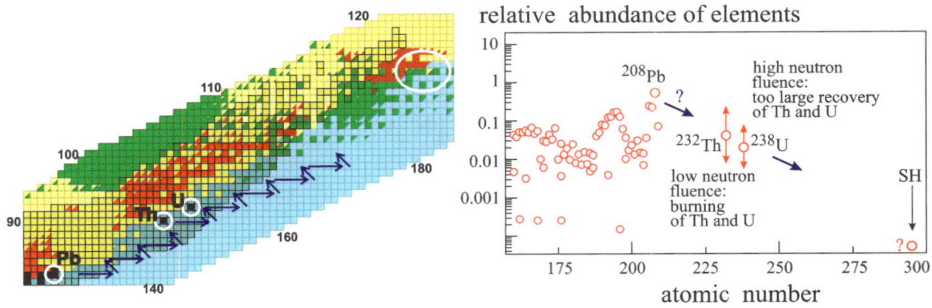


**Fig. 7** Formation of SH elements in astrophysical r-process. Unknown neutron fluence could be adjusted in such a way that relative abundances of uranium and thorium (burned and recovered from lead and lighter stable elements) keep finally their experimental values

dance in the universe. (2) SH nuclei may appear (and survive) at the last (rather cold) stage of the astrophysical r-process when the observed abundance of heavy elements (in particular, thorium and uranium to lead ratios) is also reproduced. (3) Existing (experimental) abundance of stable nuclei may be used as initial condition. During intensive neutron irradiation initial thorium and uranium material are depleted transforming to heavier elements and going to fission, while more abundant lead and lighter stable elements enrich thorium and uranium. (4) Unknown total neutron fluence may be adjusted in such a way that the ratios Y(Th)/Y(Pb) and Y(U)/Y(Pb) keep its experimental values at the end of the process. Simultaneously, for a given neutron fluence, one gets the relative yield of SH elements, Y(SH)/Y(Pb) (in accordance with our estimation, $^{291}$Cn and $^{293}$Cn are the most stable SH nuclei [22], their half-lives are about several hundred years).

We performed calculations (Fig. 7) starting from initial relative abundances of heavy elements corresponding to experimental values. The value of the neutron flux $n_0$ was fixed at $10^{24}$ cm$^{-2}$sec$^{-1}$ and the total neutron fluence was regulated by the time of exposure. At such high neutron flux the final result depends only on the total neutron fluence. After neutron irradiation the waiting time of 100 years was applied to obtain the final distribution of nuclei after all the decays. This time is still shorter than half-lives of some $\alpha$-decayed plutonium, curium and californium isotopes, and we added their yields to the yields of their daughter thorium and uranium products. At low neutron fluxes initial thorium and uranium nuclei increase their masses and charges (after neutron capture and subsequent $\beta^-$-decay), find themselves in the region of fission instability and drop out. Thus, their numbers decrease relative to lead, which, in contrast with Th and U, has an additional feed from lighter nuclei. Contribution from lead to thorium and uranium becomes noticeable only when the probability for capture of 24 neutrons is not negligible. At neutron fluence $n \sim 1.5 \cdot 10^{25}$ cm$^{-2}$ (= 15 neutrons/barn) burning of thorium and uranium is compensated by increasing contribution from lighter stable nuclei with $Z \leq 83$. However at this neutron fluence the final abundance of thorium and uranium is still too low, and only at $n \sim 2 \cdot 10^{25}$ cm$^{-2}$ the both ratios Y(Th)/Y(Pb) and Y(U)/Y(Pb) are close to the observed values. At this neutron fluence the relative to lead yield of most stable isotopes of SH element 112, namely $^{291}$Cn and $^{293}$Cn, is about $10^{-12}$ which is not extremely low and keeps hope to find them in nature (most probably in the cosmic rays).

## References


1. G.T. Seaborg, W.D Loveland, The Elements Beyond Uranium. (John Wiley & Sons, Inc., New York, 1990)
2. S.G. Nilsson, J.R. Nix, A. Sobiczewski et al., Nucl. Phys. A **115**, 545 (1968)
3. S.G. Nilsson, S.G. Thompson, C.F. Tsang, Phys. Lett. **28B**, 458 (1969)
4. U. Mosel, W. Greiner, Z. Physik **222**, 261 (1969)
5. J. Grumann, U. Mosel, B. Fink, W. Greiner, Z. Physik **228**, 371 (1969)
6. G.N. Flerov, G.M. Ter-Akopian, Rep. Prog. Phys. **46**, 817 (1983)
7. S. Hofmann, G. Münzenberg, Rev. Mod. Phys. **72**, 733 (2000)
8. K. Morita et al., J. Phys. Soc. Jpn. 76(4), 043201, 045001 (2007)
9. G.T. Seaborg, Ann. Rev. Nucl. Sci. **18**, 53 (1968)
10. Yu.Ts. Oganessian, V.K. Utyonkov, Yu.V. Lobanov et al., Phys. Rev. C **69**, 054607 (2004)
11. Yu.Ts. Oganessian, V.K. Utyonkov, Yu.V. Lobanov et al., Phys. Rev. C **70**, 064609 (2004)
12. Yu.Ts. Oganessian, V.K. Utyonkov, Yu.V. Lobanov et al., Phys. Rev. C **74**, 044602 (2006)
13. V.I. Zagrebaev, M.G. Itkis, Yu.Ts Oganessian, Phys. At. Nucl. **66**, 1033 (2003)
14. V.I. Zagrebaev, Nucl. Phys. A **734**, 164 (2004)
15. P.A. Ellison, K.E. Gregorich, J.S. Berryman et al., Phys. Rev. Lett. **105**, 182701 (2010)
16. J.M. Gates, Ch.E Düllmann, M. Schädel et al., Phys. Rev. C **83**, 054618 (2011)
17. S. Hofmann, S. Heinz, R. Mann et al., GSI Scientific Report 2010 (GSI, Darmstadt, 2010) ISSN: 0174-0814, p. 197
18. V.I. Zagrebaev, W. Greiner, Phys. Rev. C **78**, 034610 (2008)
19. V.I. Zagrebaev, W. Greiner, Phys. Rev. C **83**, 044618 (2011)
20. V.I. Zagrebaev, A.V. Karpov, I.N. Mishustin, W. Greiner, Phys. Rev. C **84**, 044617 (2011)
21. V.I. Zagrebaev, A.V. Karpov, W. Greiner, Phys. Rev. C **85**, 014608 (2012)
22. A.V. Karpov, V.I. Zagrebaev, Y. Martinez Palenzuela, L. Felipe Ruiz, W. Greiner, Int. J. Mod. Phys. E **21**, 1250013 (2012)
23. V.I. Zagrebaev, Yu.Ts. Oganessian, M.G. Itkis, W. Greiner, Phys. Rev. C **73**, 031602 (2006)
24. W. Loveland, A.M. Vinodkumar, D. Peterson, J.P. Greene, Phys. Rev. C **83**, 044610 (2011)
25. V. Zagrebaev, W. Greiner, J. Phys. G **34**, 2265 (2007)
26. M. Schädel, W. Brüchle, H. Gäggeler et al., Phys. Rev. Lett. **48**, 852 (1982)
27. M.G. Itkis, J. Äystö, S. Beghini et al., Nucl. Phys. A **734**, 136 (2004)
28. V. Zagrebaev, W. Greiner, J. Phys. G **31**, 825 (2005)
29. H. Diamond, P.R. Fields, C.S. Stevens et al., Phys. Rev. **119**, 2000 (1960)
30. H.W. Meldner, Phys. Rev. Lett. **28**, 975 (1972)
31. T.A. Thompson, A. Burrows, B.S. Meyer, Astrophys. J. **562**, 887 (2001)
32. S. Rosswog, M. Liebendörfer, F.-K. Thielemann, M.B. Davies, W. Benz, T. Piran, Astron. Astrophys. **341**, 499 (1999)

# Superheavy Nuclei: Decay and Stability

**A. V. Karpov, V. I. Zagrebaev, Y. Martinez Palenzuela and Walter Greiner**

**Abstract** Decay properties of superheavy nuclei are required for exploring the nuclei from the upper part of the nuclear map. The stability of nuclei with $Z \leq 132$ is studied with respect to $\alpha$-decay, $\beta$-decay and spontaneous fission. Performed calculations allow us to conclude that at existing experimental facilities the synthesis and detection of nuclei with $Z > 120$ produced in fusion reactions may be difficult due to their short half-lives (shorter than 1 $\mu$s). We found for the first time the region of $\beta^+$-decaying superheavy nuclei with $111 \leq Z \leq 115$ located to the "right" (more neutron-rich) to those synthesized recently in Dubna in $^{48}$Ca-induced fusion reactions. This fact may significantly complicate their experimental identification. However it gives a chance to synthesize in fusion reactions the most stable superheavy nuclei located at the center of the island of stability. Our calculations yield that the $\beta$-stable isotopes $^{291}$Cn and $^{293}$Cn with a half-life of about 100 years are the longest-living superheavy nuclei located at the island of stability.

## 1 Motivation

More than 40 years passed from the first predictions that the region of rather stable superheavy (SH) nuclei should exist around $Z{\sim}114$ and $N{\sim}184$ [1–3]. Great success was achieved during the last twenty years in the experimental study of

A. V. Karpov (✉) · V. I. Zagrebaev
Joint Institute for Nuclear Research, Dubna, Russia
e-mail: karpov@jinr.ru

V. I. Zagrebaev
e-mail: zagrebaev@jinr.ru

Y. Martinez Palenzuela · W. Greiner
Frankfurt Institute for Advanced Studies, Frankfurt, Germany
e-mail: palenzuela@fias.uni-frankfurt.de

W. Greiner
e-mail: greiner@fias.uni-frankfurt.de

W. Greiner (ed.), *Exciting Interdisciplinary Physics*,
FIAS Interdisciplinary Science Series, DOI: 10.1007/978-3-319-00047-3_6,

reactions leading to superheavy nuclei, their decay properties and structure. Up to now near-barrier fusion reactions have been used for the production of new SH elements in the "cold" [4, 5] and "hot" (using $^{48}$Ca as a projectile) [6, 7] combinations of colliding nuclei. The heaviest yet discovered element is the 118 one, synthesized in "hot" fusion reaction of $^{48}$Ca beam and $^{248}$Cf target. However, californium is the heaviest available target which has been used in these experiments for the production of element 118 [8]. Thus, to get SH elements with $Z > 118$ in fusion reactions, one should proceed to heavier than $^{48}$Ca projectiles ($^{50}$Ti, $^{54}$Cr, etc.). The corresponding cross sections for the production of the elements 119 and 120 are predicted to be smaller by about two orders of magnitude [9] as compared with $^{48}$Ca-induced fusion reactions leading to the formation of the elements 114–116. Another limitation of the fusion reactions (both "cold" and "hot") for producing superheavy elements consists in the fact that they lead to neutron-deficient isotopes having rather short life time.

The most stable SH nuclei are predicted to be located along the $\beta$-stability line in the region of more neutron-rich nuclei, which is unreachable directly by fusion reactions with stable beams. In fact, the predicted magic numbers, especially for protons, are quite different within different theoretical approaches. The magic number $Z = 114$ was predicted in earliest macro-microscopic calculations [1–3, 10] and confirmed later in Refs. [11, 12]. The fully microscopic approaches predict the proton shell closure at $Z = 120$ [13], $Z = 126$ [14], or $Z = 114, 120, 126$ [15] depending on the chosen nucleon-nucleon interaction in mean field theories. The neutron magic number $N = 184$ is almost firmly predicted by different theoretical models.

Nowadays the experimental study of heavy nuclei, in particular of superheavies, requires ideas, new theoretical predictions, and methods (reactions) that can be used for producing these nuclides. Knowledge of the decay modes and half-lives of nuclei in a very wide range of neutron and proton numbers (nuclear map) is necessary for such predictions and for the planning of the corresponding experiments. Moreover, the study of decay properties may help us to answer some principle but open questions: how far may we still move in synthesis of SH elements by the fusion reactions, where the island of stability is centered, what are the properties of the most stable SH nuclei, how to reach this region? Another field where the decay properties play a crucial role is the study of the $r$-process of nucleosynthesis in the superheavy mass region, and the related problem of a search of superheavy nuclei in nature.

## 2 Half-Lives of Heavy and SH Nuclei

This work is aimed to the analysis of the decay properties of heavy and superheavy elements with respect to $\alpha$-decay, $\beta$-decay, and spontaneous fission (SF)—the three main decay modes. All the calculations performed in this paper are based on the values of the ground-state masses obtained within the macro-microscopic approach. Here we use experimental masses for known nuclei and three sets of the ground-state masses for unknown ones, obtained by P. Möller et al. [16] (mainly these, as the most known ones), A. Sobiczewski et al. [17], and within the two-center shell-model potential [18, 19].

The $\alpha$-decay is characterized by the energy release $Q_\alpha$ and the corresponding half-life $T_\alpha$. The half-life for $\alpha$-decay can be estimated quite accurately using the well-known Viola-Seaborg formula [20]

$$\log_{10} T_\alpha\,(\text{sec}) = \frac{aZ+b}{\sqrt{Q_\alpha(\text{MeV})}} + cZ + d + h_{\log}, \tag{1}$$

where $a$, $b$, $c$, $d$, and $h_{\log}$ are adjustable parameters. We use the values of these parameters obtained in [21] $a = 1.66175$, $b = -8.5166$, $c = -0.20228$, $d = -33.9069$. The quantity $h_{\log}$ takes into account hindrance of $\alpha$-decay for nuclei with odd neutron and/or proton numbers [20]

$$h_{\log} = \begin{cases} 0, & \text{Z and N are even} \\ 0.772, & \text{Z is odd and N is even} \\ 1.066, & \text{Z is even and N is odd} \\ 1.114, & \text{Z and N are odd} \end{cases} \tag{2}$$

The phenomenological calculation of $T_\alpha$ is the most justified (as compared with $T_\beta$ and $T_{SF}$) and the most accurate. The errors arising from uncertainty in $Q_\alpha$ are much larger than the one due to the inaccuracy of phenomenological Viola-Seaborg formula.

If one moves aside the stability line, the $\beta$-processes start to play an important role. Therefore, to estimate correctly the life time of such a nucleus we have to consider the competition of $\alpha$-decay and spontaneous fission with $\beta^\pm$ decays and electron capture (EC). The decay properties of nuclei close to the $\beta$-stability line are mostly known (except for the region of superheavy nuclei). This means that we may restrict ourself to the case of nuclei far from the line of $\beta$-stability. It allows us to assume that the corresponding $Q$-values and the density of states are large enough to find in the daughter nucleus a level which is close to the ground state and which fulfils the conditions of allowed $\beta$-decays. Thus, the problem simplifies to the case of the ground-to-ground allowed $\beta$ transitions. This assumption may be not accurate enough for some specific nuclei close to the $\beta$-stability line, but this can not alter the general trend in the decay modes, which we are interested in. We should mentioned here that previous systematic calculations of the half-lives with respect to $\beta$-decay (see, e.g., [22, 23]) were performed for allowed transitions as well. The half-life with respect to all kinds of $\beta$ processes $T_\beta$ is given by

$$1/T_\beta = 1/T_{\beta^-} + 1/T_{\beta^+} + 1/T_{EC}. \tag{3}$$

The half-life with respect to the allowed $\beta$-decay is defined by the following relation [24]:

$$\log_{10}\left[f_0^b T_b\,(\text{sec})\right] = 5.7 \pm 1.1, \tag{4}$$

where $f_0^b$ is the Fermi function (which is calculated using the standard relations, see e.g. [25]), $b = \beta^{\pm}$ or $EC$. Thus, the estimation of the $\beta$-decay half-lives is reduced to the calculation of the Fermi function $f_0^b$. We use in (4) the constant value 4.7, adjusted to the corresponding experimental data. The $\beta$-decay half-lives shorter than 1000 s should be addressed to the allowed decays. Our calculations agree with the experiment within two orders of magnitude for this case. This is sufficient to estimate the $\beta$-decay half-lives in competition with $\alpha$-decay and spontaneous fission almost for all experimentally unknown nuclei.

The spontaneous fission (SF) of nuclei is a very complicated process. Knowing the multidimensional potential energy surface only is not sufficient for the accurate determination of the corresponding decay time. The most realistic calculations of the SF half-life are based on the search for the least action path in the multidimensional deformation space. Only few examples of such calculations are known [10, 27–29], that were performed in a rather restricted area of the nuclear map due to long calculation times. In Ref. [30] we propose the systematics based on idea of W.J. Swiatecki [31] that the SF half-lives are mainly determined by the height of the fission barrier. To determine the coefficients of the systematics we include in the fitting procedure not only the experimental data [32] but also the realistic theoretical predictions [27–29] for the region $100 \leq Z \leq 120$ and $140 \leq N \leq 190$

$$\begin{aligned}\log_{10} T_{SF}\,(\text{sec}) =& 1146.44 - 75.3153Z^2/A \\ &+ 1.63792\left(Z^2/A\right)^2 - 0.0119827\left(Z^2/A\right)^3 \\ &+ B_f\left(7.23613 - 0.0947022Z^2/A\right) \\ &+ \begin{cases} 0, & \text{Z and N are even} \\ 1.53897, & \text{A is odd} \\ 0.80822, & \text{Z and N are odd} \end{cases}\end{aligned} \qquad (5)$$

Here $B_f$ is the fission barrier, which is calculated as a sum of the liquid-drop barrier $B_f(LDM)$ [33] and the ground-state shell correction $\delta U(g.s.)$ [16], i.e. $B_f = B_f(LDM) + \delta U(g.s.)$. Fig. 1 shows the dependence of the SF half-life on the neutron numbers for nuclei with even atomic numbers from Uranium to $Z = 114$ element. Obviously Eq. (5) qualitatively reproduces the behavior of the half-lives in the experimentally known region. However the proposed relation substantially underestimates the abrupt decrease of the half-life for Cf, Fm, and No around $N = 160$. In the region of superheavy nuclei we get reasonable agreement with the data. The reason for larger deviation from the experimental SF half-lives for neutron-rich isotopes of Cf, Fm, and No is the influence of exit channel, caused by clusterization with two nearly double-magic tin fragments, which is a special case of this region of nuclei. This effect is not included in the relation (5), but is accounted for within the dynamical approach mentioned above. However, even in such advanced calculations, this steep decrease of the SF half-live is underestimated (see Fig. 4 in Ref. [29]). In Fig. 1 we also show the calculations of Ref. [27, 28] for the isotopes of $Z = 104 - 114$.

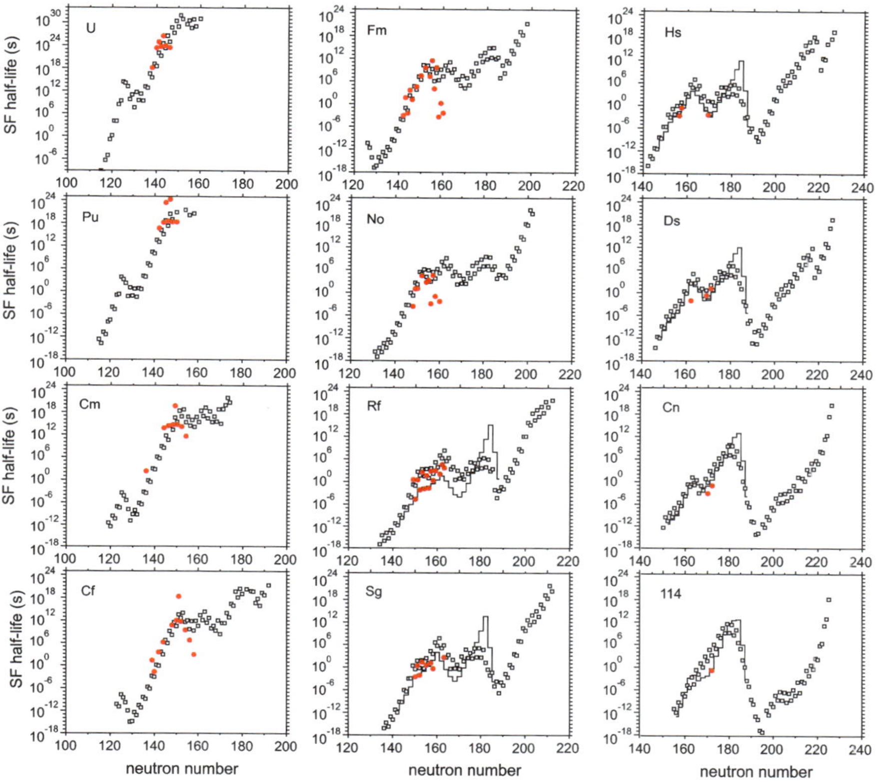


**Fig. 1** Dependence of the SF half-lives on the neutron number for the isotopes of elements from U to 114. The *open black squares* are the estimation by the phenomenological formula (5), the *full red circles* are the experimental data [26, 32], and the full lines are the calculations of Refs. [27, 28]

One may see that in this region both models give similar results for those nuclei, for which experimental data exist. However, the model of Ref. [27, 28] predicts for some nuclei a too steep decrease of the half-lives around $N \simeq 170$ and much longer times around the closed shell numbers $N = 184$.

# 3 Analysis of the Nuclear Map

Figure 2 shows upper part of the nuclear map for the total half-lives and decay modes of the nuclei with $Z \leq 132$ obtained with the ground-state masses from Ref. [16]. The known nuclei are situated along the $\beta$-stability line with a shift to the proton-rich region especially for heavy and superheavy nuclei. Almost all proton-rich nuclei with $Z \leq 118$ having half-lives sufficiently long for their experimental identification are

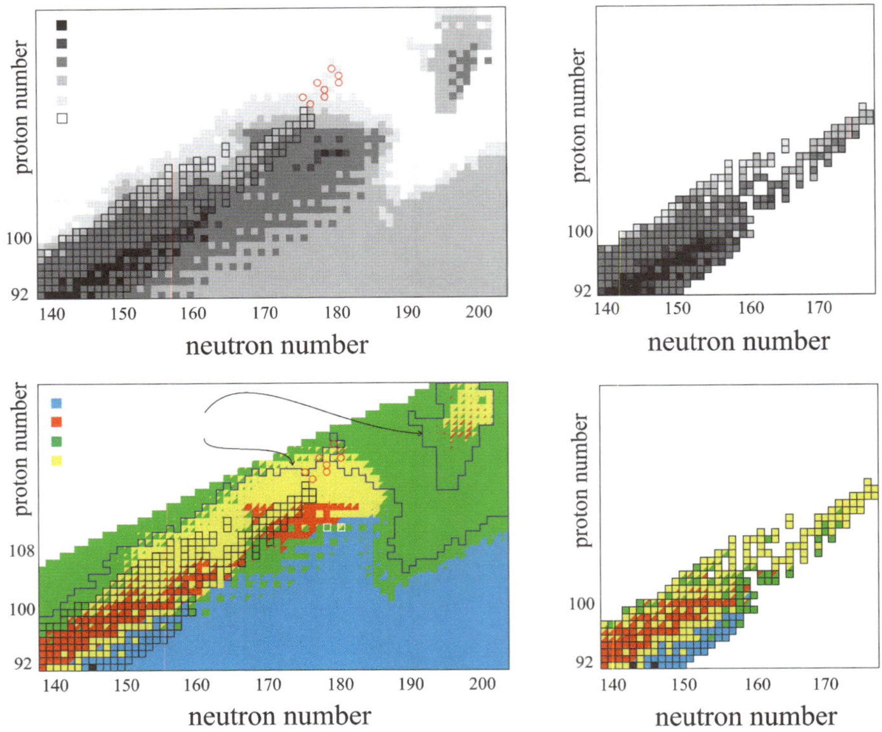


**Fig. 2** The total half-lives (*top*) and the decay modes (*bottom*) of nuclei in the *upper part* of the nuclear map. The *left panels* are calculations (performed with the ground-state masses from Ref. [16]) and the *right panels* are the experimental data taken from [26]. The contour lines on the *left bottom* panel correspond to the border of 1 μs half-life. The circles show the nuclei with $Z = 119-124$, which may be synthesized in $3n$ channel of fusion reactions $^{50}$Ti+$^{249}$Bk, $^{249}$Cf and $^{54}$Cr, $^{58}$Fe+$^{248}$Cm, $^{249}$Bk, $^{249}$Cf (see the text). The bounded cells correspond to the experimentally known nuclei. The bounded nuclei with the white color border are the most stable Copernicium isotopes $^{291}$Cn and $^{293}$Cn

already synthesized. The red circles in Figs. 2 and 3a correspond to the nuclei with $Z = 119 - 124$, which may be obtained in the $3n$ channel of the fusion reactions: $^{50}$Ti+$^{249}$Bk, $^{50}$Ti+$^{249}$Cf, $^{54}$Cr+$^{248}$Cm, $^{54}$Cr+$^{249}$Bk, $^{54}$Cr+$^{249}$Cf, $^{58}$Fe+$^{248}$Cm, $^{58}$Fe+$^{249}$Bk, and $^{58}$Fe+$^{249}$Cf. The synthesis cross section of these new superheavy nuclei with $Z > 118$ in fusion reactions is predicted to decrease substantially due to the change of the projectile from $^{48}$Ca to a heavier one [9]. Moreover, as can be seen from Figs. 2 and 3a these nuclei are very short-living. They are located at the border of 1 μs area—the critical time required to pass separator to be detected. It means that the nuclei heavier than the 120 element—even if they will be synthesized—could be hardly detected because of their very short half-lives. This conclusion is nearly model independent. Both models [see Figs. 2 and 3a] give quite similar predictions of the half-lives for the nuclei which could be synthesized in the above mentioned

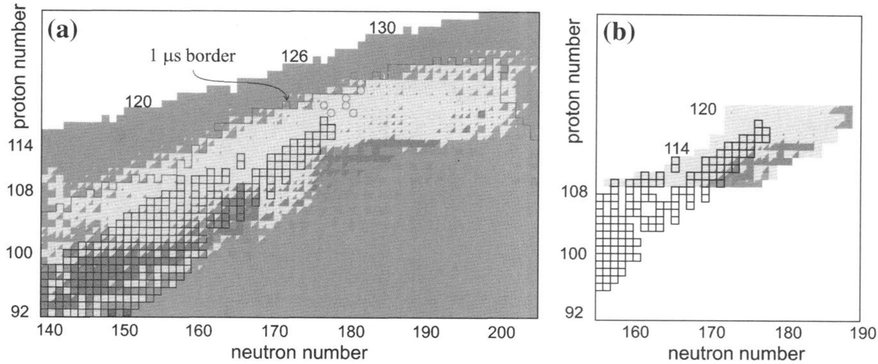


**Fig. 3** The decay modes calculated using the ground-state masses from the two-center shell model [18, 19] (**a**) and those obtained in Ref. [17] (**b**). The SH half-lives for the panel (**b**) are taken from Refs. [27, 28] (with the hindrance factor 100 for odd and odd-odd nuclei). Other notations are the same as in Fig. 2

projectile-target combinations. However the borders of 1 $\mu$s area on the neutron-rich side differ substantially for these two models. This discrepancy appears due to the extrapolation of the model parameters to the unknown region, while the results for experimentally studied nuclei are quite similar.

The discovery of new elements mentioned above (even proton-rich isotopes) is certainly of interest. However, in our opinion, the most challenging region for future studies is the region of more heavy and more neutron-rich nuclei. This is especially the island of stability of superheavy nuclei centered at $Z \sim 114$ and $N \sim 184$ (remember, however, that the microscopic meson filed theory also predicts nuclei around $Z \sim 120$ and $N \sim 184$ as a candidates for a stability island). According to our predictions (made with the masses [16]) the most long-living nuclei in the $Z \sim 114$ and $N \sim 184$ area are the $\beta$-stable isotopes of Copernicium $^{291}$Cn and $^{293}$Cn with the half-lives of about 100 years shown in Fig. 2 by the white-border squares. The main decay mode of $^{291}$Cn is predicted to be SF and $^{293}$Cn is decaying by $\alpha$-decay and SF with nearly equal probability. Because of their relatively long half-lives these isotopes—if synthesized—could be accumulated. Unfortunately these two isotopes are unreachable directly by any fusion reaction with stable ion beams. In principle, there is a chance to produce these nuclei in multi-nucleon transfer reactions [34, 35] or by multiple neutron capture processes [36]. However the corresponding cross sections are very low. A new way for the synthesis of neutron-enriched superheavy nuclei and, in particular, those from the center of the stability island may be found basing on the found area of $\beta^+$-decaying nuclei in the vicinity of the island of stability.

We found (see Figs. 2 and 3) that some isotopes of superheavy elements with $111 \leq Z \leq 115$, more neutron-rich than those synthesized recently in Dubna in the $^{48}$Ca-induced fusion reactions, also may undergo $\beta^+$-decay. Note, that such an area of $\beta^+$-decaying nuclei appears independently of the model used for the nuclear masses calculation. However, the size of this region is sensitive to the underlying shell model. The appearance of such an area of $\beta^+$-decaying nuclei in the vicinity of the island of stability becomes quite evident from the schematic Fig. 4. In this figure we consider the situation where the neutron closure $N = 184$ coincides with the region of $\beta$-stable nuclei (which is expected close by the proton number $Z = 114$). The left panel of Fig. 4 shows the typical behavior of the characteristic energies of EC, $\alpha$-decay, and SF playing the role in this region ($Q_{\beta^-}$ is negative here and not shown). In this case one may expect the following order of decay modes starting from the proton drip line up to the top of the stability island (see the right panel in Fig. 4). Due to the strong Coulomb field, the most proton-rich nuclei should undergo SF with rather short half-lives. Moving to the "right" the fission barriers increase because of increase of the neutron number (and, therefore, decrease of the Coulomb forces) as well of the stabilizing effect of the neutron shell $N = 184$. Then $\alpha$-decay starts to play a main role. Note, that most nuclei known at the moment close to $Z = 114$ (both synthesized in "cold" and "hot" fusion reactions) experience $\alpha$-decay. Approaching the island of stability the half-lives of $\alpha$-decay as well as those of SF increase by many orders of magnitude due to influence of the neutron shell $N = 184$. When these half-lives are longer than minutes and days (the typical half-lives with respect to EC of nuclei in the vicinity of the $\beta$-stability region), the EC process may dominate. Finally, the most stable nuclei (which should be $\beta$-stable) again undergo $\alpha$-decay or/and SF. This consideration of the decay modes sequence is rather natural and model independent. It explains an appearance of the area of $\beta^+$-decay found here. However, the size of this area depends on the nuclear masses and nuclear structure. It should be stressed ones more, that our calculations of $\beta$-decay half-lives are based on the assumption of

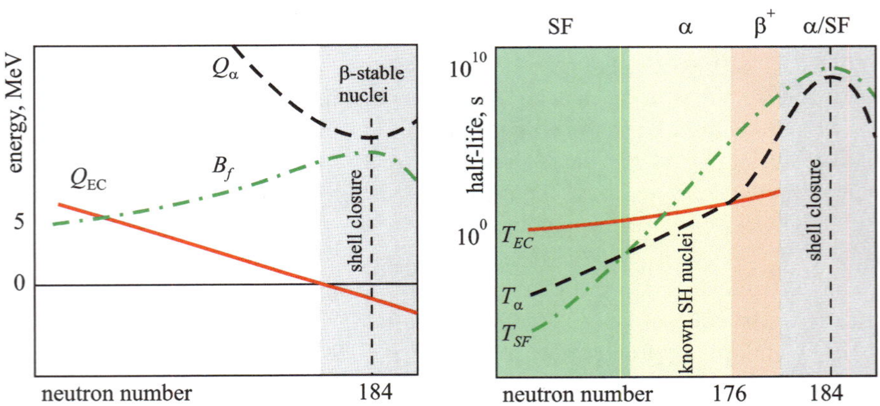


**Fig. 4** Schematic picture explaining the existence of the region of $\beta^+$-decaying nuclei in the vicinity of the stability island. (*left panel*) Dependence on the neutron number of the characteristic energies of $\beta^+$-decay ($Q_{EC}$, *solid curve*), alpha-decay ($Q_\alpha$, *dashed curve*), and spontaneous fission ($B_f$, *dash-dotted curve*). The region of $\beta$-stable nuclei and the position of the neutron shell closure ($N = 184$) are shown. (*right panel*) Expected behavior of the half-lives $T_{EC}$ (*solid curve*), $T_\alpha$ (*dashed curve*), and $T_{SF}$ (*dash-dotted curve*) from the proton-rich side up to the center of the stability island. The dominating modes of decay and the position of known SH nuclei in the vicinity of $Z = 114$ are shown

allowed $\beta$-transitions. As was said above, $\beta$-decay can be substantially suppressed, especially for nuclei close to the $\beta$-stability line (i.e. having small $Q$-values of $\beta$-decays). This means that some of the nuclei found here to have the $\beta^+$-decay as the main mode, may have much longer $\beta$-decay time, whereas the main decay mode could be $\alpha$-decay or SF. However, the gross decay-mode structure of the nuclear map (i.e. existence of the region of $\beta^+$-decaying superheavy nuclei) should remain.

Our finding indicates that the experimental identification of the nuclei to the "right" of already discovered ones may meet significant difficulties. However, the existence of the area of $\beta^+$-decay gives us the hypothetic way to reach the middle of the island of stability just in fusion processes of "stable" nuclei. In Fig. 5 several possible decay chains of the isotopes $^{291}115$ and $^{291}114$ are shown along with the corresponding values of $Q_\alpha$ and half-lives calculated with the use of nuclear masses predicted in Ref. [17] and in Ref. [16]. The SF half-lives are taken from Ref. [27, 28] (with the hindrance factor 100 for odd and odd-odd nuclei), while the values in brackets are calculated by phenomenological relations (5). The isotope $^{291}115$ may be formed after $\alpha$-decay of $^{295}117$ (the 2n evaporation channel of the $^{48}Ca$ + $^{249}Bk$ fusion reaction, cross section is 0.3 pb [9]) or after two $\alpha$-decays of $^{299}119$ (the 3n evaporation channel of the $^{48}Ca$ + $^{254}Es$ fusion reaction, cross section is

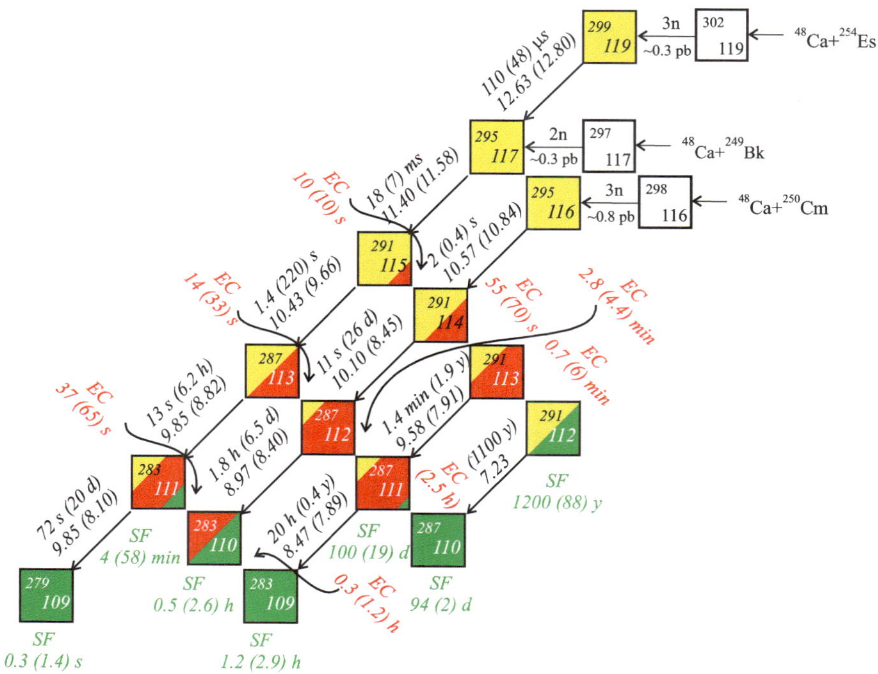


**Fig. 5** The pathway to the middle of the island of stability via a possible $\beta^+$-decay of the isotopes $^{291}115$ and $^{291}114$. Decay half-lives and $Q_\alpha$ values (in MeV) calculated with nuclear masses [17] and [16] (*in brackets*) are shown

0.3 pb [37]). The second one, $^{291}114$, is formed after $\alpha$-decay of $^{295}116$ in the 3n evaporation channel of the $^{48}Ca + ^{250}Cm$ fusion reaction with cross section of about 0.8 pb [37]. These isotopes should have rather long half-lives and, thus, they could be located already in the "red" area of the nuclear map; that is, they may be $\beta^+$-decaying nuclei. In accordance with our calculations of decay properties of SH nuclei [30], the isotopes $^{291}115$ and $^{291}114$ may experience not only $\alpha$-decay but also EC. This prediction opens a narrow pathway to the middle of the island of stability of SH nuclei by sequence of $\beta^+$ decays ending at the $^{291}Cn$ nucleus. Note that, for the moment, the proposed method is the highest in cross section method for production of the nuclei located in the middle of the first island of stability. Hopefully it may be realized in future with the progress in experimental techniques.

**Acknowledgments** This work was supported in part by DFG-RFBR collaboration and Helmholtz association. One of us (A.V.K.) is indebted to JINR for support of these studies within the young scientist grant programm. Y.M.P. is grateful to the DAAD for a PhD stipend.

## References

1. U. Mosel, B. Fink, W. Greiner, in Memorandum zur Errichtung eines gemeinsamen Ausbildungszentrums für Kernphysik der Hessischen Hochschulen Darmstadt, Frankfurt am Main, and Marburg (1966).
2. H. Meldner, Ph.D. thesis, University Frankfurt am Main (1966); see also Proceedings of International Symposium Why and How Should We Investigate Nuclides Far off the Stability Line, Lysekil, Sweden (1966).
3. S.G. Nilsson, C.F. Tsang, A. Sobiczewski, Z. Szymanski, S. Wycech, C. Gustafson, I.-L. Lamm, P. Möller, B. Nilsson, Nucl. Phys. A **131**, 1 (1969)
4. S. Hofmann, G. Münzenberg, Rev. Mod. Phys. **72**, 733 (2000)
5. K. Morita et al., J. Phys. Soc. Jpn. **76**(043201), 045001 (2007)
6. Yu. Ts, Oganessian. J. Phys. G **34**, R165 (2007)
7. Yu.Ts. Oganessian et al., Phys. Rev. Lett. 104, 142502 (2010).
8. Yu.Ts. Oganessian et al., Phys. Rev. C 74, 044602 (2006).
9. V.I. Zagrebaev, W. Greiner, Phys. Rev. C **78**, 034610 (2008)
10. U. Mosel, W. Greiner, Z. Phys. **222**, 261 (1969)
11. Z. Patyk, A. Sobiczewski, Nucl. Phys. A **533**, 132 (1991)
12. P. Möller, J.R. Nix, Nucl. Phys. A **549**, 84 (1992)
13. M. Beiner, H. Flocard, M. Vénéroni, P. Quentin, Phys. Scr. **10A**, 84 (1974)
14. S. Cwiok, J. Dobaczewski, P.H. Heenen, P. Magierski, W. Nazarewicz, Nucl. Phys. A **611**, 211 (1996)
15. K. Rutz, M. Bender, T. Bürvenich, T. Schilling, P.G. Reinhard, J.A. Maruhn, W. Greiner, Phys. Rev. C **56**, 238 (1997)
16. P. Möller, J.R. Nix, W.D. Myers, W.J. Swiatecki, At. Data Nucl. Data Tables **59**, 185 (1995)
17. I. Muntian, Z. Patyk, A. Sobiczewski, Phys. At. Nucl. 66, 1015 (2003) [Yad. Fiz. 66, 1051 (2003)].
18. J. Maruhn, W. Greiner, Z. Phys, Z. Phys, A **251**, 431 (1972)
19. V.I. Zagrebaev, A.V. Karpov, Y. Aritomo, M. Naumenko, W. Greiner, Phys. Part. Nucl. **38**, 469 (2007)
20. V.E. Viola, G.T. Seaborg, J. Inorg. Nucl. Chem. **28**, 741 (1966)
21. A. Sobiczewski, Z. Patyk, S. Cwiok, Phys. Lett. B 224, 1 (1989).

22. E.O. Fiset, J.R. Nix, Nucl. Phys. A **193**, 647 (1972)
23. P. Möller, J.R. Nix, K.-L. Kratz, At. Data Nucl. Data Tables **66**, 131 (1997)
24. C.S. Wu, S.A. Moszkowski, *Beta decay* (John Wiley & Sons, New York, 1966), p. 183
25. M.A. Preston, *Physics of the Nucleus* (Addison-Wesley Publishing Company, Inc., Reading, Massachusetts, 1962), p. 574
26. V.I. Zagrebaev, A.S. Denikin, A.V. Karpov, A.P. Alekseev, V.V. Samarin, M.A. Naumenko, V.A. Rachkov, Nuclear Map of NRV, http://nrv.jinr.ru/nrv
27. R. Smolańczuk, Phys. Rev. C **56**, 812 (1997)
28. R. Smolańczuk, J. Skalski, Phys. Rev. C **52**, 1871 (1995)
29. A. Staszczak, Z. Lojewski, A. Baran, B. Nerlo-Pomorska, K. Pomorski, In: Proceedings of the 3rd International Conference on Dynamical Aspects of Nuclear Fission, Casta-Papernicka, 30 Aug 1996, p. 22.
30. A.V. Karpov, V.I. Zagrebaev, Y. Martinez Palenzuela, L. Felipe Ruiz, W. Greiner, Int. J. Mod. Phys. E **21**, 1250013 (2012)
31. W.J. Swiatecki, Phys. Rev. **100**, 931 (1955)
32. N.E. Holden, D.C. Hoffman, Pure Appl. Chem. **72**, 1525 (2000)
33. A.J. Sierk, Phys. Rev. C **33**, 2039 (1986)
34. V.I. Zagrebaev, Yu. Ts Oganessian, M.G. Itkis, Phys. Rev. C **73**, 031602 (2006)
35. V.I. Zagrebaev, W. Greiner, Phys. Rev. C **83**, 044618 (2011)
36. V.I. Zagrebaev, A.V. Karpov, I.N. Mishustin, Phys. Rev. C **84**, 044617 (2011)
37. V.I. Zagrebaev, A.V. Karpov, Phys. Rev. C **85**, 014608 (2012)

# Stability Peninsulas at the Neutron Drip Line

**Dmitry Gridnev, V. N. Tarasov, K. A. Gridnev, S. Schramm, D. V. Tarasov and W. Greiner**

## 1 Introduction

One of the fundamental questions in nuclear physics is what combinations of neutrons and protons can build up a stable nucleus. The nuclear landscape called nuclear chart is shown in Figs. 1, 2 and 3. A large number of stable isotopes are still nuclear "terra incognita". Moving away from stable nuclei by adding either protons or neutrons, one finally reaches the particle drip lines where the nuclear binding ends. The nuclei beyond the drip lines are unbound to nucleon emission; that is, for those systems the strong interaction is unable to bind the constituent nucleons as one nucleus.

The yet unexplored parts of the nuclear chart may help answering many questions of fundamental importance: What are the limits of nuclear existence? What are the properties of nuclei with an extreme $N/Z$ ratio? There are also related important questions in the field of nuclear astrophysics. Nuclei far from stability valley are different from that around the stability line: here one finds the distortion of the known "magic" numbers, large spatial extensions of such nuclei and other interesting structural phenomena. An exciting question is whether there exist islands of stability beyond the drip lines. Such islands are neutron stars, which are bound due to gravity or the island of superheavy elements (see the talks of S. Schramm, Walter Greiner, Zagrebaev, Bandyopadhyay in this volume).

The main objective of my talk is to discuss some interesting new qualitative features of the neutron drip line that were predicted in [1–4], namely, the formation of stability peninsulas. A relatively recent experiment [5] revealed new squares on the nuclear chart, which correspond to stable isotopes $^{40}$Mg and $^{42}$Al. Figure 1 shows also comparison with existing theoretical predictions. Interestingly, Nature chooses the most optimistic scenario regarding stability of isotopes at the drip line. So far the experiment is not capable to detect the whole neutron drip line, and as can be seen

D. Gridnev(✉) · V. N. Tarasov · K. A. Gridnev · S. Schramm · D. V. Tarasov · W. Greiner
Frankfurt Institute for Advanced Studies, Goethe University, Ruth-Moufang-Str. 1, 60438 Frankfurt am Main, Germany
e-mail: gridnev@fias.uni-frankfurt.de

W. Greiner (ed.), *Exciting Interdisciplinary Physics*,
FIAS Interdisciplinary Science Series, DOI: 10.1007/978-3-319-00047-3_7,

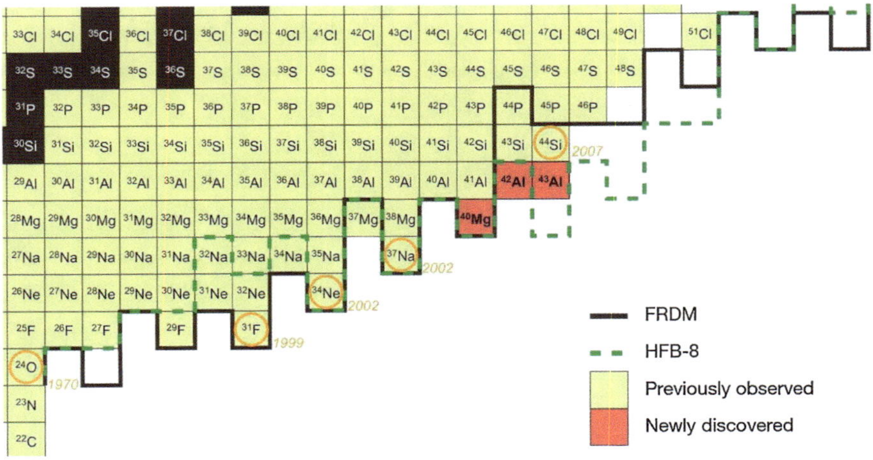


**Fig. 1** Fragment of the nuclear chart. The proton number increases vertically and the neutron number horizontally. *Yellow squares* denote previously observed nuclei. The neutron *drip lines* predicted by the FRDM and HFB-8 models are shown by the *black* and *dashed green lines*, respectively. The most recently observed *drip-line* nuclei are indicated by *orange circles* with their year of discovery, and newly discovered isotopes [5] are highlighted in *red*

in Fig. 2 it is, probably, very far from that. It is, of course, important to foresee the experimental setup that would be able to confirm or invalidate present theoretical predictions. An interesting opportunity for production of neutron rich elements is mentioned in the talk of Walter Greiner, see this volume. The suggestion is to utilize an intensive neutron flux, which can be generated by a double atomic bomb explosion or by a chain of pulsed nuclear reactors. Such flux would force nucleus capture a large number of neutrons before it undergoes beta decay.

## 2 Stability Peninsulas

In our discussion we shall focus on even-even nuclei. A reliable microscopic description of nuclei is obtained with the so-called effective forces between nucleons called Skyrme forces. Their use has become popular since the seminal papers of Vautherin and Brink [6, 7], where these forces were successfully used for systematic description of spherical and deformed nuclei. These forces generally predict very well deformations, sizes, nuclear densities, nucleon separation energies etc. Their advantage is a relatively small number of parameters and delta-functions in the interaction terms, which facilitate the calculation of the integrals considerably.

These forces can be expressed as

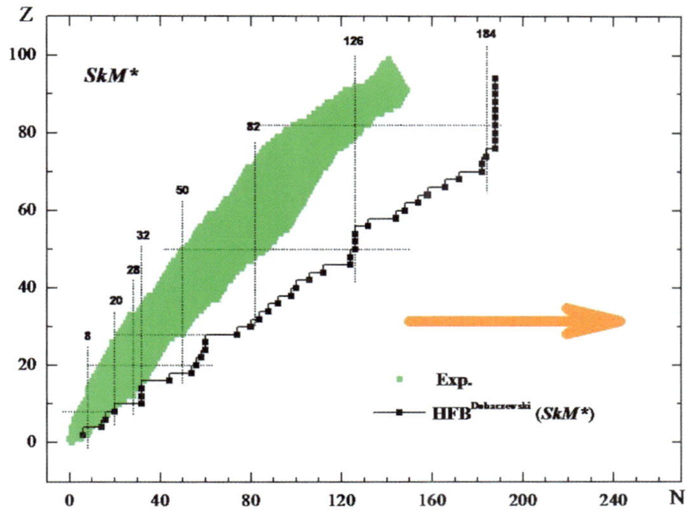


**Fig. 2** Nuclear chart. *Green* area shows experimentally observed nuclei. *Black squares* correspond to the neutron *drip line* calculated with the Hartree-Fock-Bogoliubov method [8, 9]. The *orange arrow* indicates typical direction of the calculations, when one tries to detect the *drip line*, namely, one increases the neutron number until the saturation point is reached

**Table 1** Various types of Skyrme Forces and their parameters

| Force | $t_0$ MeV fm$^3$ | $t_1$ MeV fm$^5$ | $t_2$ MeV fm$^5$ | $t_3$ MeV fm$^{3+3\alpha}$ | $x_0$ | $x_1$ | $x_2$ | $x_3$ | $W_0$ MeV fm$^5$ | $\alpha$ |
|---|---|---|---|---|---|---|---|---|---|---|
| Sly4 | −2489 | 487 | −546 | 13777 | 0.83 | −0,3 | −1 | 1.4 | 123 | 1/6 |
| SkM* | −2645 | 410 | −135 | 15595 | 0.09 | 0 | 0 | 0 | 130 | 1/6 |
| SkI2 | −1915 | 438 | 305 | 10549 | −0.21 | −1.7 | −1.5 | −0.2 | 120 | 1/4 |
| Ska | −1603 | 571 | −68 | 8000 | −0.02 | 0 | 0 | −0.3 | 125 | 1/3 |

$$V_{12} = t_0(1 + x_0 P_\sigma)\delta(\mathbf{r}) + \frac{1}{2}t_1(1 + x_1 P_\sigma)\left[\mathbf{k}'^2\delta(\mathbf{r}) - \delta(\mathbf{r})\mathbf{k}^2\right] + t_2(1 + x_2 P_\sigma)\mathbf{k}'\delta(\mathbf{r})\mathbf{k}$$
$$+\frac{1}{6}t_3(1 + x_3 P_\sigma)\rho^\gamma\delta(\mathbf{r}) + iW_0(\boldsymbol{\sigma}_1 + \boldsymbol{\sigma}_2)(\mathbf{k}' \times \delta(\mathbf{r})\mathbf{k})$$

Here $P_\sigma = (1 + \boldsymbol{\sigma}_1\boldsymbol{\sigma}_2)/2$. One can see that the force is density dependent. The parameters entering the expression are usually fixed so as to reproduce various bulk nuclear properties as well as selected properties of certain doubly magic nuclei. There is no unique set of parameters and this leads to various variants of the Skyrme force, each of which has its advantages and disadvantages. Some of these are listed in Table 1.

After fixing the parameters the Skyrme forces are used as ingredient in Hartree Fock calculations, where the ground state wave function is written in the form of a Slater determinant. One also has to introduce the pairing force, which in our case is

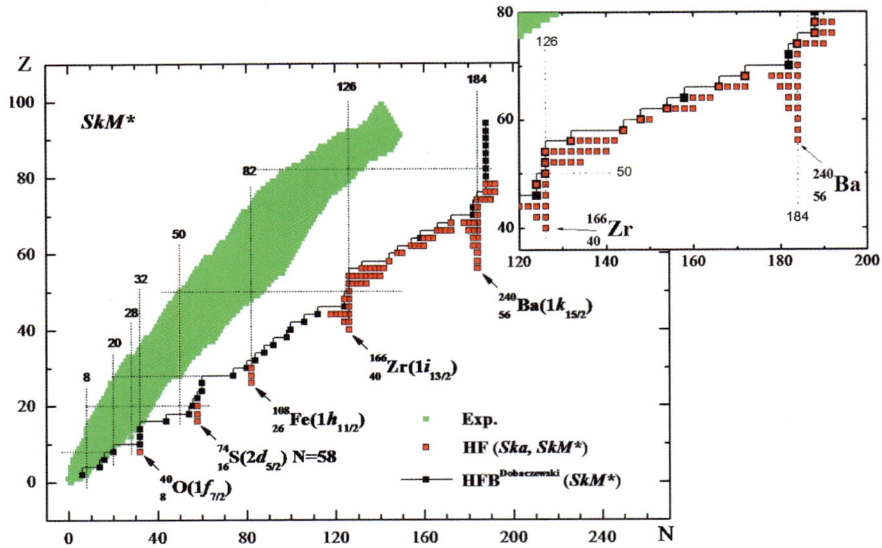


**Fig. 3** Formation of peninsulas at the neutron drip line. *Black squares* indicate the drip line obtained within HFB approach using SkM* forces. *Green area* shows experimentally known nuclei. *Red squares* are nuclei that are predicted stable against one neutron emission in our calculations using Ska and SkM* forces. One can see formation of peninsulas at "magic" numbers and "quenched magic" numbers (see the text for details)

treated in the BCS framework with a pairing constant $G_{n,p} = (19.5/2)[1 \pm 0.51(N - Z)/A]$. The Hartree-Fock equations are then solved using the basis functions of the deformed harmonic oscillator. Since we focus our attention on the neutron drip line one encounters wave functions, which correspond to small neutron separation energies, and, therefore, are very spatially extended. Clearly, one needs basis functions that match such peculiar behavior. This is done by adjusting the parameters of the harmonic oscillator on each iteration. The parameters of the oscillator are chosen in order to minimize the resulting total energy. This helpful procedure of readjusting basis functions reduces substantially the required number of basis functions as well as the required number of iterations.

In the case of spherical nuclei we solve the Hartree-Fock integral-differential equations directly without using the oscillator basis. This is a useful additional check because most of the nuclei forming the stability peninsulas are spherical. In the deformed case only bound states entered the BCS scheme. In the spherical case we also include those quasi-stable states in the continuum, which lie under the centrifugal barrier. In this procedure the spectrum is discretized by introducing a fictional wall. The states, which remain localized when the wall is moved at a large distance, are taken into account in the pairing scheme.

The standard theoretical approach in locating the neutron drip line is to take a stable nucleus with a fixed proton charge $Z$ and increase the number of neutrons $N$

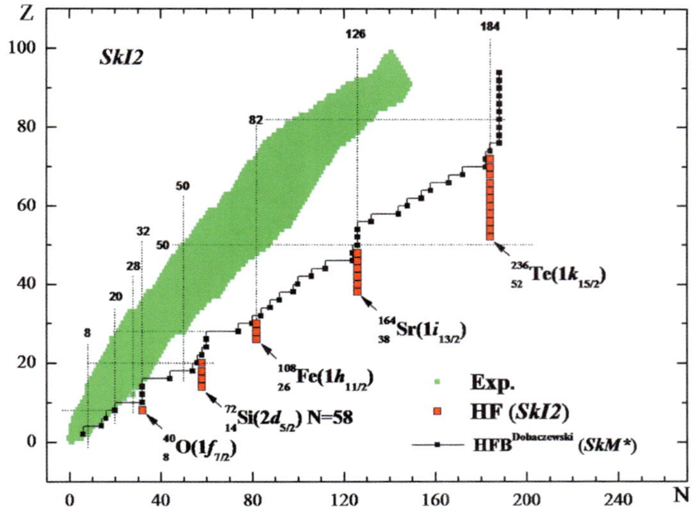


**Fig. 4** The same as in Fig. 3, except that now the forces are SkI2. These forces provide the most "optimistic" predictions regarding the stability of isotopes. Comparing to Fig. 3 one can see that the peninsulas are positioned at the same number of neutrons. In the present case the peninsulas are more extended down in $Z$

until the resulting nucleus would be "overloaded" in the sense that it gets rid of extra neutrons through decay, see Fig. 2. This method, however, implies a simple structure of the drip line, namely, that every line corresponding to a fixed number of protons on the nuclear chart crosses the neutron drip line only once. Yet, it might happen that the drip line has a more complicated structure [1–4]. In the vicinity of "magic" numbers or "quenched magic" numbers the following scenario can take place. At some point being filled with neutrons the nucleus loses its stability but then after adding more neutrons the stability is restored. This leads to formation of stability peninsulas on the nuclear chart, see Figs. 3 and 4.

The analysis of the phenomenon of stability restoration through adding neutrons has been undertaken in [1–4]. We have considered long isotope rows of the elements Pb, Zr, Ar, Kr, Rn, Gd, Ba, S, as well as many other elements. Thereby, the phenomenon of stability restoration through adding neutrons was in focus. Having found such new isotopes we also investigated their properties like masses, deformation, root mean square radii etc.

An important point is also that nuclei forming stability peninsulas are spectrally bound in the sense that there exists a well-defined ground state wave function, which minimizes the energy functional for such nuclei. At this point they become well-defined compact objects and the question about their lifetime is correctly formulated. And though for some nuclei it may be energetically favorable to get rid of two or more neutrons, a large centrifugal barrier of the last filled levels may serve as an indication that this lifetime would be large. For some nuclei the energetically favorable decay

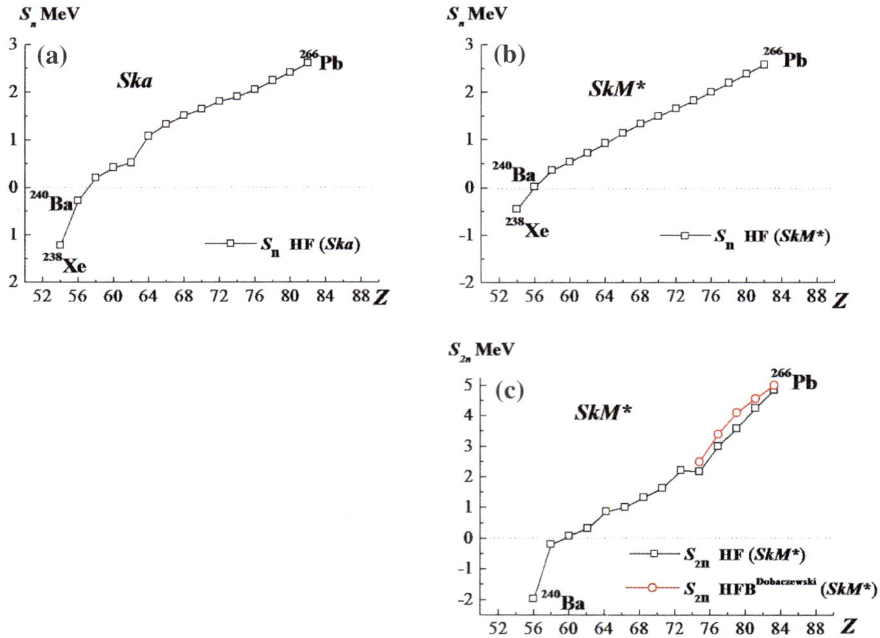


**Fig. 5** One and two neutron separation energies for the isotones with $N = 184$ calculated with Ska, SkM* forces. *Red line* corresponds to HFB calculations [8, 9]. The last element, which is stable against one and two neutrons emission, is $^{244}$Nd

would be into four or more neutrons, which enhances the lifetime considerably (such decays were not experimentally observed so far).

In searching for extensions of the neutron drip line limits we proceeded in the standard way adding as many neutrons to the nucleus as possible. Having found an unstable nucleus we did not stop and added more neutrons to see, whether the stability can be restored. Gradually, the general picture became clear. One can see in Fig. 3 that stability peninsulas are formed at neutron "magic" numbers or "quenched magic" numbers like in the case of $^{40}$O ($N = 32$) or $^{74}$S ($N = 58$), which correspond to the filled subshells $1f_{7/2}$ and $2d_{5/2}$ respectively. The stability peninsulas extend vertically along the $Z$ axis in the direction of diminishing $Z$. Figure 5 shows how the stability peninsula corresponding to $N = 184$ (a closed shell) extends in $Z$. This can be seen from one and two neutron separation energies of the isotones. Figure 6 shows the same for the closed shell with $N = 258$, which we discuss below.

In our approach we try different variants of the Skyrme forces. Let us stress that just by definition for nuclei at the neutron drip line one expects small one and two neutron separation energies. So it would be rather naive to expect that the drip line calculated with effective forces would exactly match the real one. Even for different forces the difference between separation energies is of order MeV. This is why it is important to check the results with various variants of the Skyrme forces. It turns out

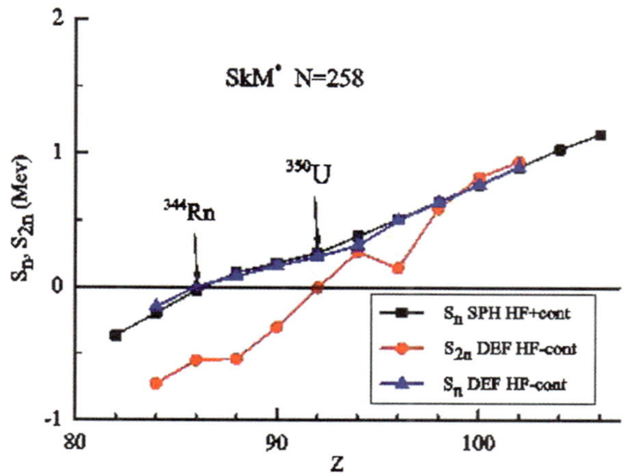


**Fig. 6** One and two neutron separation energies for the isotones with $N = 258$ calculated with SkM* forces. *Black squares* indicate calculations assuming spherical symmetry, where one solves the equations on a grid rather than expands the wave function in the harmonic oscillator basis. See also Table 2

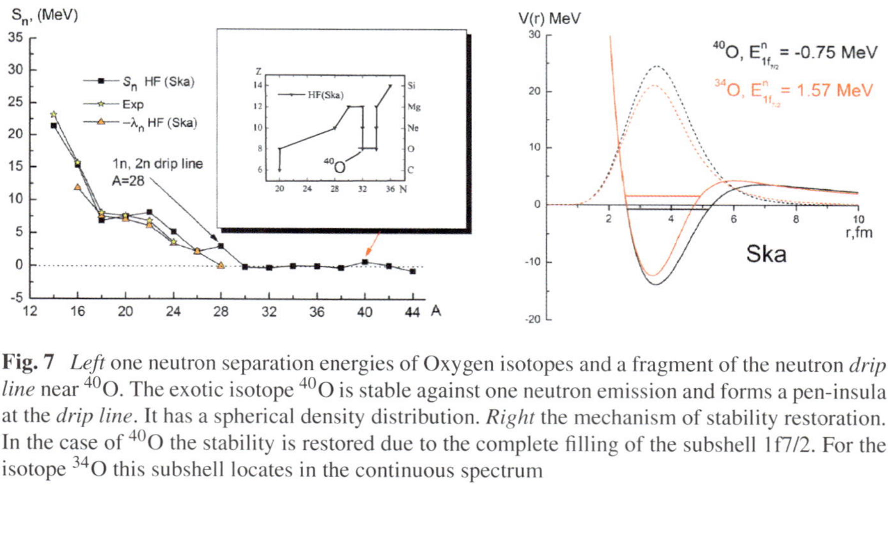


**Fig. 7** *Left* one neutron separation energies of Oxygen isotopes and a fragment of the neutron *drip line* near $^{40}$O. The exotic isotope $^{40}$O is stable against one neutron emission and forms a pen-insula at the *drip line*. It has a spherical density distribution. *Right* the mechanism of stability restoration. In the case of $^{40}$O the stability is restored due to the complete filling of the subshell 1f7/2. For the isotope $^{34}$O this subshell locates in the continuous spectrum

that the neutron numbers, where the stability peninsulas appear are the same for all forces, only the edges of these peninsulas and the degree to which they are extended depend on the specific variant of the force. This can be illustrated in Figs. 3 and 4. In view of this striking invariance with respect to the choice of the Skyrme force, we claim that such peninsulas constitute a general qualitative feature of the neutron drip line.

The mechanism working behind the formation of stability peninsulas is usually the same in all cases. When one adds neutrons to an unstable nucleus the totally filled subshell immerses from the continuum to the states with negative energy. For

**Table 2** Heaviest Uranium isotopes for spherical (1d) and axially deformed (2d) calculations

| Model | $N$(1d) | $N$(2d) | $E$tot(2d) MeV |
|---|---|---|---|
| NL-Z2 | 258 | 258 | 2021 |
| SkM* | 246 | 220 | 1987 |
| Sly6 | 184 | 206 | 1929 |
| SkI4 | 258 | 218 | 1975 |

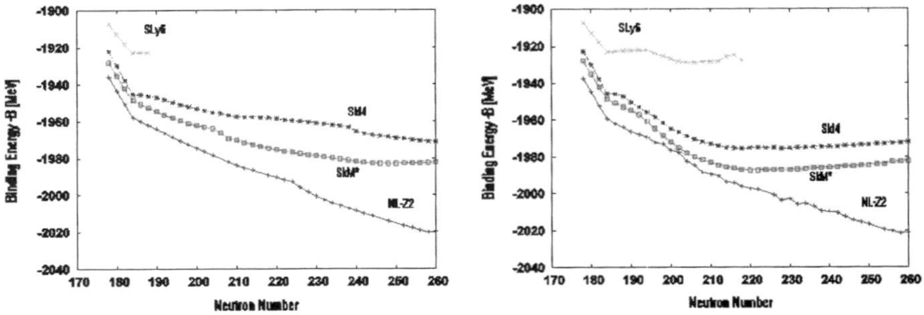


**Fig. 8** Total binding energy of Uranium isotopes for different forces and models. *Left* assuming spherical symmetry; *Right* assuming axial symmetry. SLy6, SkI4, SkM* are Skyrme forces used in Hartree-Fock calculations on a grid. NL-Z2 are relativistic mean field calculations. The global minima of these curves are presented in Table 2

example, in Fig. 7 (right) we show how this happens in the case of $^{40}$O. The shell effects are the key to the understanding of these phenomena. Below we list some of the isotopes from the stability peninsulas and the subshells responsible for the stability enhancement

$1f_{7/2}$ $^{40}$O

$2d_{5/2}$ $^{76}$Ar, $^{74}$S

$1h_{11/2}$ $^{110}$Ni, $^{108}$Fe

$1i_{13/2}$ $^{174}$Cd, $^{172}$Pd, $^{170}$Ru, $^{168}$Mo, $^{166}$Zr

$1k_{15/2}$ $^{256}$Hf, $^{254}$Yb, $^{252}$Er, $^{250}$Dy, $^{248}$Gd, $^{246}$Sm, $^{244}$Nd, $^{242}$Ce, $^{240}$Ba

An unbelievably neutron rich nucleus was proposed in [10]. It turns out that various models, not only the ones described here but also relativistic mean field calculations predict the existence of a stable extremely heavy Uranium nucleus, see Fig. 8 and Table 2. This is an effect of a closed shell corresponding to $N = 258$, which appears to be "magic". The effect is so strong that one speaks here not about the local minimum of the binding energy as a function of the neutron number like we had it previously in the case of peninsulas but rather of a global minimum. Would such nucleus ever be experimentally found it would be a truly fascinating object.

**Table 3** Nuclei with maximal binding energy for different models and different Z

| Z | NL-Z2 | SkM* | SkI4 | Sly6 |
|---|---|---|---|---|
| 92 | 258 | 220 | 218 | 206 |
| 94 | 258 | 230 | 230 | 208 |
| 96 | 258 | 258 | 230 | 208 |
| 98 | 258 | 258 | 258 | 218 |
| 100 | 258 | 258 | 258 | 220 |
| 102 | 258 | 258 | 258 | 220 |
| 104 | 258 | 258 | 258 | 232 |

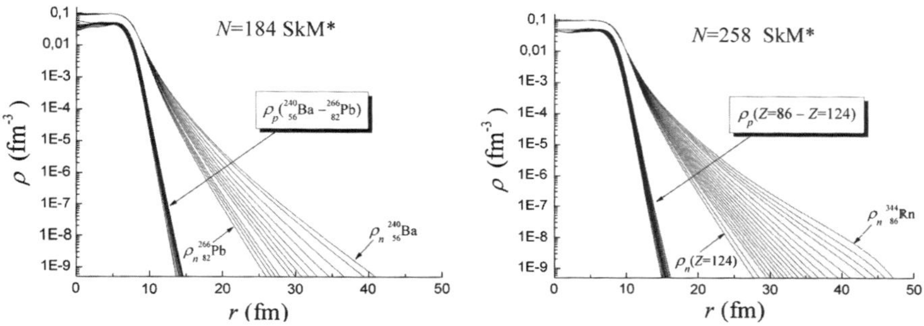


**Fig. 9** Proton and neutron density distributions for the isotones with $N = 184$ (*left*) and $N = 258$ (*right*). The calculations are performed with SkM* forces. These are closed shells and the distributions possess spherical symmetry. One can see the enormous spatial extension (halo formation) for a large number of neutrons

One can see that some models predict a stable nucleus containing 350 nucleons! To see how heavy can be the isotopes of elements neighboring to Uranium see Table 3.

Let us also mention that the wave functions near the drip line produce very spatially extended neutron densities, see Fig. 9. Here one can speak of a large halo formation. To illustrate this we shall compare proton and neutron root mean square radii, which we denote $R_p$ and $R_n$ respectively. For nuclei in the stability valley one has normally $R_n - R_p \simeq 0.1 - 0.2$ fm. For $^{40}$O we obtain $R_n - R_p \simeq 1.29$ fm and $R_n/R_p \simeq 1.44$. For $^{248}$Gd we get $R_n - R_p \simeq 0.77$ fm and $R_n/R_p \simeq 1.14$. And for $^{240}$Ba we obtain $R_n - R_p \simeq 0.94$ fm and $R_n/R_p \simeq 1.17$.

Regarding giant neutron halos it is worth mentioning a recent result [11], which concerns a universal behavior of spatially extended wave functions of three-body systems (cluster plus two neutrons). We can write the wave function of the cluster with mass $M$ and two particles with mass m in the form $\psi(x, y)$, where $x = r_2 - r_1$, $y = 2\sqrt{mM(2m + M)}(r_3 - (r_1 + r_2)/2)$, and $r_i$ are particle position vectors. It can also written in the form $\psi(\rho, \theta, \hat{x}, \hat{y})$ where $\rho = \sqrt{x^2 + y^2}$, $\theta = \arctan(|y|/|x|)$ and $\hat{x}$, $\hat{y}$ are unit vectors in the direction of $x$, $y$, respectively. The angular density distribution of a wave function can be defined as

$$D(\theta, \hat{x}, \hat{y}) = \cos^2\theta \sin^2\theta \int \rho^5 |\psi(\rho, \theta, \hat{x}, \hat{y})|^2 d\rho$$

It turns out that for a spatially extended wave function

$$D(\theta, \hat{x}, \hat{y}) \rightarrow \frac{1}{(4\pi)^2} \frac{1}{4\pi} \sin^2\theta$$

The limit becomes exact for an infinitely extended wave function. Remarkably, the resulting expression depends only on the angle $\theta$. This expression is in good agreement with the form of the so-called dineutron peak in the halo distribution, see Figs. 1 and 2 in [12].

**Acknowledgments** The author would like to express his gratitude to Walter Greiner and to local organizers of the Exciting Physics Symposium at Makutsi.

## References

1. K.A. Gridnev, D.K. Gridnev, V.G. Kartavenko, V.E. Mitroshin, V.N. Tarasov, D.V. Tarasov, W. Greiner, Eur. Phys. J. A **25**, S01, 253 (2005)
2. K.A. Gridnev, D.K. Gridnev, V.G. Kartavenko, V.E. Mitroshin, V.N. Tarasov, D.V. Tarasov, W. Greiner, Phys. Atom. Nucl. **69**, 3 (2006)
3. K.A. Gridnev, V.N. Tarasov, D.V. Tarasov, D.K. Gridnev, V.V. Pilipenko, W. Greiner, Int. J. Mod. Phys. E **15**, 673 (2006)
4. V.N. Tarasov, D.V. Tarasov, K.A. Gridnev, D.K. Gridnev, V.G. Kartavenko, W. Greiner, Int. J. Mod. Phys. E **17**, 1273 (2008)
5. T. Baumann et al., Nature **449**, 1022 (2007)
6. D. Vautherin, D.M. Brink, Phys. Rev. C **5**, 626 (1972)
7. D. Vautherin, D.M. Brink, Phys. Rev. C **7**, 296 (1973)
8. M.V. Stoitsov, J. Dobaczewski, P. Ring, S. Pittel, Phys. Rev. C **61**, 034311 (2000)
9. M.V. Stoitsov, J. Dobaczewski, W. Nazarewicz, S. Pittel, D.J. Dean, Phys. Rev. C **68**, 054312 (2003)
10. S. Schramm, D. Gridnev, D.V. Tarasov, V.N. Tarasov, W. Greiner, Int. J. Mod. Phys. E 21, 1250047 (2012)
11. D.K. Gridnev, Angular Probability Distribution of Three Particles near Zero Energy Threshold, ArXiv:0912.0414v2 (2009)
12. Yu. Ts, Oganessian, V.I. Zagrebaev, J.S. Vaagen. Phys. Rev. Lett. **82**, 4996 (1999)

# Unexpected Strong Decay Mode of Superheavy Nuclei

**D. N. Poenaru, R. A. Gherghescu and W. Greiner**

**Abstract** Calculations of half-lives of superheavy nuclei (SH) show an unexpected result: for some of them heavy particle radioactivity (HPR) dominates over alpha decay—the main decay mode of the majority of recently discovered SHs. The result is important for theory and future experiments producing heavier SHs with a substantial amount of funding. The standard identification technique by alpha decay chains will be impossible for these cases. HPR had been predicted in 1980 four years before the first experiment. The daughter is mainly the doubly magic $^{208}$Pb. We changed the concept of HPR to allow emitted particles with $Z_e > 28$ from parents with $Z > 110$ (daughter around $^{208}$Pb). We find a trend toward shorter half-lives and larger branching ratios relative to alpha decay for heavier SHs. A new table of measured masses AME11 and the theoretical LiMaZe01, KTUY05 and FRDM95 tables are used to determine Q-values.

## 1 Introduction

Superheavy (SH) elements with atomic numbers $Z = 104 - 118$ have been synthesized with cold fusion reactions [8, 9, 17] or with hot fusion induced by $^{48}$Ca projectiles [19–21]. The majority of proton-rich SH nuclides are identified through the $\alpha$ decay chains. Up to now only $\alpha$ decay, $\beta$ decay and spontaneous fission of SH nuclei have been observed (see the Ref. [11] for calculations). Review papers on theory of SHs [6, 37], and stability calculations [10] are recently published.

D. N. Poenaru (✉) · R. A. Gherghescu
Horia Hulubei National Institute of Physics and Nuclear Engineering, P O Box MG-6, 077125 Bucharest-Magurele, Romania
e-mail: Dorin.Poenaru@nipne.ro; poenaru@fias.uni-frankfurt.de

W. Greiner
Frankfurt Institute for Advanced Studies, J. W. Goethe University, Ruth-Moufang-Str. 1, 60438 Frankfurt am Main, Germany

W. Greiner (ed.), *Exciting Interdisciplinary Physics*,
FIAS Interdisciplinary Science Series, DOI: 10.1007/978-3-319-00047-3_8,

We would like to discuss the competition of $\alpha$ decay and heavy particle radioactivity (HPR) [4, 35], also named cluster radioactivity, which may be important [25] in the region of the heaviest SHs. In this binary process, from one parent nucleus, $^{A}Z$, one obtains an emitted particle, $^{A_e}Z_e$, and a daughter $^{A_d}Z_d$:

$$^{A}Z \rightarrow {}^{A_e}Z_e + {}^{A_d}Z_d \tag{1}$$

We have an indication of the possibility of extrapolating our calculations to this region from the results of the following calculations within ASAF model: the half-life for $^{128}$Sn emission from $^{256}$Fm ($Q = 252.129$ MeV) and for $^{130}$Te emission from $^{262}$Rf ($Q = 274.926$ MeV) are given by $\log_{10} T_{Fm}(s) = 4.88$ and $\log_{10} T_{Rf}(s) = 0.53$. They are in agreement with experimental values for spontaneous fission [7]: 4.02 and 0.32, respectively.

Since 1984 [34] the following HPR have been experimentally confirmed [3, 30] in heavy parent nuclei with $Z = 87 - 96$: $^{14}$C, $^{20}$O, $^{23}$F, $^{22,24-26}$Ne, $^{28,30}$Mg, and $^{32,34}$Si. The employed techniques [27] in Universities and Research Institutes are: semiconductor telescope; magnetic spectrometers (SOLENO, Enge split-pole), and solid state nuclear track detectors. The measured half-lives are in good agreement with predicted values within analytical superasymmetric fission (ASAF) model (see the reviews [5, 22, 23, 28, 29] and references therein). The shortest measured half-life of $T_c = 10^{11.01}$ s corresponds to $^{14}$C radioactivity of $^{222}$Ra and the largest branching ratio relative to alpha decay, $b_\alpha = T_\alpha/T_c$, of $10^{-8.9}$ was observed for $^{14}$C radioactivity of $^{223}$Ra. Consequently HPR in the region of heavy transfrancium nuclei is a rare phenomenon.

Very frequently the daughter nucleus was the doubly magic $^{208}_{82}$Pb$_{126}$ or one of its neighbors. This is the reason why we changed the concept of HPR, previously [32] associated to a maximum $Z_e^{max} = 28$. In the regions of SHs with $Z > 110$ we consider not only the emitted particles with atomic numbers $2 < Z_e < 29$, but also heavier ones up to $Z_e^{max} = Z - 82$, allowing to get for $Z > 110$ an atomic number of the most probable emitted HP $Z_e > 28$ and a doubly magic daughter around $^{208}$Pb.

Besides the already mentioned superasymmetric fission theory there are many other theoretical approaches of the HPR e.g. Refs. [2, 15, 23, 33, 39]. Present calculations are performed within the ASAF model, very useful for the high number of combinations parent-emitted cluster in order to check the metastability of SH parent nuclides with measured or calculated masses against many possible decay modes.

## 2 The Model

A crucial quantity for accuracy of half-life calculation is the released energy

$$Q = [M - (M_e + M_d)]c^2 \tag{2}$$

obtained as a difference of the parent and the two decay product masses in units of energy; $c$ is the light velocity. In theory the decay constant $\lambda = \ln 2/T_c$ may be expressed by a product of three model dependent quantities:

$$\lambda = \nu S P_s \tag{3}$$

where $\nu$ is the frequency of assaults on the barrier per second, $S$ is the preformation probability and $P_s$ is penetrability of external barrier, mainly of Coulomb nature. According to our method to estimate the preformation as the penetrability of internal barrier [26] we have

$$S = \exp(-K_{ov}) \tag{4}$$

A very large number of combinations parent–emitted cluster has to be considered in a systematic search for new decay modes. The numerical calculation of three-fold integrals involved in numerical models are too time-consuming. The large amount of computations can be performed in a reasonable time by using an analytical relationship for the half-life. We developed our ASAF model to fulfill this requirement. We started with Myers-Swiatecki liquid drop model [18] adjusted with a phenomenological correction.

The half-life of a parent nucleus $AZ$ against the split into a cluster $A_e Z_e$ and a daughter $A_d Z_d$

$$T = [(h \ln 2)/(2E_v)] \exp(K_{ov} + K_s) \tag{5}$$

is calculated by using the WKB quasiclassical approximation, according to which the action integral is expressed as

$$K = \frac{2}{\hbar} \int_{R_a}^{R_b} \sqrt{2B(R)E(R)}dR \tag{6}$$

with $B = \mu$, $K = K_{ov} + K_s$, and $E(R)$ replaced by $[E(R) - E_{\text{corr}}] - Q$ where $E_{\text{corr}}$ is a correction energy similar to the Strutinsky shell correction, also taking into account the fact that Myers-Swiatecki's liquid drop model (LDM) [18] overestimates fission barrier heights, and the effective inertia in the overlapping region is different from the reduced mass [32]. The turning points of the WKB integral are: $R_a = R_i + (R_t - R_i)[(E_v + E^*)/E_b^0]^{1/2}$ and $R_b = R_t E_c\{1/2 + [1/4 + (Q + E_v + E^*)E_l/E_c^2]^{1/2}\}/(Q + E_v + E^*)$ where $E^*$ is the excitation energy concentrated in the separation degree of freedom, $R_i = R_0 - R_e$ is the initial separation distance, $R_t = R_e + R_d$ is the touching point separation distance, $R_j = r_0 A_j^{1/3}$ ($j = 0, e, d$; $r_0 = 1.2249$ fm) are the radii of parent, emitted and daughter nuclei, and $E_b^0 = E_i - Q$ is the barrier height before correction. The interaction energy at the top of the barrier, in the presence of a non-negligible angular momentum, $l\hbar$, is given by:

$$E_i = E_c + E_l = e^2 Z_e Z_d / R_t + \hbar^2 l(l+1)/(2\mu R_t^2) \tag{7}$$

The two terms of the action integral $K$, corresponding to the overlapping ($K_{ov}$) and separated ($K_s$) fragments, are calculated by analytical formulas (approximated for $K_{ov}$ and exact for $K_s$ in case of separated spherical shapes within the LDM):

$$K_{ov} = 0.2196(E_b^0 A_e A_d/A)^{1/2}(R_t - R_i)\left[\sqrt{1-b^2} - b^2 \ln \frac{1+\sqrt{1-b^2}}{b}\right] \quad (8)$$

$$K_s = 0.4392[(Q + E_v + E^*)A_e A_d/A]^{1/2} R_b J_{rc} \; ; \; b^2 = (E_v + E^*)/E_b^0 \quad (9)$$

$$\begin{aligned} J_{rc} &= (c)\arccos\sqrt{(1-c+r)/(2-c)} - [(1-r)(1-c+r)]^{1/2} \\ &+ \sqrt{1-c}\ln\left[\frac{2\sqrt{(1-c)(1-r)(1-c+r)} + 2 - 2c + cr}{r(2-c)}\right] \end{aligned} \quad (10)$$

where $r = R_t/R_b$ and $c = rE_c/(Q + E_v + E^*)$. In the absence of the centrifugal contribution ($l = 0$), one has $c = 1$. We took $E_v = E_{\rm corr}$ in order to get a smaller number of parameters.

The potential barrier shape similar to that we considered within the ASAF model was calculated by using the macroscopic-microscopic method [24], as a cut through the PES at a given mass asymmetry, usually the $^{208}$Pb valley or not far from it.

Half-life calculations are very sensitive to the released energy. Even with the newly released tables of experimental masses, atomic mass evaluation 2011 (AME11) [1] many masses are still not available for new SHs. We have used not only these updated tables for 3290 nuclides (2377 measured and 913 from the systematics) ending up at $Z = 118$ but also some calculated masses, e.g. Liran-Marinov-Zeldes (LiMaZe01) [13, 14], Koura-Tachibana-Uno-Yamada (KTUY05) [12] and the finite-range droplet model (FRDM95) [16] with 1969 ($Z = 82 - 126$, $N \leq 184$) 9441 ($Z = 2 - 130$, $N \leq 200$) and 8979 ($Z = 8 - 136$) masses, respectively.

In a systematic search for HPR we calculate with the ASAF model for every parent nucleus $^A$Z the half-lives of all combinations of pairs of fragments $^{A_e}Z_e$, $^{A_d}Z_d$ with $2 < Z_d \leq Z_e^{\max}$ conserving the hadron numbers $Z_e + Z_d = Z$ and $A_e + A_d = A$.

## 3 Results

We published [25] some results obtained by using the AME11 mass tables. Here we would like to present the results obtained by using the calculated masses LiMaZe01, KTUY05 and FRDM95. When using calculated masses for parent and daughter nuclei we take into account the nuclides stable against one proton, two protons, one neutron and two neutrons spontaneous emissions.

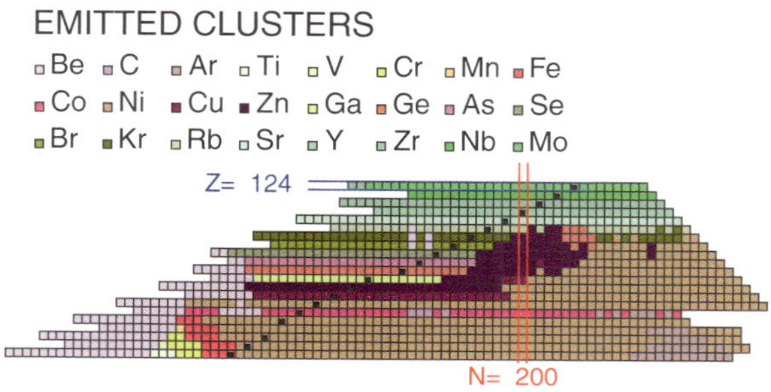


**Fig. 1** Chart of superheavy cluster emitters with atomic numbers $Z = 104 - 124$. The $Q$ values are calculated using the FRDM95 mass tables. *Black squares* mark the *Green* approximation of the line of beta stability.

The chart of cluster emitters from Fig. 1 is obtained by associating to each parent only the most probable emitted cluster. The black squares mark the Green approximation of the line of $\beta$ stability.

As may be seen from Fig. 1 besides emitted clusters with $Z_e \leq 28$, as Be, C, Ar, Ti, V, Cr, Mn, Fe, Co, and Ni, new many types of HPR with $Z_e > 28$ are present on this chart: Cu, Zn, Ga, Ge, As, Se, Br, Kr, Rb, Sr, Y, Zr, Nb, and Mo. In other words the following atomic numbers of the most probable emitted heavy particle are obtained: $Z_e = 4, 6, 18, 22, 23, 24, 25, 26, 27, 28, 29, 30, 31, 32, 33, 34, 35, 36, 37, 38, 39, 40, 41, 42$. As we previously observed [32], many of the SH nuclides are $^{8}$Be emitters, but they have a very low branching ratio $b_\alpha$. *Most frequently occurs the doubly magic* $^{78}$*Ni radioactivity*.

In few cases one has only one mass number for a given $Z_e$: $A_e = 8$ for $Z_e = 4$; $A_e = 14$ for $Z_e = 6$; $A_e = 55$ for $Z_e = 23$, and $A_e = 59$ for $Z_e = 25$. In other cases we also took only one color for every atomic number $Z_e$, despite the fact that one has various isotopes: $A_e = 50, 52, 53$ for $Z_e = 18$; $53, 54$ for 22; $58 - 60$ for 24; $62, 64, 66$ for 26; $63, 65, 71, 73, 75$ for 27; $66, 68, 70 - 78$ for 28; $73 - 80$ for 29; $74, 76-82$ for 30; $75, 77-83$ for 31; $78, 80, 82-84, 86-88$ for 32; $79-81, 83-89$ for 33; $82, 84 - 90$ for 34; $85 - 93$ for 35; $86 - 92, 94, 96, 98, 100$ for 36; $89 - 102$ for 37; $88 - 90, 92 - 96, 98, 100, 102, 104, 105$ for 38; $96 - 108$ for 39; $95 - 110$ for 40; $103 - 113$ for 41, and $100, 102, 104, 106 - 110, 112 - 115$ for $Z_e = 42$.

In Fig. 2 we compare the calculated half-lives of SHs against $\alpha$ decay based on AME11 mass tables within ASAF model with experimental data. The same quantity in Fig. 3 looks differently not showing the typical variations around the semi-magic neutron numbers of the daughter $N_d = 152, 162$. Nevertheless, the overall accuracy is not very much affected as may be seen in the Table 1.

An estimation of the accuracy gives the standard rms deviation of $\log T$ values:

$$\sigma = \left\{ \sum_{i=1}^{n} [\log(T_i/T_{\rm exp})]^2/(n-1) \right\}^{1/2} \tag{11}$$

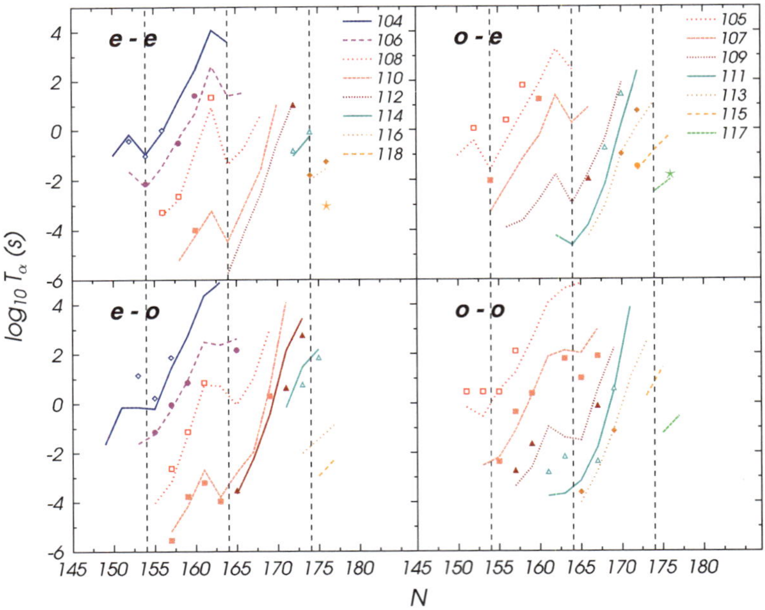


**Fig. 2** Decimal logarithm of the half-lives of SH nuclei against $\alpha$ decay versus the neutron number of the parent nucleus. Calculations performed within ASAF model. Experimental data marked with points. *Vertical dashed lines* correspond to $N = 154, 164, 174$. $Q$ values are calculated using the AME11 mass tables

As may be seen in Table 1 the best reproduction of experimental values are obtained by using AME11 mass values to calculate $Q$-values. It is followed by KTUY05 for e-e and e-o nuclei, by FRDM95 for o-e nuclei and by LiMaZe01 for o-o SH parent nuclei. Lower values of $\sigma$ for $\alpha$ decay half-lives may be obtained [31] within our UNIV (universal curve) and semFIS (semiempirical) models.

The general trend of a shorter half-life and a larger branching ratio when the atomic and mass numbers of the parent nucleus increases may be seen on Fig. 4, obtained within ASAF model by using the KTUY05 mass tables to calculate the $Q$ values in four groups of parent nuclei: even-even, even-odd, odd-even and odd-odd. This kind of plots in four groups allow to get smoother curves compared to what would be obtained if the even-odd staggering would be present. As to absolute value of $T_c$ and $b_\alpha$ there is no guarantee of reliability until some measurements will be available. More elaborate models should be used (see e.g. [36]) in order to estimate the competition of spontaneous fission.

If the calculated masses are reliable, then half-lives $T_c$ shorter than one nanosecond for SH nuclei with $Z \geq 122$ and large neutron numbers (see Fig. 5) would make difficult or even impossible any identification measurement. It would be possible to find some interesting cases with $T_c < T_\alpha$, $b_\alpha > 1$ which could be measured (see Fig. 4). For example if $Z = 122$, the neutron number of the Green approximation of

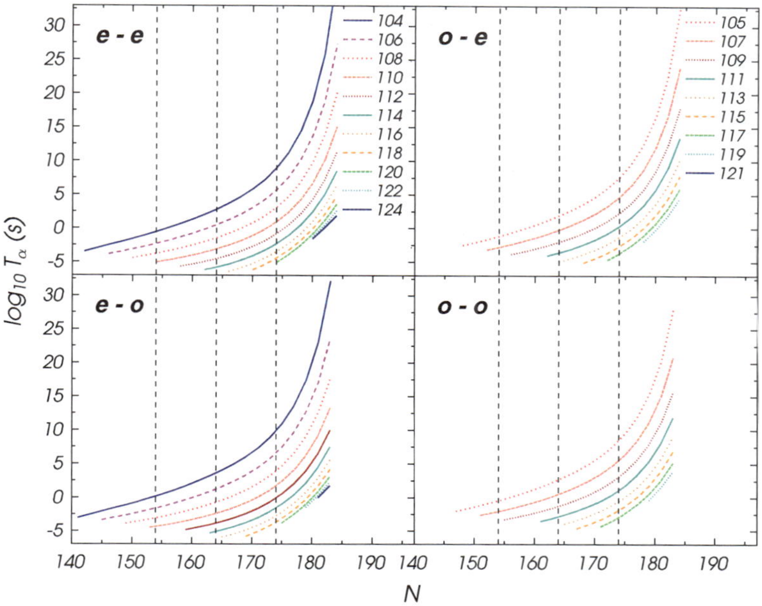


**Fig. 3** Decimal logarithm of the half-lives of SH nuclei against $\alpha$ decay versus the neutron number of the parent nucleus. Calculations performed within ASAF model. *Vertical dashed lines* correspond to $N = 154, 164, 174$. $Q$ values are calculated using the LiMaZe01 mass tables

**Table 1** Comparison of standard rms deviations from experiment of half-life calculations performed within ASAF model using different mass tables

| Parent nuclei | n | $\sigma_{AME11}$ | $\sigma_{LiMaZe01}$ | $\sigma_{KTUY05}$ | $\sigma_{FRDM95}$ |
|---|---|---|---|---|---|
| e-e | 16 | 0.582 | 1.666 | 1.264 | 1.372 |
| e-o | 20 | 0.741 | 1.627 | 1.092 | 1.559 |
| o-e | 13 | 1.072 | 2.043 | 1.637 | 1.421 |
| o-o | 19 | 0.831 | 1.092 | 1.254 | 1.135 |

the line of beta stability is $N_\beta = 202$, meaning that some neutron-deficient isotopes of 122 with $N > 188$ could have $b_\alpha > 1$. An increase in $Q$-value of $\Delta Q = 1.08$ MeV (from 273.49 to 274.57) produces a shorter half-life by 1.2 orders of magnitude, $\log_{10} T_c(s)$ from 6.58 to 5.38 for $^{81}$As emission from $^{287}$115. Similarly for $\Delta Q = 0.87$ MeV one has one order of magnitude shorter half-life for $^{85}$Se radioactivity of $^{293}$116.

The pronounced minimum of the branching ratio at $N = 186$ in Fig. 4 is the result of the strong shell effect of the assumed magic number of neutrons of the daughter $N_d = 184$ present in the KTUY05 masses. The half-life of $\alpha$ decay of a SH nucleus with $N = 186$ neutron number leading to a more stable daughter with magic neutron number $N_d = 184$ is shorter by some orders of magnitude compared to the $\alpha$ decay

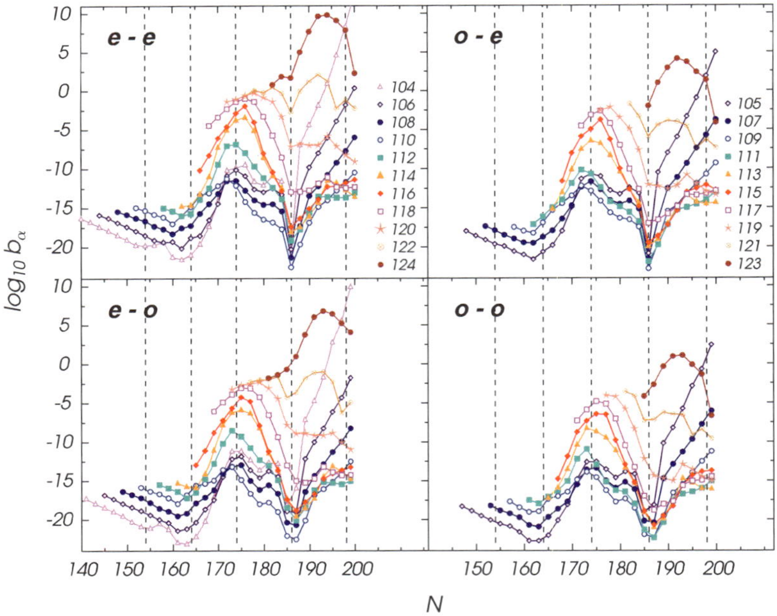


**Fig. 4** Decimal logarithm of the branching ratio relative to $\alpha$ decay for cluster emission from superheavy nuclei versus the neutron number of the parent nucleus. *Vertical dashed lines* correspond to $N = 174, 186, 198$. half-lives of superheavy nuclei against cluster radioactivities versus the neutron number of the parent nucleus. $Q$ values are calculated using the KTUY05 mass tables

of a SH with $N = 184$. Similar results were obtained using the FRDM95 masses (see Fig. 5).

Half-lives $T_c < T_\alpha$ are found for neutron-deficient SHs present on LiMaZe01 mass tables (at the top of Fig. 5). A transition from $b_\alpha < 1$ for $Z = 121$ to $b_\alpha > 1$ when $Z$ increases toward $Z = 124$ may be clearly seen in the central and bottom panels. A sharp decrease of $Q_\alpha$ values calculated with FRDM95 masses around $N = 196$ produces very high $T_\alpha$ at the bottom of Fig. 5. In Fig. 5 we can compare the absolute values of half-lives for $\alpha$ decay and HPR for SHs with $Z = 121 - 124$ calculated within ASAF model using different mass tables. The even-odd staggering is more pronounced for HPR than for $\alpha$ decay. This effect produces an alternation of $b_\alpha$ values for successive even and odd neutron numbers.

In conclusion, by changing the concept of HPR to allow spontaneous emission of heavy particles with atomic number larger than 28 from SHs with $Z > 110$ (daughter nuclei around the doubly magic $^{208}$Pb) we found that calculated half-lives $T_c$ against HPR and the branching ratios relative to $\alpha$ decay are showing a trend toward shorter $T_c$ and larger $b_\alpha$ for the heaviest SHs. It is possible to find regions of SHs where the half-life for HPR are shorter than those against $\alpha$ decay.

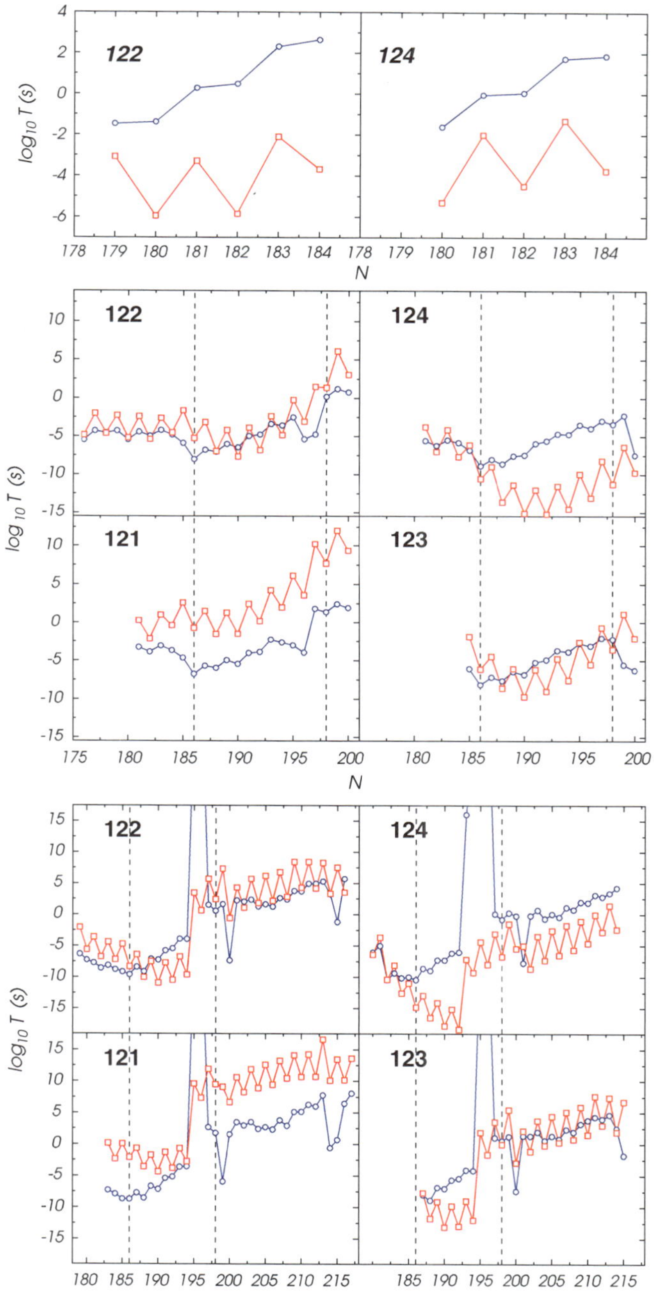


**Fig. 5** Decimal logarithm of the half-lives of superheavy nuclei with atomic numbers 122–124 against $\alpha$ decay (*blue open circles*) and HPR (*red open squares*) versus the neutron number of the parent nucleus. $Q$ values are calculated using the LiMaZe (*top*), KTUY05 (*middle*) and FRDM95 (*bottom*) mass tables

**Acknowledgments** This work was partially supported by Deutsche Forschungsgemeinschaft bibmycro and partially within IDEI Programme under contracts 43/05.10.2011 and 42/05.10.2011 with UEFISCDI, Bucharest.

## References


1. G. Audi, W. Meng: Private, communication, April (2011)
2. R. Blendowske, H. Walliser, Phys. Rev. Lett. **61**, 1930–1933 (1988)
3. R. Bonetti, A. Guglielmetti, Rom. Rep. Phys. **59**, 301–310 (2007)
4. Encyclopaedia Britannica Online, 2011. Web http://www.britannica.com/EBchecked/topic/465998/
5. W. Greiner, M. Ivaşcu, D.N. Poenaru, A. Săndulescu, Cluster radioactivities, in *Treatise on Heavy Ion Science*, vol. 8, ed. by D.A. Bromley (Plenum Press, New York, 1989), pp. 641–722
6. W. Greiner, D.N. Poenaru: Neutron Rich Long-Lived Superheavies, ed. by M. Brenner Cluster Structure of Atomic Nuclei, chap. 5, (Research Signpost, Trivandrum, India, 2010), pp. 119–146
7. D.C. Hoffman, T.M. Hamilton, M.R. Lane, Spontaneous Fission. In: Nuclear Decay Modes, chap. 10, (Institute of Physics Publishing, Bristol, 1996) pp. 393–432
8. S. Hofmann, Radiochim. Acta **99**, 405–428 (2011)
9. S. Hofmann, G. Münzenberg, Rev. Mod. Phys. **72**, 733–767 (2000)
10. P. Jachimowicz, M. Kowal, J. Skalski, Phys. Rev. C **83**, 054302 (2011)
11. A.V. Karpov, V.I. Zagrebaev, W Greiner, In these Proceedings, (2012)
12. H. Koura, T. Tachibana, M. Uno, M. Yamada, Prog. Theor. Phys. **113**, 305–325 (2005)
13. S. Liran, A. Marinov, N. Zeldes, Phys. Rev. **C62**, 047301 (2000)
14. S. Liran, A. Marinov, N. Zeldes, Phys. Rev. C 66, 024303 (2002) E-print arXiv nucl-th/0102055v1
15. R.G. Lovas, R.J. Liotta, A. Insolia, K. Varga, D.S. Delion, Phys. Rep. **294**, 265–362 (1998)
16. P. Möller, J.R. Nix, W.D. Myers, W.J. Swiatecki, Atomic Data Nucl. Data Tab. **59**, 185–381 (1995)
17. K. Morita et al., J. Phys. Soc. Jpn. **76**, 045001 (2007)
18. W.D. Myers, W.J. Swiatecki, Nucl. Phys. A **81**, 1–60 (1966)
19. Y.T. Oganessian, J. Phys. G: Nucl. Part. Phys. **34**, R165–R242 (2007)
20. Y.T. Oganessian, Radiochim. Acta **99**, 429–439 (2011)
21. Y.T. Oganessian et al., Phys. Rev. Lett. **104**, 142502 (2010)
22. D.N. Poenaru, W. Greiner (eds.), *Handbook of Nuclear Properties* (Clarendon Press, Oxford, 1996)
23. D.N. Poenaru, W. Greiner, Theories of Cluster Radioactivities, in Nuclear Decay Modes. chap. 6, (Institute of Physics Publishing, Bristol, 1996) pp. 275–336
24. D.N. Poenaru, R.A. Gherghescu, W. Greiner, Phys. Rev., C 73, 014608 (2006)
25. D.N. Poenaru, R.A. Gherghescu, W. Greiner, Phys. Rev. Lett. **107**, 062503 (2011)
26. D.N. Poenaru, W. Greiner, Phys. Scr. **44**, 427–429 (1991)
27. D.N. Poenaru, W. Greiner (eds.), *Experimental Techniques in Nuclear Physics* (Walter de Gruyter, Berlin, 1997)
28. D.N. Poenaru and W. Greiner, Cluster radioactivity, in C. Beck (ed.) Clusters in Nuclei, Vol. 1. Lecture Notes in Physics 818, chap. 1, (Springer, Berlin 2010), pp. 1–56
29. D.N. Poenaru, M. Ivaşcu, W. Greiner, in Particle Emission from Nuclei, Vol. III: Fission and Beta-Delayed Decay Modes, chap. 7, (CRC Press, Boca Raton, Florida 1989), pp. 203–235 (1989)
30. D.N. Poenaru, Y. Nagame, R.A. Gherghescu, W. Greiner, Phys. Rev. C **65**, 054308 (2002)
31. D.N. Poenaru, I.H. Plonski, W. Greiner, Phys. Rev., C 74, 014312 (2006)

32. D.N. Poenaru, D. Schnabel, W. Greiner, D. Mazilu, R. Gherghescu, Atomic Data Nucl. Data Tab. **48**, 231–327 (1991)
33. C. Qi, F.R. Xu, R.J. Liotta, R. Wyss, Phys. Rev. Lett. **103**, 072501 (2009)
34. H.J. Rose, G.A. Jones, Nature **307**, 245–247 (1984)
35. A. Sandulescu, D.N. Poenaru, W. Greiner, Sov. J. Part. Nucl. **11**, 528–541 (1980)
36. R. Smolanczuk, J. Skalski, A. Sobiczewski, Phys. Rev. C **52**, 1871–1880 (1995)
37. A. Sobiczewski, Radiochim. Acta **99**, 395–404 (2011)
38. V.M. Strutinsky, Nucl. Phys. A **95**, 420–442 (1967)
39. M. Warda, L.M. Robledo, Phys. Rev. C **84**, 044608 (2011)

# Part II
# Nuclear Structure and Reactions

# Coupled-Channel Effects in Collisions Between Heavy Ions Near the Coulomb Barrier

**C. Beck**

**Abstract** With the recent availability of state-of-the-art heavy-ion stable and radioactive beams, there has been a renew interest in the investigation of nuclear reactions with heavy ions. I first present the role of inelastic and transfer channel couplings in fusion reactions induced by stable heavy ions. Analysis of experimental fusion cross sections by using standard coupled-channel calculations is discussed. The role of multi-neutron transfer is investigated in the fusion process below the Coulomb barrier by analyzing $^{32}$S + $^{90,96}$Zr as benchmark reactions. The enhancement of fusion cross sections for $^{32}$S + $^{96}$Zr is well reproduced at sub-barrier energies by NTFus code calculations including the coupling of the neutron-transfer channels following the Zagrebaev semi-classical model. Similar effects for $^{40}$Ca + $^{90}$Zr and $^{40}$Ca + $^{96}$Zr fusion excitation functions are found. The breakup coupling in both the elastic scattering and in the fusion process induced by weakly bound stable projectiles is also shown to be crucial. In this lecture, full coupled-channel calculations of the fusion excitation functions are performed by using the breakup coupling for the more neutron-rich reaction and for the more weakly bound projectiles. I clearly demonstrate that Continuum-Discretized Coupled-Channel calculations are capable to reproduce the fusion enhancement from the breakup coupling in $^{6}$Li + $^{59}$Co.

## 1 Introduction

Heavy-ion fusion reactions at bombarding energies at the vicinity and below the Coulomb barrier have been widely studied [1–5]. In low-energy fusion reactions, the very simple one-dimensional barrier-penetration model (1D-BPM) [1, 2] is based upon a real potential barrier resulting from the attractive nuclear and repulsive

C. Beck (✉)
Institut pluridisciplinaire Hubert Curien, IN2P3-CNRS and Université de Strabourg—23, rue du Loess BP 28, 67037 Strasbourg Cedex 2, France
e-mail: christian.beck@iphc.cnrs.fr

W. Greiner (ed.), *Exciting Interdisciplinary Physics*,
FIAS Interdisciplinary Science Series, DOI: 10.1007/978-3-319-00047-3_9,

Coulomb interactions. For light- and medium-mass nuclei, one only assumes that the di-nuclear system (DNS) fuses as soon as it has reached the region inside the barrier i.e. within the potential pocket. If the system can evolve with a bombarding energy high enough to pass through the barrier and to reach this pocket with a reasonable amount of energy, the fusion process will occur after a complete amalgamation of the colliding nuclei forming the compound nucleus (CN). On the other hand, for sub-barrier energies the DNS has not enough energy to pass through the barrier.

In reactions induced by stable beams, the specific role of multi-step nucleon-transfers in sub-barrier fusion enhancement still needs to be investigated in detail both experimentally and theoretically [6–14]. In a complete description of the fusion dynamics the transfer channels in standard coupled-channel (CC) calculations [2, 8, 10, 14, 15] have to be taken into account accurately. It is known, for instance, that neutron transfers may induce a neck region of nuclear matter in-between the interacting nuclei favoring the fusion process to occur. In this case, neutron pick-up processes can occur when the nuclei are close enough to interact each other significantly [7, 8], if the Q-values of neutron transfers are positive. It was shown that sequential neutron transfers can lead to the broad distributions characteristic of many experimental fusion cross sections. Finite Q-value effects can lead to neutron flow and a build up of a neck between the target and projectile [8]. The situation of this neck formation of neutron matter between the two colliding nuclei could be considered as a "doorway state" to fusion. In a basic view, this intermediate state induced a barrier lowering. As a consequence, it will favor the fusion process at sub-barrier energies and enhance significantly the fusion cross sections. Experimental results have already shown such enhancement of the sub-barrier fusion cross sections due to the neutron-transfer channels with positive Q-values [6, 9].

In reactions induced by weakly bound nuclei and/or by halo nuclei, the influence on the fusion process of coupling both to collective degrees of freedom and to transfer/breakup channels is a key point [3–5] for the understanding of N-body systems in quantum dynamics [1]. Due to their very weak binding energies, a diffuse cloud of neutrons for $^{6}$He or an extended spatial distribution for the loosely bound proton in $^{8}$B would lead to larger total reaction (and fusion) cross sections at sub-barrier energies as compared to 1D-BPM model predictions. This enhancement is well understood in terms of the dynamical processes arising from strong couplings to collective inelastic excitations of the target (such as "normal" quadrupole and octupole modes) and projectile (such as soft dipole resonances). However, in the case of reactions where at least one of the colliding nuclei has a sufficiently low binding energy for breakup to become a competitive process, conflicting conclusions were reported [3–5, 16, 17].

Recent studies with Radioactive Ion Beams (RIB) indicate that the halo nature of $^{6,8}$He [18–22], for instance, does not enhance the fusion probability as anticipated. Rather the prominent role of one- and two-neutron transfers in $^{6,8}$He induced fusion reactions was definitively demonstrated. On the other hand, the effect of non-conventional transfer/stripping processes appears to be less significant for stable weakly bound projectiles. Several experiments involving $^{9}$Be, $^{7}$Li, and $^{6}$Li projectiles on medium-mass targets have been undertaken.

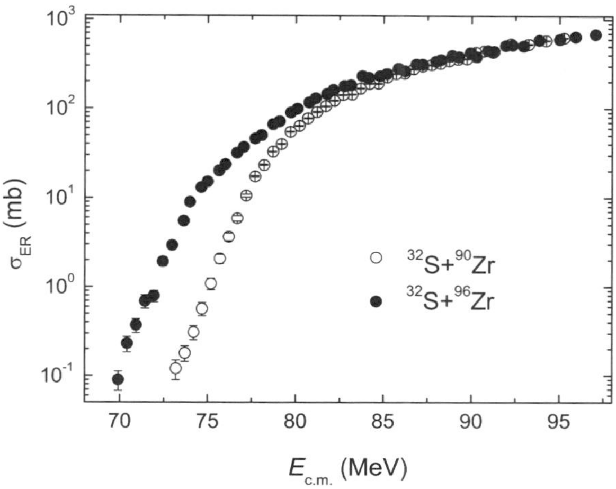


**Fig. 1** Comparison between the fusion-evaporation (ER) excitation functions of $^{32}S + ^{90}Zr$ (*open circles*) and $^{32}S + ^{96}Zr$ (*points*) as a function of the center-of-mass energy. The error bars of the experimental data taken from Ref. [23] represent purely statistics uncertainties (Courtesy of H.Q. Zhang)

**Fig. 2** Ratios of measured fusion cross sections for $^{6}Li$ and $^{7}Li$ projectiles with $^{24}Mg$, $^{28}Si$ and $^{59}Co$ targets as a function of $E_{c.m.}/V_b$. The *solid line* gives the 1D-BPM prediction while the *dotted line* shows results obtained from Wong's prescription (This figure originally shown in Ref. [24] for $^{6,7}Li + ^{59}Co$ has been adapted to display comparisons with other lighter targets [25–28])

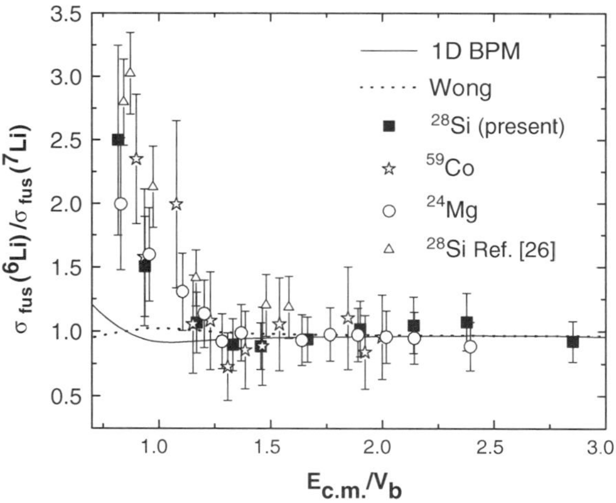


## 2 Experimental Results

In this lecture we first present the role of inelastic and transfer channel couplings in experimental data obtained in fusion reactions induced by stable $^{32}S$ projectiles [23]. The breakup coupling in both elastic scattering data and in the fusion data are also shown for weakly bound $^{6,7}Li$ projectiles [24].

### 2.1 $^{32}$S + $^{90}$Zr and $^{32}$S + $^{96}$Zr Reactions

In order to investigate the role of neutron transfers we further study $^{32}S + {}^{90}Zr$ and $^{32}S + {}^{96}Zr$ as benchmark reactions. Figure 1 displays the measured fusion cross sections for $^{32}S+{}^{90}Zr$ (open circles) and $^{32}S+{}^{96}Zr$ (points). We present the analysis of excitation functions of evaporation residues (ER) cross sections recently measured with high precision (i.e. with small energy steps and good statistical accuracy for these reactions [23]).

The differential cross sections of quasi-elastic scattering (QEL) at backward angles were previously measured by the CIAE group [13]. The analysis of the corresponding BD-QEL barrier distributions (see solid points in Fig. 3) already indicated the significant role played by neutron transfers in the fusion processes.

In Fig. 3 we introduce the experimental fusion-barrier (BD-Fusion) distributions (see open points) obtained for the two reactions by using the three-point difference method of Ref. [8] as applied to the data points of Ref. [23] plotted in Fig. 1. It is interesting to note that in both cases the BD-Fusion and BD-QEL barrier distributions are almost identical up to $E_{c.m.} \approx 85$ MeV.

### 2.2 $^{6}$Li + $^{59}$Co and $^{7}$Li + $^{59}$Co Reactions

The fusion excitation functions were measured for the $^{6,7}Li+{}^{59}Co$ reactions [24] at the VIVITRON facility of the IPHC Strasbourg and the Pelletron facility of Saõ Paulo by using $\gamma$-ray techniques. Their ratios are presented in Fig. 2 with comparisons with other lighter targets [25–28]. The theoretical curves (1D-BPM) [1, 2] and Wong [15] do not take into account the breakup channel coupling that is discussed in one of the following sections in more details.

## 3 Coupled Channel Analysis

Analysis of experimental fusion cross sections by using standard CC calculations is first discussed with the emphasis of the role of multi-neutron transfer in the fusion process below the Coulomb barrier for $^{32}S + {}^{90,96}Zr$ as benchmark reactions.

### 3.1 $^{32}$S + $^{90}$Zr and $^{32}$S + $^{96}$Zr Reactions

A new CC computer code named NTFus [29] taking the neutron transfer channels into account in the framework of the semiclassical model of Zagrebaev [10] has been developed. The effect of the neutron transfer channels yields a fairly good agreement

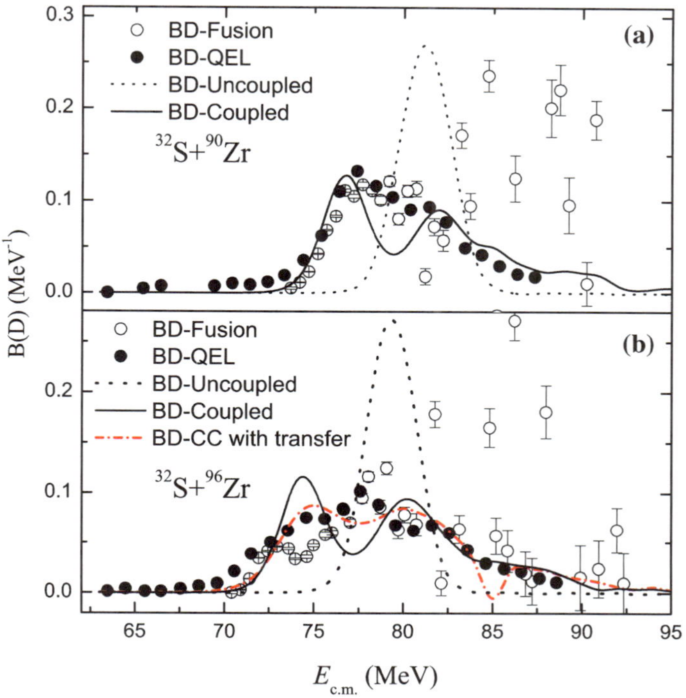


**Fig. 3** Barrier distributions (BD) from the fusion ER (*open circles*) cross sections [23], plotted in Fig. 1, and quasielastic scattering (*solid circles*) cross sections [13] for $^{32}S + ^{90}Zr$ (**a**) and $^{32}S + ^{96}Zr$ (**b**). The *dashed* and *solid black lines* represent uncoupled calculations (1D-BPM) and the CC calculations without neutron transfer coupling. The *red dash-dotted line* represents the CC calculations with neutron transfer coupling for the $^{32}S + ^{96}Zr$ reaction (Courtesy of H.Q. Zhang)

with the data of sub-barrier fusion cross sections measured for $^{32}S + ^{96}Zr$, the more neutron-rich reaction [23]. This was initially expected from the positive Q-values of the neutron transfers as well as from the failure of standard CC calculation of quasi-elastic barrier distributions without neutron-transfers coupling [13] as shown by the solid line in Fig. 3b.

By fitting the experimental fusion excitation function displayed in Fig. 1 with NTFus CC calculation [29], we concluded [30] that the effect of the neutron transfer channels produces significant enhancement of the sub-barrier fusion cross sections of $^{32}S + ^{96}Zr$ as compared to $^{32}S + ^{90}Zr$. A detailed inspection of the $^{32}S + ^{90}Zr$ fusion data presented in Fig. 1 along with the negative Q-values of their corresponding neutron transfer channels lead us to speculate with the absence of a neutron transfer effect on the sub-barrier fusion for this reaction. With the semiclassical model developed by Zagrebaev [10] we propose to definitively demonstrate the significant role of neutron transfers for the $^{32}S + ^{96}Zr$ fusion reaction by fitting its experimental excitation function with NTFUS code [29] calculations, as shown in Fig. 3.

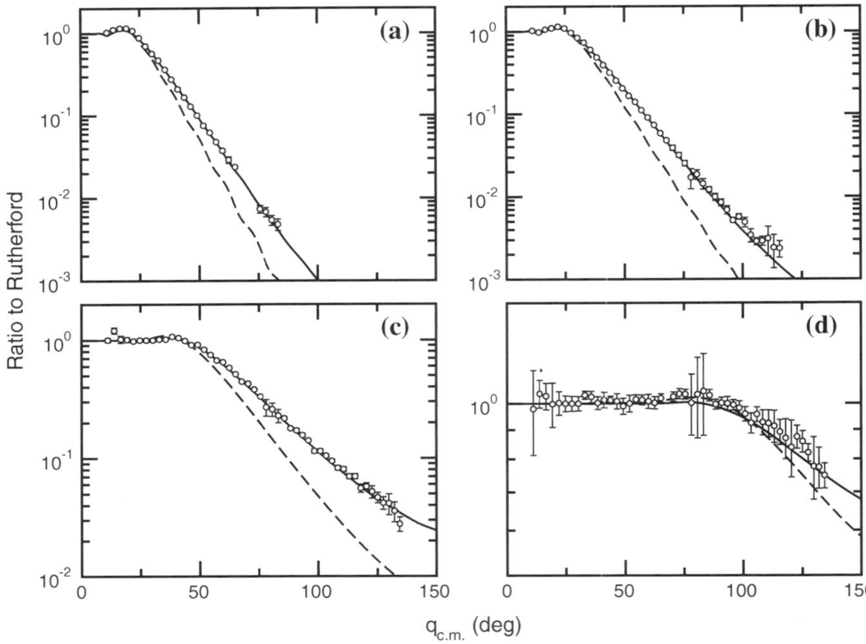


**Fig. 4** Ratios of the elastic scattering cross-sections to the Rutherford cross sections as a function of c.m. angle for the $^6$Li + $^{59}$Co system [47]. The *curves* correspond to CDCC calculations with (*solid lines*) and without (*dashed lines*) $^6\text{Li} \rightarrow \alpha + d$ breakup couplings to the continuum for incident $^6$Li energies of **a** 30 MeV, **b** 26 MeV, **c** 18 MeV and **d** 12 MeV (This figure has been adapted from the work of Ref. [47])

The new oriented object NTFUS code [29], using the Zagrebaev model [10] was implemented (at the CIAE) in C++, using the compiler of ROOT [31], following the basic equations of Ref. [32]. Let us first remind the values chosen for the deformation parameters and the excitation energies that are given in Refs. [2, 33, 34] (see Tables given in [30] for more details). The quadrupole vibrations of both the $^{90}$Zr and $^{96}$Zr are weak in energy; they lie at comparable energies. The $^{96}$Zr nucleus presents a complicated situation [35]: its low-energy spectrum is dominated by a $2^+$ state at 1.748 MeV and by a very collective [B(E3; $3^- \rightarrow 0^+$) = 51 W.u.] $3^-$ state at 1.897 MeV. CC calculations explained the larger sub-barrier enhancement as due mainly to the strong octupole vibration of the $3^-$ state in $^{36}$S + $^{96}$Zr [36]. However, the agreement is not so satisfactory below the barrier for $^{32}$S + $^{96}$Zr (see solid line of Fig. 3b), as well as for $^{40}$Ca + $^{96}$Zr [9] and, therefore, there is the need to take neutron transfers into account.

The main functions of the code NTFUS are designed to calculate the fusion excitation functions with normalized barrier distribution (based on experimental data) given by CCFULL [15], we take the dynamical deformations into account. In order to introduce the role of neutron transfers, the NTFUS code [29] applies the Zagrebaev model [10] to calculate the fusion cross sections $\sigma_{fus}(E)$ as a function of center-of-mass energy E. Then the fusion excitation function can be derived using the following formula [10]:

$$T_l(E) = \int f(B)\frac{1}{N_{tr}}\sum_k \int_{-E}^{Q_0(k)} \alpha_k(E, l, Q) \times P_{HW}(B, E + Q, l)dQdB, \quad (1)$$

and

$$\sigma_{fus}(E) = \frac{\pi\hbar^2}{2\mu E}\sum_{l=0}^{l_{cr}}(2l + 1)T_l(E), \quad (2)$$

where $T_l(E)$ are the transmission coefficients, $E$ is the energy given in the center-of-mass system, B and $f(B)$ are the barrier height and the normalized barrier distribution function, $P_{HW}$ is the usual Hill-Wheeler formula. $l$ is the angular momentum whereas $l_{cr}$ is the critical angular momentum as calculated by assuming no coupling (well above the barrier). $\alpha_k(E, l, Q)$ and $Q_0(k)$ are, respectively, the probabilities and the Q-values for the transfers of $k$ neutrons. And $1/N_{tr}$ is the normalization of the total probability taking into account the neutron transfers.

The NTFUS code [29] uses the ion-ion potential between two deformed nuclei as developed by Zagrebaev and Samarin in Ref. [32]. Either the standard Woods-Saxon form of the nuclear potential or a proximity potential [37] can be chosen. The code is also able to predict fusion cross sections for reactions induced by halo projectiles [30]; for instance $^{6}He+^{64}Zn$ [22, 38]. In the following, only comparisons for $^{32}S + ^{90}Zr$ and $^{32}S + ^{96}Zr$ are discussed.

For the high-energy part of the $^{32}S + ^{90}Zr$ excitation function, one can notice a small over-estimation of the fusion cross sections at energies above the barrier up to the point used to calculate the critical angular momentum. This behavior can be observed at rather high incident energies—i.e. between about 82 and 90 MeV (shown as the dashed line in Fig. 3a for $^{32}S + ^{90}Zr$ reaction). We want to stress that the corrections do not affect our conclusions that the transfer channels have a predominant role below the barrier for $^{32}S + ^{96}Zr$ reaction, as shown by the dotted-dashed red curve in Fig. 3b.

As expected, we obtain a good agreement with calculations not taking any neutron transfer coupling into account for $^{32}S + ^{90}Zr$ as shown by the solid line of Fig. 3a (the dashed line are the results of calculations performed without any coupling). On the other hand, there is no significant over-estimation at sub-barrier energies. As a consequence, it is possible to observe the strong effect of neutron transfers on the fusion for the $^{32}S + ^{96}Zr$ reaction at sub-barrier energies. Moreover, the barrier distribution function $f(B)$ extracted from the data contains the information of the neutron transfers. These information are also contained in the transmission coefficients, which are the most important parameters for the fusion cross sections to be calculated accurately. The $f(B)$ function as calculated with the three-point formula [8] will mimic the differences induced by the neutron transfer taking place in sub-barrier energies where the cross section variations are very small (only visible if a logarithm scale is employed for the fusion excitation function). It is interesting to note that the Zagrebaev model [10] implies a modification of the Hill-Wheeler probability and does not concern the barrier distribution function $f(B)$. Finally, the code allows us to perform each calculation by taking the neutron transfers into account or not.

The calculation with the neutron transfer effect is performed up to the channel +4n(k = 4), but we have seen that we obtain the same overall agreement with data up to channels +5n and +6n [30]. As we can see on Fig. 3b, the solid line representing standard CC calculations without the neutron transfer coupling (the dotted line is given for uncoupled calculations) does not fit the experimental data well at sub-barrier energies. On the other hand, the dotted line displaying NTFus calculations taking the neutron transfer coupling into account agrees perfectly well with the data. As expected, the Zagrebaev semiclassical model's correction applied at sub-barrier energies enhances the calculated cross sections. Moreover, it allows to fit the data reasonably well and therefore illustrates the strong effect of neutron transfers for the fusion of $^{32}$S + $^{96}$Zr at subbarrier energies.

The present full CC analysis of $^{32}$S + $^{96}$Zr fusion data [23, 30] using NTFus [29] confirms perfectly well first previous CC calculations [10] describing well the earlier $^{40}$Ca + $^{90,96}$Zr fusion data [9] and, secondly, very recent fragment-$\gamma$ coincidences measured for $^{40}$Ca + $^{96}$Zr multi-neutron transfer channels [35].

## *3.2* $^{6}$Li + $^{59}$Co *and* $^{7}$Li + $^{59}$Co *Reactions*

For reactions induced by weakly bound nuclei [16, 17, 25–28, 39–41] and exotic nuclei [18–22, 38, 42–46], the breakup channel is open and plays a key role in the fusion process near the Coulomb barrier similarly to the transfer-channel coupling described in the previous section. It is therefore appropriate to use the Continuum-Discretized Coupled-Channel (CDCC) approach [47–50] to describe the influence of the breakup channel in both the elastic scattering and the fusion process at sub-barrier energies.

Theoretical calculations (including CDCC predictions given in Refs. [47, 49] indicate only a small enhancement of total fusion for the more weakly bound $^{6}$Li below the Coulomb barrier (see curves of Fig. 2), with similar cross sections for both $^{6,7}$Li + $^{59}$Co reactions at and above the barrier [24]. It is interesting to notice, however, that the same conclusions have been reached for other targets such as $^{24}$Mg [26] and $^{28}$Si [25, 27, 28] as can be clearly seen in the plot of Fig. 2. These results are consistent with rather low breakup cross sections measured for the $^{6,7}$Li + $^{59}$Co reactions even at incident energies larger than the Coulomb barrier [39–41]. But the coupling of the breakup channel is extremely important for the CDCC analysis of the angular distributions of the elastic scattering [47] as shown in Fig. 4 for $^{6}$Li + $^{59}$Co. The curves show the results of calculations with (solid lines) and without (dashed lines) $^{6,7}$Li $\rightarrow \alpha + d, t$ breakup couplings. The main conclusion is that effect of breakup on the elastic scattering is stronger for $^{6}$Li than $^{7}$Li.

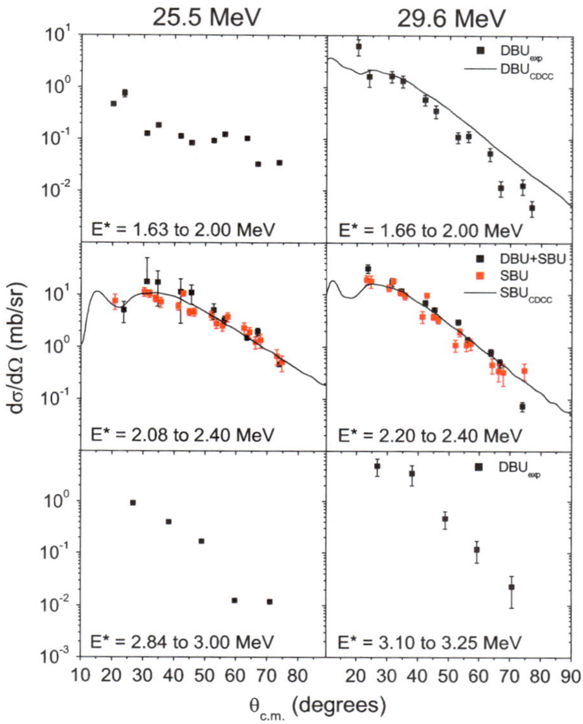


**Fig. 5** Experimental [39–41] and theoretical CDCC [30] angular distributions for the SBU and DBU projectile breakup processes (see text for details) obtained at $E_{lab} = 25.5$ MeV and 29.6 MeV for $^{6}Li + {}^{59}Co$. The chosen experimental continuum excitation energy ranges are given (Courtesy of F.A. Souza)

A more detailed investigation of the breakup process in the $^{6}Li + {}^{59}Co$ reaction with particle coincidence techniques is now proposed to discuss the interplay of fusion and breakup processes. Coincidence data compared to three-body kinematics calculations reveal a way how to disentangle the contributions of breakup, incomplete fusion and/or transfer-reemission processes [39–41].

Figure 5 displays experimental (full rectangles) and theoretical angular distributions (solid lines) for the sequential (SBU) and direct (DBU) projectile breakup processes at the two indicated bombarding energies for the $^{6}Li + {}^{59}Co$ reaction. In the CDCC calculations the $\alpha + d$ binning scheme is appropriately altered to accord exactly with the measured continuum excitation energy ranges. For this reaction it was not necessary to use a sophisticated four-body CDCC framework. The CDCC cross sections [47] are in agreement with the experimental ones [16, 40, 41], both in shapes and magnitudes within the uncertainties. The relative contributions of the $^{6}Li$ SBU and DBU to the incomplete fusion/transfer process has been discussed in great details in Refs. [39–41] by considering the corresponding lifetimes obtained by using a semi-classical approach fully described in a previous publication [39].

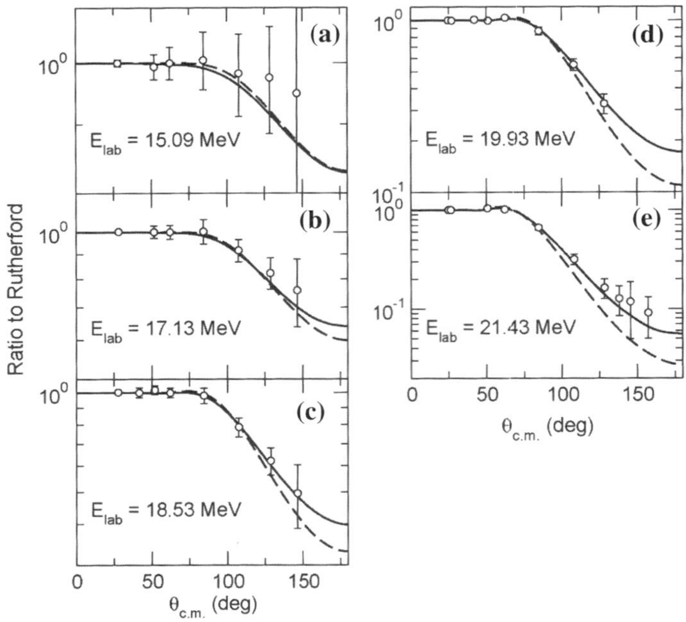


**Fig. 6** Ratios of the elastic scattering cross-sections to the Rutherford cross sections as a function of c.m. angle for the $^7$Be + $^{58}$Ni system [52] for incident $^7$Be energies of **a** 15.09 MeV, **b** 17.13 MeV, **c** 18.53 MeV, **d** 19.93 MeV and **e** 21.43 MeV. The *solid* and *dashed curves* denote full and no coupling to the continuum (This figure has been adapted from the work of Ref. [17])

We conclude that the flux diverted from complete fusion to incomplete fusion would arise essentially from DBU processes via high-lying continuum (non-resonant) states of $^6$Li; this is due to the fact that both the SBU mechanism and the low-lying DBU processes from low-lying resonant $^6$Li states occur at large internuclear distances [39–41]. Work is in progress to study incomplete fusion for $^6$Li + $^{59}$Co within a newly developed 3-dimensional classical trajectory model [51].

## 3.3 Coupled-Channel Calculations for Reactions Induced by Halo Nuclei

As far as exotic halo projectiles are concerned we have initiated a systematic study of $^8$B and $^7$Be induced reactions data [52] with an improved CDCC method [48]. Figure 6 displays the analysis of the elastic scattering for the $^7$Be + $^{58}$Ni system [52]. The curves correspond to CDCC calculations with (solid lines) and without (dashed lines) $^7$Be $\rightarrow \alpha$ + $^3$He breakup couplings to the continuum. The $^6$Li and $^7$Be calculations were similar, but with a finer continuum binning for $^7$Be. As compared to $^7$Be + $^{58}$Ni (similar to $^{6,7}$Li + $^{58,64}$Ni) the CDCC analysis of $^8$B + $^{58}$Ni reaction

[48] while exhibiting a large breakup cross section (consistent with the systematics) is rather surprizing as regards the consequent weak coupling effect found to be particularly small on the near-barrier elastic scattering.

Recently, the scattering process of $^{17}$F from $^{58}$Ni target was investigated [43] slightly above the Coulomb barrier and total reaction cross sections were extracted from the Optical-Model analysis. The small enhancement as compared to the reference (tightly bound) system $^{16}$O + $^{58}$Ni is here related to the low binding energy of the $^{17}$F valence proton. This moderate effect is mainly triggered from a transfer effect, as observed for the $2n$-halo $^{6}$He [18, 19] and the $1n$-halo $^{11}$Be [42] in contrast to the $1p$-halo $^{8}$B + $^{58}$Ni reaction where strong enhancements are triggered from a breakup process [45].

## 4 Summary, Conclusions and Outlook

We have investigated the fusion process (excitation functions and extracted barrier distributions [23]) at near- and sub-barrier energies for the two neighbouring reactions $^{32}$S + $^{90}$Zr and $^{32}$S + $^{96}$Zr. For this purpose a new computer code named NTFus [29] has been developed by taking the coupling of the multi-neutron transfer channels into account by using the semiclassical model of Zagrebaev [10].

The effect of neutron couplings provides a fair agreement with the present data of sub-barrier fusion for $^{32}$S + $^{96}$Zr. This was initially expected from the positive Q-values of the neutron transfers as well as from the failure of previous CC calculation of quasi-elastic barrier distributions without coupling of the neutron transfers [13]. With the agreement obtained by fitting the present experimental fusion excitation function and the CC calculation at sub-barrier energies, we conclude that the effect of the neutron transfers produces a rather significant enhancement of the sub-barrier fusion cross sections of $^{32}$S + $^{96}$Zr as compared to $^{32}$S + $^{90}$Zr. At this point we did not try to reproduce the details of the fine structures observed in the fusion barrier distributions. We believe that to achieve this final goal it will first be necessary to measure the neutron transfer cross sections to provide more information on the coupling strength of neutron transfer because its connection with fusion is not yet fully understood [35].

In the second part of this lecture, we have studied the breakup coupling on elastic scattering and fusion by using the CDCC approach with a particular emphasis on a very detailed analysis of the $^{6}$Li + $^{59}$Co reaction. The CDCC formalism, with continuum–continuum couplings taken into account, is probably one of the most reliable methods available nowadays to study reactions induced by exotic halo nuclei, although many of them have added complications like core excitation and three-body structure. The respective effects of transfer/breakup are finally outlined for reactions induced by $1p$-halo, $1n$-halo and $2n$-halo nuclei.

The complexity of such reactions, where many processes compete on an equal footing, necessitates kinematically and spectroscopically complete measurements [53], i.e. ones in which all processes from elastic scattering to fusion are measured simultaneously, providing a technical challenge in the design of broad range detection systems. A full understanding of the reaction dynamics involving couplings to the breakup and nucleon-transfer channels will need high-intensity RIB and precise measurements of elastic scattering, fusion and yields leading to the breakup itself. A new experimental program with SPIRAL beams and medium-mass targets is getting underway at GANIL.

**Acknowledgments** I would like to thank A. Diaz-Torres, N. Keeley, F.A. Souza and A. Richard for very fruitful discussions on many theoretical aspects of this lecture.

## References

1. A.B. Balantekin, N. Takigawa, Rev. Mod. Phys. **70**, 77 (1998); arXiv:nucl-th/9708036 (1997)
2. M. Dasgupta, D.J. Hinde, N. Rowley, A.M. Stefanini, Annu. Rev. Nucl. Part. Sci. **48**, 401 (1998)
3. J.F. Liang, C. Signorini, Int. J. Mod. Phys. E **14**, 1121 (2005); arXiv:nucl-ex/0504030 (2005)
4. L.F. Canto, P.R.S. Gomes, R. Donangelo, M.S. Hussein, Phys. Rep. **424**, 1 (2006)
5. N. Keeley, R. Raabe, N. Alamanos, J.-L. Sida, Prog. Part. Nucl. Phys. **59** 579 (2007); arXiv:nucl-th/0702038 (2007)
6. R. Pengo, D. Evers, K.E.G. Lobner, U. Quade, K. Rudolph, S.J. Skorka, I. Weidl, Nucl. Phys. A **411**, 256 (1983)
7. P.H. Stelson, Phys. Lett. B **205**, 190 (1988)
8. N. Rowley, G.J. Thompson, M.A. Nagarajan, Phys. Lett. B **282**, 276 (1992)
9. H. Timmers, D. Ackermann, S. Beghini, L. Corradi, J.H. He, G. Montagnoli, F. Scarlassara, A.M. Stefanini, N. Rowley, Nucl. Phys. A **633**, 421 (1998)
10. V.I. Zagrebaev, Phys. Rev. C **67**, 061601 (2003)
11. A.M. Stefanini, F. Scarlassara, S. Beghini, G. Montagnoli, R. Silvestri, M. Trotta, B.R. Behera, L. Corradi, E. Fioretto, A. Gadea, Y.W. Wu, S. Szilner, H.Q. Zhang, Z.H. Liu, M. Ruan, F. Yang, N. Rowley, Phys. Rev. C **73**, 034606 (2006)
12. A.M. Stefanini, B.R. Behera, S. Beghini, L. Corradi, E. Fioretto, A. Gadea, G. Montagnoli, N. Rowley, F. Scarlassara, S. Szilner, M. Trotta, Phys. Rev. C **76**, 014610 (2007)
13. F. Yang, C.J. Lin, X.K. Wu, H.Q. Zhang, C.L. Zhang, P. Zhou, Z.H. Liu, Phys. Rev. C **77**, 014601 (2008)
14. S. Kalkal, S. Mandal, N. Madhavan, E. Prasad, S. Verma, A. Jhingan, R. Sandal, S. Nath, J. Gehlot, B.R. Behera, M. Saxena, S. Goyal, D. Siwal, R. Garg, U.D. Pramanik, S. Kumar, T. Varughese, K.S. Golda, S. Muralithar, A.K. Sinha, R. Singh, Phys. Rev. C **81**, 044610 (2010)
15. K. Hagino, N. Rowley, A.T. Kruppa, Comput. Phys. Commun. **123**, 143 (1999); arXiv:nucl-th/9903074 (1999)
16. C. Beck, Nucl. Phys. A **787**, 251 (2007); arXiv:nucl-ex/0701073 (2007); arXiv:nucl-th/0610004 (2006)
17. C. Beck, N. Rowley, P. Papka, S. Courtin, M. Rousseau, F.A. Souza, N. Carlin, R. Liguori Neto, M.M. de Moura, M.G. Del Santo, A.A.P. Suaide, M.G. Munhoz, E.M. Szanto, A. Szanto de Toledo, N. Keeley, A. Diaz-Torres, K. Hagino, Nucl. Phys. A **834**, 440 (2010); arXiv:0910.1672 (2010)
18. A. Di Pietro, P. Figuera, F. Amorini, C. Angulo, G. Cardella, S. Cherubini, T. Davinson, D. Leanza, J. Lu, H. Mahmud, M. Milin, A. Musumarra, A. Ninane, M. Papa, M.G. Pellegriti, R. Raabe, F. Rizzo, C. Ruiz, A.C. Shotter, N. Soic, S. Tudisco, Phys. Rev. C **69**, 044601 (2004)
19. A. Navin, V. Tripathi, Y. Blumenfeld, V. Nanal, C. Simenel, J.M. Casandjian, G. de France, R. Raabe, D. Bazin, A. Chatterjee, M. Dasgupta, S. Kailas, R.C. Lemmon, K. Mahata, R.G. Pilay, E.C. Pollacco, K. Ramachandran, M. Rejmund, A Shrivastava, J.L. Sida, E. Tryggestad, Phys. Rev. C **70**, 044601 (2004)

20. A. Chatterjee, A. Navin, A. Shrivastava, S. Bhattacharyya, M. Rejmund, N. Keeley, V. Nanal, J. Nyberg, R.G. Pillay, K. Ramachandran, I. Stefan, D. Bazin, D. Beaumel, Y. Blumenfeld, G. de France, D. Gupta, M. Labiche, A. Lemasson, R. Lemmon, R. Raabe, J.A. Scarpaci, C. Simenel, C. Timis, Phys. Rev. Lett. **101**, 032701 (2008)
21. A. Lemasson, A. Shrivastava, A. Navin, M. Rejmund, N. Keeley, V. Zelevinsky, S. Bhattacharyya, A. Chatterjee, G. de France, B. Jacquot, V. Nanal, R.G. Pillay, R. Raabe, C. Schmitt, Phys. Rev. Lett. **103**, 232701 (2009)
22. V. Scuderi, A. Di Pietro, P. Figuera, M. Fisichella, F. Amorini, C. Angulo, G. Cardella, E. Casarejos, M. Lattuada, M. Milin, A. Musumarra, M. Papa, M.G. Pellegriti, R. Raabe, F. Rizzo, N. Skukan, D. Torresi, M. Zadro, Phys. Rev. C **84**, 064604 (2011)
23. H.Q. Zhang, C.J. Lin, F. Yang, H.M. Jia, X.X. Xu, F. Jia, Z.D. Wu, S.T. Zhang, Z.H. Liu, A. Richard, C. Beck, Phys. Rev. C **82**, 054609 (2010); arXiv: 1005.0727 (2010)
24. C. Beck, F.A. Souza, N. Rowley, S.J. Sanders, N. Aissaoui, E.E. Alonso, P. Bednarczyk, N. Carlin, S. Courtin, A. Diaz-Torres, A. Dummer, F. Haas, A. Hachem, K. Hagino, F. Hoellinger, R.V.F. Janssens, N. Kintz, R. Liguori Neto, E. Martin, M.M. Moura, M.G. Munhoz, P. Papka, M. Rousseau, A. Sanchez i Zafra, O. Stezowski, A.A. Suaide, E.M. Szanto, A. Szanto de Toledo, S. Szilner, J. Takahashi, Phys. Rev. C **67**, 054602 (2003); arXiv:nucl-ex/0411002 (2004)
25. M. Sinha, H. Majumdar, P. Basu, S. Roy, R. Bhattacharya, M. Biswas, M.K. Pradhan, S. Kailas, Phys. Rev. C **78**, 027601 (2008); arXiv:nucl-ex:0805.0953 (2008)
26. M. Ray, A. Mukherjee, M.K. Pradhan, R. Kshetri, M. Saha Sarkar, R. Palit, I. Majumdar, P.K. Joshi, H.C. Jain, B. Dasmahapatra, Phys. Rev. C **78**, 064617 (2008); arXiv:nucl-ex:0805.0953 (2008)
27. A. Pakou, K. Rusek, N. Alamanos, X. Aslanoglou, M. Kokkoris, A. Lagoyannis, T.J. Mertzimekis, A. Musumarra, N.G. Nicolis, D. Pierroutsakou, D. Roubos, Eur. Phys. J. A **39**, 187 (2009)
28. M. Sinha, H. Majumdar, P. Basu, Su Roy, R. Bhattacharya, M. Biswas, M.K. Pradhan, R. Palit, I. Mazumdar, S. Kailas, Eur. Phys. J. A **44**, 403 (2010)
29. H.Q. Zhang, H.Q. Zhang, C.L. Zhang, H.M. Jia, C.J. Lin, F. Yang, Z.H. Liu, Z.D. Wu, F. Jia, X.X. Xu, A. Richard, A.K. Nasirov, G. Mandaglio, M. Manganaro, G. Giardina, K. Hagino, AIP Conf. Proc. **1235**, 50 (2010)
30. A. Richard, C. Beck, H.Q. Zhang, EPJ Web Conf **17**, 08005 (2011); arXiv:1104.5333 (2011)
31. http://root.cern.ch, website of ROOT
32. V.I. Zagrebaev, V.V. Samarin, Phys. Atom. Nucl. **67**(8), 1462 (2004)
33. S. Raman, C.W. Nestor, P. Tikkanen, At. Data Nucl. Data Tables **78**, 1 (2001)
34. T. Kebedi, R.H. Spear, At. Data Nucl. Data Tables **89**, 77 (2005)
35. L. Corradi, S. Szilner, G. Pollarolo, G. Colo, P. Mason, E. Farnea, E. Fioretto, A. Gadea, F. Haas, D. Jelavic-Malenica, N. Marginean, C. Michelagnoli, G. Montagnoli, D. Montanari, F. Scarlassara, N. Soic, A.M. Stefanini, C.A. Ur, J.J. Valiente-Dobon, Phys. Rev. C **84**, 034603 (2011)
36. A.M. Stefanini, L. Corradi, A.M. Vinodkumar, F. Yang, F. Scarlassara, G. Montagnoli, S. Beghini, M. Bisogno, Phys. Rev. C **62**, 014601 (2000)
37. J. Blocki, J. Randrup, W.J. Swiatecki, C.F. Tsang, Ann. Phys. **105**, 427 (1977)
38. M. Fisichella, V. Scuderi, A. Di Pietro, P. Figuera, M. Lattuada, C. Marchetta, M. Milin, A. Musumarra, M.G. Pellegriti, N. Skukan, E. Strano, D. Torresi, M. Zadro, J. Phys.: Conf. Ser. **282**, 012014 (2011)
39. F.A. Souza, C. Beck, N. Carlin, N. Keeley, R. Liguori Neto, M.M. de Moura, M.G. Munhoz, M.G. Del Santo, A.A.P. Suaide, E.M. Szanto, A. Szanto de Toledo, Nucl. Phys. A **821**, 36 (2009); arXiv:0811.4556 (2008)
40. F.A. Souza, N. Carlin, C. Beck, N. Keeley, A. Diaz-Torres, R. Liguori Neto, C. Siqueira-Mello, M.M. de Moura, M.G. Munhoz, R.A.N. Oliveira, M.G. Del Santo, A.A.P. Suaide, E.M. Szanto, A. Szanto de Toledo, Nucl. Phys. A **834**, 420 (2010); arXiv:0910.0342 (2010)
41. F.A. Souza, N. Carlin, C. Beck, N. Keeley, A. Diaz-Torres, R. Liguori Neto, C. Siqueira-Mello, M.M. de Moura, M.G. Munhoz, R.A.N. Oliveira, M.G. Del Santo, A.A.P. Suaide, E.M. Szanto, A. Szanto de Toledo, Eur. Phys. J. A **44**, 181 (2010); arXiv:0909.5556 (2009)

42. A. Di Pietro, G. Randisi, V. Scuderi, L. Acosta, F. Amorini, M.J.G. Borge, P. Figuera, M. Fisichella, L.M. Fraile, J. Gomez-Camacho, H. Jeppesen, M. Lattuada, I. Martel, M. Milin, A. Musumarra, M. Papa, M.G. Pellegriti, F. Perez-Bernal, R. Raabe, F. Rizzo, D. Santonocito, G. Scalia, O. Tengblad, D. Torresi, A.M. Vidal, D. Voulot, F. Wenander, M. Zadro, Phys. Rev. Lett. **105**, 022701 (2010)
43. M. Mazzocco, C. Signorini, D. Pierroutsakou, T. Glodariu, C. Boiano, F. Farinon, A. Di Pietro, P. Figuera, D. Filipescu, L. Fortunato, A. Guglielmetti, G. Inglima, M. La Commara, M. Lattuada, P. Lotti, C. Mazzocchi, P. Molini, A. Musumarra, A. Pakou, C. Parascandolo, N. Patronis, M. Romoli, M. Sandoli, V. Scuderi, F. Soramel, L. Stroe, D. Torresi, E. Vardaci, A. Vitturi, Phys. Rev. C **82**, 054604 (2010)
44. Z. Kohley, F. Liang, D. Shapira, R.L. Varner, C.J. Gross, J.M. Allmond, A.L. Caraley, E.A. Coello, F. Favela, K. Lagergren, P.E. Mueller, Phys. Rev. Lett. **107**, 202701 (2011)
45. E.F. Aguilera, J.J. Kolata, Phys. Rev. C **85**, 014603 (2012)
46. M.J. Rudolph, Z.Q. Gosser, K. Brown, S. Hudan, R.T. de Souza, A. Chbihi, B. Jacquot, M. Famiano, J.F. Liang, D. Shapira, D. Mercier, Phys. Rev. C **85**, 024605 (2012)
47. C. Beck, N. Keeley, A. Diaz-Torres, Phys. Rev. C **75**, 054605 (2007); arXiv:nucl-th/0703085 (2007)
48. N. Keeley, R.S. Mackintosh, C. Beck, Nucl. Phys. A **834**, 792 (2010)
49. A. Diaz-Torres, I.J. Thompson, C. Beck, Phys. Rev. C **68**, 044607 (2003); arXiv:nucl-th/0307021 (2003)
50. C. Beck, N. Rowley, P. Papka, S. Courtin, M. Rousseau, F.A. Souza, N. Carlin, F. Liguori Neto, M.M. De Moura, M.G. Del Santo, A.A.I. Suade, M.G. Munhoz, E.M. Szanto, A. Szanto De Toledo, N. Keeley, A. Diaz-Torres, K. Hagino, Int. J. Mod. Phys. E **20**, 943 (2011); arXiv:1009.1719 (2010)
51. A. Diaz-Torres, D.J. Hinde, J.A. Tostevin, M. Dasgupta, L.R. Gasques, Phys. Rev. Lett. **98**, 152701 (2007); arXiv:nucl-th/0703041 (2007)
52. E.F. Aguilera, E.F. Aguilera, E. Martinez-Quiroz, D. Lizcano, A. Gomez-Camacho, J.J. Kolata, L.O. Lamm, V. Guimaraes, R. Lichtenthaler, O. Camargo, F.D. Becchetti, H. Jiang, P.A. DeYoung, P.J. Mears, T.L. Belyaeva, Phys. Rev. C **79**, 021601 (2009)
53. P. Papka, C. Beck, Clusters in nuclei, vol. 2. Lect. Notes Phys. **848**, 299 (2012)

# Collinear Cluster Tri-Partition as a Probe of Clustering in Heavy Nuclei

**D. V. Kamanin, A. A. Alexandrov, I. A. Alexandrova, N. A. Kondtatyev, E. A. Kuznetsova, O. V. Strekalovsky, V. E. Zhuchko, Yu. V. Pyatkov, W. von Oertzen, Yu. E. Lavrova, A. N. Tyukavkin, O. V. Falomkina, N. Jacobs, V. Malaza and Yu. V. Ryabov**

## 1 Introduction

Nuclear fission, a process where a heavy nucleus decays into two fragments of intermediate mass (e.g., Ba + Kr) has been identified by Hahn and Strassmann in 1938. It was discovered by chemical analysis while irradiating natural uranium with thermal neutrons [1]. Shortly afterwards Petrzhak and Flerov [2] observed spontaneous fission of the $^{238}$U isotope. The energy release in the fission process was immediately calculated by all leading physicists at that time to be very large, typically 200–205 MeV (e.g., Meitner and Frisch [3]). The large value is due to the larger binding energy per nucleon ($E_B/N$) in the mass range around mass $A = 54$ (iron, $E_B/N = 8.2$ MeV), as compared to the value at the end of the periodic table, $E_B/N = 7.2$ MeV. This fact could have been noticed four years before these discoveries, because of the existence of the liquid-drop model and the nuclear-mass

D. V. Kamanin (✉) · A. A. Alexandrov · I. A. Alexandrova · N. A. Kondtatyev · E. A. Kuznetsova · O. V. Strekalovsky · V. E. Zhuchko · Y. V. Pyatkov · O. V. Falomkina
Joint Institute for Nuclear Research, Dubna 141980, Russia
e-mail: kamanin@jinr.ru

Y. V. Pyatkov · Y. E. Lavrova · A. N. Tyukavkin
National Nuclear Research University MEPHI, Moscow 115409 , Russia

W. von Oertzen
Helmholtz-Zentrum Berlin, Glienickerstr. 100, 14109 Berlin, Germany

O. V. Falomkina
Physics Faculty, Computer Methods in Physics Division,
Lomonosov Moscow State University, Moscow 119899 , Russia

N. Jacobs · V. Malaza
Faculty of Military Science, Military Academy, University of Stellenbosch,
Saldanha 7395, South Africa

Y. V. Ryabov
Institute for Nuclear Research RAN, Moscow 117312, Russia

W. Greiner (ed.), *Exciting Interdisciplinary Physics*,
FIAS Interdisciplinary Science Series, DOI: 10.1007/978-3-319-00047-3_10,

formula of Bethe and Weizsäcker [4]. However, the large collective motion through a large deformation (today called super-deformation) was considered to be unlikely. The fission of heavy low-excited nuclei into three fragments of comparable masses, so called "true ternary fission", has been intensively investigated soon after the discovery of fission. Swiatecki [5] has shown within the framework of the liquid-drop model that fission into three heavy fragments is energetically more favorable than binary fission for all nuclei with fission parameters $30.5 < Z^2/A < 43.3$. In 1963 Strutinsky [6] has calculated the equilibrium shapes of the fissioning nucleus and has shown that along with the ordinary configuration with one neck, there is the possibility of more complicated elongated configurations with two and even three necks; at the same time it was stressed that such configurations are much less probable. Later, Diehl and Greiner [7, 8] have shown a preference for prolate over oblate saddle point shapes for the fission of a nucleus into three fragments of similar size. Such prescission configurations could lead to an almost collinear separation of the decay partners, at least in a sequential fission process. Actually, the Coulomb interaction in the total potential energy is the smallest for the linear arrangements of the three fragments. Furthermore, results demonstrating a decisive role of shell effects in the formation of the multi-body chain-like nuclear molecules were obtained also by Poenaru et al. [9]. On the experimental side there have been multiple attempts to find the true ternary fission in low-energy fission by means of counting techniques and in radiochemical studies. The schemes of the spectrometric experiments were based on the assumption of comparable angles between all three fragments emitted [10, 11]. Masses of the fragments were calculated in this case based on experimental values of the energies and angles. Contradictory results have been obtained; these were treated as showing the absence of fission fragments in the vicinity of mass fifty both in binary and ternary fission [12]. At the same time almost collinear ternary decays of excited heavy nuclear systems were known from the experiments in refs. [13, 14] at the early stage of our work. Bearing in mind the results mentioned above, we came to the conclusion, that collinear tri-partition of low excited heavy nuclear systems would be a promising field of research. In our first experiments dedicated to this problem [15, 16] some indications of such processes were already observed. At least one of the decay products detected was a magic nucleus. By analogy with the known cluster decay (or lead radioactivity), the process has been called "collinear cluster tri-partition" (CCT).

## 2 "Ni"-Bump and Its Internal Structure

We report here some results of three different experiments (marked Ex1, Ex2, Ex3 below) devoted to the search for collinear cluster tri-partition of $^{252}$Cf (sf) and performed in the Flerov Laboratory of Nuclear Reactions (FLNR) of the Joint Institute for Nuclear Research (JINR) in Dubna. The TOF-E (time-of-flight vs. energy) method for the measurements of two FF masses in coincidence with detectors placed at 180° was used in all three experiments. In this method, the fragment velocities

V, obtained by means of TOF and the energy $E$ are measured for each detected fragment individually. Only two fragments were actually detected in each fission event (in two detectors, at 1800) and their total mass, the sum $M_s$ will serve as a sign of a multi-body decay, if it is significantly smaller than the mass of the initial system ("missing mass" method).

In the first experiment (Ex1) performed at the FOBOS [17] set-up about $13 \times 10^6$ coincident binary fission events have been collected. It has the highest statistics among all three experiments discussed here. The TOF of the fragment has been measured over a flight path of 50 cm between the "start" detector which is based on micro-channel plates (MCP) placed next to the $^{252}$Cf source and the "stop" detectors formed by position sensitive avalanche counters (PSAC). The energies of those coincident fragments, which passed through the PSACs, were measured in the Bragg ionization chambers.

In the second experiment (Ex2), due to the low yield of the process under study, a multi-arm configuration containing five big and one small standard FOBOS modules in each arm was used. In order to select the CCT events accompanied almost isotropic neutron emission [18] the "neutron belt" consisting of 140 $^3$He-filled neutron counters in the moderator was assembled in a plane perpendicular to the symmetry axis of the spectrometer which serves as the mean fission axis at the same time. The centre of the belt coincides with the location of the fission source.

The experiment Ex3 has been performed using the Correlation Mosaic $E$–$T$ Array (COMETA). It is a double arm time-of-flight spectrometer which includes a MCP based "start" detector with the $^{252}$Cf source inside, similar to that used in Ex1. Two mosaics of eight PIN diodes each and a "neutron belt" comprising 28 $^3$He-filled neutron counters are used. Each PIN diode ($2 \times 2$ cm of surface area) provides both energy and timing signals.

The most pronounced manifestation of the CCT as a missing mass event is a bump (Fig. 1) in the two dimensional of the mass-mass correlation plot [19]. In this distribution of the fission fragment masses the bump occurs in one of the spectrometer arms with dispersive media ($M_1$), whereas it is absent in the analogues variable for the second arm ($M_2$). The bump is marked by the arrow in Fig. 1a. We see two great bumps due to binary fission; the pronounced vertical and horizontal intensities are due to binary fission fragments scattered from the entrance support grid for the windows of the gas detectors. The FF mass correlation plot similar to that obtained in Ex1 (Fig. 1a) is shown in Fig. 2a. Projections of this distribution both on the $M_1$ axis and on the $M_s = \text{const}$ directions are presented in Fig. 2b, c. They are compared with the analogous spectra from the experiments Ex1 including the result from the $^{235}$U($n_{th}$, f) reaction [19]. The bump in the projected FF mass correlation data in Fig. 2b is centered on mass (68–70) amu, associated with magic isotopes of Ni. This bump will be called below as the "Ni"-bump. The bump marked by the arrow in Fig. 2a looks less pronounced as compared to that obtained in Ex1 (Fig. 1a). This can be partially explained by a worse mass resolution due to the wide-aperture avalanche counter used as "start" detectors in Ex2, instead of the MCP based detector in Ex1. Projections for Ex2 are shown in the "difference" version, i.e. as a difference of the

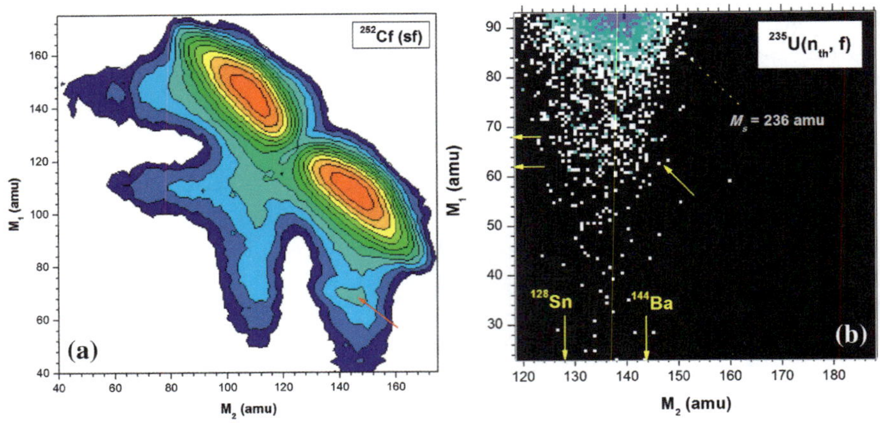


**Fig. 1** **a** Contour map (in logarithmic scale, the steps between the lines are approximately a factor 2.5) of the mass-mass distribution of the collinear fragments of $^{252}$Cf (sf), detected in coincidence in the two opposite arms of the FOBOS spectrometer. The specific bump in arm1 is indicated by an *arrow*. **b** The region of the mass distribution for the FFs from the reaction $^{235}$U($n_{th}$, f) around the bump. The bump is bounded by magic clusters (marked by corresponding symbols near the axes). The *tilted arrow* shows a valley between the ridges $M_1 + M_2 = 210$ amu of $M_s = $ const. See text for details

tail regions in arm1 and in arm2, respectively. Overall a good agreement is observed in the position of the peaks in Fig. 2b, c for all three experiments.

The methodically quite different experiment Ex3 shows results, which confirm our previous results concerning the structures in the missing mass distributions. In this case there is no tail due to scattering from material in front of the $E$-detectors. Figure 3 shows the region of the mass distribution for the FFs from $^{252}$Cf (sf) around the "Ni"-bump ($M_1 = 68$–$80$ amu, $M_2 = 128$–$150$ amu). The structures are seen in the spectrometer arm facing the source backing only. No additional selection of the fission events has been applied in this case; the experiment has no background from scattered FFs. A rectangular-like structure below the locus of binary fission is bounded by magic nuclei (their masses are marked by the numbered arrows) namely $^{128}$Sn (1), $^{68}$Ni (2), $^{72}$Ni (3). Two tilted diagonal lines with $M_s = 196$ amu and $M_s = 202$ amu (marked by number 4) start from the partitions 68/128 and 68/134, respectively. In experiment Ex1 [19], Fig. 6, similar sub-structures have been seen for masses $M_s = 204, 208, 212, 214$ amu where they were revealed indirectly by the applying of the second derivative filter, but in the absolutely statistically reliable distribution ("Ni"-bump) processed. Bearing in mind essential difference in the geometry of blocking mediums in Ex1 and Ex3 to be decisive for the relative experimental yields of the CCT modes with different angular distributions between the fragments forming the fork flying in the same direction the preference of lighter partitions standing behind the tilted ridges in Ex3 is not strange.

Positions of the points in the lower part of Fig. 3 do not contradict to possible existence of all the ridges revealed in Ex1 if the following magic partitions are assigned to their beginnings: 70/134, 68/140, 68/144, 70/144.

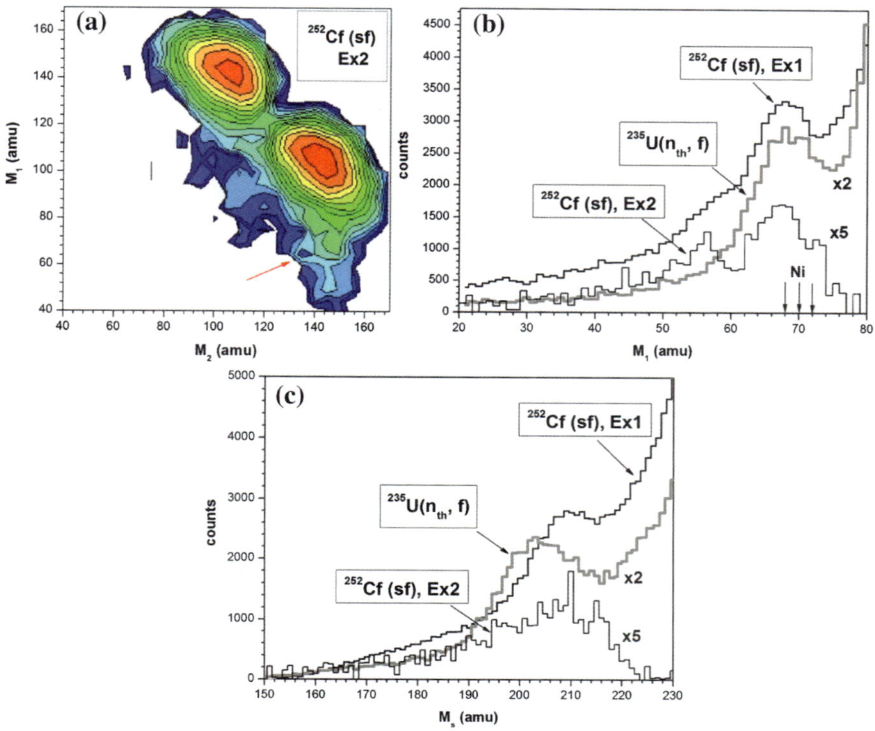


**Fig. 2** Ex2. **a** Contour map of the mass-mass distribution (logarithmic scale, with lines approximately a factor 1.5) from a coincidence in the two opposite arms of Ex2. The bump in the spectrometer arm (arm1) facing the backing of the Cf source is marked by the *arrow*. **b** Projections onto the $M_1$ axis for comparison with the experiments Ex1, and with the results of the $^{235}U(n_{th}, f)$ reaction (Fig.1b) [1]. **c** Projections onto the direction $M_s = M_2 + M_1$. Ex1 is presented by two curves marked by the *arrows* 1 and 2 (*dotted*) for the arm1 and arm2, respectively

Thus, comparison of Ex1 and Ex3 which are absolutely different both by the detectors and mass calculation procedures used as well as the statistics collected delivers strong confirmation of the existence of tilted ridges $M_s = \text{const}$ linked with magic partitions. As can be inferred from Fig. 3, the yield of the FFs with the mass 128 amu, which is extremely low in conventional binary fission, is clearly seen. It means that scattered binary fragments in any case cannot give rise to this structure. A part of the plot just below the locus of the binary FFs is shown in a larger scale in the insert. The structure is bounded by the magic nuclei of $^{80}$Ge, $^{78}$Ni, $^{132}$Sn, $^{144}$Ba (their masses are marked by the arrows 5, 6, 7, 8, respectively).

The observations presented point to the fact that the CCT decay occurs in a variety of modes (mass combinations), which could not be distinguished in Ex1 without additional gating due to the large background from scattered FFs. Likely due to the difference in the parameters of the blocking mediums the yield of the "Ni"-bump in Ex3 does not exceed $10^{-3}$ per binary fission i.e. much less then in Ex1 and Ex2. At

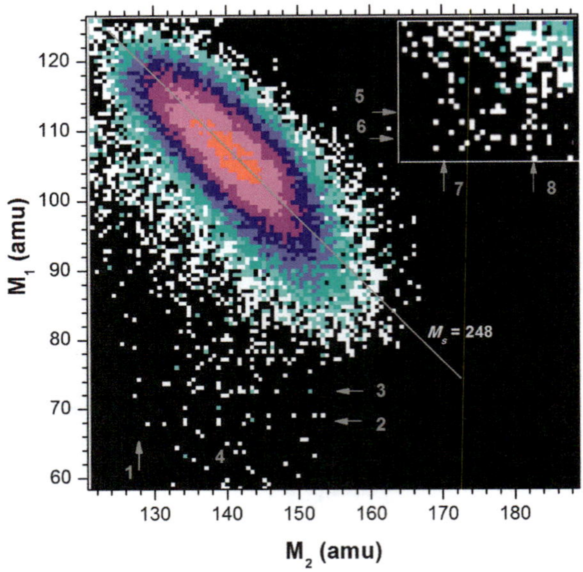


**Fig. 3** Results of Ex3: The region of the mass-mass distribution for the FFs from $^{252}$Cf (sf) around the CCT bump (Figs. 1a, 2a). No additional gates were applied. A part of the plot just below the locus of binary FFs produces the rectangular structure seen before. It is shown in the *insert* in a larger scale

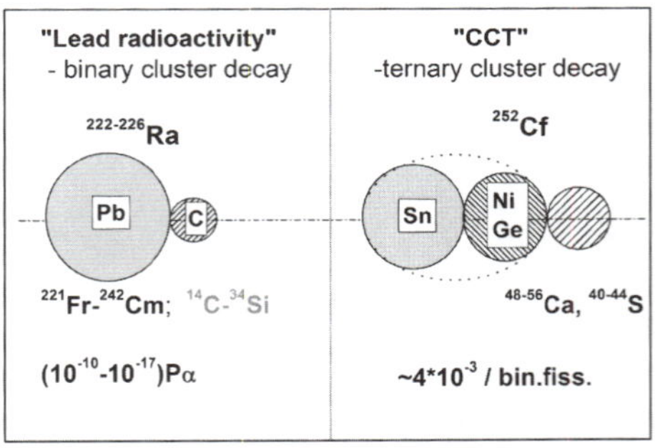


**Fig. 4** Cluster scheme for the comparison of the lead radioactivity with collinear cluster tri-partition

the same time with the absence of scattered FFs in Ex3, allowed the observation of the internal structure, without any additional cleaning of the FF mass distribution.

We would like to stress that one of the decay modes manifesting itself via tilted ridges $M_S = \text{const}$ can be treated as a new type of cluster decay as compared to the well-known heavy ion or lead radioactivity. Key features of both are summed up in Fig. 4. The relatively high CCT yield can be understood if one assumes collective motion through hyper-deformed pre-scission shapes of the mother systems, which is supported by the fact that the linear arrangement realizes the lowest Coulomb potential energies of three clusters. We also emphasize that the $Q$-values for ternary fission are by 25–30 MeV more positive, again due to the formation of magic fragments, as in binary fission. The ternary fission process must be considered to proceed sequentially, with two neck ruptures in a short time sequence characteristic for binary fissions.

## 3 Light Ion Accompanied CCT Modes

Specific CCT modes were observed in the data from the reaction $^{235}U(n_{th},f)$ [20]. The experiment was performed at the mini-FOBOS setup. This is a two arm time-of-flight-fission fragment-energy spectrometer based on standard detector modules of the $4\pi$ FOBOS spectrometer [17].

The discussed data are related to almost collinear tri-partition of $^{236}U^*$ nucleus due to the limited spectrometer aperture (the maximum deviation of the lightest fragment from the fission axis cannot exceed $14^0$). As was shown in [20] there is a possibility to discriminate fission events distorted due to scattering of the fragments involved by means of special gating. Corresponding procedure is based on the experimental values of the time-drift of the track formed by the detected fragment in the ionization chamber. We have find [20] that almost in all events selected the light ion (light charged particle) is registered in coincidence with a fragment from the light group of mass distribution of fission fragments observed in conventional binary fission.

The obtained experimental information can be generalized in the context of the following scenario of the two-stage decay of $^{236}U^*$: Being sufficiently elongated, the system clusterizes, forming the di-nuclear system (Fig. 5) from two magic clusters. Upon further elongation, the deformed light magic cluster (Mo) clusterizes with partitioning of the light charged particle (carbon nucleus) and the magic remainder (Kr). The process of collinear cluster tri-partition according to a similar scenario in $^{236}U^*$ may take place not only in valleys of mass asymmetric but also mass symmetric shapes [21]. To the best of our knowledge, the described effect was not

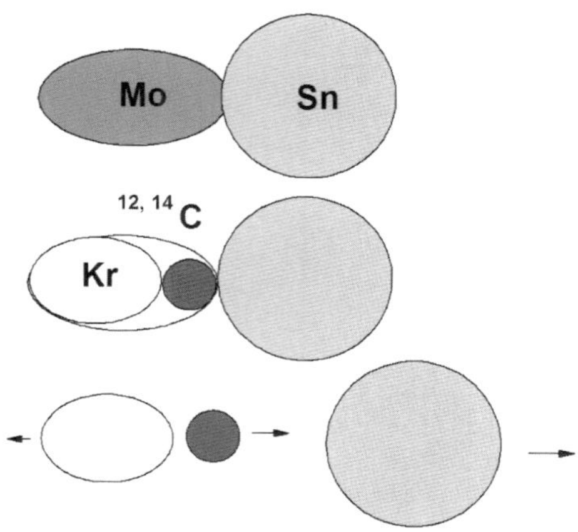


**Fig. 5** Illustration of the scenario of collinear cluster tri-partition accompanied by the emission of the light ion

observed earlier in works on the polar emission of light charged particles, which is probably associated with the excessive thicknesses of the $dE$ detectors used to identify the charge of light charged particles [22].

## 4 New Aspects from Neutron Gated Data

The experiments with coincident neutrons were motivated by the expectation that the center fragment is connected to an isotropic neutron source of increased (as compared to binary fission) multiplicity linked with the CCT. For this reason a selection of the fission events with an increased number of detected neutrons was studied.

Corresponding results obtained at the COMETA setup are presented in Fig. 6.

A rectangular structure bounded by the magic clusters is seen. This structure is invisible in the initial ungated distribution, because it is located very close to the centre of the conventional binary fission events, as can be seen from the comparison of Fig. 6 with Fig. 1a. The structure manifests itself exclusively thanks to the difference of the neutron sources for the fragments appearing in both binary fission and CCT, respectively. These two decay modes must differ in the neutron multiplicity or/and in their angular distributions of the emitted neutrons in order to provide the higher registration efficiency for neutrons linked with the CCT channel. The value

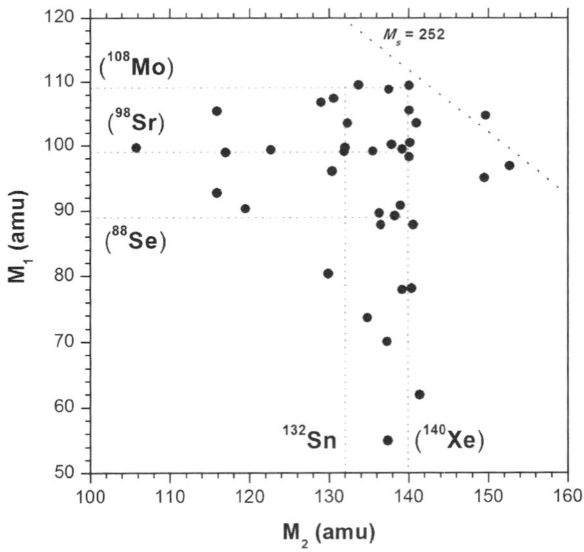


**Fig. 6** Ex3 ($n = 3$, and $w1$): Results obtained at the COMETA setup: mass-mass distribution of the FFs from $^{252}Cf$ (sf) under the condition that three neutrons ($n = 3$) were detected in coincidence and an additional selection with the gate w1 in the (V1-E1) distribution below the loci of conventional binary fission was applied [18]

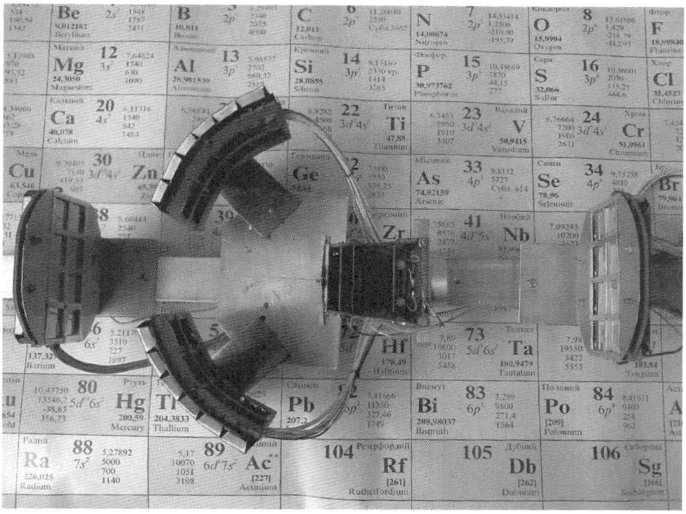


**Fig. 7** Photo of the COMETA-2 setup

of the excitation energy at the scission point known from the experiment let come to conclusion that the neutron source provided the structures in Fig. 6 differs from this in binary fission just by the angular distribution of the emitted neutrons.

## 5 To a Unified Model of Ternary Decays of Low Excited Nuclei

For the moment three different types of ternary decays of low excited nuclei are known, namely, conventional ternary fission, polar emission and CCT. It seems there is a deep link between the polar emission and CCT, at least with the CCT accompanied by a light charged particle. It would be extremely interesting to compare all three ternary decays in the frame of the unified experimental approach. We are planning to do this by means of step by step increasing of the aperture of the COMETA spectrometer and the first step has been already done. Recently COMETA-2 set up (Fig. 7) was put into operation at the FLNR of the JINR.

It contains four mosaics of Si semiconductor detectors of eight diodes each and the micro channel plates based "start" detector with the $^{252}$Cf inside. The FFs detectors are surrounded by the "neutron belt" which was used previously at the COMETA spectrometer. Processing of the data of the test run is in progress.

## 6 Conclusions

1. A new bright phenomenon is observed—CCT.
2. CCT is due to the pre-formation of at least two magic clusters, deformed as well. The CCT modes based on these combinations are more preferable.
3. There is an open field for the systematic study of CCT since it may bring us new knowledge on the fission process and clustering in cold nuclear matter.

## References


1. O. Hahn, F. Strassmann, Naturwissenschaften **27**, 89 (1939)
2. K.A. Petrzhak, G.N. Flerov, J. Phys., USSR **3**, 275 (1940)
3. L. Meitner, O. Frisch, Nature **143**, 239 (1939)
4. C.F. von Weizsäcker, Z. Phys. **96**, 431 (1935)
5. W.J. Swiatecki, in *Proceeding of the Second UN Conference on the Peaceful Uses of Atomic Energy, Geneva*, vol. 15, p. 651 (1958)
6. V.M. Strutinsky et al., Nucl. Phys. **46**, 639 (1963)
7. H. Diehl, W. Greiner, Phys. Lett. B **45**, 35 (1973)
8. H. Diehl, W. Greiner, Nucl. Phys. A **229**, 29 (1974)
9. D.N. Poenaru et al., Phys. Rev. C **59**, 3457 (1999)
10. M.L. Muga, Phys. Rev. Lett. **11**, 129 (1963)
11. P. Schall, Phys. Lett. B **191**, 339 (1987)
12. F. Gïnenwein, Nucl. Phys. A **734**, 213 (2004)
13. S.A. Karamian et al., Jad. Fiz. **5**, 959 (1963)
14. P. Glässel et al., Z. Phys. A **310**, 189 (1983)
15. Yu.V. Pyatkov et al., in *Proceeding of the International Conference of Nuclear Physics "Nuclear Shells—50 Years"* (World Scientific, Singapore, 2000), p. 301
16. Yu.V. Pyatkov et al., in *Proceeding of the International Symposium on Exotic Nuclei (EXON-2001)* (World Scientific, Singapore, 2002), p. 181
17. H.-G. Ortlepp et al., Nucl. Instrum. Methods A **403**, 65 (1998)
18. Yu.V. Pyatkov et al., Eur. Phys. J. A **48**, 94 (2012)
19. Yu.V. Pyatkov et al., Eur. Phys. J. A **45**, 29 (2010)
20. Yu.V. Pyatkov, Bulletin of the Russian academy of sciences. Physics **75**, 949 (2011)
21. V.V. Pashkevich et al., Int. J. Mod. Phys. E **18**, 907 (2009)
22. E. Piasecki et al., Nucl. Phys. A **255**, 387 (1975)

# Pairing Influence in Binary Nuclear Systems

**R. A. Gherghescu, D. N. Poenaru and W. Greiner**

**Abstract** A specialized macroscopic-microscopic method is applied to calculate the deformation energy and penetrability for binary nuclear configurations typical for fusion processes. The deformed two-center shell model is used to obtain the single particle energy levels for the transition region of two partially overlaped projectile and target nuclei. The macroscopic part is obtained with the Yukawa-plus-exponential potential. The microscopic shell and pairing corrections are obtained with the Strutinsky and BCS approaches and the cranking formulae yield the inertia tensor. Finally the WKB method is used to calculate the penetrabilities for sub-barrier fusion reactions.

## 1 Introduction

When the total deformation energy is calculated along the distance between centers for fusion configurations, some valleys appear for different mass asymmetries. These valleys can be obtained as the result of multidimensional minimization of the action integral within the space of deformation. In order to take as many as possible deformation parameters into account, one has to calculate all the terms in the total deformation energy with the appropriate binary model able to describe the stages of the fusion process. Such a model has been pioneered and improved by the Frankfurt school in the group of Greiner and collaborators [1, 2]. The importance of the deformed valleys in the potential energy surfaces is that they provide promising

R. A. Gherghescu (✉) · D. N. Poenaru
Horia Hulubei National Institute of Physics and Nuclear Engineering, P.O. Box MG-6, 077125 Bucharest-Magurele, Romania
e-mail: radu@theory.nipne.ro; radu@fias.uni-frankfurt.de

R. A. Gherghescu · D. N. Poenaru · W. Greiner
Frankfurt Institute for Advanced Studies, University, Ruth-Moufang-Str. 1, 60438 Frankfurt am Main, Germany

W. Greiner (ed.), *Exciting Interdisciplinary Physics*,
FIAS Interdisciplinary Science Series, DOI: 10.1007/978-3-319-00047-3_11,

doors towards complete sub-barrier fusion towards synthesis of superheavy nuclei. Sub-barrier fusion reactions have low cross section values, but the final compound nucleus could be reached in a more stable point, close to the ground state. When dynamics is studied one has to introduce the influence of the mass tensor. We use the results from the pairing calculations for the occupation probabilities. In this way the mass tensor components contain the binary character of the process, since the pairing parameters are calculated with the two-center shell model levels.

## 2 The Binary Macroscopic-Microscopic Method

Fusionlike configurations are used for the total deformation energy calculation. A typical shape is displayed in Fig. 1, where $b_T$, $a_T$ and $b_P$, $a_P$ are the small and large semiaxes of the target and projectile respectively, $z_s$ is the position of the separation plane and $R$ is the distance between centers. All these geometrical parameters form the space of deformation, and furtheron one shall work with $\chi_T = b_T/a_T$, $\chi_P = b_P/a_P$, $b_P$ and $R$ as degrees of freedom.

The microscopic part starts with the binary Hamiltonian:

$$H = -\frac{\hbar}{2m_0}\Delta + V(\rho, z) + V_{\Omega s} + V_{\Omega^2} \tag{1}$$

where the potentials are deformation dependent. The deformed two-center oscillator potential for target and projectile regions $v_T$ and $v_P$ reads:

$$V(\rho, z) = \begin{cases} V_T(\rho, z) = \frac{1}{2}m_0\omega_{\rho T}^2\rho^2 + \frac{1}{2}m_0\omega_{zT}^2(z + z_T)^2 & \text{for } v_T \\ V_P(\rho, z) = \frac{1}{2}m_0\omega_{\rho P}^2\rho^2 + \frac{1}{2}m_0\omega_{zP}^2(z - z_P)^2 & \text{for } v_P \end{cases} \tag{2}$$

where $z_T$ and $z_P$ are the centers of the target and projectile. Angular momentum dependent potentials, $V_{\Omega s}$ and $V_{\Omega^2}$ are constructed to comply to the $V(\rho, z)$ - dependence and hermiticity of the operators, so that:

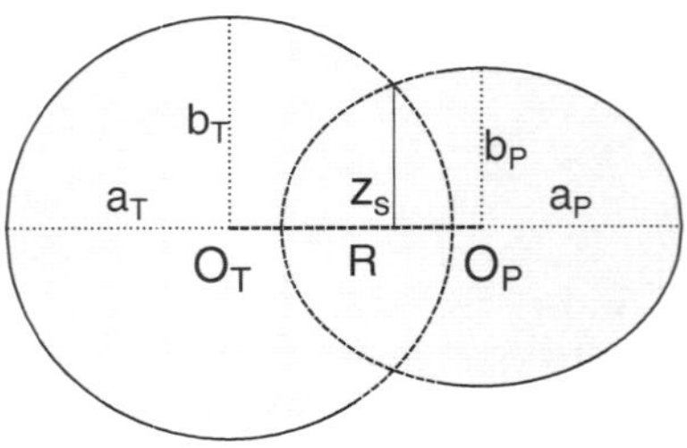


**Fig. 1** Binary fusionlike configuration for two ellipsoidal target and projectile nuclei. The free geometrical parameters (the semiaxes of the two ellipsoids and the distance between centers) are marked

$$V_{so} = \begin{cases} -\left\{ \frac{\hbar}{m_0\omega_{0T}} \kappa_T(\rho, z), (\nabla V^{(r)} \times \mathbf{p})\mathbf{s} \right\} & v_T\text{–region} \\ -\left\{ \frac{\hbar}{m_0\omega_{0P}} \kappa_P(\rho, z), (\nabla V^{(r)} \times \mathbf{p})\mathbf{s} \right\} & v_P\text{–region} \end{cases} \tag{3}$$

and similarly for the $l^2$ term. The matrix diagonalization of $H$ generates the level scheme of the fusion configuration, for spheroidally deformed nuclei, at any given distance $R$ between centers and intermediary independent $b_P$, $\chi_T$ and $\chi_P$. The level scheme sequence from the touching point up to complete overlapping is input data for the Strutinsky method [3], and calculations are performed separately for protons and neutrons. The shell correction energy is obtained as the difference between the simple sum of level enegies and the smoothed part of the same scheme:

$$\delta E_{\text{shell}} = \sum_i E_i - \tilde{U} \tag{4}$$

where the summation is performed for all occupied levels. The main part of the calculation consists in obtaining the smoothed term $\tilde{U}$. A smoothed–level distribution density $\tilde{g}(\varepsilon)$ is defined by averaging the actual distribution over a finite interval $\gamma$ (here equal to 1.2 in $\hbar\omega$ units). If the level energies in units of $\hbar\omega$ are denoted with $\varepsilon_i$, one can write the integral which replaces the discrete sum and one obtains the smoothed distribution:

$$\begin{aligned} \tilde{g}(\varepsilon) &= \frac{1}{\gamma} \int_{-\infty}^{\infty} \zeta\left(\frac{\varepsilon - \varepsilon'}{\gamma}\right) g(\varepsilon') d\varepsilon' \\ &= \frac{1}{\gamma} \sum_{i=1}^{\infty} \zeta\left(\frac{\varepsilon - \varepsilon_i}{\gamma}\right) \end{aligned} \tag{5}$$

This work utilizes a smoothing function $\zeta$ of the form:

$$\zeta(x) = \frac{1}{\sqrt{\pi}} \exp(-x^2) f_m(x) \tag{6}$$

where $x = (\varepsilon - \varepsilon')/\gamma$ and the smoothing function $f$ is taken as a polynomial sum:

$$f_m(x) = \sum_{k=0}^{m} a_{2k} H_{2k}(x) \tag{7}$$

$H_n(x)$ are the Hermite polynomials, and the maximum degree $m$ (here 3) is taken such as $d\tilde{U}/d\gamma$ = constant (the plateau condition). The maximum level is chosen such as $|x_i| \geq 3$. Beyond this limit the contribution of more remoted levels is negligeable. Once the density of the smooth levels $\tilde{g}(\varepsilon)$ is obtained by this smearing procedure, the smoothed part of the energy is given by:

$$\tilde{u} = \tilde{U}/\hbar\omega = \int_{-\infty}^{\tilde{\lambda}} \tilde{g}(\varepsilon)\varepsilon d\varepsilon \tag{8}$$

where the Fermi level $\tilde{\lambda}$ for smoothed distribution is obtained from the conservation of the total number of delocalized valence electrons:

$$N_e = \int_{-\infty}^{\tilde{\lambda}} \tilde{g}(\varepsilon) d\varepsilon \tag{9}$$

By substituting the above expression for $\tilde{g}(\varepsilon)$ one obtains:

$$N_e = \frac{2}{\sqrt{\pi}} \sum_{1}^{\infty} \int_{-\infty}^{x_{iF}} f_m(x_i^2) \exp(x_i^2) dx_i \tag{10}$$

where $x_{iF} = (\tilde{\lambda} - \varepsilon_i)/\gamma$. The summation is in fact reduced to the levels around the Fermi limit. The latter equation yields the Fermi level for smoothed distribution $\tilde{\lambda}$, and is solved numerically. We consider a set of doubly degenerate energy levels $\{\varepsilon_i\}$ expressed in units of $\hbar\omega_0^0$. Calculations for neutrons are similar with those for protons, hence for the moment we shall consider only protons. In the absence of a pairing field, the first $Z/2$ levels are occupied, from a total number of $n_t$ levels available. Only few levels below $(n)$ and above $(n')$ the Fermi energy are contributing to the pairing correlations. Usually $n' = n$. If $\tilde{g_s}$ is the density of states at Fermi energy obtained from the shell correction calculation $\tilde{g_s} = dZ/d\varepsilon$, expressed in number of levels per $\hbar\omega_0^0$ spacing, the level density is half of this quantity: $\tilde{g_n} = \tilde{g_s}/2$.

We can choose as computing parameter, the cut-off energy (in units of $\hbar\omega_0^0$), $\Omega \simeq 1 \gg \tilde{\Delta}$. Let us take the integer part of the following expression

$$\Omega \tilde{g_s}/2 = n = n' \tag{11}$$

When from calculation we get $n > Z/2$ we shall take $n = Z/2$ and similarly if $n' > n_t - Z/2$ we consider $n' = n_t - Z/2$.

The gap parameter $\Delta = |G| \sum_k u_k v_k$ and the Fermi energy with pairing corellations $\lambda$ (both in units of $\hbar\omega_0^0$) are obtained as solutions of a nonlinear system of two BCS equations

$$n' - n = \sum_{k=k_i}^{k_f} \frac{\varepsilon_k - \lambda}{\sqrt{(\varepsilon_k - \lambda)^2 + \Delta^2}} \tag{12}$$

$$\frac{2}{G} = \sum_{k=k_i}^{k_f} \frac{1}{\sqrt{(\varepsilon_k - \lambda)^2 + \Delta^2}} \tag{13}$$

where $k_i = Z/2 - n + 1$; $k_f = Z/2 + n'$.

The pairing interaction $G$ is calculated from a continuous distribution of levels

$$\frac{2}{G} = \int_{\tilde{\lambda}-\Omega}^{\tilde{\lambda}+\Omega} \frac{\tilde{g}(\varepsilon)d\varepsilon}{\sqrt{(\varepsilon-\tilde{\lambda})^2+\tilde{\Delta}^2}} \tag{14}$$

where $\tilde{\lambda}$ is the Fermi energy deduced from the shell correction calculations and $\tilde{\Delta}$ is the gap parameter, obtained from a fit to experimental data, usually taken as $\tilde{\Delta} = 12/\sqrt{A}\hbar\omega_0^0$. Both $\Delta_p$ and $\Delta_n$ decrease with increasing asymmetry $(N-Z)/A$. From the above integral we get

$$\frac{2}{G} \simeq 2\tilde{g}(\tilde{\lambda}) \ln\left(\frac{2\Omega}{\tilde{\Delta}}\right) \tag{15}$$

Real positive solutions of BCS equations are allowed if

$$\frac{G}{2}\sum_k \frac{1}{|\varepsilon_k-\lambda|} > 1 \tag{16}$$

i.e. for a pairing force (G-parameter) large enough at a given distribution of levels.

As a consequence of the pairing correlation, the levels situated bellow the Fermi energy are only partially filled, while those above the Fermi energy are partially empty; there is a given probability for each level to be occupied by a quasiparticle

$$v_k^2 = \frac{1}{2}\left[1 - \frac{\varepsilon_k-\lambda}{\sqrt{(\varepsilon_k-\lambda)^2+\Delta^2}}\right] \tag{17}$$

or a hole

$$u_k^2 = 1 - v_k^2 \tag{18}$$

Only the levels in the near vicinity of the Fermi energy (in a range of the order of $\Delta$ around it) are influenced by the pairing correlations. For this reason, it is sufficient for the value of the cut-off parameter to exceed a given limit $\Omega \gg \tilde{\Delta}$, the value in itself having no significance. The shell and pairing corrections calculated for the splitting of $^{236}$Pu in two $^{118}$Ag nuclei are displayed in Fig. 2 along the reduced distance between centers. One observes large fluctuations of the proton and neutron shell corrections, in counterphase with the corresponding pairing energy.

The macroscopic part is obtained using the Yukawa-plus-exponential method, specialized to binary processes. The Coulomb term $E_C$ [4] and the nuclear surface term $E_Y$ [5] are computed as:

$$E_C = \frac{2\pi}{3}(\rho_{eT}^2 F_{C_T} + \rho_{eP}^2 F_{C_P} + 2\rho_{eT}\rho_{eP} F_{C_{TP}}) \tag{19}$$

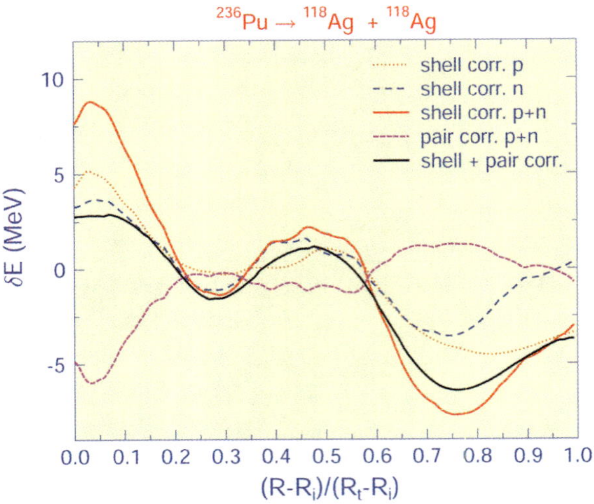


**Fig. 2** Shell and pairing corrections for neutrons and protons, and their sum for the symmetric splitting of $^{236}$Pu

and

$$E_Y = \frac{1}{4\pi r_0^2}[c_{sT}F_{EYT} + c_{sP}F_{EY_P} + 2(c_{sT}c_{sP})^{1/2}F_{EY_{TP}}] \tag{20}$$

where $\rho_{ei}$ is the charge density and $c_{si}$ the surface coefficient. $F_{C_i}$ and $F_{EY_i}$ are shape dependent integrals.The peculiarity resides in the last term in both formulas, $F_{C_{TP}}$ and $F_{EY_{TP}}$, which account for the interaction between non-overlapped parts of the overlapping configuration. Details about these terms are given in [6].

The total deformation dependent macroscopic energy is calculated as the sum of the Coulomb and surface terms:

$$E_{\text{macro}} = (E_C - E_C^{(0)}) + (E_Y - E_Y^{(0)}) \tag{21}$$

where $E_C^{(0)}$ and $E_Y^{(0)}$ are the values for the corresponding spherical compound nucleus. Finally the deformation energy is computed as the sum of the macroscopic part and the shell correction:

$$E_{\text{def}} = E_{\text{macro}} + E_{\text{shell}} \tag{22}$$

## 3 Dynamics

In order to obtain the penetrabilities for different reaction channels, the action integral must be computed. Besides the usual deformation energy, the nuclear inertia tensor,

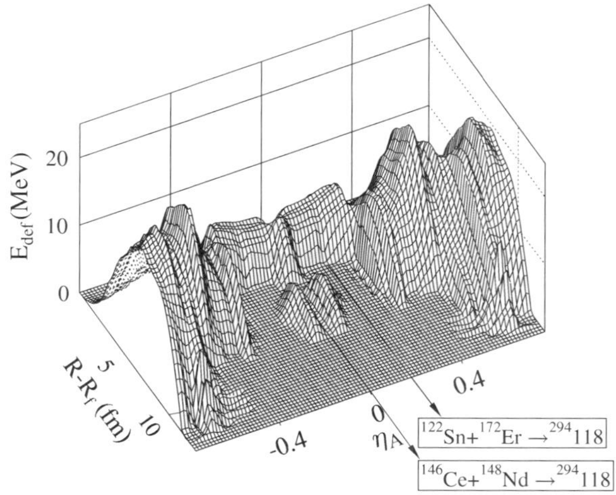


**Fig. 3** Potential energy surface of $^{294}118$ as a function of fusion reaction mass asymmetry $\eta_A$ and distance between centers $R$, after multidimensional minimization over $(b_P, \chi_T, \chi_P)$

which accounts for the reaction of the nucleus to the deformation along a given degree of freedom, is to be computed. This work uses the cranking approach to obtain the mass tensor components within the four-dimensional space of $(b_P, \chi_T, \chi_P, R)$. According to the cranking model, after including the BCS pairing correlations [7], the inertia tensor is given by [8]:

$$B_{ij} = 2\hbar^2 \sum_{\nu\mu} \frac{\langle \nu|\partial H/\partial \beta_i|\mu\rangle\langle\mu|\partial H/\partial\beta_j|\nu\rangle}{(E_\nu + E_\mu)^3}(u_\nu v_\mu + u_\mu v_\nu)^2 + P_{ij} \qquad (23)$$

where $H$ is the single-particle Hamiltonian allowing to determine the energy levels and the wave functions $|\nu\rangle$, $u_\nu$, $v_\nu$ are the BCS occupation probabilities, $E_\nu$ is the quasiparticle energy, and $P_{ij}$ gives the contribution of the occupation number variation when the deformation is changed (terms including variation of the gap parameter, $\Delta$, and Fermi energy, $\lambda$, $\partial\Delta/\partial\beta_i$ and $\partial\lambda/\partial\beta_i$).

The penetrability $P$ for a given fusion path is calculated as usual:

$$P = \exp(-K_{ov}) \qquad (24)$$

where $K_{ov}$ is the overlapping action integral. The barriers are supposed to be tunneled at the level of the final compound nucleus ground state energy. This is the minimum value of the kinetic energy in this work where sub-barrier fusion reactions are intended to take place at the lowest energy. $K_{ov}$ is calculated numerically as:

$$K_{ov}(b_P, \kappa_T, \kappa_P; R) = \frac{2}{\hbar}\int_{(\mathrm{fus})} [2B(R)_{b_P,\kappa_T,\kappa_P} E_{\mathrm{def}}(R)_{b_P,\kappa_T,\kappa_P}]^{1/2} dR \qquad (25)$$

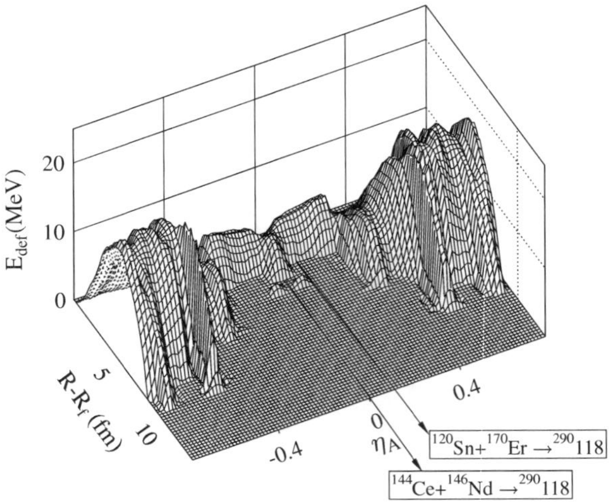


**Fig. 4** Potential energy surface of $^{290}118$ as a function of fusion reaction mass asymmetry $\eta_A$ and distance between centers $R$, after multidimensional minimization over $(b_P, \chi_T, \chi_P)$

Since $K_{ov}$ is calculated for every set $(b_P, \kappa_T, \kappa_P)$ at every point $R$, the penetrability appears as a multidimensional lot. The final value of $P$ for every channel reaction is the result of action integral $K_{ov}$ minimization over the whole range of $(b_P, \kappa_T, \kappa_P, R)$. The multidimensional minimization of the action integral is performed over the grid in the space of $I(b_P, \kappa_T, \kappa_P; R)$, where I is the integrand.

# 4 Results and Discussion

The algorithm has been applied to the synthesis of $^{294,290}118$ nuclei. For every superheavy system the entire possible range of mass asymmetry has been taken into account. Target-projectile pairs start from symmetry $\eta_A = 0$ $(A_T \simeq A_P)$ up to the asymmetry value, where there is still the possibility to have a stable target (a few tens of minutes halflife). For superheavy production, especially in the neutron-rich region like $^{294}118$, there is always the problem of having both partners between the drip lines. One has to stress that most of the reactions used in this work are most likely to be obtained only with the help of radioactive beams, since due to the large necessary neutron number, for certain mass asymmetries it is impossible to have both partners around the stability line. Once the static barriers are obtained from the minima on the potential energy surface, the mass asymmetry is completed with finding the charge asymmetry, by repeating the calculations for all possible $(Z_T, Z_P)$ for the same $(A_T, A_P)$. At the end the mass tensor and penetrability are calculated for all $(A_T, Z_T)$-$(A_P, Z_P)$ reaction channels by preserving $R$ as the main free variable.

### *4.1* $^{294}$*118*

As the result of multidimensional minimization in the $(b_P, \chi_T, \chi_P)$ space along $R$, the first potential energy surface (PES) has been obtained for the neutron rich superheavy system $^{294}$118 in Fig. 3, as a function of the reduced distance between centers $R - R_f$ and the mass asymmetry $\eta_A = (A_T - A_P)/A$. Two maxima appear between symmetric reactions and $\eta_A \simeq 0.15$, both pf them lower than any other energy height for larger mass asymmetry. Two out of the possible reactions are marked on the figure: the symmetric one with $^{146}$Ce as a projectile and proton magic projectile $^{122}$Sn-reaction. These reactions have been computed as having the largest values of penetrability after minimization of the action integral.

### *4.2* $^{290}$*118*

Four neutrons poorer, $^{290}$118 displays a PES without the pronounced ridge at symmetry, see Fig. 4. This superheavy nucleus is calculated as spherical. The lowest barrier height has been computed for $^{144}$Ce+$^{146}$Nd, having also the highest penetrability value. As mass asymmetry increases, the entrance point is more advanced and barriers are very narrow, but slightly higher.

## 5 Conclusions

A binary configuration model has been used within a large number of degrees of freedom to calculate the barriers and penetrabilties towards the synthesis of $^{294,290}$118 superheavy isotopes. Dynamical multidimensional minimization of the action integral yielded the penetrabilities by WKB method. The barriers are larger and higher, and penetrabilities are lower as the system is neutron richer. The highest log $P$ for every superheavy isotope are calculated for spherical projectiles.

**Acknowledgments** This work was partially supported by Deutsche Forschungsgemeinschaft and partially within IDEI Programme under contracts 43/05.10.2011 and 42/05.10.2011 with UEFISCDI, Bucharest.

## References

1. J. Maruhn, W. Greiner, Z. Phys. **251**, 431 (1972)
2. R.A. Gherghescu, Phys. Rev. C **67**, 014309 (2003)
3. V.M. Strutinsky, Nucl. Phys. A **95**, 420 (1967)
4. K.T. Davies, A.J. Sierk, J. Comp. Phys. **18**, 311 (1975)
5. H.J. Krappe, J.R. Nix, A.J. Sierk, Phys. Rev. C **20**, 992 (1979)
6. D.N. Poenaru, W. Greiner, *in Nuclear Decay Modes* (IOP Publishing, Bridstol, 1996)
7. J. Bardeen, L. Cooper, J. Schrieffer, Phys. Rev. C **108**, 1175 (1957)
8. M. Brack et al., Rev. Mod. Phys. **44**, 320 (1972)

# Chiral Symmetry in Real Nuclei

**Obed Shirinda and Elena Lawrie**

**Abstract** Identifying the presence of a symmetry may have a profound impact on the nuclear structure features. We are discussing whether strongly broken chiral symmetry is likely to occur in nuclei, how weakly broken chirality is exhibited, and whether we can rely on the chirality fingerprints in order to identify chiral symmetry in real nuclei.

## 1 Symmetries, Nuclear Chiral Symmetry and Some Open Questions

Nuclear structure theory is undertaking a formidable challenge, which is to understand and describe the nature of hundreds of nuclear states in a single nucleus, without a complete knowledge of the nuclear force, and for a quantum system built of many (but not an infinite number of) nucleons. In order to simplify the task nuclear structure models are used. These are descriptions of the nucleus, based on simplified assumptions. For instance the liquid drop model assumes that the nuclear matter behaves like a liquid drop, while the shell model description is based on an independent motion of the nucleons in a mean field potential. The application of such models, even the most successful ones, is to some extend limited, since any nuclear phenomenon not covered by the initial assumptions remains beyond the model description.

There is however a very powerful and very helpful additional consideration, i.e. the symmetries of the nuclear system. The presence of a symmetry poses strict requirements on the model formalism and usually has direct consequences for the nuclear structure. Consider the following example. In some cases the assumptions of

O. Shirinda · E. Lawrie(✉)
iThemba LABS, National Research Foundation, P.O. Box 722, Somerset West 7129, South Africa
e-mail: obed@tlabs.ac.za

E. Lawrie
e-mail: elena@tlabs.ac.za

W. Greiner (ed.), *Exciting Interdisciplinary Physics*,
FIAS Interdisciplinary Science Series, DOI: 10.1007/978-3-319-00047-3_12,

a model brake a basic symmetry. For instance the assumption of the deformed shell model that a nucleus is deformed, brakes the spherical symmetry and defines a special direction in space (the axis of symmetry of the nuclear matter) not equivalent to the other directions. However, the description of the nuclear system should not depend on the orientation of the coordinate system, thus the wavefunctions are defined in such a way that the broken symmetry is restored. This restoration of the broken symmetry, among other things, leads in this specific case to the requirement that the ground state rotational bands in deformed even-even nuclei can have states with even spins only. Nuclear structure properties that result from symmetry requirements are very important in nuclear structure. They reflect basic principles, and do not depend on the particularities of the model, e.g., on the exact shape of the nuclear potential, on the assumed residual interactions, on the number of nucleons, etc. Thus, investigating symmetries and identifying their impact on the nuclear structure is of extreme importance for all nuclear models.

It was shown that one can form a nuclear system for which the chiral symmetry is broken. Such nuclear system has to be defined in the angular momentum space [1]. One such system is shown in Fig. 1. It is built by three mutually orthogonal angular momenta, and since one could arrange these angular momenta in two different ways, (corresponding to a right-handed and to a left-handed systems), the chiral symmetry is broken. One restores the chiral symmetry by forming a combination of the left- and right-handed wavefunctions. The wavefunctions corresponding to the restored chirality describe two identical rotational bands. Therefore it was suggested that a nuclear system possessing chiral symmetry should exhibit two partner bands with degenerate properties [1].

In order for a chiral system to form the total angular momentum of the nucleus should be "aplanar", with significant projections along all three nuclear axes [1]. The simplest nuclear chiral system is then formed in an odd-odd nucleus with triaxial shape, where the angular momenta of the odd proton, the odd neutron, and the collective rotation are predominantly aligned along the three major nuclear axes. In order to get such an alignment of the angular momenta, the proton and the neutron

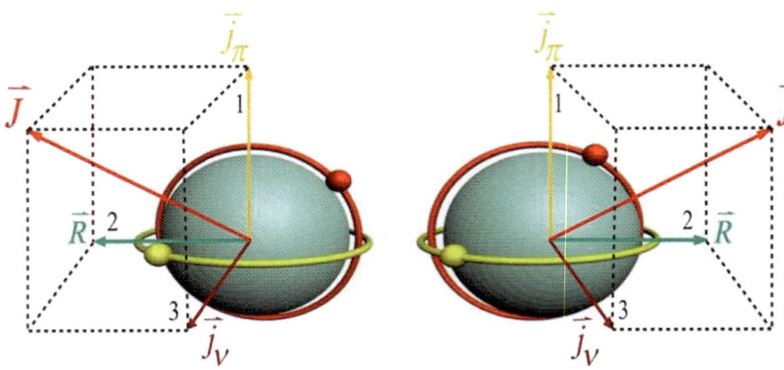


**Fig. 1** A nuclear chiral system built by the angular momenta of the odd proton ($\mathbf{j}_\pi$), the odd neutron ($\mathbf{j}_\nu$) and the collective rotation ($\mathbf{R}$), from [2]

should have predominantly a particle and a hole nature, in which case the particle angular momentum will be predominantly aligned along the short nuclear axis, while the angular momentum of the hole will be predominantly aligned along the long nuclear axis. If the nuclear shape is triaxial, rotation around the intermediate axis will be dominant.

When dealing with a broken symmetry it is important to also keep in mind the possibility of a weak symmetry breaking. Consider the example of the breaking of the spherical symmetry described above. In this case weak breaking could occur if the potential energy surface has a soft minimum with respect to the quadrupole deformation, and although the wavefunction includes mostly components corresponding to a deformed nuclear shape, it also contains some contributions describing a spherically symmetric nuclear shape. Chiral symmetry might be weakly broken, if there is an overlap between the left-handed and the right-handed systems. Such overlap could occur for instance if the wavefunctions of the left- and right-handed systems include a contribution describing a planar orientation of the three angular momenta, e.g., the angular momentum of the collective rotation lies in the plane defined by the proton and neutron angular momenta. In the case of weakly broken chiral symmetry the degeneracy of the chiral partner bands is lifted and the two bands show a similarity.

Therefore, a strongly broken chiral symmetry should result in a pair of degenerate $\Delta I = 1$ rotational bands with the same parity [1]. It is very interesting whether such perfect nuclear chiral systems can exist in nature. Apart from degeneracy, these chiral partner bands should also show staggering in the $B(M1)$ transition probabilities [3], and a smooth dependence of the energy staggering parameter with spin [4]. According to the particle-rotor model calculations such a pair of bands can show large discrepancies at low excitation energies, and become degenerate at higher spins, where the chiral symmetry becomes strongly broken [1].

Searches for (near-)degenerate partner bands were carried out in many laboratories. Several pairs of bands were suggested as possibly linked to chiral symmetry. Such partner bands were discovered in the nuclei with mass numbers around 130, where the odd proton and the odd neutron occupy $h_{11/2}$ orbitals near the bottom and top of the $h_{11/2}$ shell, respectively. Other possible chiral candidates were found in the 100 mass region, where the chiral structures involve an odd proton hole in the $g_{9/2}$ shell, while the odd neutron particle occupies an orbital near the bottom of the $h_{11/2}$ shell. A short review of these experimental results can be found in [5]. The latest chiral mass region is A = 80 [6]. At iThemba LABS chiral candidates were studied in the 190 mass region [7–9]. The Tl isotopes in this mass region have rotational bands built on the $\pi h_{9/2} \otimes \nu i_{13/2}^{-1}$ configuration.

A common feature of all chiral candidates is that none of them shows really close near-degeneracy in the partner bands. Partner bands with energy discrepancy of as much as 500 keV were suggested as possible chiral partners. One of the suggested best chiral candidate, $^{128}$Cs, exhibits energy discrepancy of about 200 keV [10, 11]. An example of experimental results obtained at iThemba LABS is shown in Fig. 2. The four-quasiparticle bands in $^{194}$Tl is one of the best examples of near-degeneracy found in any chiral candidate, with energy difference remaining less than 110 keV through the whole spin range of the bands, and reaching 37 keV at $I = 21$. Obviously,

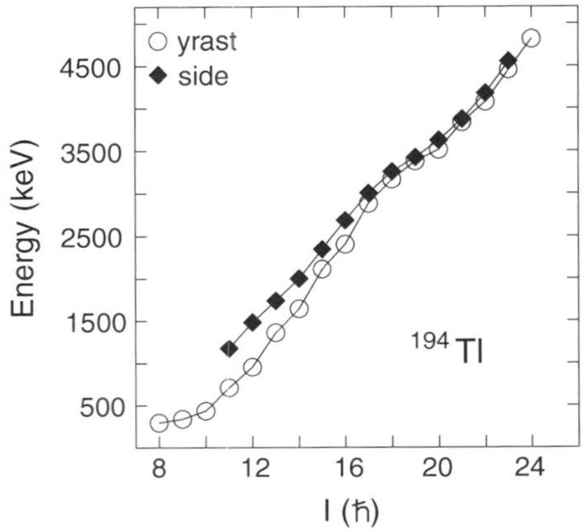


**Fig. 2** The excitation energies as a function of the spin for the partner bands in $^{194}$Tl. The two-quasiparticle bands are assigned the $\pi h_{9/2} \otimes \nu i^{-1}_{13/2}$ configuration ($I < 18$), while the four-quasiparticle bands are associated with the $\pi h_{9/2} \otimes \nu i^{-3}_{13/2}$ configuration ($I > 18$)

a close near-degeneracy in the excitation energies of the partner bands within certain spin range leads also to near degeneracy in several other properties of the bands, such as moments of inertia, alignments, etc. Thus, in addition to the energy near degeneracy usually the near degeneracy of the $B(M1)$ and $B(E2)$ reduced transition probabilities is also tested.

The experimental data published so far shows that strongly broken chiral symmetry is not yet discovered in nuclei. All partner bands found to date show some divergence in their properties. This raises several questions, such as: (i) is it at all possible to observe strongly broken chiral symmetry in nuclear systems?; (ii) assuming that weakly broken chiral symmetry is present, what features can be expected in the partner bands and how can the chiral symmetry be identified?; (iii) are the presently accepted fingerprints of chirality a reliable tool to identify real nuclear chiral systems?

In order to address these questions we have investigated nuclear systems with chiral geometry using the two-quasiparticle-plus-triaxial-rotor-model (TQTRM) [12, 13]. The particle-rotor model was applied in the original work on chiral symmetry [1], and the calculations showed almost perfect degeneracy of the $\pi h_{11/2} \otimes \nu h^{-1}_{11/2}$ chiral partner bands at high spins. Thus we used a similar model, which however has the options to include pairing, to include residual proton-neutron interaction, to describe the nucleon configurations within a configuration space of one or more orbitals originating in one or more spherical shells, etc. The additional more sophisticated features in the model offer the possibility of a more realistic description of the nuclear chiral system.

## 2 Nuclear Chiral System—Ideal and Realistic Descriptions

The TQTRM calculations were performed for chiral systems in the 100, 130 and 190 mass regions, built on two-quasiparticle configurations in odd-odd nuclei. Standard parameters were used for the Nilsson potential [14] and for the pairing. The quadrupole deformation was set to 0.15, which is suitable for the nuclei of interest. The moment of inertia was approximated with an irrotational-flow model formula, with a fixed energy of the $2^+$ state of the even-even core. The parameters $u_0$ and $u_1$ of the proton-neutron interaction [12, 13] were set to the previously optimized values of $u_0 = -7.2$ and $u_1 = -0.8$ MeV for the $\pi h_{11/2} \otimes \nu h^{-1}_{11/2}$ configuration in the A = 130 [15] and $u_0 = -4.95$ MeV and $u_1 = -0.55$ MeV for the $\pi h_{9/2} \otimes \nu i^{-1}_{13/2}$ configuration in the A = 190 [16, 17] mass regions. When calculating the $B(M1)$ reduced transition probabilities $g_R = Z/A$ were used, while the free spin $g_s$ factors were attenuated by a factor of 0.6. Results for the 130 mass region will be discussed here, since they are representative for all three mass regions.

## 2.1 The Degeneracy of the Partner Bands as a Fingerprint of Chiral Symmetry

It is known that best chiral systems in odd-odd nuclei are formed when the odd particle and the odd hole occupy the lowest-energy and the highest-energy orbitals of the corresponding high-j shell, and when the triaxiality of the nuclear shape is maximum, $\gamma \sim 30°$ [18]. Our calculations are consistent with this conclusion. Therefore, here calculations for which the Fermi surfaces for the odd proton and the odd neutron are situated at the bottom and top of the $h_{11/2}$ shell will be discussed.

Calculations done with what we call in the following "restricted configuration" will be examined first. These involve configuration spaces for the odd proton and the odd neutron restricted to only one orbital each, i.e. the lowest- and the highest-energy orbital originating from the $h_{11/2}$ shell, respectively. Monopole pairing was included, the single particle orbitals in the deformed mean field were described within several spherical shells and the proton-neutron interaction was considered too. This more sophisticated description seemed to yield almost perfect near degeneracy of the chiral partner bands. The difference in the excitation energy, $\Delta$E, of the two partner bands for such restricted configuration and for $\gamma = 30°$ and $\gamma = 20°$ are shown in Fig. 3. For $\gamma = 30°$ the energy difference is about 1–2 keV for $I > 15$, showing almost perfect degeneracy in the partner bands. In addition other properties of the partner bands, such as the projections of the total and the individual[1] angular momenta along the three major nuclear axes, the $B(M1)$ and $B(E2)$ reduced transition probabilities, etc., show similarly almost perfect degeneracy for this spin range [19]. Although the TQTRM has no prior assumption of chiral symmetry the simultaneous degeneracy in all calculated properties of the partner bands indicates that a symmetry is present.[2] The calculated average angles between the angular momenta of the odd proton, the odd neutron and the collective rotation are very close to 90° for the spin range where degeneracy is reached (see Fig. 4), indicating a chiral geometry. Furthermore

---

[1] the angular momenta of the odd proton, the odd neutron and the collective rotation

[2] Strictly speaking the TQPRM calculations only indicate that a symmetry is present, without identifying its nature. It is customarily assumed, that it is the chiral symmetry.

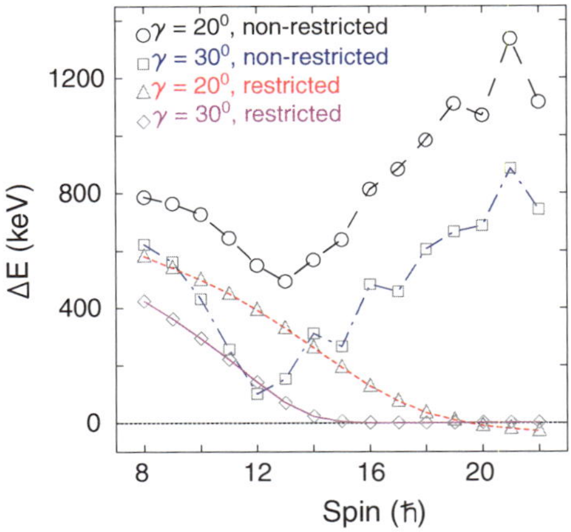


**Fig. 3** Calculated difference in the excitation energies, $\Delta E$, of the partner bands for the restricted and the non-restricted configurations, and for $\gamma = 20°$ and $\gamma = 30°$

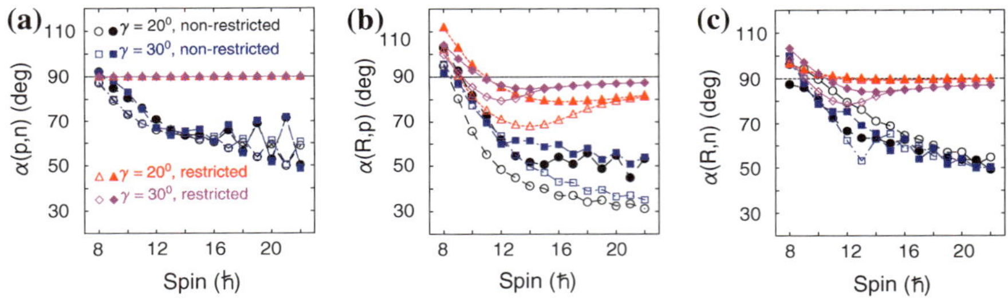


**Fig. 4** Expectation values of the angles between the angular momenta of the proton (**p**), neutron (**n**) and collective rotation (**R**) for the yrast (*open symbols*) and side (*filled symbols*) bands

the probabilities for vanishing projections of the total angular momentum along the three major nuclear axes, shown in Fig. 5, show negligible values (less than 2 %) in the spin range where the degeneracy is reached. Therefore the probability for non-chiral (i.e. planar) contributions to the wavefunctions is negligible in this spin range.

Therefore a nearly perfect degenerate pair of partner bands, associated with the presence of strongly broken symmetry, is found within the TQPRM calculations, when the nucleon configuration is restricted to only one orbital for each, the odd proton and the odd neutron. This result suggests that one can investigate the properties of perfect chiral structures within this description. One disadvantage however, is that such a restricted configuration is not a realistic description of the nucleus. In order to investigate how the properties of the partner bands change when a realistically large configuration space is used, the same calculations were performed, but this time the configuration was "non-restricted", i.e. the proton and the neutron configuration spaces contained the five orbitals closest to the corresponding Fermi level, each.

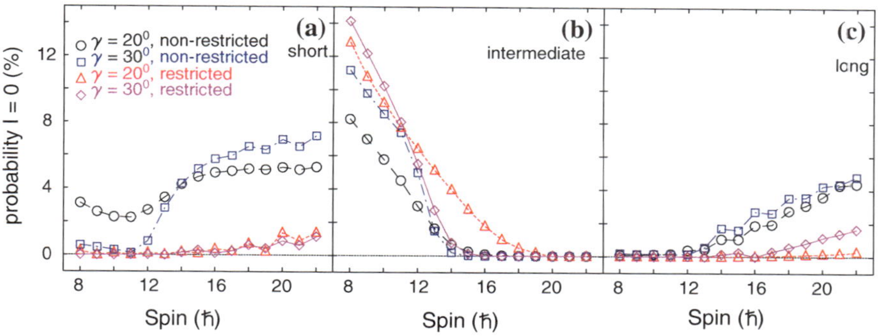


**Fig. 5** Calculated probability for vanishing projection of the total angular momentum along the (**a**) short, (**b**) intermediate, and (**c**) long axes, respectively, and for the yrast band

These orbitals were typically orbitals originating from the same high-j shell, in this case from the $h_{11/2}$ shell.

Results from these calculations are shown in Fig. 3. The involvement of a larger configuration space lifts the degeneracy in the excitation energies of the partner bands. The divergence in the excitation energies is much larger than when a restricted configuration and non-optimal, $\gamma = 20°$, value for the $\gamma$ deformation is used. Simultaneously the degeneracy is lifted from the other properties of the partner bands, such as the projections of the total and individual angular momenta, $B(M1)$ and $B(E2)$ reduced transition probabilities, etc. (more details can be found in [19]). The divergence is considerable, seeming to indicate that even the best realistic chiral systems cannot reach near-degeneracy.

This conclusion needs further investigation, in particular it is important to examine whether the chiral geometry of the three individual angular momenta persists for the non-restricted configuration. As seen in Fig. 4, the expectation values of the angles between the three individual angular momenta remain larger than 30° for the whole spin range, and furthermore the contribution to the wavefunction from components with vanishing projections of the total angular momentum remains less than 8 % for high spins, see Fig. 5. Thus it seems that the three-dimensional, chiral geometry of the three angular momenta is dominant, although small non-chiral components are present too. Therefore a realistic description of a system in chiral geometry seems to involve small non-chiral components, i.e. this is a weak symmetry breaking), which leads to a loss of the degeneracy in the chiral partner bands. These bands show certain similarities, which do not persist for a large spin range, and the discrepancies in the properties could be well pronounced (e.g., $\Delta E$ could differ by several hundreds of keV). Therefore, the use of this fingerprint to identify chiral symmetry in nuclei could become very difficult.

## 2.2 Staggering of $B(M1)$ Transition Probabilities and Energy Staggering as Fingerprints of Chiral Symmetry

It was suggested that a chiral system should show a characteristic staggering in the $B(M1)$ reduced transition probabilities [3], with high values for even spins for the intra-band transitions. In a few following works [20–22] $B(M1)$ staggering, not necessarily with the same phase, was found. It was also suggested that the presence of strong $B(M1)$ staggering can be used to identify strongly broken chiral symmetry, since the staggering is sensitive to the optimal value of the $\gamma$ deformation [20]. Our TQTRM calculations performed with restricted configurations, and yielding almost perfect degeneracy in the chiral partner bands, are consistent with such conclusion. Staggering in the $B(M1)$ values was observed in our calculations too. High values were found for even spins for the intra-band $B(M1)$ transition probabilities. The magnitude of staggering was maximal for the optimal value of the $\gamma$ deformation. These results, however, hold true for the restricted configuration calculations only. A more realistic description, including a non-restricted configuration, shows important differences. Firstly, it seems that the magnitude of the $B(M1)$ staggering is not particularly sensitive to the optimal $\gamma$ deformation, for instance the $B(M1)$ staggerings for $\gamma = 30°$ and $\gamma = 20°$ seem very similar for the intra-band transitions, see Fig. 6. This also shows that such staggering may be present also for weakly broken chiral symmetry. Secondly, the phase of the $B(M1)$ staggering for non-restricted configuration is found to be opposite to that for a restricted configuration. This is a very curious change. One could think that such change could be caused by an additional phenomenon. Another look at the expectation values of the angles between the individual angular momenta, see Fig. 4, may possibly explain this change in the phase of the staggering. One observes that the angles between the angular momenta

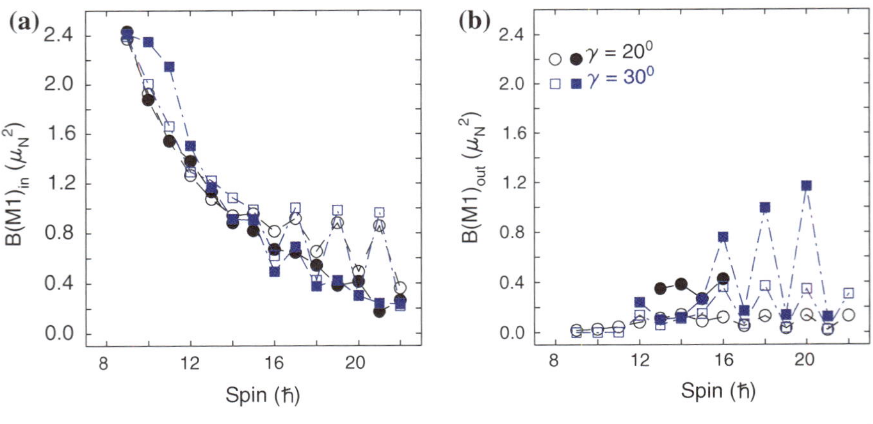


**Fig. 6** Calculated $B(M1)$ reduced transition probabilities for the (**a**) intra-band, and (**b**) inter-band transitions. *Open* and *filled symbols* in panel (**a**) denote transitions in the yrast and the side bands, respectively. In panel (**b**) *open symbols* denote side → yrast, while *filled symbols* yrast → side bands transitions, respectively

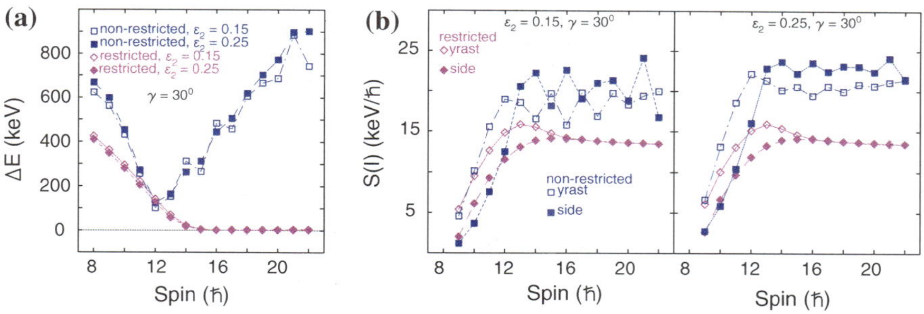


**Fig. 7** **a** Calculated difference in the excitation energies of the partner bands for quadrupole deformations of 0.15 and 0.25. **b** Energy staggering parameter, $S(I) = [E(I) - E(I-1)]/(2I)$, for the calculations done with quadrupole deformation of 0.15. **c** As (**b**) but for quadrupole deformation of 0.25

of the odd proton and the neutron show pronounced staggering for the non-restricted configuration, while a smooth behaviour is present for the restricted configuration. A staggering in these angles will inevitably lead to a staggering in the $B(M1)$ values. Thus, it seems that for a non-restricted configuration there is an addition source of staggering in the $B(M1)$ reduced transition probabilities. That also means that the observation of a staggering in the $B(M1)$ values may not necessarily be caused by a chiral symmetry, but also by an entirely different phenomenon. It seems that the phase of the $B(M1)$ staggering should be examined if one wants to understand the nature of the observed staggering.

A lack of energy staggering in the partner bands was suggested as another possible fingerprint of chiral symmetry [4]. This suggestion is based on the argument, that the energy staggering is caused by the Coriolis interaction, which should be vanishing for a chiral system for which the rotational angular momentum and the angular momenta of the odd proton and the odd neutron are orthogonal. Our calculations show that for a strongly broken chiral system, (obtained in our calculations with a restricted configuration), these three angular momenta are nearly orthogonal to each other, see Fig. 4. In this case no energy staggering is found in the partner bands. However, the realistic description of a chiral system yields much smaller expectation values for the angles between the individual angular momenta, indicating that a considerable Coriolis interaction able to cause a well pronounced energy staggering is present. Thus a real chiral system may in fact show energy staggering. Furthermore the lack of energy staggering is obviously not a prove of the orthogonality of the individual angular momenta. It could simply result from a larger moment of inertia, as shown in Fig. 7, where the same calculations but with a larger quadrupole deformation of 0.25 are presented. One can notice that the energy staggering for the larger nuclear deformation is vanishing, while the divergence in the excitation energies of the partner bands remains considerable.

Thus, it seems that none of the previously identified fingerprints of chiral symmetry is reliable and easily applied for real chiral systems. This leads to a few important conclusions. First, realistic chiral systems are associated with weakly broken chiral symmetry, for which the expected perfect degeneracy in the partner bands is lifted, and only some similarity in the properties of the partner bands remains. Thus, partner bands which show divergence in the excitation energy of several hundreds of keV, and in a similar way divergence in other properties including the $B(M1)$ and $B(E2)$ transition probabilities, may in fact possess chiral symmetry. Second, none of the other previously accepted fingerprints of chirality is reliable and easily applied for real chiral systems. The staggering in the $B(M1)$ transition probabilities may not be necessarily caused by chiral symmetry. Chiral bands may, or may not, show energy staggering. It seems alternative ways of testing for the presence of chiral symmetry in nuclei are needed.

Another method of identifying chiral geometry in nuclei could be to measure substantial triaxiality of the nuclear shape. Techniques for model-independent measurements of nuclear $\gamma$ deformation are already developed [23] and may possibly be used in the search of chiral symmetry in nuclei.

**Acknowledgments** This work is based upon research supported by the National Research Foundation, South Africa. We thank Ingemar Ragnarsson for making available the two-quasiparticle-plus-triaxial-rotor model codes and for numerous fruitful discussions.

## References

1. S. Frauendorf, J. Meng, Nucl. Phys. A **617**, 131 (1997)
2. J. Meng, S.Q. Zhang, J. Phys. G **37**, 064025 (2010)
3. T. Koike, K. Starosta, I. Hamamoto, Phys. Rev. Lett. **93**, 172502 (2004)
4. C. Vaman et al., Phys. Rev. Lett. **92**, 032501 (2004)
5. T. Koike, Nucl. Phys. A **834**, 36c (2010)
6. S.Y. Wang et al., Phys. Lett. B **703**, 40 (2011)
7. E.A. Lawrie et al., Phys. Rev. C **78**, 021305(R) (2008)
8. E.A. Lawrie et al., Eur. Phys. J. A **45** (2010)
9. P.L. Masiteng et al., Acta Phys. Pol. B **40**, 657 (2009)
10. T. Koike, K. Starosta, C.J. Chiara, D.B. Fossan, D.R. LaFosse, Phys. Rev. C **67**, 044319 (2003)
11. E. Grodner et al., Phys. Rev. Lett. **97**, 172501 (2006)
12. P.B. Semmes, I. Ragnarsson, in *Proceedings of International Conference on High-Spin Physics and Gamma-Soft nuclei*, Pittsburg, 1990, ed. by J.X. Saladin, R.A. Sorenson, C.M. Vincent (World Scientific, 1991), pp. 500
13. P.B. Semmes, I. Ragnarsson, in *Proceedings of "Future Directions in Nuclear Physics with 4π Gamma Detection Systems of the New Generation"*, ed. by J. Dudek, B. Haas, (AIP Conf. Proc. 259, 1992), pp. 566
14. T. Bengtsson, I. Ragnarsson, Nucl. Phys. A **436**, 14 (1985)
15. N. Tajima, Nucl. Phys. A **572**, 365 (1994)
16. R.A. Bark et al., Phys. Lett. B **406**, 193 (1997)
17. R.A. Bark et al., Phys. Lett. B **416**, 453(E) (1998)
18. S.Q. Zhang, B. Qi, S.Y. Wang, J. Meng, Phys. Rev. C **78**, 044307 (2007)
19. E.A. Lawrie, O. Shirinda, Phys. Lett. B **689**, 66 (2010)
20. B. Qi, S.Q. Zhang, S.Y. Wang, J.M. Yao, J. Meng, Phys. Rev. C **79**, 041302(R) (2009)
21. Ch. Droste et al., Eur. Phys. J. A **42**, 79 (2009)
22. S.Q. Zhang, B. Qi, S.Y. Wang, J. Meng, Phys. Rev. C **75**, 044307 (2007)
23. Julian Srebrny, Douglas Cline, Int. J. Mod. Phys. E **20**, 422 (2011)

# The Fascinating $\gamma$-Ray World of the Atomic Nucleus: The Evolution of Nuclear Structure in $^{158}$Er and the Future of $\gamma$-Ray Spectroscopy

**Xiaofeng Wang and Mark A. Riley**

**Abstract** The rare-earth nucleus $^{158}$Er exhibits a number of beautiful structural changes as it evolves with increasing excitation energy and angular momentum. After undergoing Coriolis induced alignments of high-$j$ neutron and proton pairs, a dramatic prolate collective to oblate non-collective transition takes place via the mechanism of band termination. At the highest spins, a spectacular return to collective rotation is observed in the form of triaxial strongly deformed bands. This latter suggestion is based on a comparison of transition quadrupole moments ($Q_t$) between experiment and theory, and long standing predictions that such heavy nuclei will possess nonaxial shapes on their path towards fission. These exciting discoveries in $^{158}$Er have benefited greatly from the progression of $\gamma$-ray detector developments through recent decades. The new $\gamma$-ray energy tracking technique and the next generation detector arrays utilizing this technique, e.g., GRETINA ($1\pi$) and GRETA ($4\pi$), are briefly discussed.

## 1 The Evolution of Nuclear Structure in $^{158}$Er

In 1937 Bohr and Kalckar [1] proposed that we could learn about the structure of nuclei by detecting their $\gamma$-ray emissions. Indeed, to this day $\gamma$-ray spectroscopic studies continue to revolutionize our understanding of the atomic nucleus revealing an extremely rich system that displays a wealth of static and dynamical facets. One of the most fundamental and fascinating topics of research in $\gamma$-ray spectroscopy is the response of atomic nuclei to increasing angular momentum and excitation energy [2], often referred to as high spin nuclear physics.

X. Wang · M. A. Riley (✉)
Department of Physics, Florida State University, Tallahassee, FL32306, USA
e-mail: xwang3@nucmar.physics.fsu.edu

M. A. Riley
e-mail: mriley@nucmar.physics.fsu.edu

W. Greiner (ed.), *Exciting Interdisciplinary Physics*,
FIAS Interdisciplinary Science Series, DOI: 10.1007/978-3-319-00047-3_13,

In the field of high spin nuclear physics, the rare-earth region has always been one of the most favored domains since nuclei here can accommodate the highest values of angular momentum. The $^{158}$Er ($N = 90$) nucleus has become a textbook example in terms of the evolution of nuclear structure with increasing excitation energy and angular momentum [3–5]. The discoveries in $^{158}$Er illustrated in Fig. 1 have benefited much from the progression of detector techniques. On the other hand, the excitement generated by these observations and discoveries in other nuclei have also pushed the advancement of $\gamma$-ray detector systems.

As displayed in Fig. 1, many fascinating phenomena have been observed in $^{158}$Er. For example, as the angular momentum increases, this nucleus exhibits Coriolis-induced alignments of both neutron and proton pairs along the yrast line (see Fig. 2). It was among the first in which backbending was discovered [6] ($I \sim 14$), but it was also the first nucleus where the second ($I \sim 28$) and third ($I \sim 38$) discontinuities along the yrast line were identified [7, 8]. At spins 40–50$\hbar$, the yrast line is crossed by a very different structure, where the $^{158}$Er nucleus undergoes a dramatic shape transition from a prolate collective rotation to non-collective oblate configurations [9–11]. This transition manifests itself as favored, fully aligned band termination. In $^{158}$Er, three terminating states, $46^+$, $48^-$, and $49^-$, have been observed [11]. Band termination occurs when the valence nucleons outside the $^{146}$Gd spherical core are fully aligned with the axis of collective rotation [12–14]. A schematical illustration of band termination is shown in Fig. 3. It represents a clear manifestation of mesoscopic physics, since the underlying finite-particle basis of the nuclear angular momentum generation is revealed [4, 5]. Experimentally there is a huge drop in intensity of $\gamma$ rays above the terminating state. For example, relative to the decay from the $46^+$ terminating state in $^{158}$Er, the feeding transitions above this state observed in the 1300–2000 keV energy range are lower in intensity by at least one order of magnitude, see Fig. 4a.

## 2 The Return of Collective Rotational Band Structures at Spins Beyond Band Termination in $^{158}$Er

It had been a goal for decades to establish the nature of the states in the rare earth nuclei well beyond the very favored band-termination states (in the spin range of 40–50$\hbar$). In 2007, a new frontier of discrete-line $\gamma$-ray spectroscopy in the spin 50–70$\hbar$ range (the so-called "ultrahigh-spin regime") was opened. Four rotational structures in $^{158}$Er and $^{157}$Er (two in each nucleus), displaying high dynamic moments of inertia and possessing very low intensities ($\sim 10^{-4}$ of the respective channel intensity), were identified and extended up to spin $\sim 65\hbar$ [15]. These structures bypass the well-known "band-terminating" states, marking a return to collectivity at spins beyond band termination. As shown in Figs. 1 and 4, the ultrahigh spin bands have properties significantly different from the low-lying collective prolate bands (normal deformed) or the non-collective oblate states. These bands were proposed to be

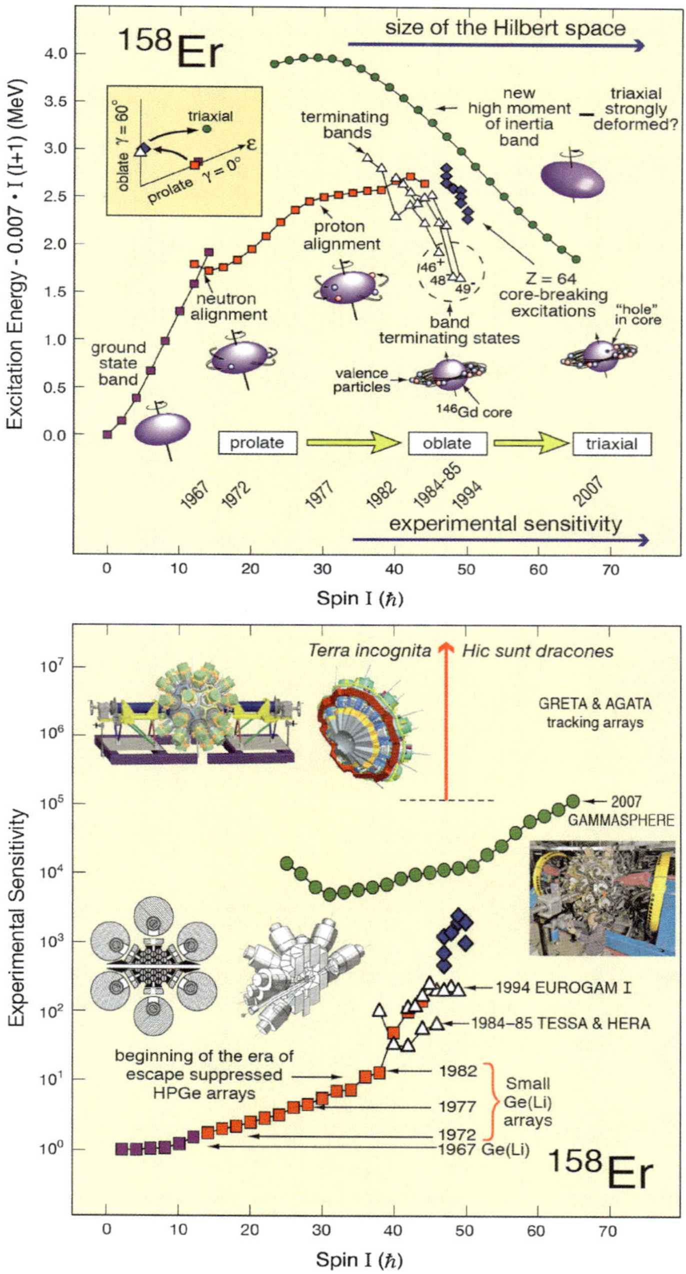


**Fig. 1** *Top* the evolution of nuclear structure in $^{158}$Er with excitation energy and angular momentum (spin). The inset illustrates the changing shape of $^{158}$Er with increasing spin within the standard $(\varepsilon, \gamma)$ deformation plane. *Bottom* the experimental sensitivity of detection is plotted as a function of spin showing the progression of $\gamma$-ray detector techniques with time that are associated with nuclear structure phenomena in $^{158}$Er

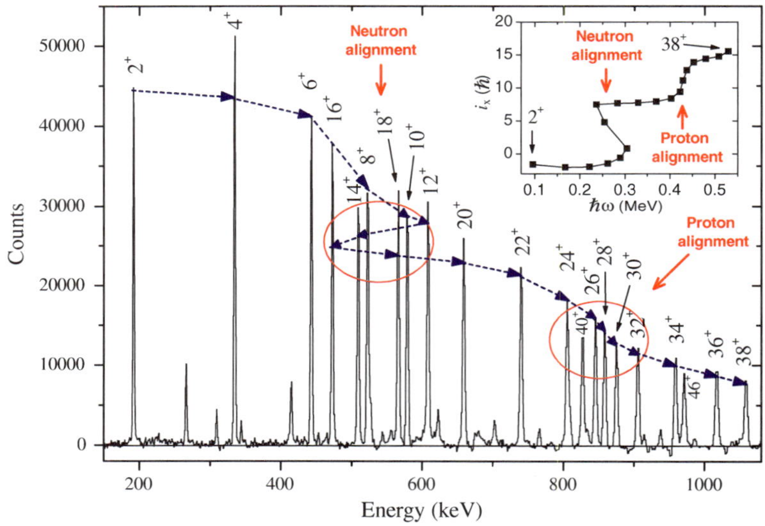


**Fig. 2** Coincidence $\gamma$-ray spectrum representative of the yrast band in $^{158}$Er. *Inset* experimental aligned spin (i.e., angular momentum in the intrinsic reference frame of a nucleus), $i_x$, as a function of the rotational frequency, $\hbar\omega$, for the same band. The observed neutron and proton alignments are marked in both the spectrum and the aligned spin plot. Note that the $i_x$ line bends back (towards lower rotational frequency) with increasing spin when the neutron alignment occurs, hence the name "backbending"

triaxial strongly deformed (TSD) structures [15], consistent with the predictions of the early cranking calculations of Bengtsson and Ragnarsson [12] and Dudek and Nazarewicz [13]. Thus, a new chapter in the story of $^{158}$Er began.

It is worth mentioning that a triaxial nuclear shape has distinct short, intermediate, and long principal axes, as shown in Fig. 5. This shape is commonly described using the parameters ($\varepsilon_2$, $\gamma$) of the Lund convention [16], where $\varepsilon_2$ and $\gamma$ represent the eccentricity from sphericity and triaxiality, respectively. At high spin, collective rotation about the short axis, corresponding to a positive $\gamma$ value ($0° < \gamma < 60°$), usually has the lowest excitation energy based on moment of inertia considerations [17, 18]. Thus, this mode is expected to be favored over rotation about the intermediate axis ($-60° < \gamma < 0°$). In $^{158}$Er, configurations with $\varepsilon_2 \sim 0.34$ and a positive value of $\gamma = 20°$–$25°$ were predicted to be low in energy theoretically and, were thus initially adopted to interpret the collective bands at ultrahigh spin [15]. However, it was imperative to determine their deformation experimentally in order to elucidate their character further and to test the above predictions.

A follow-up experiment to measure the transition quadrupole moments was carried out using the Doppler Shift Attenuation Method (DSAM) [19]. A 215 MeV $^{48}$Ca beam was delivered by the ATLAS facility at Argonne National Laboratory in the USA and bombarded a 1 mg/cm$^2$ $^{114}$Cd target backed by a 13 mg/cm$^2$ $^{197}$Au layer. A 0.07 mg/cm$^2$ $^{27}$Al layer between Cd and Au was used to prevent the migration

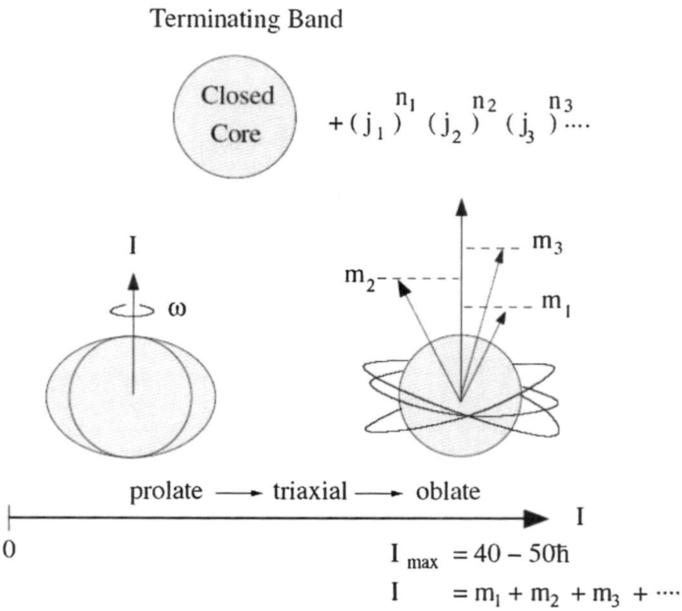


**Fig. 3** A schematical illustration of the band termination phenomenon. See text for details. Adopted from Ref. [14]

of the target material into the backing. The emitted $\gamma$ rays were detected by the Gammasphere spectrometer [20]. More details of this DSAM experiment can be found in Refs. [21, 22]. Compared with the previous thin-target experiment [15], the enhanced statistics of the DSAM experiment allowed the observation of a new collective band at ultrahigh spin in $^{158}$Er (band 3), which was estimated to carry an intensity of only ~10 % of the strongest collective band at ultrahigh spin (band 1). In spite of their extremely low intensities, an analysis of fractional Doppler shifts $F(\tau)$ was conducted for the three bands in $^{158}$Er (see Ref. [21] for details). The transition quadrupole moments ($Q_t$) of bands 1 and 2 in $^{158}$Er have been published in Ref. [21], while the preliminary result for band 3 was reported in Ref. [22].

The transition quadrupole moments $Q_t$ of the three collective bands at ultrahigh spin in $^{158}$Er have been experimentally determined to be ~9–11 eb, as shown in Fig. 6. This result demonstrates that they are all associated with strongly deformed shapes, since the low spin yrast band in $^{158}$Er has a measured $Q_t$ of ~6 eb [23]. However, the measured $Q_t$ values appear too large for the energetically favored positive-$\gamma$ (rotation about the short axis) triaxial shape (TSD1: $\varepsilon_2 \sim 0.34$). Rather, they are more compatible with a negative-$\gamma$ (rotation about the intermediate axis) triaxial deformed minimum (TSD2: $\varepsilon_2 \sim 0.34$) or a positive-$\gamma$ minimum with larger deformation (TSD3: $\varepsilon_2 \sim 0.43$) within the current cranked Nilsson-Strutinsky (CNS) theoretical framework [24].

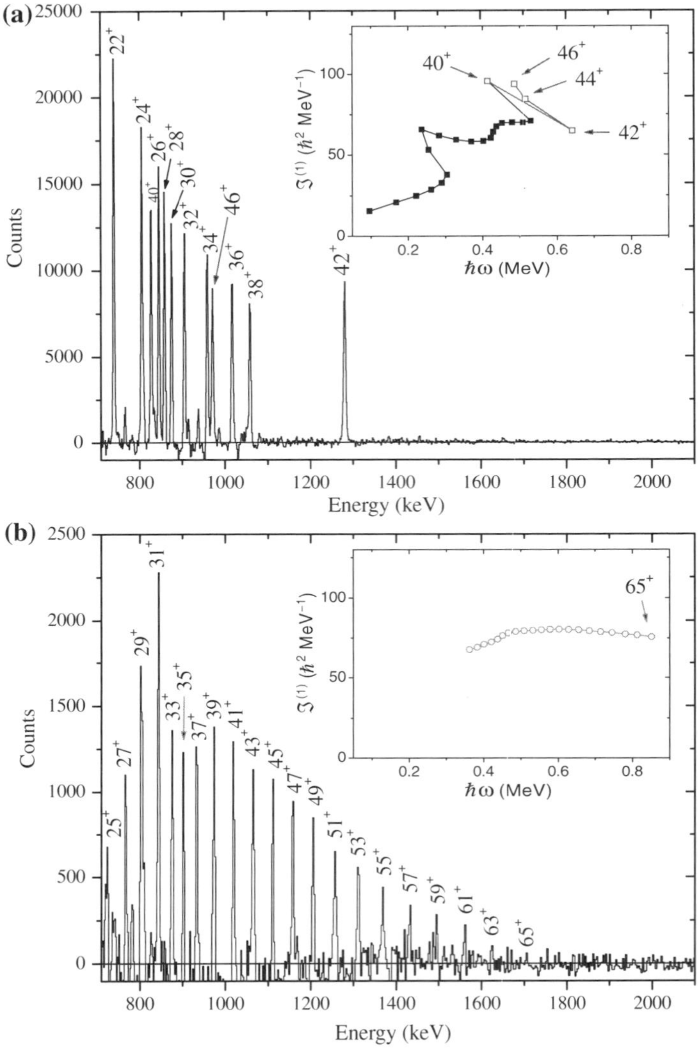


**Fig. 4** **a** Sample spectrum highlighting the high spin $\gamma$-ray transitions of the yrast band in $^{158}$Er from the thin target data [15] (in coincidence with the $44^+ \rightarrow 42^+$ transition and any one of the transitions between the $38^+$ and the $22^+$ states in the yrast band). The huge drop in intensity of $\gamma$ rays above the $46^+$ terminating state (marked in *blue*) is evident, see text for details. **b** Coincidence spectrum representative of the strongest collective band at ultrahigh spin (band 1) observed in $^{158}$Er from the same data as **a**. The spins assigned are tentative and the parity is not known for band 1. In both **a** and **b**, transitions are marked with the states (spin$^{\text{parity}}$) which they decay from. *Insets* kinematic moments of inertia, $\Im^{(1)}$, as a function of rotational frequency, $\hbar\omega$, for the yrast sequence and band 1 (see also Sect. 2) in $^{158}$Er. The $40^+$, $42^+$, $44^+$, and $46^+$ states are based on a configuration that terminates at the $46^+$ non-collective oblate state, while band 1 (collective) extends up to a spin of $65\hbar$. Note that the collective band beyond band termination has a behavior of $\Im^{(1)}$ significantly different from the structures shown in **a** (Color figure online)

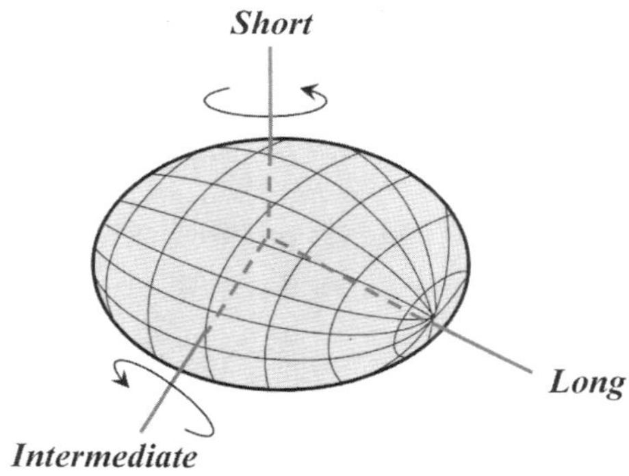


**Fig. 5** A schematical illustration of a triaxial nuclear shape. The distinct short, intermediate, and long principal axes are marked

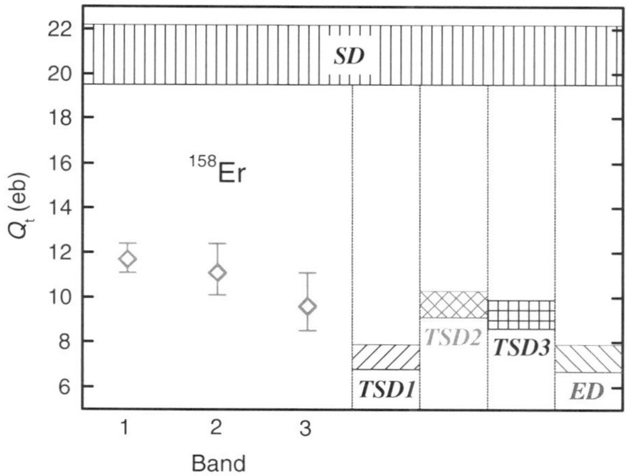


**Fig. 6** Measured transition quadrupole moments $Q_t$ (plotted with *diamonds*) of the three bands at ultrahigh spin in $^{158}$Er, compared with the theoretical $Q_t$ values (*horizontal shaded areas*) associated with the minima of interest calculated in the CNS model [21] (*ED* enhanced deformed; *SD* superdeformed). See text for the definitions of the three TSD minima

This puzzling discrepancy in our Er work has recently motivated another theoretical study in which calculations using the 2-dimensional tilted axis cranking (TAC) method [25] based on a self-consistent Skyrme-Hartree-Fock (SHF) model were conducted for configurations associated with triaxial shapes at ultrahigh spin in $^{158}$Er [26]. In this TAC/SHF work, it is shown that the negative-$\gamma$ minimum (rotation about the intermediate axis) becomes only a saddle point when titled cranking is considered and, thus, the previously mentioned TSD2 minimum may be removed from consideration. The calculated $Q_t$ value for a candidate positive-$\gamma$ triaxial minimum with large deformation ($\sim$10.5 eb) also agrees well with the experimental values, nevertheless, this minimum does not become yrast until spin $\sim 70\hbar$. Thus, the question "where are the band structures associated with the most energetically favored TSD minimum?" still remains open.

## 3 An Extended Study of Collective Bands at Spins Beyond Band Termination in the Rare Earth Region

Several questions naturally arise from the above striking observations of collective band at spins beyond band termination in $^{158}$Er. For example, is observation of such structures a general feature of the light rare-earth nuclei? How do properties of these minima with exotic shapes change with $Z$ and $N$? In fact, the recent discoveries in $^{158}$Er have triggered a comprehensive project to explore this phenomenon in the light rare earth nuclei with a mass of 150–165. The primary aim of this ongoing project is to bridge our physics understanding of the evolution of nuclear structure between the well known SD (superdeformed) domain of the Gd, Tb, and Dy nuclei ($^{152}$Dy [27], for example) and the lower spin TSD/Wobbling regime of the Lu nuclei ($^{163}$Lu [28], for example) and the neighbors ($^{167}$Ta [29], for example) in the nuclear landscape, as illustrated in Fig. 7.

Our research plan for this project is currently conducted along two directions: in the Er isotopes and in the $N = 90$ isotones. So far, collective bands with similar characteristics to the $^{158}$Er case have also been found in the Er isotopes, e.g., $^{154}$Er [30, 31] and $^{160}$Er [32], as well as in the $N = 90$ isotones, e.g., $^{160}$Yb [33] and $^{157}$Ho [22]. A new DSAM experiment of $^{160}$Yb has been approved and is expected to be performed soon at the ATLAS facility at Argonne National Laboratory.

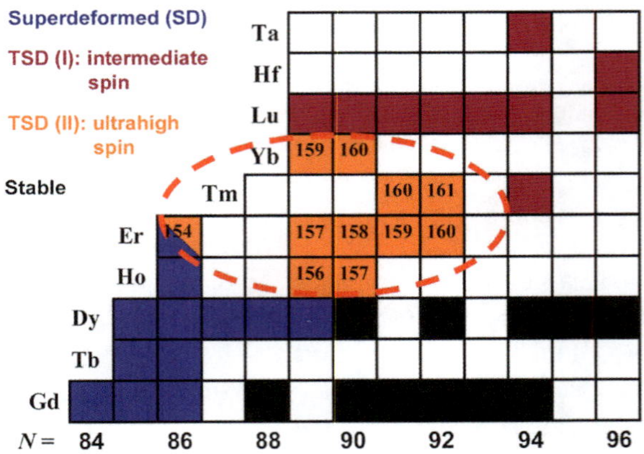


**Fig. 7** Section of the nuclear landscape highlighting nuclei in the SD domain of Gd, Tb, and Dy and the TSD/Wobbling (*TSD* triaxial strongly deformed) regime of Lu and its neighbors as well as those covered by the present project. Nuclei in the thickened, *dashed circle* are covered by the present project. See text for details

## 4 The $\gamma$-Ray Tracking Technique and the New Generation of Detector Arrays

Every major advance in $\gamma$-ray detector techniques and systems has resulted in significant progress in studying the nuclear structure of rapidly rotating nuclei, as illustrated, for example, in $^{158}$Er (Fig. 1). The current state-of-the-art $4\pi$ $\gamma$-ray detector arrays, for example, Gammasphere at Argonne National Laboratory (ANL) in the USA [20] which consists of more than 100 large volume HPGe detectors each surrounded by a Compton-suppression shield, have pushed this particular detector technology to its limit, with an efficiency for a 1 MeV $\gamma$ ray of about 10 % and a peak-to-total ratio of 55 % [20, 34].

Current generation radioactive beam facilities in the USA, Europe, and Japan are beginning to offer tantalizing glimpses of new physics in the terra-incognita of rare isotopes with extreme proton to neutron ratios, while a new generation of high-intensity radioactive beam facilities, for example, FRIB at Michigan State University (MSU), are being constructed or planned worldwide. To fully exploit the scientific discovery potential of these facilities a new generation of high-efficiency detectors with improved position resolution is required. The next major step in $\gamma$-ray spectroscopy is to abandon the concept of a physical suppression shield, which greatly reduces the overall possible efficiency, and to move towards the goal of a $4\pi$ Ge ball utilizing the technique of $\gamma$-ray energy tracking in electrically segmented Ge crystals [35, 36].

With the new tracking technique, the position and energy of $\gamma$-ray interaction points are identified in the detector segments. Since most $\gamma$ rays interact more than once within the crystal, the energy-angle relationship of the Compton scattering formula is used to track the path of a given $\gamma$ ray. The full $\gamma$-ray energy is obtained by summing only the interactions belonging to that particular $\gamma$ ray. In this way, there are no vetoed Compton scatters in the suppression shields and scattered $\gamma$ rays

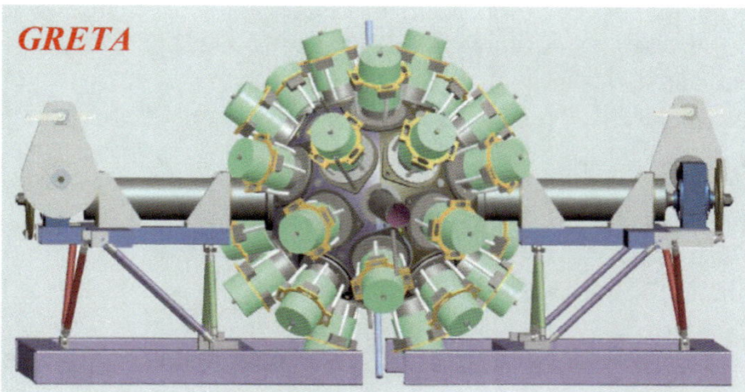


**Fig. 8** The planned $4\pi$ GRETA (Gamma Ray Energy Tracking Array) spectrometer. Adapted from Ref. [41]. Note that a very similar system, AGATA, is planned in Europe

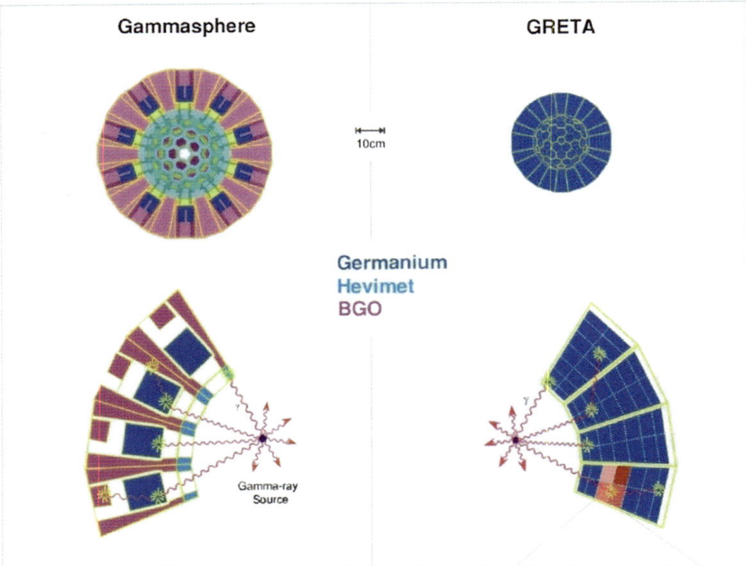


| | Gammasphere | GRETA |
|---|---|---|
| **Efficiency (1 MeV):** | **~ 10%** | **~ 60%** |
| **Efficiency (15 MeV):** | **~ 0.5%** | **~ 12%** |
| **Peak/Total (1 MeV):** | **~ 55%** | **~ 85%** |
| **Position resolution:** | **~ 20 mm** | **~ 2 mm** |

**Fig. 9** Comparison of GRETA with Gammasphere, demonstrating the advantages of a new generation $\gamma$-ray detector array utilizing the energy tracking technique. Adapted from Ref. [42]

between crystals are recovered. Thus, a $4\pi$ $\gamma$-ray energy tracking array, for example, GRETA [35, 36] in the USA (see Fig. 8), will have a high overall efficiency, ~60 % for a single 1 MeV $\gamma$ ray. Other key benefits of a tracking array include good peak-to-total ratio (~85 %), high counting rate (~50 kHz) capability per crystal, excellent position resolution (~2 mm), the ability to handle high multiplicities without summing, the ability to pick out low-multiplicity events hidden in a high background environment, and high sensitivity for linear polarization measurements. A back-to-back comparison of GRETA with Gammasphere on several key technical parameters is given in Fig. 9. During the last few years, $\gamma$-ray energy tracking technology has been shown to be feasible. GRETINA [37], a $1\pi$ detector system, has been constructed in the USA, while the AGATA [38–40] demonstrator has been constructed in Europe. We will focus on GRETINA and GRETA below.

In March 2011, the construction of GRETINA was completed at Lawrence Berkeley National Laboratory (LBNL). From April to July in 2011, a series of engineering runs, in-beam tests of GRETINA, were carried out to help with the debugging, characterization, and initial optimization of GRETINA. Some photos and results from the first engineering run, of which the primary aim was to test GRETINA under high multiplicity conditions using a reaction populating high spin

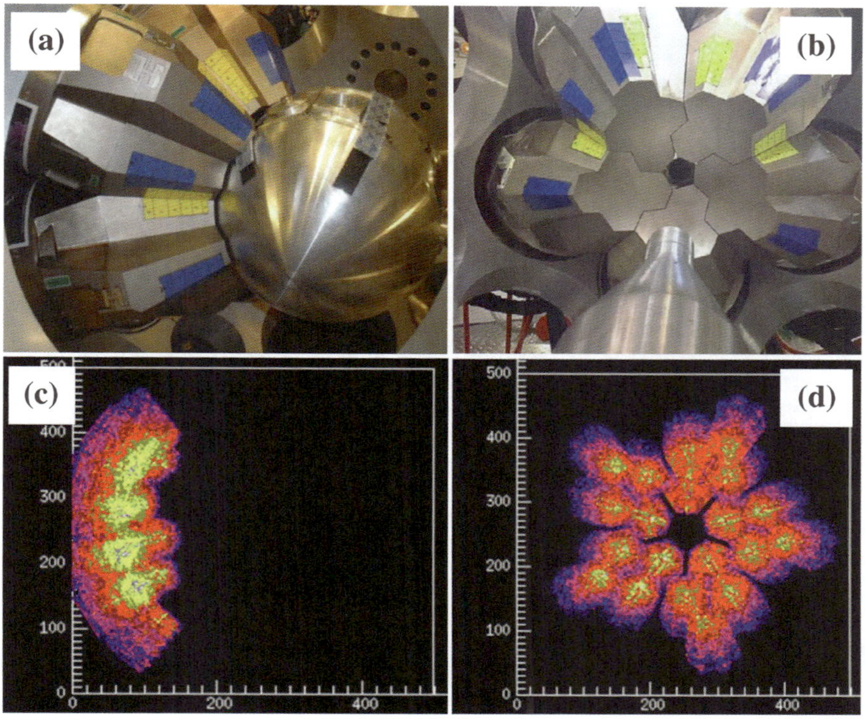


**Fig. 10** Photos and images from the first engineering run of GRETINA in April 2011 at LBNL where reactions leading to $^{158}$Er were used to test the new GRETINA system. **a** Side view photo of GRETINA; **b** Front view photo of GRETINA; **c** Interaction points analysis side view; and **d** Interaction points analysis front view. Adopted from Ref. [43]

states in $^{158}$Er, are displayed in Fig. 10. From the fall of 2011 to present, a series of commissioning runs with GRETINA coupled to the Berkeley Gas Separator (BGS) have been conducted. It is hoped that these runs will provide the opportunity both to obtain physics results on the spectroscopy of super heavy elements (SHE) and to test GRETINA under "battle conditions".

In the summer of 2012, the first scientific campaign of GRETINA will begin. It is first to be stationed at the NSCL facility at MSU for about one year and then will move to the ATLAS facility at ANL. The GRETINA detector array is thus about to enter its operations phase. However, the momentum in developing the $\gamma$-ray energy tracking technology to its ultimate potential must continue towards GRETA which is a full $4\pi$ spectrometer. As addressed in the 2007 NSAC Long Range Plan [41], GRETA will improve the power of GRETINA by a factor of 10–100 for most experiments. This $4\pi$ $\gamma$-ray energy tracking spectrometer (along with AGATA in Europe) will revolutionize $\gamma$-ray spectroscopy in the same way that Gammasphere and Euroball [44] did. These next generation $\gamma$-ray spectrometers in conjunction with a suite of specially designed auxiliary detector systems will be essential to fully exploit the compelling scientific opportunities of the new radioactive beam facilities and will bring nuclear physics into a new era.

**Acknowledgments** The authors want to thank all the collaborators involved in the recent experiments on $^{158}$Er (and neighboring nuclei) including: J. Simpson, E. S. Paul, R. V. F. Janssens, A. O. Evans, P. J. Nolan, A. Pipidis, I. Ragnarsson, P. J. Twin, A. Aguilar, A. D. Ayangeakaa, A. J. Boston, H. C. Boston, D. B. Campbell, M. P. Carpenter, C. J. Chiara, R. M. Clark, M. Cromaz, I. G. Darby, P. Fallon, U. Garg, D. J. Hartley, C. R. Hoffman, D. T. Joss, D. S. Judson, F. G. Kondev, T. Lauritsen, K. Lagergren, I. Y. Lee, N. M. Lumley, A. O. Macchiavelli, J. Matta, J. Ollier, M. Petri, D. C. Radford, J. M. Rees, J. P. Revill, L. L. Riedinger, S. V. Rigby, J. F. Sharpey-Schafer, F. S. Stephens, C. Teal, J. Thomson, C. Unsworth, D. Ward, and S. Zhu. In addition, the authors would like to thank everyone who has been involved in the GRETINA/GRETA and AGATA projects. Your brilliant efforts have created the next step in the evolution of $\gamma$-ray spectroscopy and this is greatly appreciated by the nuclear physics community. This work has been supported in part by the U.S. National Science Foundation under grants No. PHY-0756474 (FSU), PHY-0554762 (USNA), and PHY-0754674 (UND), the U.S. Department of Energy, Office of Nuclear Physics, under contracts No. DE-AC02-06CH11357 (ANL), DE-AC02-05CH11231 (LBNL), DE-AC05-00OR22725 (ORNL), DE-FG02-94ER40834 (UMD), and DE-FG02-96ER40983 (UTK), the United Kingdom Science and Technology Facilities Council, the Swedish Science Research Council, and by the State of Florida.

## References

1. N. Bohr, F. Kalckar, Kgl. Dan. Vid. Selsk. Math. Phys. Medd. **14**, 10 (1937)
2. A. Bohr, B.R. Mottelson, *Nuclear Structure*, vol. II (Benjamin, New York, 1975)
3. K.S. Krane, *Introductory Nuclear Physics* (Wiley, New York, 1988)
4. S.G. Nilsson, I. Ragnarsson, *Shapes and Shells in Nuclear Structure* (Cambridge University Press, Cambridge, 1995)
5. K. Heyde, *Basic Ideas and Concepts in Nuclear Physics* (Institute of Physics, Bristol, 1999)
6. H. Beuscher et al., Phys. Lett. B **40**, 449 (1972)
7. I.Y. Lee et al., Phys. Rev. Lett. **38**, 1454 (1977)
8. J. Burde et al., Phys. Rev. Lett. **48**, 530 (1982)
9. J. Simpson et al., Phys. Rev. Lett. **53**, 648 (1984)
10. P.O. Tjøm et al., Phys. Rev. Lett. **55**, 2405 (1985)
11. J. Simpson et al., Phys. Lett. B **327**, 187 (1994)
12. T. Bengtsson, I. Ragnarsson, Phys. Scr. **T5**, 165 (1983)
13. J. Dudek, W. Nazarewicz, Phys. Rev. C **31**, 298 (1985)
14. I. Ragnarsson et al., Phys. Scr. **34**, 651 (1986)
15. E.S. Paul et al., Phys. Rev. Lett. **98**, 012501 (2007)
16. G. Andersson et al., Nucl. Phys. A **268**, 205 (1976)
17. Z. Szymański, *Fast Nuclear Rotation* (Clarendon Press, Oxford, 1983)
18. B.G. Carlsson et al., Phys. Rev. C **78**, 034316 (2008)
19. S. Devons, G. Manning, D.S.P. Bunbury, Proc. Phys. Soc. A **68**, 18 (1955)
20. R.V.F. Janssens, F.S. Stephens, Nucl. Phys. News **6**, 9 (1996)
21. X. Wang et al., Phys. Lett. B **702**, 127 (2011)
22. X. Wang et al., J. Phys. Conf. Ser. **381**, 012065 (2012)
23. S.L. Shepherd et al., Phys. Rev. C **65**, 034320 (2002)
24. B.G. Carlsson, I. Ragnarsson, Phys. Rev. C **74**, 011302(R) (2006)
25. S. Frauendorf, Rev. Mod. Phys. **73**, 463 (2001)
26. Y. Shi et al., Phys. Rev. Lett. **108**, 092501 (2012)

27. P.J. Twin et al., Phys. Rev. Lett. **57**, 811 (1986)
28. S.W. Ødegård et al., Phys. Rev. Lett. **86**, 5866 (2001)
29. D.J. Hartley et al., Phys. Rev. C **80**, 041304(R) (2009)
30. K. Lagergren et al., Phys. Rev. Lett. **87**, 022502 (2001)
31. J.P. Revill et al., J. Phys. Conf. Ser. **381**, 012066 (2012)
32. J. Ollier et al., Phys. Rev. C **80**, 064322 (2009)
33. A. Aguilar et al., Phys. Rev. C **77**, 021302(R) (2008)
34. I.Y. Lee, Nucl. Phys. A **520**, c641 (1990)
35. M.A. Deleplanque et al., Nucl. Instr. Meth. A **430**, 292 (1999)
36. I.Y. Lee, M.A. Deleplanque, K. Vetter, Rep. Prog. Phys. **66**, 1095 (2003)
37. I.Y. Lee et al., Nucl. Phys. A **746**, 255c (2004)
38. J. Simpson, R. Kruecken, Nucl. Phys. News **13**, 15 (2003)
39. J. Simpson, Acta Phys. Polon. B **36**, 1383 (2005)
40. S. Akkoyun et al., Nucl. Instr. Meth. A **668**, 26 (2012)
41. *2007 NSAC Long Range Plan*, http://science.energy.gov/~/media/np/nsac/pdf/docs/nuclear_science_low_res.pdf
42. *GRETA White Paper*, http://grfs1.lbl.gov/DocDB/0001/000139/001/GRETA_WP_Jan06_4.pdf
43. *GRETINA Newsletter*, vol. 9, http://www.physics.fsu.edu/Gretina/GRETINA_Newsletter_9.pdf
44. J. Simpson, Z. Phys. A **358**, 139 (1997)

# High Energy-Resolution Experiments with the K600 Magnetic Spectrometer at Intermediate Energies

**Iyabo Usman**

**Abstract** The K600 magnetic spectrometer of iThemba LABS is a unique facility world-wide for the study of nuclear structure and reactions mechanisms at intermediate energies using light-ion projectiles. In particular, high energy-resolution experiments on nuclei in the region of giant resonances present a powerful tool to extract information about the dominant processes leading to equilibration. Proton inelastic scattering on nuclei spanning the periodic table from light to heavy-mass ($^{12}$C to $^{208}$Pb) has established fine structure as a global phenomenon for the Isoscalar Giant Quadrupole Resonance (ISGQR). In addition, it has been possible to extract level densities in the nuclear continuum very reliably for use in astrophysical calculations. Experimental results and corresponding theoretical predictions elucidating the origin of fine structure will be presented.

## 1 K600 Magnetic Spectrometer

The iThemba LABS K600 QDD magnetic spectrometer consists of five active elements, namely, a quadrupole magnet, two dipole magnets and two trim coils (K and H) as shown in Fig. 1. Vertical focusing at the focal-plane is achieved with the quadrupole magnet at the entrance of the spectrometer. The two trim coils, which are shaped pole-face current windings located inside the dipoles, are used to achieve kinematic correction and final focusing at the focal-plane. The collimator carousel, which defines the solid angle acceptance of the spectrometer, is situated in front of the quadrupole magnet at a distance of 735.5 mm from the target centre and can hold six collimators. Situated at the turning axis of the spectrometer is the 524 mm diameter scattering chamber which houses the target ladder at the centre. The target

I. Usman (✉)
iThemba Laboratory for Accelerator Based Sciences,
Somerset West, Cape Town 7129 , South Africa
e-mail: zzv@tlabs.ac.za

W. Greiner (ed.), *Exciting Interdisciplinary Physics*,
FIAS Interdisciplinary Science Series, DOI: 10.1007/978-3-319-00047-3_14,

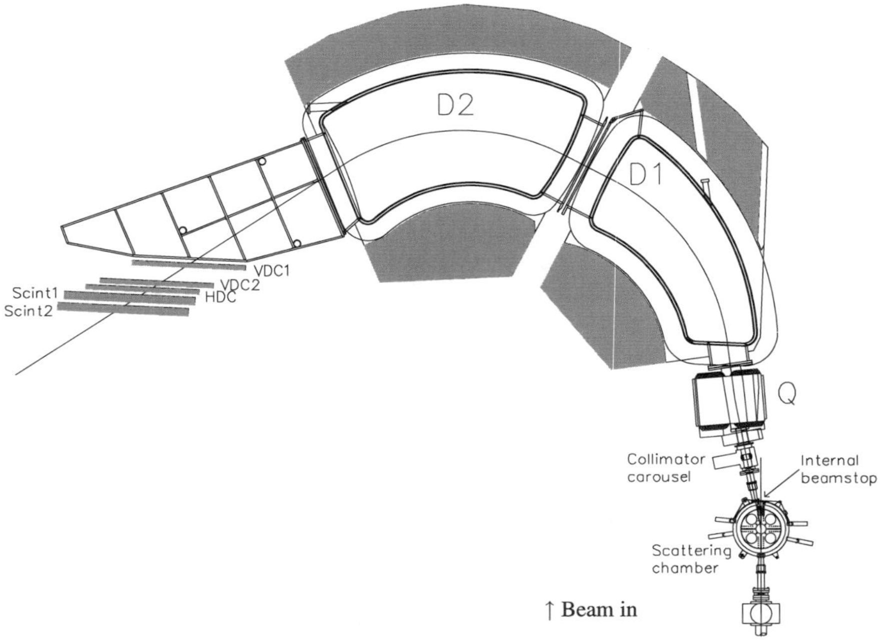


**Fig. 1** The layout of the K600 Magnetic Spectrometer showing its main components and the focal plane detector package. The K600 is positioned at 11° scattering angle

ladder consists of six target positions around which is a turntable that can carry a small Faraday cup or detectors. Behind the second dipole magnet is the focal plane position-sensitive detector package. It can be positioned in one of the three dispersion modes: low, medium, and high dispersion with dispersion values 6.2 cm/%, 8.1 cm/% and 9.8 cm/%, respectively [1], by varying the ratio of the fields of the two dipole magnets D1 and D2. For the isoscalar giant quadrupole resonance experiments, the spectrometer settings were optimized for the medium dispersion mode for protons of incident energy 200 MeV, covering an excitation energy range of approximately $E_x$ = 6–30 MeV. The required excitation energy range for each nucleus studied could be recorded within a single measurement, while at the same time providing sufficient energy resolution across the focal plane. It is important to note that the high energy-resolution capability of this facility is a pre-requisite for the analysis of the inelastically scattered protons of these investigations. The solid angle can be defined by a selection of six collimators housed in a collimator carousel upstream from the spectrometer quadrupole magnet. The character of the background for (p, p′) scattering was found to be sensitive to collimator size and geometry.

## 2 Fine Structure of Nuclear Giant Resonances

Nuclear giant resonances are simple, collective, particle-hole excitations of the nucleus. Such resonances are elementary modes of nuclear excitation which involve the coherent motion of the nucleons in the nucleus. These oscillations are classified by their angular momentum as monopole ($L = 0\hbar$), dipole ($L = 1\hbar$), quadrupole ($L = 2\hbar$) etc. resonances [2]. Each type is subdivided according to isospin ($\Delta T$) and spin ($\Delta S$) transfers. The isoscalar ($\Delta T = 0$) modes are vibrations in which neutrons and protons move in phase. Modes in which neutrons and protons move out of phase are called isovector ($\Delta T = 1$). Similar oscillations may take place in the spin space. Nucleons with spin-up and spin-down can move either as in phase ($\Delta S = 0$) modes or as out of phase ($\Delta S = 1$) modes. Typical probes for giant resonance studies include:

1. $\gamma$-rays and electrons for the excitation of the Isovector Giant Dipole Resonance (IVGDR),
2. $\alpha$-particles, protons and electrons for the excitation of the Isoscalar Giant Monopole Resonance (ISGMR) and the Isoscalar Giant Quadrupole Resonance (ISGQR), and
3. charge exchange reactions like (p,n), (n,p) or ($^3$He,t) for the Gamow-Teller (GT) resonance.

The initial observations [3, 4] have blossomed into a *bona fide* subfield of nuclear physics, one that is pursued in nearly every major intermediate-energy nuclear physics facility throughout the world. Light ions such as protons, tritons, deuterons and alpha particles have been preferred for studying the nuclear structure and reaction mechanisms because they are relatively easy to accelerate. Each one of these light ions has some unique property that makes it useful for a certain type of study. For example, since both the spin and isospin of an alpha particle are zero, alpha-particle scattering is ideal for exciting a nucleus without changing the isospin. Proton inelastic scattering can be used to excite both isoscalar and isovector giant resonances due to its spin and isospin value of one-half.

Some of the gross features of giant resonances such as centroid energies and collectivity, measured in terms of sum rules, have been studied and are well understood in microscopic models [5, 6]. Yet, the widths have not been fully understood due to the limitations in the experimental methods. Also, in recent years, high energy-resolution proton inelastic-scattering experiments revealed that giant resonances additionally carry fine structure. An early example of fine structure was seen in inelastic electron scattering experiments [7] investigating the giant quadrupole resonance in $^{208}$Pb, carried out using the DALINAC electron accelerator of the TU Darmstadt, Germany. In addition, the fine structure of Gamow-Teller (GT) resonances in medium-heavy nuclei ($40 \le A \le 90$) has been investigated using the ($^3$He,t) reaction at the Grand-Raiden magnetic spectrometer of RCNP Osaka, Japan [8]. Even so, the fine structure of giant resonances, which carries unique information on the underlying physical nature and the dominant decay mechanisms of the resonances, is still a relatively unexplored topic especially in light nuclei ($A \leq 40$). Thus, the problem of determining the dominant damping mechanisms of giant multipole resonances in light nuclei needs thorough investigation in order to reveal their characteristic features. Therefore, an understanding of the decay modes, in particular the systematic behaviour of the escape width relative to the spreading width, is an essential ingredient for the complete description of the nuclear giant quadrupole resonance across the periodic table. High energy-resolution results obtained at intermediate proton beam energies, using the k600 magnetic spectrometer in combination with dispersion matching techniques, provides the most suitable approach to study fine structure of the ISGQR in nuclei. The necessary experimental techniques were discussed in [9]. With these techniques the possibility exists to perform (p, p′) experiments at forward angles with an energy resolution much better than the energy spread of the incident scattering beam.

## 3 Experiments and Data Analysis

The fine structure of the ISGQR in medium-mass and heavy nuclei has been investigated with high-resolution (p, p′) scattering at iThemba LABS [10, 11] with the aim of extracting information about their dominant decay processes. Recently, these studies have been extended to the low-mass region $12 \leq A \leq 40$ [12, 13]. The present work focuses on the case of $^{40}$Ca where considerable fragmentation of the $E2$ strength has been observed in electron [14, 15], proton [16] and $\alpha$-scattering [17] studies, albeit measured with varying energy resolution.

The experiments were carried out with a 200 MeV proton beam produced by the Separated Sector Cyclotron of iThemba LABS. The proton beams were inelastically scattered off a natural Ca target with an areal density of 3.0 mg/cm$^2$ and momentum-analyzed with the K600 magnetic spectrometer. Dispersion matching techniques [9] were used in order to exploit the high energy-resolution capability of the spectrometer. Energy resolutions $\Delta E = 35$–$40$ keV Full Width at Half Maximum (FWHM) were achieved. The scattering angles $\theta_{\rm Lab} = 7°$, $11°$ and $15°$ were selected to lie below, at and above the maximum of the cross sections for $\Delta L = 2\hbar$ transitions populating the ISGQR. The acceptance of the spectrometer allowed to take data for excitation energies between 6 and 30 MeV with a single field setting. Details of the data analysis are described elsewhere [12].

The excitation energy spectrum of the $^{40}$Ca(p, p′) reaction in Fig. 2 taken at $\theta_{\rm Lab} = 11°$, which corresponds to the maximum cross-section for the ISGQR, reveals a broad resonance at a mean energy of $E_{\rm x} \approx 18$ MeV with strength distributed between approximately 12 and 22 MeV. Intermediate structure is visible through peaks around 12, 14, 16, 17 and 18 MeV consistent with previous experimental work [14–17].

Pronounced fine structure is observed up to about 20 MeV excitation. At the larger scattering angle of $\theta_{\rm Lab} = 15°$, the overall structure is reduced in magnitude relative to the background but distinctive features of the fine structure persist demonstrating that the observed fluctuations are still mainly caused by $\Delta L = 2\hbar$ transitions and

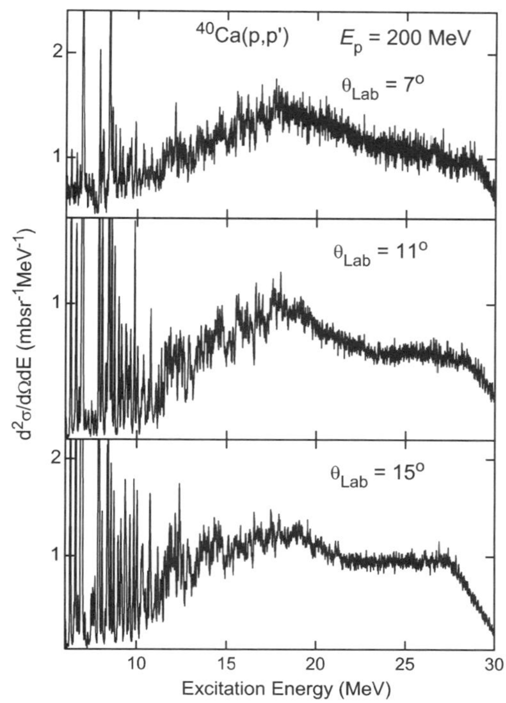


**Fig. 2** Excitation energy spectra for $^{40}$Ca at scattering angles $\theta_{\mathrm{Lab}} = 7°, 11°$ and 15°. Note that the ISGQR is expected to be most strongly excited at $\theta_{\mathrm{Lab}} = 11°$ in the region $E_x = 10$–20 MeV

possibly instrumental background. At the smaller scattering angle $\theta_{\mathrm{Lab}} = 7°$, the fine structure changes considerably indicating the presence of other multipoles. The similarity of structures seen in the 11° and 15° spectra and the difference to the 7° spectrum has also been demonstrated by a cross-correlation analysis [13].

## 3.1 Extraction of Characteristic Energy Scales

A variety of methods have been put forward to extract information on characteristic energy scales of the fine structure including a doorway-state analysis [18], the local scaling dimension [19, 20], the entropy index method [21, 22] and a wavelet analysis [10, 23]. A comparison for representative cases indicates that wavelet analysis is a particularly promising tool [24], since it provides simultaneously a quantitative measure of the fine structure and information on the localization in the excitation spectrum. Characteristic scales can be extracted from the power spectra of wavelet transforms, which allow a direct comparison between experiment and theory. A systematic study of the ISGQR [11] shows that the observed scales in medium-mass to heavy nuclei originates from a collective damping mechanism induced by the coupling of elementary one-particle, one-hole (1p-1h) to low-lying surface vibrations [25].

### 3.1.1 Wavelet Analysis

There are basically two classes of wavelet transform: Continuous Wavelet Transforms (CWT) and Discrete Wavelet Transforms (DWT). The Continuous Wavelet Transform (CWT) is defined as the sum over all time of the signal multiplied by a scaled, shifted version of the wavelet function $\Psi$, which can be expressed as

$$C(a,b) = \int S(t) \frac{1}{\sqrt{a}} \Psi\left(\frac{t-b}{a}\right) dt, \tag{1}$$

where $a$ is the scale and $b$ is the position and are equivalent to $\delta E$ and $E_{\mathrm{x}}$, respectively, in the analysis of energy scale. By folding the original energy spectrum $\sigma(E)$ with a chosen wavelet function $\Psi$, the coefficients are obtained as

$$C(E_{\mathrm{x}}, \delta E) = \frac{1}{\sqrt{\delta E}} \int \sigma(E)\, \Psi\left(\frac{E_{\mathrm{x}} - E}{\delta E}\right) dE. \tag{2}$$

The parameters excitation energy $E_{\mathrm{x}}$ and bin size $\delta E$ can be varied continuously or in discrete steps $j$, where $\delta E = 2j$, $j = 1, 2, 3, \ldots$, and $E_{\mathrm{x}} = \delta E$. Complex wavelets produce a complex CWT analysis, allowing the phase of the result to be examined. This is given by

$$\Psi(x) = \frac{1}{\sqrt{\pi f_{\mathrm{b}}}} \exp(2\pi i f_{\mathrm{c}}) \exp\left(-\frac{x^2}{f_{\mathrm{b}}}\right), \tag{3}$$

where $f_{\mathrm{c}}$ is the wavelet centre frequency and $f_{\mathrm{b}}$ controls the wavelet band width.

However, we have made an effort to develop a physically more realistic Lorentzian mother wavelet. This development was motivated by the underlying physics which suggest that the structures of a giant resonance should have a Lorentzian shape [5]. The complex Lorentzian mother-wavelet would ideally be the sum of an infinite series of Lorentzian functions of width $\Gamma$ and spacing $f_{\mathrm{c}}$ localized by a Gaussian envelope. However, in practice it was found sufficient to sum on each side of the centre Lorentzian function only sixteen other Lorentzian functions with the resulting real part of the new mother wavelet being given by

$$\Psi(x) = \sum_{n=-16}^{+16} \frac{\left(\frac{\Gamma}{2}\right)^2}{(x - (x_{\mathrm{o}} + n f_{\mathrm{c}}))^2 + \left(\frac{\Gamma}{2}\right)^2} \exp\frac{1}{2}\left(-\frac{x^2}{f_{\mathrm{b}}}\right). \tag{4}$$

Here, the parameters $x_{\mathrm{o}}$ and $\Gamma$ are the centre position and FWHM of the Lorentzian function, respectively. The parameter $f_{\mathrm{c}}$ represents the spacing or wavelength of he wavelet. By setting $f_c = 2\Gamma$, there is an obvious similarity with the cosine function which was successfully used in analysis using the complex Morlet mother wavelet [10, 13]. However, an advantage of the new Lorentzian mother wavelet is that the localization of wavelet transform of the data can be varied by changing the width of the Gaussian envelope.

Application of this newly developed Lorentzian mother wavelet to $^{40}$Ca(p, p′) data in the energy region of the ISGQR is shown in Fig. 3. By plotting the real part of the complex coefficients in a two-dimensional distribution of energy scales *versus* excitation energy, the positions of the structures within the original energy spectrum can be identified. Maxima of the wavelet coefficients at certain scale values over the energy region of ISGQR (or parts of it) indicate characteristic scales. This is illustrated in Fig. 3b, where the maximum wavelet-energy scale was restricted to 3.0 MeV in order to reveal scales related to intermediate structure. Indication of fragmented scales can be observed between 12 and 14 MeV with a scale around 500 keV which changes in magnitude at higher excitation energies. Also, a scale at around 1 MeV can be observed between 14 and 18 MeV excitation energy which is less visible at lower excitation energies. These discontinuities in a particular energy scale over the excitation energy region are referred to as fragmentation of scales. This phenomenon was not observed in medium and heavy-nuclei where the energy

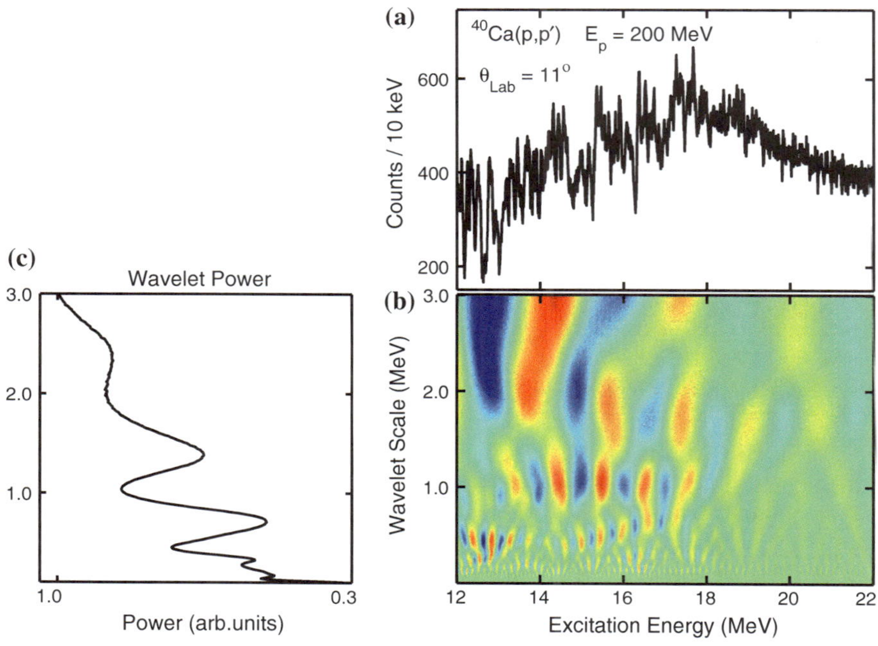


**Fig. 3** Excitation energy spectrum (**a**), wavelet coefficients (**b**) and power spectrum (**c**) for $^{40}$Ca(p, p′) at $\theta_{Lab} = 11°$ using the Lorentzian mother wavelet for scales up to 3.0 MeV

scales are compact and evenly distributed over the excitation energy region of the ISGQR [10, 11, 24].

In order to obtain a quantitative measure for the characteristic energy scales, the absolute values of the complex coefficients are projected onto the wavelet scale axis. The resulting power spectrum obtained is shown in Fig. 3c. Three peaks at scales values $<1$ MeV arising from the fine structure can be identified while two scales at approximately 1 and 2 MeV correspond to the intermediate structure.

## 4 Theoretical Calculations

The importance of the theoretical model of nuclear giant resonance excitation is to understand the origin and nature of the fine structure of the characteristic scales found in experiments by applying the most suitable microscopic models. A great deal of theoretical work has been directed towards the understanding of the origin of fine structure and damping mechanisms in light nuclei. Such models include Extended Theory of Finite Fermi Systems (ETFFS) [26], Quasi-particle Phonon Model (QPM) [27], Extended Time Dependent Hartree-Fock (ETDHF) [28], Random Phase Approximation (RPA) [29] and Second-RPA (SRPA) [30]. The main focus of this study is on the low-mass region $A \leq 40$ for which the SRPA treatment has been applied and is elaborated upon in Sect. 4.1.

### *4.1 Second Random Phase Approximation (SRPA)*

One of the methods of a microscopic calculation for the description of giant resonances or the nuclear response is the Random Phase Approximation (RPA) [29], in which giant resonances appear as a coherent superposition of one particle-one hole ($1p$-$1h$) excitations in closed-shell nuclei or two quasiparticle excitations in open-shell nuclei built on a correlated ground state. While it is well suited for the description of the mean energies and total transition probabilities, it fails to account for the energy distribution of the response function, for instance, the width of giant resonances. Thus, a theoretical prediction obtained in the framework of models that take coupling to complex degrees-of-freedom into consideration is essential.

An extension of the $1p$-$1h$ RPA by inclusion of $2p$-$2h$ excitations seems to be the most applicable candidate for incorporating damping effects which go beyond the mean-field description, and hence provides a more complete theory of small-amplitude collective motion. This was done within the framework of Second Random-Phase Approximation (SRPA) [30]. Such an approach is appropriate for the description of fine structure of giant resonances in light nuclei due to the fact that it improves the description of anharmonic effects and is also suited for incorporating higher-order effects in the ground state wave-function [31]. The different contributions to the damping of the single-particle and collective response in light nuclei can

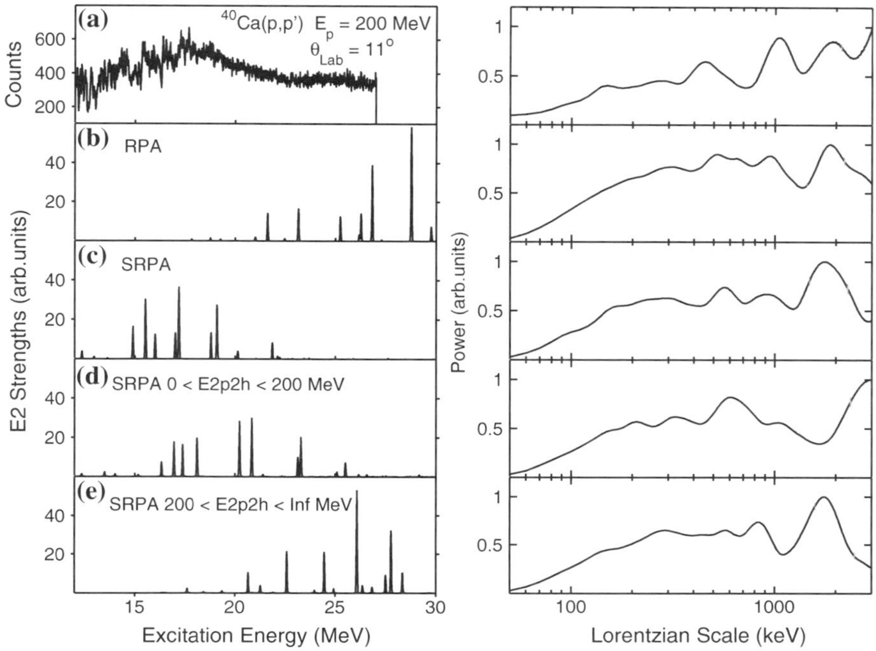


**Fig. 4** Experimental (**a**) and theoretical $E2$ strength functions for the RPA (**b**), SRPA (full calculation) (**c**), low-lying collective couplings in SRPA (**d**) and high-lying collective couplings in SRPA (**e**) in $^{40}$Ca. The *right hand side* indicates the corresponding power spectra

be disentangled. In addition, SRPA has been used successfully for the description of damping of giant resonances in [31, 32].

A large-scale SRPA calculation with a realistic interaction was performed for $^{40}$Ca using $V_{\mathrm{UCOM}}$, a realistic potential derived from the Argonne V18 interaction by renormalizing it within the Unitary Correlation Operator Method [33, 34]. More details on the SRPA approach can be found in Refs. [13, 35–38]. The strength functions and power spectra resulting from RPA and SRPA are shown in Fig. 4b–e. [] A striking result is that the ISGQR in $^{40}$Ca appears fragmented already at the RPA level. Thus, our RPA results suggest that Landau damping plays an important role in the case of $^{40}$Ca (but not in heavier nuclei), contrary to predictions using phenomenological effective interactions and density functionals.

## 5 Extraction of Level Densities

Another important aspect of proton inelastic scattering data with high energy-resolution is the use of it as a direct measurement of level densities even in the excitation energy region of giant resonances. Level densities are of fundamental interest not only as a test of the understanding of nuclear dynamics, but it also serve as a key ingredient of large reaction-network codes in modeling stellar energy production and nucleosynthesis. The phenomenon of fluctuating cross-sections in compound nucleus reactions in the region of overlapping resonances was first studied theoretically in the early 1960s [39] using the assumption of random phases between the scattering amplitudes. In the higher excitation-energy region, the levels still do not overlap but states are unresolved because of the limited energy resolution.

The level density of $2^+$ states in $^{40}$Ca extracted by means of a self-consistent procedure based on a fluctuation analysis [40] in the excitation energy interval between 10 and 20 MeV is discussed. As a starting point the measured excitation energy spectra for $^{40}$Ca(p, p′) at $\theta_{\mathrm{Lab}} = 11°$ and 15° are background subtracted using the DWT analysis with a BIOR6.8 wavelet. More detailed discussion can be found in [41]. Shown in Fig. 5 are the experimental level densities deduced from both spectra in comparison to model calculations. The theoretical results considered include the phenomenological BackShifted Fermi Gas (BSFG) and microscopic models. Two different sets of values were taken for the BSFG parameters $\Delta$, the ground-state energy correction accounting for pairing and shell effects, and the level density parameter $a$ describing the exponential increase with energy. Rauscher et al. [42] provide a fit to stable nuclei across the nuclear chart which is used for astrophysical network calculations of the $s$-process, including extra parameters for an improvement of the description in local mass areas. Von Egidy and Bucurescu [43] performed a global fit with parameters dependent on experimental masses only. The latter approach has been recently improved by a modification of the spin-cutoff parameter [44], which is in accordance with shell-model Monte Carlo calculations [45]. A current microscopic approach is based on a Hartree-Fock Bogoliubov (HFB) plus combinatorial model [46]. This has been improved in Ref. [47] to include beyond rotational also vibrational degrees-of-freedom. We also include a comparison with HF-BCS model results [48].

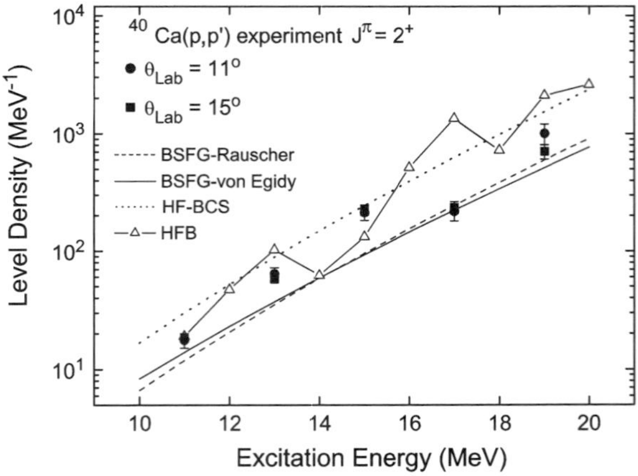


**Fig. 5** Level density of $2^+$ states in $^{40}$Ca extracted from the (p, p′) data (*filled symbols*) compared to model predictions: BSFG-Rauscher [42] (*dashed line*), BSFG-von Egidy [44] (*solid line*), HF-BCS [48] (*dotted line*) and HFB [46, 47] (*open triangles*, the connecting line is to guide the eye only)

Both BSFG parameterizations provide very similar results for $2^+$ states in $^{40}$Ca in the excitation energy range considered $E_x$ = 10–20 MeV. The corresponding level density parameter $a \approx 5.3$ MeV$^{-1}$, which is unusually low because of the double shell closure, provides a reasonable description of the energy dependence in the experimental results, but the magnitudes are about a factor of two too low. The energy dependence of the HF-BCS calculation is again very similar to the data and the BSFG results, but the predicted level densities are about 50 % too high. Since an implicit assumption of all three models is the equipartition of states with positive and negative parity for a given spin, the theoretical results were divided by a factor of two for the comparison in Fig. 5. In detail, the correspondence of fluctuations around the average increase of the level densities extracted from the data and the HFB calculations is limited. For example, at $E_x = 17$ MeV the data find a local minimum while the HFB result predicts a maximum. However, the average dependence on energy as well as the absolute magnitude are described well by the HFB calculations despite the fact that no renormalization (cf. Eq. (9) of Ref. [47]) to the experimental level scheme at low energy and neutron resonance spacings was included.

**Acknowledgments** The author would like to acknowledge the K600 collaboration comprising groups from iThemba LABS, Wits University, TU Darmstadt and RCNP Osaka. This work has been supported by the DFG under contracts SFB 634 and NE 679/2-2, and by the South African NRF.

## References

1. R. Neveling, F.D. Smit, H. Fujita, R.T. Newman, *K600 Usermanual* (iThemba LABS, Somerset West, 2008)
2. G.F. Bertsch, Nature **280**, 23 (1979)
3. R. Pitthan, Th. Walcher, Phys. Lett. B **36**, 563 (1971)
4. M.B. Lewis, F.E. Bertrand, Nucl. Phys. A **196**, 337 (1972)
5. M.N. Harakeh, A. van der Woude, *Giant Resonances: Fundamental High-Frequency Modes of Nuclear Excitation* (Oxford University, Oxford, 2001)
6. P.F. Bortignon, A. Bracco, R.A. Broglia, *Giant Resonances: Nuclear Structure at Finite Temperature* (Harwood Academic, Amsterdam, 1998)
7. G. Kühner, D. Meuer, S. Müller, A. Richter, E. Spamer, O. Titze, W. Knüpfer, Phys. Lett. B **104**, 189 (1981)
8. Y. Kalmykov, T. Adachi, G.P.A. Berg, H. Fujita, K. Fujita, Y. Fujita, K. Hatanaka, J. Kamiya, K. Nakanishi, P. von Neumann-Cosel, V.Yu. Ponomarev, A. Richter, N. Sakamoto, Y. Sakemi, A. Shevchenko, Y. Shimbara, Y. Shimizu, F.D. Smit, T. Wakasa, J. Wambach, M. Yosoi, Phys. Rev. Lett. **96**, 012502 (2006)
9. Y. Fujita, K. Hatanaka, G.P.A. Berg, K. Hosono, N. Matsuoka, S. Morinobu, T. Noro, M. Sato, K. Tamura, H. Ueno, Nucl. Instr. Methods B **126**, 274 (1997)
10. A. Shevchenko, J. Carter, R.W. Fearick, S.V. Förtsch, H. Fujita, Y. Fujita, Y. Kalmykov, D. Lacroix, J.J. Lawrie, P. von Neumann-Cosel, R. Neveling, V.Yu. Ponomarev, A. Richter, E. Sideras-Haddad, F.D. Smit, J. Wambach, Phys. Rev. Lett. **93**, 122501 (2004)
11. A. Shevchenko, O. Burda, J. Carter, G.R.J. Cooper, R.W. Fearick, S.V. Förtsch, H. Fujita, Y. Fujita, Y. Kalmykov, D. Lacroix, P. von Neumann-Cosel, R. Neveling, V.Yu. Ponomarev, A. Richter, E. Sideras-Haddad, J. Wambach, Phys. Rev. C **79**, 044305 (2009)
12. I. Usman, Ph.D. Thesis, University of the Witwatersrand (2009)

13. I. Usman, Z. Buthelezi, J. Carter, G.R.J. Cooper, R.W. Fearick, S.V. Förtsch, H. Fujita, Y. Fujita, Y. Kalmykov, P. von Neumann-Cosel, R. Neveling, P. Papakonstantinou, A. Richter, R. Roth, A. Shevchenko, E. Sideras-Haddad, F.D. Smit, Phys. Lett. B **698**, 191 (2011)
14. H. Diesener, U. Helm, G. Herbert, V. Huck, P. von Neumann-Cosel, C. Rangacharyulu, A. Richter, G. Schrieder, A. Stascheck, A. Stiller, J. Ryckebusch, J. Carter, Phys. Rev. Lett. **72**, 1994 (1994)
15. H. Diesener, U. Helm, G. Herbert, V. Huck, P. von Neumann-Cosel, C. Rangacharyulu, A. Richter, G. Schrieder, A. Stascheck, S. Strauch, J. Ryckebusch, J. Carter, Nucl. Phys. A **696**, 292 (2001)
16. K. Schweda, J. Carter, A.A. Cowley, H. Diesener, R.W. Fearick, S.V. Förtsch, J.J. Lawrie, P. von Neumann-Cosel, J.V. Pilcher, A. Richter, F.D. Smit, G.F. Steyn, S. Strauch, Phys. Lett. B **506**, 247 (2001)
17. K. van der Borg, M.N. Harakeh, A. van der Woude, Nucl. Phys. A **365**, 243 (1981)
18. J. Winchenbach, K. Pingel, G. Holzwarth, G. Kühner, A. Richter, Nucl. Phys. A **410**, 237 (1983)
19. H. Aiba, M. Matsuo, Phys. Rev. C **60**, 034307 (1999)
20. H. Aiba, M. Matsuo, S. Nishizaki, T. Suzuki, Phys. Rev. C **68**, 054316 (2003)
21. D. Lacroix, P. Chomaz, Phys. Rev. C **60**, 064307 (1999)
22. D. Lacroix, A. Mai, P. von Neumann-Cosel, A. Richter, J. Wambach, Phys. Lett. B **479**, 15 (2000)
23. W.D. Heiss, R.G. Nazmitdinov, F.D. Smit, Phys. Rev. C **81**, 034604 (2010)
24. A. Shevchenko, et al., Phys. Rev. C **77**, 024302 (2008)
25. G.F. Bertsch, P.F. Bortignon, R.A. Broglia, Rev. Mod. Phys. **55**, 287 (1983)
26. S. Kamerdzhiev, J. Lisantti, P. von Neumann-Cosel, A. Richter, G. Tertychny, J. Wambach, Phys. Rev. C **55**, 2101 (1997)
27. V.G. Solov'ev, V.I. Kisin, *Theory of Atomic Nuclei: Quasiparticles and Phonons* (Institute of Physics, Bristol, 1992)
28. D. Lacroix, S. Ayik, P. Chomaz, Phys. Rev. C **63**, 064305 (2001)
29. G. Brown, *Unified Theory of Nuclear Models* (North-Holland, Amsterdam, 1964)
30. S. Drozdz, V. Klent, J. Speth, J. Wambach, Nucl. Phys. A **451**, 11 (1986)
31. S. Drozdz, S. Nishizaki, J. Speth, J. Wambach, Phys. Rep. C **197**, 1 (1990)
32. C. Yannouleas, M. Dworzecka, J. Griffin, Nucl. Phys. A **397**, 239 (1983)
33. R. Roth, H. Hergert, P. Papakonstatinou, T. Neff, H. Feldmeier, Phys. Rev. C **72**, 034002 (2005)
34. R. Roth, T. Neff, H. Feldmeier, Prog. Part. Nucl. Phys. **65**, 50 (2010)
35. P. Papakonstantinou, R. Roth, Phys. Lett. B **671**, 356 (2009)
36. P. Papakonstantinou, R. Roth, Phys. Rev. C **81**, 024317 (2010)
37. R. Roth et al., Phys. Rev. C **73**, 044312 (2006)
38. N. Paar, P. Papakonstantinou, H. Hergert, R. Roth, Phys. Rev. C **74**, 014318 (2006)
39. T. Ericson, Adv. Phys. **9**, 425 (1963)
40. P.G. Hansen, B. Jonson, A. Richter, Nucl. Phys. A **518**, 13 (1990)
41. I. Usman, Z. Buthelezi, J. Carter, G.R.J. Cooper, R.W. Fearick, S.V. Förtsch, H. Fujita, Y. Kalmykov, P. von Neumann-Cosel, R. Neveling, I. Poltoratska, A. Richter, A. Shevchenko, E. Sideras-Haddad, F.D. Smit, J. Wambach, Phys. Rev. C **84**, 054322 (2011)
42. T. Rauscher, F.K. Thielemann, K.-L. Kratz, Phys. Rev. C **58**, 1613 (1997)
43. T. von Egidy, D. Bucurescu, Phys. Rev. C **72**, 044311 (2005)
44. T. von Egidy, D. Bucurescu, Phys. Rev. **80**(80), 054310 (2009)
45. Y. Alhassid, S. Lui, H. Nakada, Phys. Rev. Lett. **99**, 162504 (2007)
46. S. Hilaire, S. Goriely, Nucl. Phys. A **779**, 63 (2006)
47. S. Goriely, S. Hilaire, A.J. Koning, Phys. Rev. C **78**, 064307 (2008)
48. P. Demetriou, S. Goriely, Nucl. Phys. A **695**, 95 (2001)

# Activities at iThemba LABS Cyclotron Facilities

**R. M. Bark, J. Cornell, J. J. Lawrie and Z. Z. Vilakazi**

**Abstract** iThemba Laboratory for Acceleratory Based Sciences (iThemba LABS) is a multi-disciplinary cyclotron facility. Chief among its activities is the operation of a $k = 200$ separate sector cyclotron (SSC) which provides proton beams of energies up to 200 MeV. These beams are used for fundamental nuclear physics research in the intermediate energy region, isotope production and medical physics applications. Details on developments regarding the new flagship project at iThemba LABS are also presented.

## 1 Introduction

iThemba Laboratory for Accelerator Based Sciences is a multi-disciplinary research centre, operated by the National Research Foundation (NRF). It provides accelerator and ancillary facilities for: Research and training in the physical, biomedical and material sciences; treatment of cancer patients with energetic neutrons and protons and related research; production of radioisotopes and radiopharmaceuticals for use in nuclear medicine, research and industry and related research. At the heart of iThemba LABS accelerator complex is the variable-energy, separated-sector cyclotron, which provides beams with a maximum energy of 200 MeV for protons. As shown in Fig. 1, these are directed to vaults for the production of radioisotopes, proton and neutron therapy and nuclear physics experiments. Light ions, preaccelerated in the first solid-pole injector cyclotron with a $K$-value of 8 are used for therapy and radioisotope production. For radioisotope production and neutron therapy a 66 MeV proton beam is used while for proton therapy and nuclear physics research, 200 MeV beams are

R. M Bark · J. Cornell · J. J. Lawrie · Z. Z. Vilakazi (✉)
iThemba Laboratory for Accelerator Based Sciences, Old Faure Road, 722, Somerset West 7129, South Africa
e-mail: zzv@tlabs.ac.za

W. Greiner (ed.), *Exciting Interdisciplinary Physics*,
FIAS Interdisciplinary Science Series, DOI: 10.1007/978-3-319-00047-3_15,

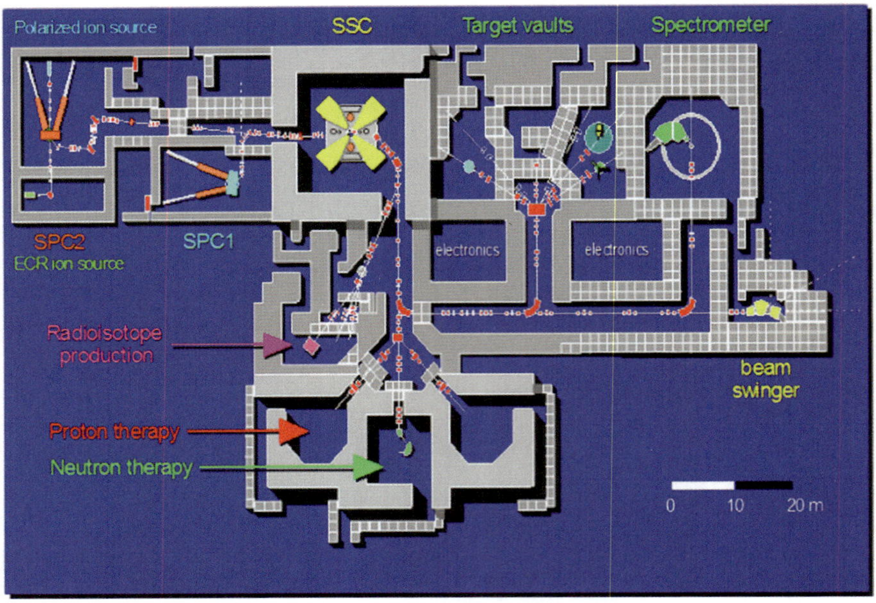


**Fig. 1** Layout of iThemba LABS cyclotrons and beam lines

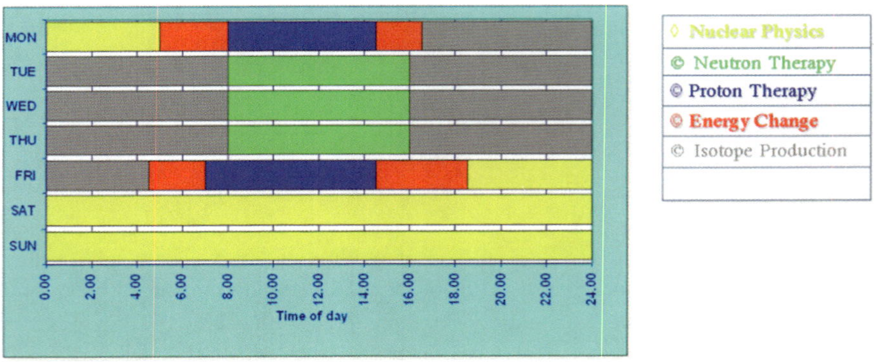


**Fig. 2** Present weekly schedule of iThemba LABS SSC beamtime

used. The second solid-pole injector cyclotron with a $K$-value of 10 is used for pre-acceleration of heavy ions and polarized protons from these two external sources.

Beams are delivered to the different users groups for 24 h per day and seven days per week as shown in Fig. 2 below. A 66 MeV proton beam is available for radionuclide production and neutron therapy from Monday evening until midday Friday. Patients are treated during day time and between treatments the beam is switched to the radionuclide production vault and the intensity increased to 250 $\mu$A. On weekends, a 200 MeV beam is used either for proton therapy or nuclear physics

**Table 1** Beams delivered at iThemba LABS

| Element | Mass | Energy range (MeV) | | Element | Mass | Energy range (MeV) | |
|---|---|---|---|---|---|---|---|
| | | from | to | | | from | to |
| H | 1 | 11.5 | 227 | Si | 28 | 141 | 141 |
| He | 4 | 25 | 200 | Cl | 37 | 205 | 250 |
| B | 11 | 55 | 60 | Ar | 40 | 280 | 280 |
| C | 12 | 58 | 400 | Zn | 64 | 165 | 280 |
| C | 13 | 75 | 82 | Kr | 84 | 450 | 530 |
| N | 14 | 140 | 400 | Kr | 86 | 396 | 462 |
| O | 16 | 73 | 400 | I | 127 | 730 | 730 |
| O | 18 | 70 | 110 | Xe | 129 | 750 | 790 |
| Ne | 20 | 110 | 125 | Xe | 136 | 750 | 750 |
| Al | 27 | 150 | 349 | | | | |

research using beams of light and heavy ions, as well as polarized protons, pre-accelerated in a second solid-pole injector cyclotron (SPC2).

# 2 Accelerators and Beam Deliver Systems of iThemba LABS

## *2.1 The Beam Delivery Injection Systems*

The light-ion injector cyclotron (SPC1) has maximum proton energy of 8 MeV. Beam produced in the internal source are accelerated with two 90° dees operated at a maximum voltage of 60 kV. The RF-system can be tuned over the frequency range 8.6–26 MHz. The beam is extracted with an electrostatic channel and two active magnetic channels. For the 66 MeV proton beams, a 600 μA beam at energy of 3.14 MeV is extracted. Flat-topping systems which superimpose a fifth harmonic on the main dees have been installed. The injector cyclotron for heavy ions and polarized protons (SPC2) is in many respects similar to SPC1, except that the ion sources are external and the beam is injected axially with a spiral inflector. Heavy ions with mass numbers up to that of Xenon are delivered by an electron-cyclotron resonance ion source (ECRIS). Proton beams from a polarized ion source are also accelerated with this injector cyclotron. Listed in Table 1 are particle species and beam energies which have been accelerated using the internal PIG (Penning or Philips Ion Gauge) source of the light-ion injector cyclotron and the Minimafios ECR ion source of the heavy-ion injector cyclotron. The ECR source has been in use since 1994.

## 2.2 The Separated Sector Cyclotron

This is in the form of the separated-sector cyclotron which has four separate magnet sectors with an overall diameter of 13 m. The RF-system consists of two lambda-half resonators that operate in the frequency range 7–26 MHz. The dee voltage is 220 kV at a power level of 80 kW per resonator. Although it has been designed for maximum proton energy of 200 MeV it has delivered proton beams of 227 MeV. The beam is inflected with two bending magnets and a magnetic channel and extracted with two septum magnets. Typically, the cyclotron delivers a 250 $\mu$A beam of 66 MeV protons for radioisotope production with 99.8 % transmission through the machine. A flat-topping system, operating on the third harmonic, has been installed in one of the vacuum chambers.

## 2.3 Ion Source and Beam Line Developments

As part of the ongoing process of infrastructure upgrade; two new ion sources have been installed at iThemba LABS. The first was a process of commissioning the 18 GHz Grenoble GTS ECR ions source (ECRIS). Thus far, 14 GHz and 18 GHZ microwave generators and the hexapole magnet coils have been delivered. Some components have still to be ordered. With this source $^{129}Xe^{37+}$-ions were made be available for acceleration to 2.2 GeV in the separated-sector cyclotron. The other ion source being installed is the 14.5 GHz ECRIS donated by the Hahn Meitner Institute (HMI). This source has already been installed in the injection beam line of the heavy ion injector cyclotron. The first beams from the source were delivered in 2008. With this source beam currents will increase dramatically.

# 3 Radio-Isotope Production

During the past 20 years the Radionuclide production Group has established a comprehensive production programme that is based on a 66 MeV proton beam. iThemba LABS was for many years the only producer of accelerator-based radio nuclides in South Africa. Radiopharmaceuticals produced from these radionuclides are supplied to more than 50 nuclear medical centres throughout South Africa. Beam currents of up to 100 $\mu$A have during the past 20 years been used for routine production of a range of radionuclides. Furthermore, iThemba LABS is one of a few facilities that utilises proton beam energies significantly above 30 MeV for large-scale production of radioisotopes. High-energy proton accelerators have a definite use for the production of radioisotopes, e.g. $^{68}$Ge: $(p, 4n)$ nuclear reaction on $^{71}$Ga—increases production yield of $^{68}$Ge significantly when added to the yield of the $(p, 2n)$ reaction on $^{69}$Ga. In the vertical target station beam currents of up to 250 $\mu$A are used to produce $^{82}$Sr and $^{68}$Ge in tandem. These long lived isotopes are for the international export market.

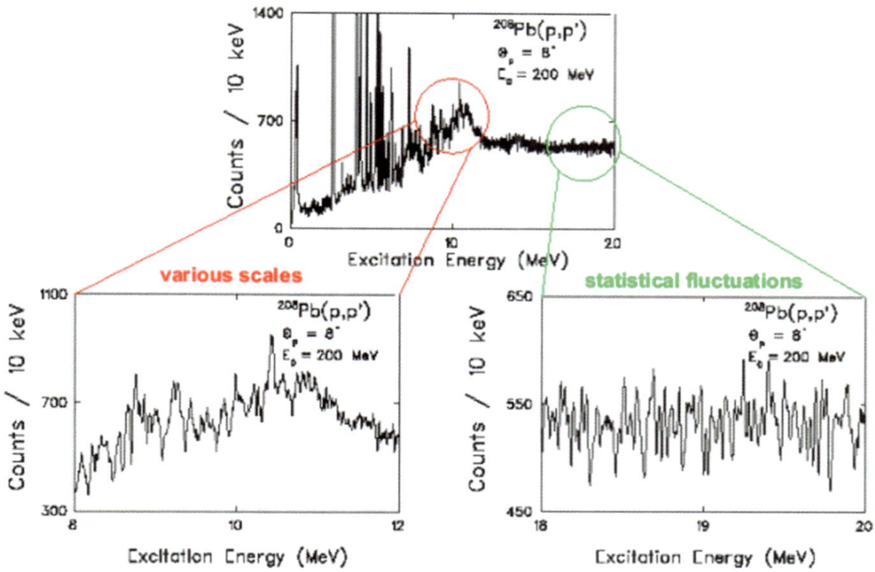


**Fig. 3** Fine structure of ISGQR from high resolution proton inelastic scattering processes

# 4 Nuclear Physics Research

Two main research areas are covered at iThemba LABS; these are light ion and heavy-ion nuclear structure investigations. The former uses the $k = 600$, kinematically corrected magnetic spectrometer for light particles. It has an angular acceptance of 70 mrad and a resolution of 26 keV for 200 MeV proton beams. Among the experimental investigations conducted are high resolution knock-out reactions, namely $(p, 2p)$ and (p, $2\alpha$); decays of giant resonances on $(p, px)$ for $x = n, p, \alpha$ and studies of mixed symmetry states (see Fig. 3).

Gamma ray nuclear spectroscopy using heavy ions is undertaken at the AFRODITE gamma-ray spectrometer array (see [1] and references there in) composed of 9 Compton suppressed clover detectors and 7 segmented planar Ge detectors. The detector performances are characterized by efficiencies of 1.6 % at 1.33 MeV and 6 % at 100 keV, respectively. Recent investigations carried out using this detector include the study of chiral systems across the nuclear chart, the search for tetrahedral structures in mass 160 and 230 regions, and a focus on excited $0^+$ states in the rare-earth region.

## 5 Proton and Neutron Therapy

No attempt will be made to give exhaustive details on hadron therapy since articles and reviews on this subject have been published over the years. That notwithstanding, it would be worth pointing out though that it is well established fact that compared with conventional radiations (photons, electrons), neutrons have similar physical characteristics to X-rays and are more effective for treating radio-resistant tumours, usually large and/or slow-growing, such as salivary gland tumours and advanced prostate cancer. To this end, the iThemba LABS neutron facility has been operational since 1988 and to date in close to 1,600 patients have been treated.

The rationale for proton therapy relates primarily to improved physical dose selectivity (details are contained in a review of the ICRU journal [2]). The success of this treatment modality is evidenced by the proliferation of proton therapy centres. Since 1954, a total of 53,439 patients have been treated. At iThemba LABS, close to 500 patients have been treated since the first beams (for proton therapy) were delivered in 1993.

## 6 Future Scientific Developments

**The Radioactive Isotope project**

In the past, nuclear accelerators accelerated stable atoms—those occurring naturally —to energies required for nuclear or material physics research. Often the goal was to create artificial radioactive nuclei. The study of these nuclei allows theories of nuclear forces to be tested but the tests are limited to those nuclei that can be produced with a stable beam—these are the neutron *deficient* nuclei. The neutron *rich* nuclei are largely unknown and represent a severe gap in our knowledge of nuclear structure and nuclear forces.

To create neutron rich nuclei, two methods are available. The Projectile Fragmentation (PF) method accelerates a beam of heavy-ions and breaks them up by passing them through a target. If protons are knocked from the atoms of the beam by the target, a new *neutron rich* radioactive beam is produced. However, to penetrate the target and maintain focus, extremely high energies are required, necessitating the construction of an expensive accelerator. Furthermore, because the beam is relativistic, it cannot easily be used for many studies of interest.

The second way of producing radioactive beams is the Isotope Separation OnLine (ISOL) method. Here, a driver accelerator delivers a high-intensity, possibly low-energy, beam which strikes a very thick production target. The radioactive ions produced in this target are initially trapped within it, but are liberated by heating the target to a temperature of several hundred degrees. The radioactive species diffuse out of the target, are ionized, selected, charge-bred, and injected into a second (post-) accelerator for transport to an experimental station. Thus in this method two accelerators are required.

At iThemba LABS, a new 70 MeV proton accelerator would be ideal as a driver accelerator while the existing SSC is already suitable, with a upgraded injection accelerator and improvements to the vacuum and control system, as a post accelerator. The addition of a Radioactive Ion Beam (RIB) production target and ion-source would complete the system.

To be able to be used for both RIB production and for isotope production and neutron therapy, a negative-ion cyclotron is required one for which it is possible to extract two beams simultaneously.

**Research with the New Facility**

The physics research conducted at the new facility will broadly fall in line with the goals of the five and ten year research plan. It is in the second stage that radioactive beams will become available, pushing the South African nuclear physics and material science research to the international forefront. In nuclear medicine new opportunities will arise with the use of Positron Emission Tomography (PET). The development of the facility itself will create numerous opportunities for research within the field of accelerator physics.

In general terms, the research opportunities are described below:

**Nuclear Physics**

Stage 1

Let us begin by highlighting some aspects of the research programme on the SSC, which comprises pure and applied nuclear physics.

The pure research programme tests models of the nucleus under extreme conditions. Special programmes are underway to test predictions of nuclei having exotic shapes resembling pyramids (tetrahedrons) or stretched pears (hyperdeformation) while another programme seeks to test whether nuclei can be right-handed or left-handed (chirality). Some of the research questions some of the basic assumptions of the low-lying modes of excitation of the nucleus. Are they actually vibrations as has been assumed or are they really due to differing configurations of paired nucleons? At higher excitation energies, giant vibrations of protons relative to neutrons occur. The formation of these vibrations is studied at the laboratory using the K600 spectrometer. Soon this instrument will be able to operate at zero degrees and related questions arising about the formation of the elements in supernova explosions will be answered. At the other extreme, research has also been conducted into possible ways of artificially forming superheavy elements—elements so heavy that they are unlikely to have been formed even in supernova explosions

More and more nations have a footprint in space. South Africa is no exception with the proclamation of the South African National Space Agency and the launch of its second satellite, built by SunSpace in Stellenbosch, in 2009. One of the major difficulties faced by any space application is the increase in radiation dose to equipment and personnel when one no longer has the protective shield of the earths atmosphere. iThemba LABS is in an ideal position to test and calibrate detectors and electronic components for radiation hardness with neutron and proton beams of up to 200 MeV energy. Neutrons at iThemba LABS are already used by the European Radiation Dosimetry Group on a regular basis to develop and calibrate detectors for space applications, such as for use on the International Space Station. There have already been enquiries from SunSpace, Nelson Mandela University, Stellenbosch University and a private company to perform radiation hardness screening of components on a routine and "on demand" basis.

*Most of these areas of research will benefit greatly from extended beamtime, by allowing experiments that would otherwise be practically impossible, since high statistics are required to resolve weakly populated structures while for other experiments, extended beamtime is required to set up delicate instrumentation.*

Stage 2

On the chart of the nuclides, shown in Fig. 3, the known nuclei, those that occur naturally (black) or those that have been artificially produced (yellow), comprise approximately half of all the species that are predicted to exist.

The still unknown nuclei, (green) lie predominantly on the neutron rich side of the naturally occurring isotopes. Indeed even of the known neutron rich nuclei, very little other than their existence and a few basic properties are known.

Thus the neutron rich region, "terra incognita", is also to be regarded as a "final frontier" in low-energy nuclear physics. It is in this region that the nuclear shell model—the "standard model" of the nucleus—has been subjected to the most stringent of tests and has been found wanting. The shell model correctly predicts the location of especially stable nuclei in naturally occurring isotopes, with shell gaps at the so-called magic numbers of protons and neutrons. However, the little evidence that we already have, from radioactive beam laboratories overseas, points to the dissolution of the shell gaps in neutron rich nuclei. The shell model, which has been the basis of our understanding of nuclear structure for over 40 years, has been exposed as inadequate and requiring the inclusion of charge and isospin exchange forces in addition to the spin-orbit force. This evidence has been gleaned only from very light neutron rich nuclei—the effect of exchange forces and tensor forces in heavier neutron rich nuclei is largely untested.

The heavier neutron rich nuclei are also expected to show modified behaviour due to the existence of a neutron skin. This is also expected to influence the shell structure but it should be possible to observe new modes of the giant vibrations that scientists at iThemba LABS have been studying with stable beams. These new modes are called the pygmy resonances—in effect a vibration of the neutron skin around the core.

**Astrophysics**

The ramifications of our poor knowledge of nuclear forces extend beyond the realms of strictly nuclear physics and into astrophysics. It is in stars that the elements were created but for elements heavier than iron, it is more accurate to say that they were created in exploding stars, or supernova. In supernova large fluxes of neutrons necessarily created neutron rich nuclei. To understand the precise details of this process, the so-called "r-process", and to be able to explain the abundances of the naturally occurring elements, it is necessary to have an understanding of neutron rich nuclei, and of the location of the shell gaps. Unfortunately, since the shell model has been shown to be deficient, our understanding of the origin of the elements remains imprecise.

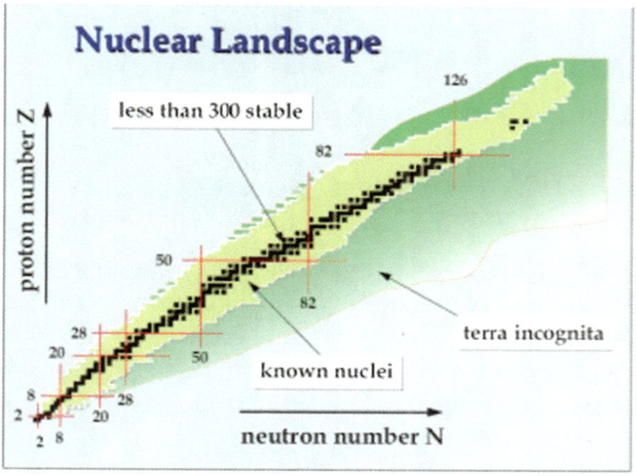


**Fig. 4** The "nuclear landscape", showing the unknown region in *green* (from RIA taskforce report)

*A radioactive beam facility, delivering neutron rich nuclei from the fission of uranium, will allow South African researchers, and indeed international collaborators, an unprecedented opportunity of understanding the fundamental forces of nature and the origin of the elements.*

**Material Science**

In condensed mater and solid states research the use of radioactive nuclei as probes for the study of structural and/or electronic/magnetic lattice environment in materials (metals, insulators or semiconductors) on surfaces and interfaces is of critical importance for technical development. Most of these studies are concentrated on the investigation of defects and impurities in semiconductors like Si, Ge, III-V or II-VI compounds. The main topics of such investigations are implantation induced lattice damage and its annealing behaviour, the lattice site of the implanted ion after annealing, the interaction between impurities or impurities and intrinsic defects, the electronic and optical properties of the implanted species and the identification of defects and impurities. For example the microscopic insight into the structure and the thermodynamic properties of materials formed by interacting defects can be obtained by detecting the hyperfine interaction between the nuclear moments of the radioactive dopants and the electromagnetic fields present at the site of the radioactive nucleus.

The understanding and the control of diffusion profiles of intrinsic and extrinsic defects particularly in semiconductors is significantly enhanced using radioactive tracer diffusion. In this process a thin layer of material containing the tracer is deposited on the surface of the sample under study; alternatively the radioactive tracer can be implanted using highly selective and pure isotopes accelerated from suitable ion sources. Subsequently the system is heated for a predetermined time and temperature ranges. After cooling back to room temperature, the sample is sectioned into thin slices step by step and either the content of the radioactive tracer in each slice or in the unsectioned part of the sample is measured. From this measurement

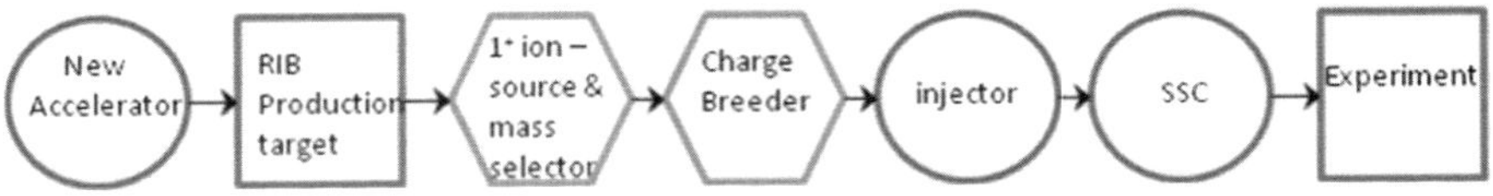


**Fig. 5** Stages in the production of radioactive ion beams

the diffusion profile of the tracer is determined. By repeating this experiment for several diffusion times and/or temperatures the diffusion parameters of the tracer, especially its activation enthalpy is determined. From these values and the shape of the diffusion profile, conclusions about the diffusion mechanism can be inferred. For this type of work the use of radioactive beams with half-lives of minutes is desirable.

An important part of materials science research is the study of the behaviour of materials under radiation conditions. It has been recognised that bombardment with energetic particles offered a unique method to create controlled populations of defects and atom displacement in solids. These phenomena produce chemical effects in solid materials since they affect the way atoms of various species are arranged in space and thus, irradiation can promote or inhibit phase changes. One of the main interests is the study of Pt-thin films coated systems subjected to radiation effects as a state variable which can induce phase and structural changes with particular interest for ordering transition. There is a set of superstructures (superlattices) that have been predicted to appear at particular temperature ranges, and the knowledge of the interactions of Pt with thin films of other compatible metals is of utmost interest. These effects are very important for many applications of the Pt coated systems as catalyst, gas sensors and optical devices.

**Production of Radioactive Beams**

Although there are numerous Radioactive Beam Facilities around the world, the technology for the production of RIBs is still under development, particularly with regard to increasing beam intensities. The intention of the iThemba LABS proposal is to draw on past international experience through collaborative efforts. The present proposal is similar to the SPES proposal of Legnaro, INFN, Italy, which makes use of a 70 MeV proton machine, originally developed for radioisotope production. Since a ~70 MeV proton beam will be used for the neutron and isotope programmes, production of RIBs will be most cost effective if it can be produced by the same beam. This beam energy is in fact ideal for the ISOL technique, and could, for example, produce neutron rich species using proton-induced fission on a production target of uranium carbide. Alternatively, the proton beam could be used to produce neutron deficient beams, using $(p, xn)$ reactions. These beams are also of interest for the material sciences.

A schematic diagramme showing the different stages of possible RIB production at iThemba LABS is given in Fig. 5 below.

Radioactive ions from the production target will be extracted with a source directly connected to the production target, and ionized to the $1^+$ charge state. A number of ionization techniques are available, including surface ionization, plasma ionization with a FEBIAD ion source, or using laser ionization (RILIS). Already a workshop on Lasers and Accelerators has been held at the University of Stellenbosch where it was clearly shown that the expertise to build RILIS ion sources exists in South Africa. Ionization to the $1^+$ charge state alone is insufficient to select the desired radioactive species; final selection is achieved by using a mass analyser.

Before injection of the beam into an injector accelerator, (which will deliver the beam for injection into the SSC), the beam has to be "charge bred" with a second ion-source to reach the final charge state that is needed to obtain the specified final energy of the beam from the SSC. This charge breeding can be performed by an ECR ion-source, with the final charge state determining the ultimate beam energy. For light ions, beam energies of up to $\sim$40 MeV per nucleon should be possible. For isotopes in the mass range 80–130, such as the fission products, the SSC should be able to deliver beam energies of between 5 and 7 MeV per nucleon without severely limiting RIB intensities.

The new facility will require expenditure on

I. A high-intensity, negative-ion cyclotron
II. New target stations for isotope production
III. Target stations and handling facilities for radioactive beam production
IV. Upgrades to existing infrastructure for nuclear and materials research
V. Civil engineering to accommodate the new facilities.

The overall cost of the project is estimated to be of the order of R950M expended over a seven year period.

## 7 Conclusions and Outlook

The production of radioactive beams is intellectually challenging and will require considerable development and design effort. We propose to use this as an important opportunity for human resource development by strongly linking to South African Universities and other research institutions. Certainly these developments will ensure that South Africa remains at the forefront of nuclear sciences for the foreseeable future.

## References

1. M. Lipoglavšek et al., Phys. Lett. B **593**, 61 (2004)
2. D. T. L. Jones, H. D. Suit et al, Prescribing, Recording, and Reporting Proton Beam Therapy, ICRU Report 78. J. ICRU **7**(2) (2007), ISSN 1473–6691

# Part III
# High-Energy Nuclear Physics

# New Forms of High Energy Density Matter

**Larry McLerran**

**Abstract** I will discuss new forms of matter that might be seen in ultra-relativistic heavy ion collisions. The Quark Gluon Plasma is matter in thermal equilibrium at very high temperature and low to intermediate baryon density. This is a deconfined plasma of quarks and gluons. At high baryon density and low to moderate temperature, there are a number of different possible phases of Quarkyonic Matter. Quarkyonic matter is at energy densities large compared to the natural scale of strong interactions $(200\,\mathrm{MeV})^4$. The quarks in the Fermi sea behave as quasi free quarks, but Fermi surface and thermal excitations are confined into baryons and mesons. Chiral symmetry is broken in a translational invariant way. In the wave function of a high energy hadron, the states that control the high energy limit of scattering are highly coherent and very dense gluons. This is the Color Glass Condensate. A short time after the collision of two high energy hadrons, very strong longitudinal color electric and color magnetic fields are formed. This is the Glasma. At later times the Glasma fields evaporate into gluons and thermalize. During this thermalization, gluon Bose condensates might be formed. Thermalization might occur through coherent effects. Although the intrinsic strength of interaction might be weak, the Glasma may be a realization of a strongly interacting Quark Gluon Plasma.

## 1 Introduction

This talk will address the following questions:

- What are the possible forms of high energy density matter?
- How might such matter be produced and studied?
- How does such matter determine the high energy limit of strong interactions?

L. McLerran (✉)
Brookhaven National Laboratory and RIKEN BNL Center, Upton , NY 11973, USA
e-mail: mclerran@bnl.gov

W. Greiner (ed.), *Exciting Interdisciplinary Physics*,
FIAS Interdisciplinary Science Series, DOI: 10.1007/978-3-319-00047-3_16,

The issues in this subfield of nuclear physics are similar in spirit to the very successful search for superheavy nuclei carried out by the group at Dubna led by Oganessian and Itkis [1], and by groups at LBL and GSI. This work was guided by the early theoretical work of Greiner and colleagues [2].

Some of this talk will be a discussion of results that have been in the literature for some time. There will be some commentary about very new and speculative results concerning how matter might thermalize in heavy ion collisions and the possible formation of Bose Einstein condensates. The subjects that will be covered are:

- The phase diagram of thermalized strongly interacting matter.
- The strongly interacting QGP and heavy ion lore.
- The Color Glass Condensate as high density and coherent fields in the high energy hadron wave function.
- The Glasma as strong colored fields produced in heavy ion collisions, and the early time evolution of the Glasma.
- The thermalization of the Glasma as a strongly interacting non-thermalized QGP.
- The possible formation of transient Bose condensates during the thermalization of the Glasma.

I would like to briefly comment on what we mean by "very high energy density" matter. The density of matter inside a proton is

$$\varepsilon_P \sim Mc^2/R^3 \tag{1}$$

where $M$ is the proton mass, and $R \sim$ 1 fm is its radius. This energy density scale is roughly 1 $\text{GeV}/\text{fm}^3$. The scale corresponding baryon number density is $\rho_B \sim$ 1 $\text{Baryon}/\text{fm}^3$. High energy densities and baryon densities will mean densities of this order of magnitude or larger. This corresponds to temperatures of $T_{\text{proton}} \sim 10^{13}$ degrees Kelvin, which is about 6 orders of magnitude larger than that of the solar interior. Fermi energies corresponding to this scale of baryon density are typically of order $\mu_{\text{baryon}} - Mc^2$ of order 100's of MeV's. In the later analysis that follow I will use relativistic units where Planck's constant and the speed of light are set to 1, and baryon Fermi energies will include the rest mass energy of the proton.

## 2 Matter in Thermal Equilibrium

My current understanding of the phase diagram of QCD is summarized in Fig. 1.

There are several landmark features of this phase diagram:

At low temperature and density, there is Hadronic Matter. This is an interacting gas of quarks and gluons confined into mesons and baryons. The baryons are massive as a consequence of the breaking of Chiral Symmetry. This symmetry is approximate, and follows from the near masslessness of light quarks. If the symmetry were realized, baryons would be massless.

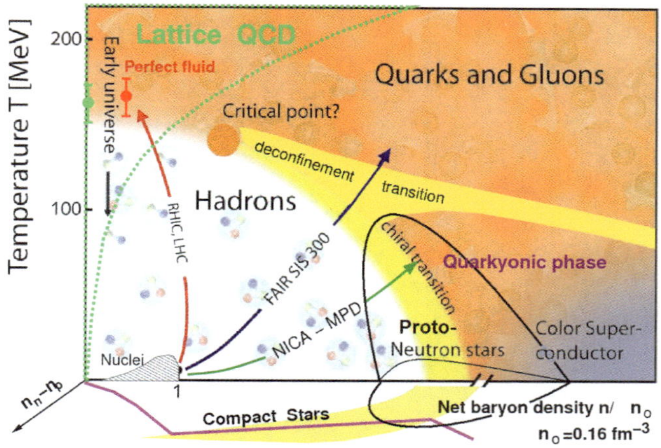


**Fig. 1** A conception of the phase diagram of QCD

At high temperatures, there is the Quark Gluon Plasma, which is a gas of de-confined quarks and gluons. In this phase, chiral symmetry is restored. At asymptotically high temperature, the quarks and gluons interact weakly, due to the fact that strong interactions become weak at short distances.

At high baryon density and low temperatures, there is Quarkyonic Matter [3]. This matter has a Fermi sea of almost free quarks. In the Fermi sea, the dominant quark interactions are at short distances where the interaction is weak. Near the Fermi surface, or for thermal excitations, quarks and gluons can interact at long distances. In the QGP, these interactions are cutoff due to Debye screening of gluons. In Quakrkonic matter, Debye screening arises from quarks, and there are a variety of arguments that this should be ineffective until much higher baryon density than is typical of Quarkyonic matter. By definition, high baryon density means baryons are present. However in the hadron phase, baryons are strongly suppressed, $\sim e^{-M_B/T}$ due to the large mass of baryons. These qualitative arguments can be made rigorous in the large number of color limit of QCD, $N_c \to \infty$. In addition, the chiral properties of QCD may be different in the Quarkyonic phase, and there may be a variety of phase transitions associated with non-translationally invariant breaking of chiral symmetry [4, 5]. This breaking of translational invariance is because the quark pairs that form a condensate must come from the top of the Fermi sea, and therefore carry net momentum. This momentum means the condensate has a typical DeBroglie wavelength, $\lambda \sim 1/2E_F$, and therefore breaks the continuous translational invariance.

In the high density phase, there may will be color superconductivity, associated with the formation of quark Cooper pairs [6, 7].

The transitions between the QGP, Quarkyonic Matter and Hadronic Matter involve large changes in the number of degrees of freedom. At low temperatures and densities, there are three pion degrees of freedom. In the Quarkyonic phase, there are three color degrees of freedom for each quark, two flavor degrees of freedom and two spins for total of 12. In the QGP, there are two spins and eight colors for gluons, and three colors, two spins, two flavors and two for quark and antiquark degrees of freedom, or about 40.

It is possible that there are true phase transition lines that separate the QGP, Quarkyonic Matter and the Hadron phase. In reality, these may be approximate. Corresponding to these transitions, there may be a triple point associated with the place where there is coexistence of these three phases. There may be a critical end point, if there are true phase transitions, where a line of first order transitions vanishes.

Much is known about the transition at finite temperature and zero baryon number density. This has been learned from lattice gauge theory computations. One can see the effects of chiral symmetry restoration and deconfinement clearly in such computations and one can also compute energy density and pressure reliably from the underlying theory of strong interactions [8, 9].

# 3 Matter as it Might Appear in Heavy Ion Collisions

In Fig. 2, the various stages of head on heavy ion collisions are shown. Initially two Lorentz contracted nuclei approach one another in the center of mass frame. The gluons in these nuclei are very dense and form a Color Glass Condensate. The Lorentz contracted nuclei pass through one another in a very short time corresponding to the thinness of their Lorentz contracted wavefunction. In the asymptotic high energy limit, this makes a singularity at the instant of collision. After the collision, very strong color longitudinal color electric and color magnetic fields are formed, which subsequently decay. During this decay, the matter thermalizes and perhaps forms a Bose condensate. After thermalization, the decay products form a thermalized Quark Gluon Plasma. At the latest time after the system cools, a hadron gas is formed.

## *3.1 The Color Glass Condensate*

The wave function of a high energy hadron is controlled by a high density state of gluons, the CGC [10]. As shown in Fig. 3, the quark and gluon distribution functions

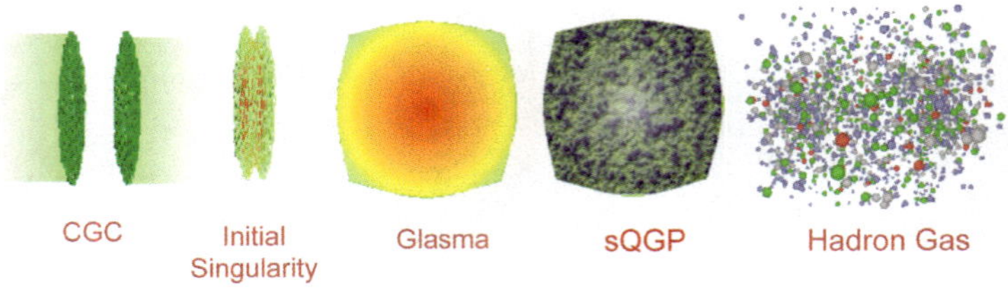


**Fig. 2** An artistic conception of heavy ion collisions

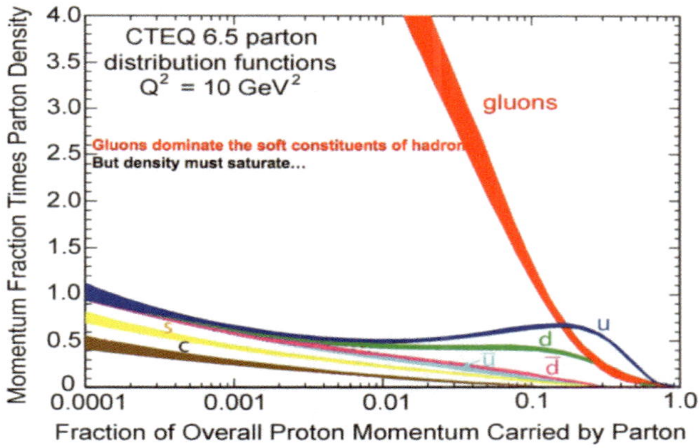


**Fig. 3** The quark and gluon distributions inside a hadron

rise rapidly as the fraction momentum $x = E_{\text{gluon}}/E_{\text{hadron}}$ is decreased. The gluons dominate the wave function at small values of $x$. For a gluon with fixed energy $\sim\Lambda_{QCD}$, the smallest values of $x$ corresponding to the highest gluon numbers are achieved at the highest energy. Since the hadron cross section grows very slowly with increasing energy, this means the gluon density grows rapidly. Therefore the typical separation between gluons is small compared to the QCD scale, and one can compute hadronic properties using weak coupling methods.

The gluon phase space density, $dN/d^2p_Td^2r_T$ at small values of $x$ is dimensionless and should be of order $1/\alpha_S \gg 1$ in this limit. This means that the gluons are highly coherent as is the case in ordinary Bose condensates. The gluons which are produced nearly at rest in the center of mass frame ("zero rapidity gluons") can be described as classical fields on account of this high phase space density. Because they are produced by gluons of much higher energy, whose time scale of evolution is Lorentz dilated, these classical fields will also appear as static, compared to natural time scales. This means that the gluons have properties similar to spin glasses, that is there is an ensemble of classical fields that can be summed over incoherently with a weight function. These observations are the basis of the name Color Glass Condensate, and there is a very well developed theory to construct the ensemble of classical fields. This theory is based on functional renormalization group equations, which are derived from first principles in the theory of strong interactions, QCD. The density of gluons per unit area has the dimensions of a momentum squared, so that there is a physical scale called the saturation momentum. The high energy limit is when $Q^2_{\text{sat}} \gg \Lambda^2_{QCD}$.

The typical gluon fields in the CGC are a Lorentz boosted Coulomb field, corresponding to Lienard-Wiechart potentials. Their polarizations and color are random, and their density is determined by the renormalization group equations for the CGC. Two sheets of CGC are represented in the left hand of Fig. 4.

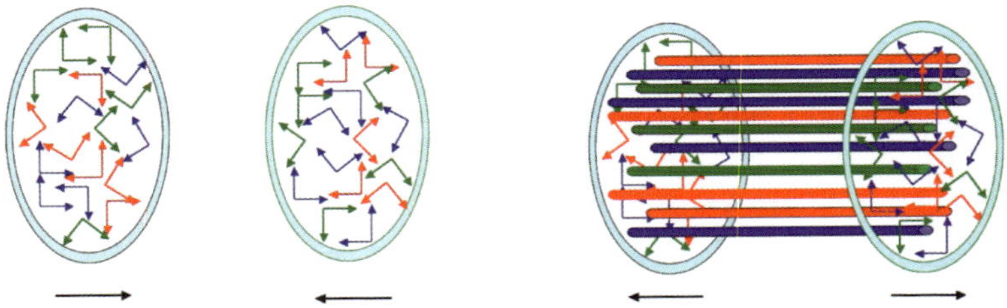

**Fig. 4** The CGC fields of a high energy hadron

## *3.2 The Glasma and the Collisions of High Energy Hadrons*

The collision of two sheets of CGC are shown in Fig. 4. On the right hand side, we see the longitudinal color electric and color magnetic fields formed in the collisions. These fields are the initial condition for the Glasma [11, 12]. The word Glasma comes from the fact that these fields are formed from the CGC fields, but at the same time eventually evolve into a thermalize QGP.

The longitudinal color electric and color magnetic fields are made because during the time the sheets of CGC pass through one another, they acquire a local color electric and color magnetic charge density. There is a symmetry in the high energy limit between electric and magnetic fields, so that one needs both electric and magnetic charge. There is an equal and opposite density induced on each sheet, and therefore longitudinal color electric and magnetic fields are induced. These fields can make a maximal local topological charge density, $\rho_{\mathrm{Top}} \sim \mathbf{E}_a \cdot \mathbf{B}^a$ that may have signatures [13].

When first proposed, it was thought that the Glasma would decay away by the classical field equations. This however does not happen: Small fluctuations in the classical field grow with time, indicating the onset of classical turbulence [14, 15]. Such fluctuations arise from the underlying quantum mechanical structure of the hadron wavefunction. These fluctuations break the boost invariance of the original classical fields, and lead to an isotropic distribution of gluons. This happens in a parametrically short time after the collision. To see this, note that the classical field is $A_{\mathrm{classical}} \sim 1/g$, and the quantum fluctuations are $A_{\mathrm{quantum}} \sim 1$. It takes a time of order $t \sim \ln(1/g)/Q_{\mathrm{sat}}$ for the system become isotropic, because of the exponential growth of fluctuations.

An interacting system would generate flow, and therefore may be thought of as a non-thermalized QGP. It is strongly interacting because of the coherent nature of the Glasma fields. One can ask whether or not the Glasma remains strongly interacting as the system proceeds to thermalization.

### 3.3 Thermalization of the Glasma

How can the Glasma become thermalized [16, 17]? We will assume that the initial phase space distribution is isotropic due to Glasma instabilities, and is of the form

$$\frac{dN}{d^3x d^3p} \sim \frac{Q_{\rm sat}}{\alpha_s E} F(E/Q_{\rm sat}) \tag{2}$$

A thermal distrbution on the other hand would be of the form

$$\frac{dN}{d^3x d^3p} \sim \frac{1}{e^{E/T} - 1} \sim T/E \tag{3}$$

for $E \ll T$. The distribution we start with for a Glasma therefore looks like the low energy coherent part of a thermal distribution cut off at an energy scale $Q_{\rm sat}$ but with a temperature $Q_{\rm sat}/\alpha_s$. The entire distribution sits at a scale $E \sim \alpha_s T$. In a thermal system it is well known that the dynamics is strongly interacting due to coherence and non-perturbative.

Thermal systems with over occupied phase space often develop Bose condensates. To see how this might occur, suppose we have a thermal distribution function with a chemical potential that allows for over occupation of phase space,

$$f \sim \frac{1}{e^{(E-\mu)/T} - 1} \tag{4}$$

The maximum occupation number one can have is when $\mu = m$ where $m$ is the particle mass. Then for $T \gg m$, $\rho_{\rm max} \sim T^3$ and $\varepsilon_{\rm max} \sim T^4$. The ratio $\rho_{\rm max}/\varepsilon_{\rm max}^{3/4}$ is a fixed number of order one. On the other hand for the Glasma distribution, this is $\rho_{\rm max}/\varepsilon_{\rm max}^{3/4} \sim 1/\alpha_s^{1/4}$ which for weak coupling is generically larger than one.

Where do the excess particles go? They end up in a Bose condensate. The correct distribution function for over occupied phase space is

$$f_{\rm thermal} = \rho_{\rm cond}\delta^{(3)}(p) + \frac{1}{e^{(E-m)/T} - 1} \tag{5}$$

The problem that we recently tried to solve was how does a Yang-Mills system with an initial over occupied phase space thermalize [16]. We assumed massless particle dynamics and took the distribution functions to be of the form

$$f(p,t) = \frac{\Lambda_s(t)}{\alpha_s p} g(p/\Lambda(t)) \tag{6}$$

The quantity $\Lambda_s$ is the coherence scale when the distribution function is of order $1/\alpha_s E$. The scale $\Lambda$ is the ultraviolet cutoff scale. Initially $\Lambda_s(0) = \Lambda(0) \sim Q_{\rm sat}$. At thermalization $\Lambda_s(t_{\rm therm}) \sim \alpha_s \Lambda(t_{\rm therm})$. Even though scattering may be quite strong during the thermalization of the system, it will take some time for the separation of scales to be established. One can therefore have a non-thermally equilibrated strongly interacting QGP, which is the thermalizing Glasma.

We looked at the equations for particle number transport, and could argue that after integrating out the momentum, we had equations for $\Lambda(t)$ and $\Lambda_s(t)$. One of the equations was energy conservation. Another followed from the transport equations themselves, which requires from dimensional reasoning that the scattering time be of the order of the time, if the system is not close to thermal equilibrium. An amusing feature of the transport equations formulated in terns of these two scales is that the coupling entirely disappears. The system is strongly interacting due to coherence during the entire time between formation and thermalization!

The behavior in time of the scales turned out to be power law, suggesting some scale invariant underlying dynamics. This leads to a thermalization time that parametrically involves inverse powers of coupling times $1/Q_{\mathrm{sat}}$. We could compute the fractional momentum space isotropy and found it to remain approximately constant during thermalization,

$$\frac{\langle p_T^2\rangle}{\langle p_L^2\rangle} \sim \text{constant} \tag{7}$$

The constancy of this ratio is a consequence of the power law evolution of the dimensional scales, and can be extracted from analyzing scattering.

The issue of formation of a Bose condensate is a delicate one that deserves further study. In Yang-Mills theory, such a condensate can decay away due to inelastic particle production. Nevertheless, while the system maintains over occupation, prior to thermalization, one might still have a condensate. This is known to occur in scalar field theory. In scalar field theory, there are classical processes that change particle number. The simulations of Epelbaum and Gelis [18] show that for an over occupied

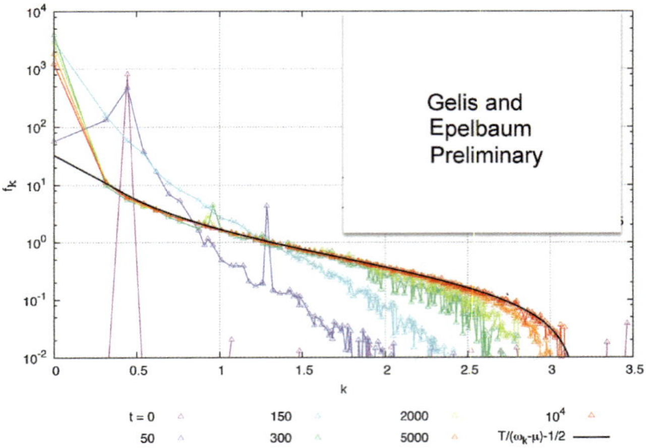


**Fig. 5** Thermalization and the formation of a Bose condensate in scalar field theory

initial distribution, one generates a Bose condensate that is long lived. This is shown in Fig. 5. One begins with a distribution that is peaked at some momentum, and then studies its evolution. One clearly sees the formation of a zero momentum mode, corresponding to a condensate, and thermalization of the system with a maximally occupied distribution, $\mu = m$, where the mass of the particle $m$ is determined by a dispersion relation extracted from the computation. The Bose condensate survives long after thermalization!

Corresponding computations for Quantum Chromodynamics are complicated because of gauge invariance and proper specification of initial conditions. There are preliminary computations that suggest the formation of a Bose condensate [19].

If such a scenario for thermalization as described above is correct, then this will require a revision of many of our ideas about the phenomenology of the formation of a sQGP in heavy ion collisions. First, the sQGP should be thought of as including both the Glasma and thermalized QGP in its definition. Both are strongly interacting plasmas that can produce flow. Jet quenching might be enhanced from scattering from the highly coherent Glasma. One significant difference between the thermalized QGP and the Glasma is that the early Glasma has an under-abundance of quarks relative to gluons. Initially, the number of gluons is $\sim 1/\alpha_s$, but the number of quarks is of order 1. Only at thermalization, do the densities of quarks and gluons approach one another.

Finally, signatures of the existence of a Bose condense state of gluons should be sought both experimentally and theoretically. Perhaps a signature might be found in the decay properties of a condensate. A condensate displaced from equilibrium should oscillate in time producing energetic particles. However, the particle's spatial momentum should be zero.

**Acknowledgments** L. McLerran wishes to acknowledge the organizers of this meeting, in particular Walter Greiner, Michael Itkis and Richard Newman. The research of L. McLerran and R. Venugopalan is supported under DOE Contract No. DE-AC02-98CH10886.

## References

1. Y.T. Oganessian, V.K. Utyonkov, Y.V. Lobanov, F.S. Abdullin, A.N. Polyakov, I.V. Shirokovsky, Y.S. Tsyganov, G.G. Gulbekian et al., Phys. Rev. Lett. **83**, 3154 (1999)
2. For a review see: W. Greiner, K. Rutz, M. Bender, T. Bürvenich et al., Phys. Rev. C **56**, 238 (1997)
3. L. McLerran, R.D. Pisarski, Nucl. Phys. A **796**, 83 (2007) [arXiv:0706.2191 [hep-ph]]
4. L.Y. Glozman, R.F. Wagenbrunn, Phys. Rev. D **77**, 054027 (2008) [arXiv:0709.3080 [hep-ph]]
5. T. Kojo, Y. Hidaka, L. McLerran, R.D. Pisarski, Nucl. Phys. A **843**, 37 (2010) [arXiv:0912.3800 [hep-ph]]
6. M.G. Alford, K. Rajagopal, F. Wilczek, Phys. Lett. B **422**, 247 (1998) [hep-ph/9711395]
7. R. Rapp, T. Schäfer, E.V. Shuryak, M. Velkovsky, Phys. Rev. Lett. **81**, 53 (1998) [hep-ph/9711396]
8. Y. Aoki, Z. Fodor, S.D. Katz, and K.K. Szabo, Phys. Lett. B **643**, 46 (2006) [hep-lat/0609068]
9. G. Boyd, J. Engels, F. Karsch, E. Laermann, C. Legeland, M. Lutgemeier, B. Petersson, Nucl. Phys. B **469**, 419 (1996) [hep-lat/9602007]

10. E. Iancu, R. Venugopalan, in Quark Gluon Plasma 3, eds. R.C. Hwa and X.N. Wang (World Scientific, Singapore, 2004) pp. 249-363 [hep-ph/0303204]
11. A. Kovner, L.D. McLerran, H. Weigert, Phys. Rev. D **52**, 3809 (1995) [hep-ph/9505320]
12. A. Kovner, L.D. McLerran, H. Weigert, Phys. Rev. D textbf52, 6231 (1995) [hep-ph/9502289]
13. D.E. Kharzeev, L.D. McLerran, H.J. Warringa, Nucl. Phys. A **803**, 227 (2008) [arXiv:0711.0950 [hep-ph]]
14. P. Romatschke, R. Venugopalan, Phys. Rev. Lett. **96**, 062302 (2006) [hep-ph/0510121]
15. K. Dusling, F. Gelis, R. Venugopalan, Nucl. Phys. A **872**, 161 (2011) [arXiv:1106.3927 [nucl-th]]
16. J.-P. Blaizot, F. Gelis, J.-F. Liao, L. McLerran, R. Venugopalan, Nucl. Phys. A **873**, 68 (2012) [arXiv:1107.5296 [hep-ph]]
17. A. Kurkela, G.D. Moore, JHEP **1111**, 120 (2011) [arXiv:1108.4684 [hep-ph]]
18. T. Epelbaum, F. Gelis, Nucl. Phys. A **872**, 210 (2011) [arXiv:1107.0668 [hep-ph]]
19. J. Berges, S. Schlichting, D. Sexty, Phys. Rev. D **86**, 074006 (2012) [arXiv:1203.4646 [hep-ph]]

# Antinuclei Produced in Relativistic Collisions: Results and Expectations

**Thorsten Kollegger and Reinhard Stock**

**Abstract** Relativistic nuclear collisions create fireball volumes of energy density reaching up to 50 $\mathrm{GeV/fm^3}$. The hadronizing system thus spreads out over the entire known and unknown spectrum of hadron-resonance species. It features a considerable yield of antibaryons, hyperons and antihyperons, as well as corresponding light nuclei, antinuclei and hypernuclei and metastable clusters. A vision emerges, of establishing an extension of the nuclear isotope chart, to antinuclei and into a further, novel dimension in addition to isospin, the strangeness coordinate. Production of light nuclear isotopes is well studied already from first Bevalac and AGS investigations. With recent increase of energy at RHIC and LHC the coverage has moved up to the anti-helium isotopes. We discuss the systematics of antinucleus production, and its interpretation in terms of the coalescence, and statistical equilibrium models. It is argued that substantial LHC and experiment upgrade would be required, to turn from initial, low statistics observations, to a precision coverage of the anti-$A = 4$ region, perhaps extending to anti-$^6$Li with a devoted LHC experiment.

## 1 Introduction

Deuterons and antideuterons were first observed in p+p collisions once the collision energy $\sqrt{s}$ moved up to the 30 GeV domain at the SPS and AGS [1]. Both the very fact of formation, and the observed yields, were qualitatively understood (or, at least, made plausible) to arise from a "coalescence" of neutrons and protons (and corresponding antinucleons) in the final phase space [2]. The essential quantity in forming

T. Kollegger · R. Stock (✉)
FIAS, University of Frankfurt, 60438 Frankfurt/M, Germany
e-mail: stock@ikf.uni-frankfurt.de

R. Stock
Institut für Kernphysik, University of Frankfurt, 60438 Frankfurt/M, Germany

W. Greiner (ed.), *Exciting Interdisciplinary Physics*,
FIAS Interdisciplinary Science Series, DOI: 10.1007/978-3-319-00047-3_17,

a light nucleus with mass number $A$ was expected to be the $A$th power of the nucleon phase space density, and its overlap with the internal wave function of nucleus $A$. Likewise for antinuclei $\bar{A}$, referring to antinucleon density. Thus cluster/anticluster formation could be expected to grow vastly if one proceeded from p+p to nucleus–nucleus collisions B+B, at a given energy: perhaps simply by $(B/2)^A$ at sufficiently high energy. This expectation turned out to be qualitatively fulfilled when the yields of light nuclei were first measured in heavy ion collisions at the Berkeley Bevalac [3–5], in the energy domain of 1 GeV/nucleon.

Since then the phenomena of light nuclear/antinuclear species production have been further pursued, culminating in the recent observation of anti-$^4$He by the STAR experiment at RHIC [6] and augmented further by a venture into truely exotic direction by detecting the anti-hyper-tritium species [7]. The nucleosynthesis of light clusters, among a primordial gas of protons and neutrons, is familiar from "The First Three Minutes" [8] of the cosmological big bang evolution. The present cold intergalactic gas exhibits a high ratio of $^4$He to protons, of about 0.25, that are "frozen in" from the primordial hot dynamics, during the multi-second era. However, a closer look at this "explosive nucleosynthesis" phenomenon [9, 10] reveals a physics different, in detail, from nuclear collisions. The abundance ratios of various light nuclei result from a system of coupled rate equations that is dominated by the breakup of emerging bound nuclear states, by a $10^9$ times denser field of primordial photons, at a temperature in the 10 MeV domain. This photon field results from the primordial, almost complete particle–antiparticle annihilation that took place at about 5 $\mu$s big-bang time, immediately following the cosmological parton to hadron phase transition. We recall that this global annihilation brought an end to the initial, almost complete matter–antimatter symmetry (up to a tiny fraction of $10^{-9}$ of matter over antimatter excess), that existed until the cosmological parton to hadron transition. Only the tiny matter excess survives.

Quite on the contrary, an ultra-relativistic nucleus–nucleus collision B+B starts from a maximal matter–antimatter asymmetry, with net baryon number $2B$ that is strictly conserved throughout the collision. Thus there is, further, no such annihilation photon field in the final state, which would break up lightly bound nuclear species like e.g. deuterium. Conserved net baryon number results in a substantial formation rate of light nuclei (including hypernuclei from concurrent hyperon formation), as is observed from Bevalac to SPS energy. Moreover, however, a significant fraction of the initial center of mass energy (about 70 %) gets converted to a newly created quark–antiquark and gluon population, in the collision "fireball". Toward top RHIC and LHC energies, this zero net baryon number quark-gluon matter (the so-called Quark-Gluon Plasma of QCD) becomes dominant at mid-rapidity (in the center of phase space).

Upon hadronization (due to cooling) essentially no matter–antimatter annihilation takes place, unlike in the universe, owing to the rapid expansion of the fireball which is about 30 orders of magnitude smaller than the universe, at hadronization time which is thus much smaller, about $10^{-22}$ s instead of 5 $\mu$s. Thus nucleons and antinucleons (more generally: baryons and antibaryons) are about equally abundant in the center (at mid-rapidity) of the expanding fireball, once the collision energy moves up to the 200 GeV at RHIC, and beyond. This gives rise to the approximate equality of, e.g., the $^4$He and anti-$^4$He production rates, as was recently observed by the STAR experiment [6] at RHIC (next section).

This is exciting new physics, chiefly for the following reasons:

1. Quantitative analysis of the various nuclear, antinuclear and hypernuclear yields, along with all other measured hadronic yields, within the framework of the statistical hadronization model [11, 12] shows a surprise: All hadronic species including the light nuclei/antinuclei and hypernuclei are produced with a "hadro-chemical" equilibrium yield distribution that corresponds to the fireball temperature, energy and entropy density prevailing at hadronization, at $T = 165 \pm 5$ MeV. All the observed species share in this equilibrium, "democratically", i.e. according to their statistical weights. This also includes all the nuclear/antinuclear/hypernuclear species [12]. The implied equilibrium among all conceivable open production channels, upon hadronization from the partonic phase, remains a key unresolved problem of QCD dynamics of hadron formation, a quantum mechanical paradox. Moreover, the recent success of the statistical model in predicting also the nuclear/antinuclear yields opens up another puzzle. The yields, and yield ratios, of composite, well extended nuclei appear to survive the hadronic expansion cascade evolution that sets in after hadron/nucleus formation, at rather high matter density initially. And this in spite of very substantial, and very different breakup cross sections, implied by the widely different nuclear wave functions. However, we note that the intensity of light nuclear cluster formation, relative to the emission rate of unclustered nucleons, represents a measure of the total entropy per baryon, created at hadronization; and likewise for antinuclei/antinucleons [13]. Is entropy strictly conserved during the hadron/resonance stage of fireball expansion (as was first conjectured by Bertsch and Cugnon [14] decades ago)? A profound challenge to current microscopic hadron transport models.
2. Walter Greiner has proposed a mechanism of spontaneous antiquark creation [15] in the superhigh energy density of the collisional fireball, of more that 50 $\text{GeV/fm}^3$ in the RHIC-LHC domain. This unprecedented energy density could become "overcritical" for the population of the negative energy (antiparticle) Dirac Sea—in analogy to the proposal [16] of spontaneous QED vacuum excitation in the overcritical Coulomb field occurring with $Z > 172$ of fused cold heavy nuclei, leading to positron emission. The strong interaction counterpart would result in enhanced antinucleon and antinucleus production from an energy-overcriticality of the collisional fireball. A formidable challenge, recalling that at top RHIC energy only every about 100.000th event features one anti-helium at midrapidity—not yet a territory for precision measurements.

## 2 Data

Before focusing on the data concerning anti-deuteron, anti-$^3$He and anti-$^4$He production at RHIC, a short glance is in order at experimental technique. NA49 has covered [17] "coalescence" at the SPS energies, $6.3 < \sqrt{s} < 17.3$ GeV, and STAR chiefly at top RHIC energy [6, 7]. Both experiments employ tracking in large TPC volumes, as illustrated for STAR in Fig. 1, in combination with time of flight (TOF) determination outside the TPC volume. One then combines the track-by-track specific ionization signal from the TPC with the TOF signal, as is shown in Fig. 2. These observables ideally complement each other, ionization ($\mathrm{d}E/\mathrm{d}x$) measuring $Z$, and TOF the mass $A$ (more precisely $\mathrm{A}^2$). Figure 2 indicates that the combined $(Z, A)$ resolution is perfectly satisfactory, to separate the light nuclear/antinuclear species. We note that the experiment ALICE at the LHC will shortly join this physics, with data at $\sqrt{s} = 7.2$ TeV, employing the same technique.

The principal experimental problem for an uninhibited study of antinuclear production is the smallness of the yield, more specifically the fall-off with increasing mass $A$. This is illustrated in Fig. 3 by STAR results [6] for $A = 3$ and 4 nuclei and antinuclei. We note that at top RHIC energy, and at midrapidity, the pbar yield closely approaches the p yield. We assume the corresponding n and nbar yields to equal the $p$, $\bar{p}$ yields (see below). Thus the $^4$He and anti-$^4$He are closely similar, but note the absolute count numbers: only 18 anti-$^4$He could be identified from about $10^9$ central Au+Au collision events recorded at $\sqrt{s} = 200$ GeV, which contained a total of about $10^{12}$ analyzed tracks.

The final data summary of the STAR study [6] is shown in Fig. 4. From the elementary proton and antiproton yields the $A$ dependence moves down extremely steeply, each incremental step in $A$ occurring with a yield suppression of almost 3 orders of magnitude. On the narrow logarithmic scale a slight cross section excess for nuclei over their antinuclei is hardly visible. In coalescence model terminology this excess arises from taking the $A$th power of the elementary nucleon/antinucleon densities in phase space, to arrive at the $A$ yield. This aspect of the model is further illustrated in Fig. 5 which gives the antinucleus to nucleus yield ratios of the data shown in Fig. 4. At top RHIC energy the $\bar{p}/p$ ratio is about 0.8. Coalescence reduces

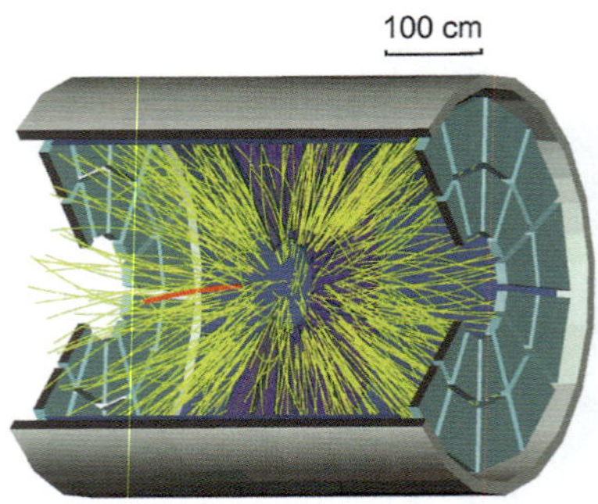


**Fig. 1** Tracking in the STAR Time Projection Chamber (TPC) [6]

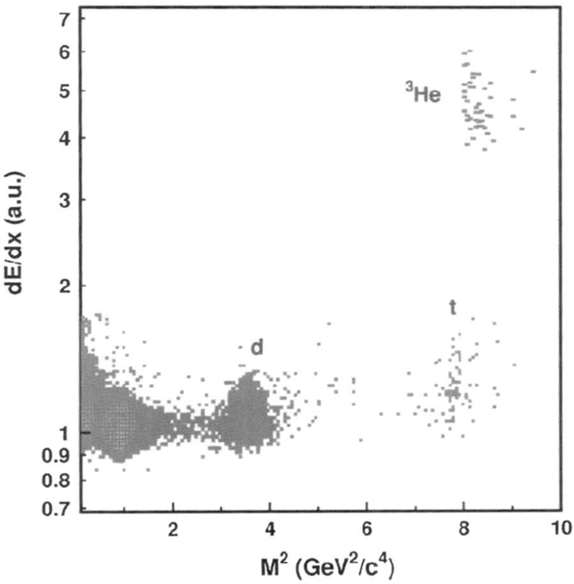


**Fig. 2** Separation of p, d, t and $^3$He by combination of specific ionization (TPC) and time of flight in NA49 [19]

this ratio by successive powers $A$ (dashed line), well corresponding to the STAR data.

A final remark concerning Fig. 4. The indicated exponential decay of the yields may hide smaller second order dependences. We note here that the extremely fast fall-off with increasing $A$ is, on the one hand, well compatible with the coalescence picture: it requires the phase space densities of $A$ nucleons or antinucleons to overlap with the internal wave function of nucleus $A$. This consideration reflects in the famous coalescence yield formula [4] for the yield of nuclear species $A(Z, N)$:

$$E(A)\frac{\mathrm{d}^3N(A)}{\mathrm{d}^3p(A)} = B(A)\left(\frac{E(p)\mathrm{d}^3N(p)}{\mathrm{d}^3p(p)}\right)^Z\left(\frac{E(n)\mathrm{d}^3N(n)}{\mathrm{d}^3p(n)}\right)^{(A-Z)} \tag{1}$$

Besides the phase space densities it contains the essential "coalescence factor" $B(A)$ which accounts [18] for the relation of the coherent fraction of the fireball freeze-out volume, to the internal wave function of cluster $A$. Under conditions of constant collisional energy, as considered in Figs. 4 and 5, the well-known dramatic difference between the weakly bound deuteron, and the strongly bound $^4$He momentum space wave functions may cause specific deviations from a simple universal exponential decay with $A$. Thus the indications of a slightly above-trend $^4$He and $^4\overline{\mathrm{He}}$ yield, in Fig. 4, require a more detailed theoretical analysis before we could claim

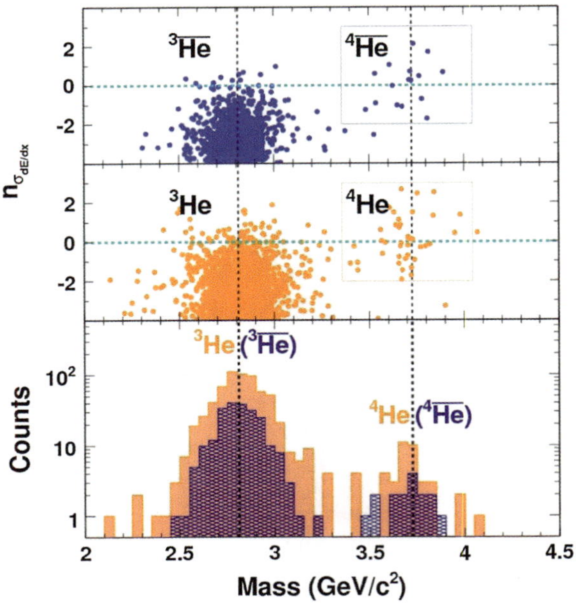


**Fig. 3** Separation of helium and anti-helium isotopes at top RHIC energy in STAR [6]

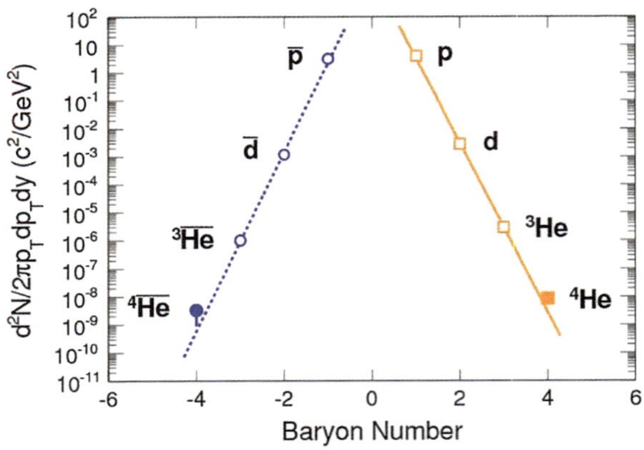


**Fig. 4** Light nucleus/antinucleus production in central Au+Au collisions at $\sqrt{s} = 200$ GeV in STAR [6]

indication of an additional creation mechanism, perhaps along the lines explored in Ref. [15].

A last remark on coalescence concerns the unknown neutron phase space density required in Eq. 1. It is well known that the initial neutron excess in heavy nuclear collision partners gets transformed in the fireball chemical equilibration, the excess down quarks being transferred mostly to a slight excess of $\pi^-$ over $\pi^+$. The final $p$

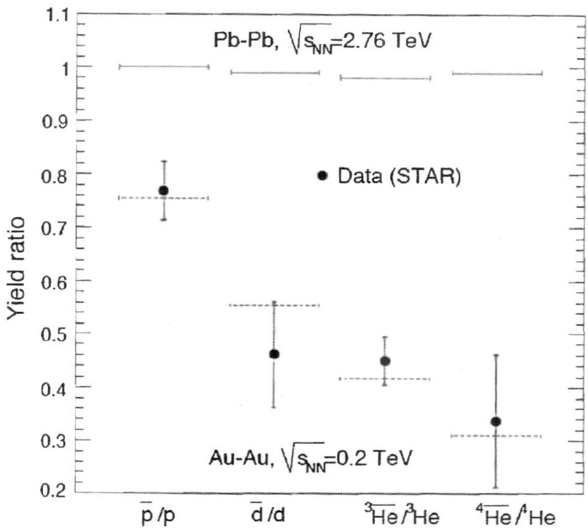


**Fig. 5** Antinucleus to nucleus ratios at top RHIC energy. Data from Fig. 4

to $n$ ratio of the produced baryons and light nuclei can, however, be measured by the yield ratio of $^3$He to $^3$H (tritium). These two wave functions are nearly identical, a deuteron plus an additional proton or neutron in the $1s$ shell. The energy dependence of the t/$^3$He ratio is shown in Fig. 6 [17, 19], to in fact closely approach unity with increasing energy. In Fig. 7, finally, we show the coalescence factor $B(2)$ from (1), for anti-deuteron production [20]. It reflects, primarily, the widening of momentum space due to increasing temperature and radial flow [18], which both saturate toward high $\sqrt{s}$.

# 3 Statistical Model Analysis

We note first that the coalescence formula (1) can be written in the form of the "mass action law", with $B$ as the equilibrium constant. The other terms are the hadrochemical equilibrium concentrations that are also the subject of the statistical hadronization model pioneered by Hagedorn [21]. At hadron formation the energy-and entropy densities, as well as the net charges of the fireball, are shared in equilibrium among the produced hadrons according to their statistical weights (the Grand Canonical Ensemble [11, 12]). This equilibrium property is not yet fundamentally understood, along with the QCD confinement process that drives hadronization. In its most recent realization the statistical model has been applied to light nuclei/antinuclei production in Au+Au collisions by Andronic et al. [12]. The most striking point of this investigation lies in the fact that the nuclear/antinuclear yields share in the hadrochemical

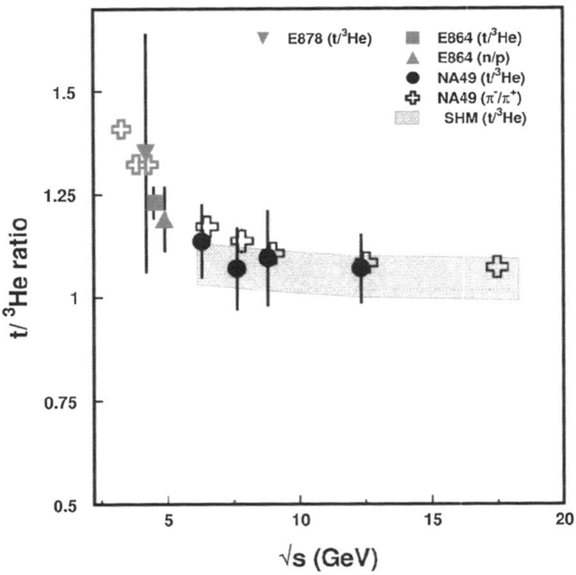


**Fig. 6** The t/$^3$He yield ratio as a measure of the n/p ratio, at AGS and SPS [17]

equilibrium that is established at hadronization, $T = 165$ MeV. The nuclei are not a secondary effect as indicated by the term "coalescence", they are established before the onset of the final dynamical phase, the hadron-resonance expansion. Of course this is clearly the case for the produced (anti-)nucleons, of which the (anti-)clusters are a part. However from the cosmological synthesis example we are used to presume that nuclei form much later, at the time of final decoupling from strong interaction, i.e. at the end of the hadron cascade expansion. Not so.

The following consideration could be helpful. Hadronic freeze-out at $T =$ 165 MeV fixes the total entropy, and the entropy per baryon in the system. Actually its value is born even earlier in the dynamics, at about 1–2 fm/c when the partonic system first settles to local equilibrium, such that hydrodynamics can describe the further evolution, until hadronization. This appears to occur, according to our inferences from the observed hydrodynamic directed flow phenomena [22]. Such an expansion is isentropic, it preserves the initially generated, low values of $S/A$. A crossover hadronization transition also conserves this quantity. Decades ago it was pointed out by Siemens and Kapusta [13] that $S/A$ is proportional to $-\ln R(d, p)$, $R$ being the ratio of clustering to nonclustering; more explicitely the ratio of bound nucleon pair number to total nucleon number [5, 22]. Quite intuitively, clustering decreases the entropy. Thus this clustering is, or gets fixed at hadronization (hadrochemical freeze-out at $T = 165$ MeV). Agreement with the data thus indicates that the subsequent hadron/resonance expansion phase is also isentropic [14] thus preserving the extent of clustering, along with the other parts of the hadronic mass distribution. The

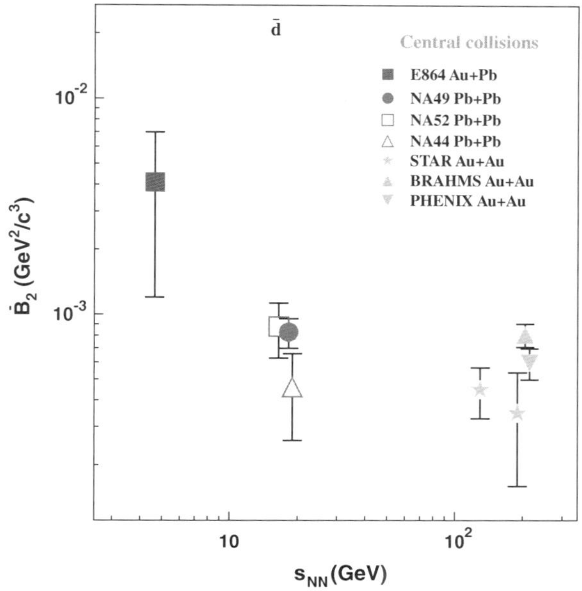


**Fig. 7** The coalescence factor $B(2)$ for anti-deuteron production [20]

properties of a cascade expansion mechanism that could preserve the weakly bound nuclear species, at work here, are still unexplored.

A summary of all mid-rapidity cluster predictions of Ref. [12] is shown in Fig. 8. Nuclear yields ($^3$He, $^4$He) decrease exponentially with incident energy, due to growing specific entropy and a widening of nucleonic phase space (see Fig. 6) due to increasing flow, until some saturation occurs toward RHIC energy. Antinuclear yields increase steeply, along with the antinucleon production rate, until they meet with the nuclear yields at top RHIC energy. Further toward LHC energy we observe a slight increase of yields, probably due to an increase in fireball volume, the only parameter of the Grand Canonical Ensemble that does not saturate like $T$ and $\mu(B)$. The various hypernuclear yields first approach a maximum (due to strangeness saturation), then fall off and slightly re-rise like the nuclear yields.

## 4 Conclusions

We have seen that all existing data can be well understood within the framework of the statistical model. Nuclei, antinuclei and hypernuclei thus behave like all other created hadrons. They equilibrate according to their masses and quantum numbers. We can thus exploit the predictive power of the SM to extrapolate into unknown and exotic domains.

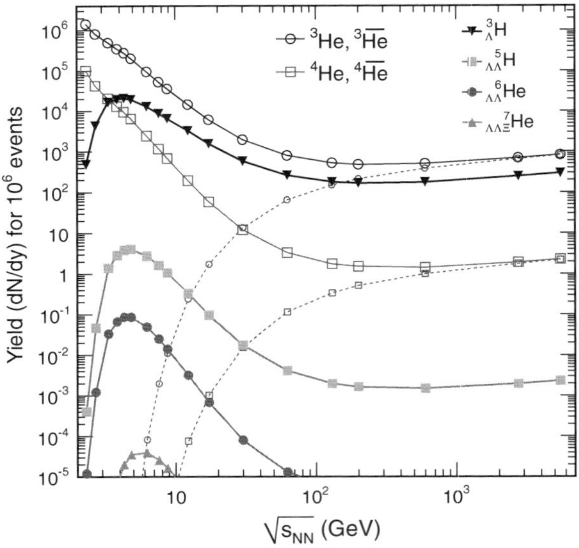


**Fig. 8** Light cluster production predicted by the statistical model [12]

Antinuclei are becoming tractable from top RHIC energy onward. Each incremental step in (anti-)atomic mass number A costs a little less than 3 orders of magnitude in the yields. The present state of the art just about covers anti-$^4$He. Novel QCD mechanisms [15] might just begin showing up here. What about anti-$^6$He or anti-$^6$Li? We extrapolate to a multiplicity of a few times $10^{-11}$ per unit rapidity at LHC energy: beyond the possible? LHC heavy ion luminosity may increase by a further order of magnitude at most, as might be possible with the ALICE experiment data acquisition rate (under a selective trigger scheme)—not too promising. However, note that anti-Li ions, with $Z = -3$, have an enormous specific ionization. One might think of a devoted LHC experiment with a detection scheme that is blind to the bulk hadrons with $Z = \pm 1$, put into a long solenoid magnet (the only possibility at colliders) with a weak magnetic field, just efficient for charge 3 separation, but covering a wider rapidity interval?

As Fig. 8 indicates, new exotic hypernuclear species which are of key interest even from the perspective of current nuclear physics, should be investigated at $\sqrt{s}$ about 5 GeV, corresponding to about 15 GeV per nucleon in fixed target experiments. This is a task for the future FAIR facility at GSI.

The future is rich in formidable challenges. This includes a challenge to future theory. All the above mentioned theoretical tools that apparently give a fair account of our data at hadronization are not QCD but classical physics: from relativistic hydrodynamics to the grand canonical Gibbs ensemble. What is the quantum mechanics that must be acting behind the scene?

## References


1. S.T. Butler, C.A. Pearson, Phys. Rev. **129**, 836 (1963)
2. A. Schwarzschild, C. Zupancic, Phys. Rev. **129**, 854 (1963)
3. H.H. Gutbrod, A. Sandoval, P.J. Johansen, A.M. Poskanzer, J. Gosset, W.G. Meyer, G.D. Westfall, R. Stock, Phys. Rev. Lett. **37**, 667 (1976)
4. J. Gosset, H.H. Gutbrod, W.G. Meyer, A.M. Poskanzer, A. Sandoval, R. Stock, G.D. Westfall, H.H. Gutbrod, Phys. Rev. C **16**, 629 (1977)
5. K.G.R. Doss, H.A. Gustafsson, H.H. Gutbrod, B. Kolb, H. Lohner, B. Ludewigt, A.M. Poskanzer, T. Renner, Phys. Rev. C **32**, 116 (1985)
6. H. Agakishiev et al. [STAR Collaboration], Nature **473**, 353 (2011), erratum ibid. 475, 412 (2011) [arXiv:1103.3312[nucl-ex]] (Erratum)
7. B.I. Abelev [STAR Collaboration], Science **328**, 58 (2010) [arXiv:1003.2030 [nucl-ex]]
8. S. Weinberg, *The First Three Minutes* (Basic Books, New York, 1977), p. 269
9. A. Mekjian, Phys. Rev. Lett. **38**, 640 (1977)
10. A.Z. Mekjian, Phys. Rev. C **17**, 1051 (1978)
11. F. Becattini, M. Gazdzicki, A. Keranen, J. Manninen, R. Stock, Phys. Rev. C **69**, 024905 (2004) [hep-ph/0310049]
12. A. Andronic, P. Braun-Munzinger, J. Stachel, H. Stocker, Phys. Lett. B **697**, 203 (2011) [arXiv:1010.2995 [nucl-th]]
13. P.J. Siemens, J.I. Kapusta, Phys. Rev. Lett. **43**, 1486 (1979)
14. G. Bertsch, J. Cugnon, Phys. Rev. C **24**, 2514 (1981)
15. W. Greiner, Int. J. Mod. Phys. E **5**, 1 (1996)
16. U. Müller, T. de Reus, J. Reinhardt, B. Müller, G. Soff, W. Greiner, Phys. Rev. A **37**, 1449 (1988)
17. V.I. Kolesnikov [NA49 Collaboration], J. Phys. Conf. Ser. **110**, 032010 (2008) [arXiv:0710.5118 [nucl-ex]]
18. R. Scheibl, U.W. Heinz, Phys. Rev. C **59**, 1585 (1999) [nucl-th/9809092]
19. V. I. Kolesnikov, private communication
20. T. Anticic [NA49 Collaboration], Phys. Rev. C **85**, 044913 (2012) [arXiv:1111.2588 [nucl-ex]]
21. R. Hagedorn, Nuovo Cim. Suppl. **3**, 147 (1965)
22. R. Stock, In *Landolt-Börnstein I 21A: Elementary particles 7*, (Springer Verlag, 2008) [arXiv:0807.1610[nucl-ex]]

# RHIC and LHC Phenomena with a Unified Parton Transport

**Ioannis Bouras, Andrej El, Oliver Fochler, Felix Reining, Florian Senzel, Jan Uphoff, Christian Wesp, Zhe Xu and Carsten Greiner**

**Abstract** We discuss recent applications of the partonic pQCD based cascade model BAMPS with focus on heavy-ion phenomeneology in hard and soft momentum range. The nuclear modification factor as well as elliptic flow are calculated in BAMPS for RHIC end LHC energies. These observables are also discussed within the same framework for charm and bottom quarks. Contributing to the recent jet-quenching investigations we present first preliminary results on application of jet reconstruction algorithms in BAMPS. Finally, collective effects induced by jets are investigated: we demonstrate the development of Mach cones in ideal matter as well in the highly viscous regime.

## 1 Introduction

In collisions of heavy ions at ultrarelativistic energies at the Relativistic Heavy-Ion Collider (RHIC) and the Large Hadron Collider (LHC) a new state of matter, the quark-gluon plasma (QGP), has been created. Although the QGP is not available for direct observation, its properties can be deduced from the measurement of the produced hadrons in the final state.

The large values of the measured hadronic elliptic flow $v_2$ [1–3], which is the second coefficient of the Fourier series of the azimuthal particle multiplicity, suggests that equilibration of quarks and gluons occurs on a very short time scale $\lesssim$1 fm/c. This also suggests that the shear viscosity over entropy density ratio $\eta/s$ of the QGP is very small, which means that the QGP behaves like a nearly perfect fluid. All these

I. Bouras · A. El · O. Fochler · F. Reining · F. Senzel · J. Uphoff · C. Wesp · C. Greiner (✉)
Institut für Theoretische Physik, Johann Wolfgang Goethe-Universität, Max-von-Laue-Str. 1, 60438 Frankfurt am Main, Germany
e-mail: carsten.greiner@th.physik.uni-frankfurt.de

Z. Xu
Department of Physics, Tsinghua University, Beijing 100084, China
e-mail: xuzhe@mail.tsinghua.edu.cn

W. Greiner (ed.), *Exciting Interdisciplinary Physics*,
FIAS Interdisciplinary Science Series, DOI: 10.1007/978-3-319-00047-3_18,

conclusions can be drawn from comparison of experimental results with hydrodynamic calculations. However, an understanding of the mechanism of fast thermalization can not be achieved in the scope of hydrodynamic models. The early pre-equilibrium dynamics of the QGP must be studied in the scope of the kinetic theory.

In contrast to the hydrodynamic approach, kinetic transport theory is a microscopic theory and thus allows to study processes of soft and hard processes simultaneously. This is in particular important for detailed understanding of further properties of the quark-gluonic medium, such as the suppression of jets and heavy-quarks. Suppression of jets, also known as jet quenching, is quantified by comparing the hadron multiplicities measured in heavy-ion collisions with appropriately scaled multiplicities from $p + p$ collisions [4–6]. In addition, very exciting jet-associated particle correlations were observed [7], which might be the result of a conical emission off propagating shock waves in form of Mach Cones. These Mach Cones might be induced by high-energy partons traversing the expanding medium [8]. Observations of these effects is consistent with the picture of a nearly perfect fluidity of the QGP.

The kinetic transport model BAMPS (Boltzmann Approach to Multiparton Scatterings) [9] has been developed to provide a unified description of dynamics of the early QGP stage of heavy-ion collisions (HIC) including perturbative QCD based elastic and inelastic processes. BAMPS has been applied to provide explanation of fast thermalization on a very short time scale $\leq 1$ fm/c [10] as well as a small value of $\eta/s \approx 0.08 - 0.2$ for $\alpha_s = 0.6 - 0.3$ [11, 12]. Furthermore recent calculations with BAMPS provide results on elliptic flow [13, 14] and jet quenching [15] at RHIC energies, which is for the first time done in a consistent and fully pQCD–based microscopic transport model. In addition, BAMPS has been used in certain works as a reference for hydrodynamic calculations. This opens the possibility to study hydrodynamic phenomena for arbitrary viscosity.

In these proceedings we discuss application of BAMPS to describe a number of phenomena observed in the recent heavy-ion experiments. In Sect. 2 we introduce calculations of the nuclear modification factor $R_{AA}(p_T)$ for RHIC and LHC conditions. In Sect. 3 BAMPS results on elliptic flow and suppression of charm and bottom quarks for RHIC and LHC energies are introduced. In Sect. 4 for the first time in the framework of BAMPS we introduce preliminary results on application of jet reconstruction algorithms. Finally in Sect. 5 the formation and propagation of shock waves in form of Mach Cones are discussed for a wide range of viscosity to entropy density ratio $\eta/s$.

## 2 Nuclear Modification Factor and Elliptic Flow from Partonic Transport Simulations

As established in [13–15] the partonic medium in BAMPS simulations of ultra relativistic heavy ion collisions features a small ratio of the shear viscosity to the entropy, $\eta/s$, and develops a strong collectivity with an integrated $v_2$ that is in good

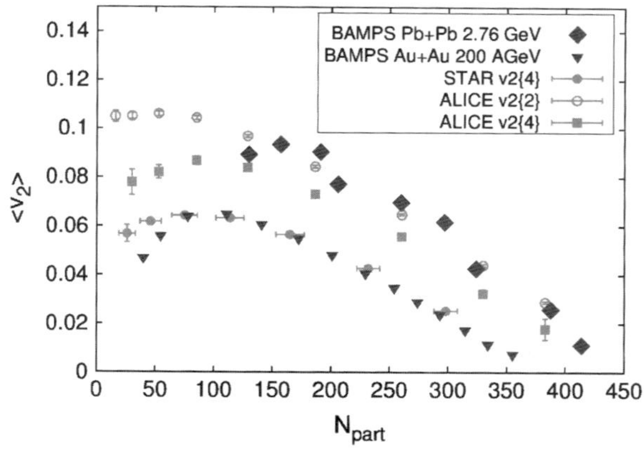


**Fig. 1** Integrated partonic $v_2$ from BAMPS as a function of $N_{\mathrm{part}}$ for Pb+Pb at 2.76 ATeV ($|y| < 0.8$) and Au+Au at 200 AGeV ($|y| < 0.5$) compared to the measured $v_2$ of charged particles from ALICE [16] and from STAR [17]

agreement with experimental results over a large centrality range for a fixed coupling of $\alpha_s = 0.3$ and a kinetic freeze-out energy density $\varepsilon_c = 0.6\,\mathrm{GeV/fm^3}$. These parameters are used for all calculations that are presented in this section.

Figure 1 shows the integrated partonic $v_2$ as a function of centrality from simulations of Au+Au at 2.76 ATeV and of Au+Au at 200 AGeV compared to experimental data from the ALICE experiment [16] at LHC and from the STAR experiment [17] at RHIC. Using the same set of parameters ($\alpha_s = 0.3$, $\varepsilon_c = 0.6\,\mathrm{GeV/fm^3}$) that have been fixed to the RHIC data [13, 14] also the integrated elliptic flow at LHC can be described over a large range in centrality. Accordingly the simulated differential $v_2$ of Pb+Pb collisions at LHC energies shows no significant deviation from the Au+Au results at 200 AGeV in the low to intermediate $p_T$ region which is also in agreement with experimental findings [18].

One of the main virtues of the transport model BAMPS is that it allows for the investigation of different observables within a consistent framework and consequently also the nuclear modification factor, $R_{AA}$, is studied using the same parameters that provide a description of the elliptic flow ($\alpha_s = 0.3$, $\varepsilon_c = 0.6\,\mathrm{GeV/fm^3}$). Figure 2 shows the nuclear modification factor $R_{AA}$ obtained from BAMPS simulations of central, 0–5 %, Pb+Pb collisions at 2.76 ATeV. The results are both shown on the partonic level for gluons and light quarks and on the hadronic level for neutral pions based on AKK fragmentation functions [19]. The suppression of high-$p_T$ particles in simulations with BAMPS is distinctly stronger than the experimentally observed suppression, which is also observed in simulations of $R_{AA}$ at RHIC energies [20]. Additionally the rise towards larger transverse momenta that is present in the LHC data is not reproduced.

The strong quenching observed in BAMPS calculations is due to the energy loss in $2 \rightarrow 3$ interactions that include an effective implementation of the LPM effect [21] via a mean free path-dependent cutoff [9, 15]. The strong quenching is then

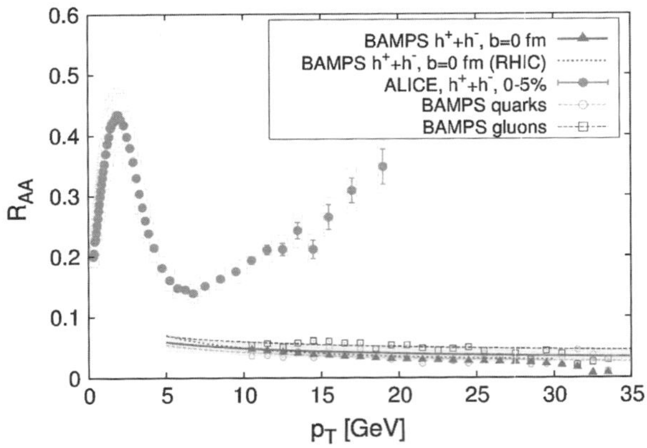


**Fig. 2** Nuclear modification factor $R_{AA}$ of charged hadrons, gluons and quarks from BAMPS simulations of Pb+Pb at $b = 0$ fm compared to results from ALICE for 0–5 % central Pb+Pb collisions [18]. Lines indicate $R_{AA}$ computed from fits to the simulated parton spectra, while symbols indicate $R_{AA}$ computed directly from the parton spectra as obtained from BAMPS. For comparison the $R_{AA}$ of charged hadrons from simulations of Au+Au at 200 AGeV and $b = 0$ fm is also shown

caused by [20] a) a strong energy loss that is caused by a complex interplay of the Gunion-Bertsch matrix element and the effective implementation of the LPM effect [20], b) a conversion of quark into gluon jets in $2 \rightarrow 3$ interactions and c) a small difference in the energy loss of gluons and quarks caused by the iterative computation of interaction rates required by the inclusion of the LPM cutoff.

Thus, while the collectivity of the medium can be well described within the current approach, the quenching of high-$p_T$ particles is overestimated. Future studies will therefore focus on the implementation of a running coupling for light quarks and gluons and also systematically explore the modeling of the LPM effect. These modifications are qualitatively expected to bring the results for the nuclear modification factors into better agreement with experimental data.

## 3 Elliptic Flow and Suppression of Heavy Quarks

Heavy quarks are a good probe to study the properties of the QGP. They are well calibrated in a sense that they are produced entirely in the early stage of the heavy ion collision due to their large mass [23] and are also tagged during hadronization due to flavor conservation. Whereas heavy quarks at RHIC can only be measured indirectly via heavy flavor electrons, at LHC for the first time it is possible to reconstruct $D$ mesons and, therefore, receive information only about charm quarks.

The heavy flavor electron data from RHIC [22, 26, 27] and the heavy flavor electron, muon and $D$ meson data from LHC [24, 28] show that the suppression of

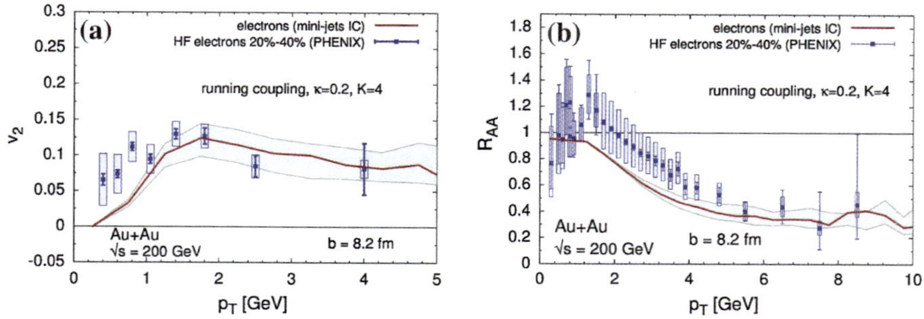


**Fig. 3** Elliptic flow $v_2$ (*left*) and nuclear modification factor $R_{AA}$ (*right*) of heavy flavor electrons for Au+Au collisions at RHIC with an impact parameter of $b = 8.2$ fm together with data [22]. The elastic cross section of $gQ \to gQ$ is multiplied with the factor $K = 4$ to mimic the influence of radiative processes

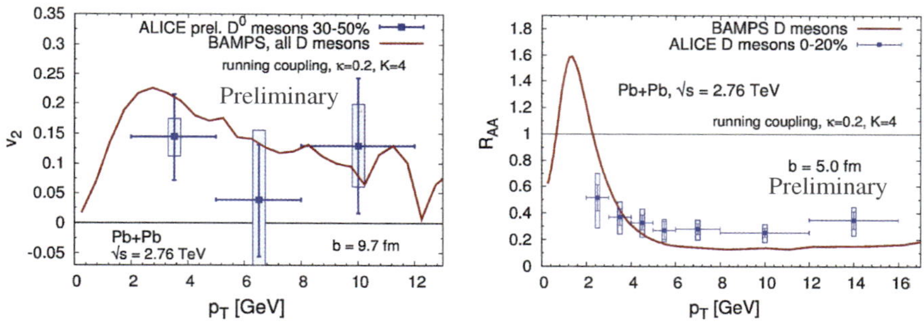


**Fig. 4** Preliminary BAMPS results on elliptic flow $v_2$ (*left*) and nuclear modification factor $R_{AA}$ (*right*) of $D$ mesons at Pb+Pb collisions at LHC with an impact parameter $b$ together with data [24, 25]. The cross section of $gQ \to gQ$ is multiplied with the factor $K = 4$

heavy quarks is on the same order as for light quarks. From the theory perspective it was thought that radiative processes involving heavy quarks are suppressed due to the dead cone effect [29, 30], which means that gluon radiation at small angles is suppressed and, therefore, the energy loss is smaller compared to light partons. Elliptic flow $v_2$ measurements of particles associated with open heavy flavor also show that heavy quarks interact strongly with the other particles of the medium. Whether these observations can be explained by collisional or radiative energy loss or other effects is currently in debate.

The elliptic flow $v_2$ and the nuclear modification factor $R_{AA}$ are important observables for heavy quarks. Although those particles are rare probes, both observables are experimentally accessible for fragmentation and decay products of heavy quarks such as $D$ mesons or heavy flavor electrons. The $R_{AA}$ reflects how much energy heavy quarks lose in the QGP. The $v_2$ is large if heavy quarks interact often with the medium and pick up its collective flow.

All the calculations for heavy quarks in this section are done with a running coupling and an improved Debye screening. The latter means that the screening mass of the $t$ channel of elastic scatterings is determined such that the energy loss matches the energy loss of a heavy quark calculated within the hard thermal loop approach. More information how this matching is done can be found in Refs. [31–34].

Detailed studies in BAMPS [33, 35–38] show that elastic energy loss of heavy quarks alone is not compatible with the experimental data at RHIC and LHC. However, elastic energy loss explains a significant portion of the overall suppression. If we employ a running coupling and improved Debye screening the experimental data for both $v_2$ and $R_{AA}$ for both RHIC and LHC can be explained if the elastic cross section is multiplied with the artificial factor $K = 4$. This indicates that radiative energy loss should be three times larger than the elastic energy loss. However, this must be checked in a forthcoming study. First results on implementing radiative energy loss of heavy quarks in BAMPS look promising [35].

Figure 3 compares our results of the $v_2$ and $R_{AA}$ at RHIC to the heavy flavor electron data from Ref. [22]. The agreement with the experimental data is very good for both observables if one employs a factor $K = 4$ for the elastic cross section to mimic the effect of radiative energy loss. At LHC for the first time it is possible to reconstruct $D$ mesons and, therefore, distinguish between charm and bottom quarks. In Fig. 4 our results on $D$ mesons is compared to data from ALICE. For the same parameters, that describe the RHIC data, a good agreement is also found at LHC. The suppression of $D$ mesons at LHC is slightly larger than the data. This can be due to a different relation between collisional and radiative processes at LHC compared to RHIC or due to the fact that we represent the rather large centrality class $0-20\,\%$ by only one impact parameter. We note that muon data from charm and bottom quarks at forward rapidity is also well described for the same parameters [35].

## 4 Jet Reconstruction Within BAMPS

Another observable to determine the parton energy loss inside a heavy-ion medium is the reconstruction of full dijets. The initial hard scattering processes of the approaching nucleons lead to back-to-back parton pairs, which gain a high amount of virtuality during these scattering processes. In the subsequent evolution of partons, they try to decrease their virtuality by splitting processes like $\mathrm{q} \rightarrow \mathrm{qg}$ or $\mathrm{g} \rightarrow \mathrm{gg}$, which can be described by the DGLAP evolution equation [39–41]. These fragmentation processes lead to particle showers with a broad angle and momentum distribution. In order to provide a description of the energy loss mechanism inside the created medium, jet reconstruction methods [42–44] are used. They combine single shower particles to a common "full jet" based on their distance $\Delta R = \sqrt{(\Delta y)^2 + (\Delta\phi)^2}$ to the jet axis.

In p + p collisions, where no medium creation is expected, these splitting processes already lead to an imbalance in the momenta of the reconstructed jets with the two highest transverse momenta. These jets are associated with the initial back-to-back

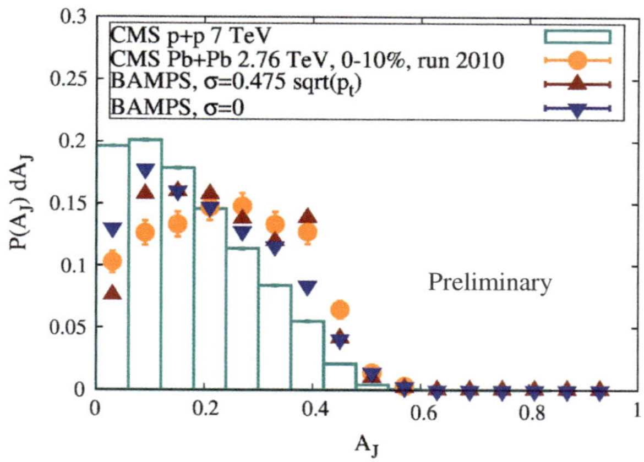


**Fig. 5** Momentum imbalance $A_J$ in central Pb + Pb collisions at the LHC with and without gaussian smearing for $\alpha_s = 0.3$ [45]

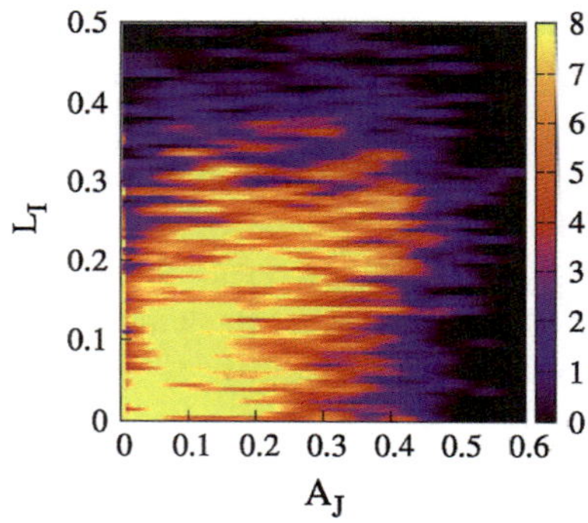


**Fig. 6** Distribution of length imbalance $L_i$ depending on $A_J$

parton pair and momentum asymmetry is caused by stochastic distributed vacuum splitting processes out of the considered jet cone. Experimental results [45–47] in $\sqrt{s_{NN}} = 2.76$ TeV Pb + Pb collisions at the LHC showed an enhancement of this momentum imbalance in central HIC in comparison to p + p-collisions. As a measure of this enhancement the momentum imbalance $A_J$,

$$A_J = \frac{p_{\mathrm{t;Leading}} - p_{\mathrm{t;Subleading}}}{p_{\mathrm{t;Leading}} + p_{\mathrm{t;Subleading}}}, \tag{1}$$

is defined, where $p_{\mathrm{t;Leading}}$ ($p_{\mathrm{t;Subleading}}$) is the reconstructed transverse momentum of the jet with the highest (second highest) transverse momentum. The additional suppression of balanced events in HIC are supposed to be the result of different in-medium energy loss of the two partons within the created bulk medium, which is a consequence of a non-central spatial production point of the initial dijet pair.

In this section we present our first preliminary results on momentum imbalance simulated within the transport model BAMPS. For the initial momentum spectra of the partons we use a distribution sampled according to a parametrized parton distribution [48], starting at $p_{t;0} = 100$ GeV. Because BAMPS describes only scattering processes of particles on the mass-shell, it is necessary to model the initial splitting processes of the virtual partons properly for reproducing the findings in p+p collisions. Therefore the shower routines of PYTHIA [49] are used to model the virtual splitting processes. Because the medium modification of the created parton showers is to be evaluated within the BAMPS framework, it is necessary to switch off hadronization processes and terminate the splitting processes within PYTHIA prematurely. Therefore, the standard PYTHIA global termination criterion in the virtuality $Q_0 = 1$ GeV is replaced by an energy-dependent minimum virtuality scale $Q_0 = \sqrt{\frac{E_{\text{parton}}}{\tau}}$ depending on the individual parton energy and a global shower time $\tau$. Throughout this section the shower time is assumed as $\tau = 0.2$ fm. Calculations within a static medium showed that the energy loss of the reconstructed jets is, for realistic values of $\tau$, nearly independent of the used shower time. The initial spatial production points of the parton pairs are determined by a Glauber modelling of the initial nucleus-nucleus collisions based on a Woods-Saxon density profile.

The so created parton showers are evolved within an offline recorded BAMPS background event. At every timestep the shower particles can interact with medium particles which then become shower particle by their own. With this procedure it is possible to clearly discriminate between shower and background particles.

In the following we compare our simulations with the experimental data measured by CMS. All event trigger conditions by CMS ($p_{\text{t;Leading}} > 120$ GeV, $p_{\text{t;Subleading}} > 50$ GeV, $\Delta\phi > \frac{2\pi}{3}$ and $|\eta_{\text{jet}}| < 2$) and an effective handling of the detector response and background fluctuations were used. For that an independent Gaussian smearing of the reconstructed jet momenta is applied. The width $\sigma$ is chosen in such a way that the smeared hadronic PYTHIA events without shower termination can reproduce the measured p+p data by CMS [45].

Figure 5 shows the calculated $A_J$ distribution for central $\sqrt{s_{NN}} = 2.76$ TeV Pb+Pb collisions (0–10 %, which corresponds to a mean impact parameter $b = 3.4$ fm) with and without smearing of the reconstructed jet momenta. As one can see, already the "true" jet momenta lead to an increase in the momentum asymmetry, though it is insufficient to reproduce the measured experimental data at $A_J > 0.3$. Therefore it can be assumed that the background fluctuations of the medium and the detector response play significant roles in explaining the strong imbalance in dijet momenta.

One advantage of simulations within a full 3+1D transport model is the availability of microscopic particle informations like space and momentum coordinates at every timestep. With this information it is possible to further investigate the processes leading to the observed momentum imbalance. The imbalance of the in-medium path lengths of the parton pair is studied introducing the length imbalance observable $L_i$

$$L_i = \frac{L_{\text{long}} - L_{\text{short}}}{L_{\text{long}} + L_{\text{short}}} . \tag{2}$$

Lower values of $L_i$ correspond to equal paths of the partons inside the medium and thus more central production of the partons. The in-medium path lengths of the initial partons ($L_{\text{long}}/L_{\text{short}}$) are determined by their spatial production point, their initial transverse momentum direction and their distance to the Wood-Saxon surface. Figure 6 shows the distribution of the length imbalance in bins of $A_J$. We observe that the length imbalance seems to be correlated to the underlying momentum asymmetry. One can state that the different transverse momenta of the reconstructed jets are mainly caused by the different in-medium path length of the two initial partons and hence a different energy loss. This suggests that there are events in which the parton pairs are produced in a more peripheral region so that one parton has to travel a longer distance through the medium than the other one, before leaving the collision zone.

We showed that the observation of a momentum imbalance by CMS is in agreement with simulated BAMPS events. Therefore we showed that the consideration of background fluctuations and detector responses plays significant role. This momentum imbalance is caused by a different in-medium path length of the two initial partons. Recent experimental results by CMS [47] with a lower cone radius $R = 0.3$ and lower subleading jet trigger $p_{\text{t;SubleadingJet}} > 30$ GeV show a broader and flatter $A_J$ distribution which can only be explained within BAMPS by usage of a higher cone radius ($R = 0.5$). This implies further investigations of the influence of bulk particles on the momentum of the reconstructed dijets. In addition, to understand the excess on energy loss of single hadrons simulated within BAMPS, it is highly necessary to study the relation between the momentum imbalance $A_J$ and the nuclear modification factor $R_{AA}$.

## 5 Transition from Ideal to Viscous Mach Cones in BAMPS

Highly energetic partons propagating through the hot and dense QGP rapidly lose their energy and momentum as the energy is deposited in the medium. Measurements of two- and three-particle correlations in heavy-ion collisions show a complete suppression of the away-side jet, whereas for lower $p_T$ a double peak structure is observed in the two-particle correlation function [7]. One possible and promising origin of these structures is assumed to be the interaction of fast partons with the soft matter which generates collective motion of the medium in form of Mach cones [8, 50].

For this purpose we investigate the propagation and formation of Mach cones in the microscopic transport model BAMPS (Boltzmann Approach of MultiParton Scatterings) [9] in the limit of vanishing mass and very small shear viscosity over entropy density ratio $\eta/s$ of the matter. Two different scenarios for the jet are used. In addition, by adjusting $\eta/s$, the influence of the viscosity on the profile of the Mach

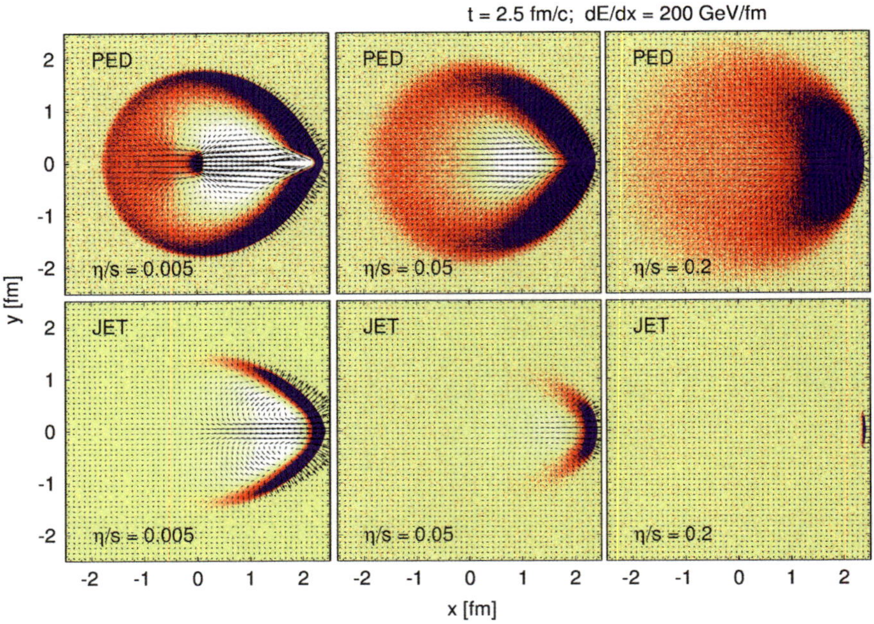


**Fig. 7** Transition from ideal to viscous Mach cones. Shape of a Mach cone shown for different jet scenarios and different viscosity over entropy density ratios, $\eta/s = 0.005$, 0.05 and 0.5. The energy deposition is $dE/dx = 200$ GeV/fm. The *upper panel* shows the pure energy deposition scenario (PED); the *lower panel* shows the propagation of a highly energetic jet (JET) depositing energy and momentum in $x$-direction. Depicted are the LRF energy density within a specific range; as an overlay we show the velocity profile with a scaled arrow length. The results are a snapshot of the evolution at $t = 2.5$ fm/c

cone and the corresponding two-particle correlation is explored for the first time. The results presented are based on a recent publication [51].

Shock waves are phenomena which have their origin in the collective motion of matter. In a simplified one-dimensional setup shock waves have already been studied within the framework of BAMPS for the perfect fluid limit [52, 53]. Furthermore BAMPS calculations have demonstrated that the shock profile is smeared out when viscosity is large. It was also found that a clear observation of the shock within the short time available in HIC requires a small viscosity.

In the following we study the evolution of "Mach cone"-like structures with different scenarios of the jet-medium interaction by using the parton cascade BAMPS. We focus on investigation of Mach cone evolution in absence of any other effects - i.e. we neglect such effects as initial fluctuations or expansion, which are however relevant in HIC. We use a static box with $T_{\text{med}} = 400$ MeV and binary collisions with an isotropic cross section. Furthermore, we keep the mean free path $\lambda_{\text{mfp}}$ of the medium particles constant in all spatial cells by adjusting the cross section according

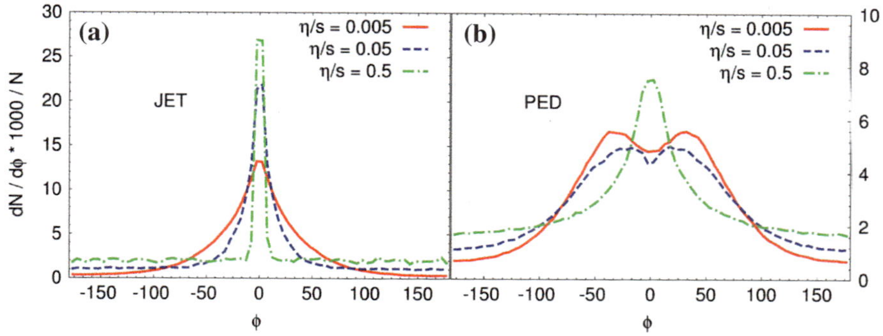


**Fig. 8** Two-particle correlations $dN/(Nd\phi)$ for different viscosities extracted from calculations shown in Fig. 7. The results are shown in the for the JET **(a)**, and PED **(b)** scenario for $dE/dx = 200$ GeV/fm

to $\sigma = 1/(n\lambda_{\text{mfp}})$, where $n$ is the particle density. The related shear viscosity for isotropic binary collisions is given by $\eta = 0.4\, e\lambda_{\text{mfp}}$ [54].

The Mach Cones studied here are induced by two different sources. The first of them we refer to as the pure energy deposition scenario (PED) [55]. This is simulated by a moving source depositing momentum end energy isotropically according to the thermal distribution $f(x, p) = \exp(-E/T)$. The second source we refer to as JET. This is simulated by a highly massless particle (jet) which has only momentum in $x$-direction, i.e. $p_x = E_{\text{jet}}$. After each timestep the energy of the jet particle is reset to its initial value. For both scenarios the sources are initialized at $t = 0$ fm/c at the position $x = -0.1$ fm and propagate in $x$-direction with $v_{\text{source}} = 1$, i.e. with the speed of light.

In Fig. 7 we show the Mach Cone structure for both PED scenario (upper panel) and JET scenario (lower panel) with $\eta/s = 0.005, 0.05$ and $0.5$ from left to right, respectively. We show a snapshot at $t = 2.5$ fm/c. The energy deposition rate is fixed to $dE/dx = 200$ GeV/fm. In both scenarios, PED and JET, for $\eta/s = 0.005$ (left panel), we observe a conical structure, but with obvious differences. The PED case with the isotropic energy deposition induces a spherical shock into back region; this structure is missing in the JET scenario because of the high forward peaked momentum deposition. Another difference is that in the JET scenario a clearly visible head shock appears. This in turn is missing in the PED scenario. Furthermore a (anti)-diffusion wake is induced by the JET (PED) scenario.

Adjusting the shear viscosity over entropy density ratio $\eta/s = 0.05 - 0.5$ we observe a smearing out of the Mach cone structure. For a sufficient high $\eta/s = 0.5$ the conical structure in both scenarios disappears. This is true for shock fronts as well as for the (anti-) diffusion wake. The difference between the PED and the JET case is that as $\eta/s$ increases, in the PED scenario the resulting "Mach cone" solution covers approximately the same spatial region regardless of a value of $\eta/s$, while in the JET case the structure is concentrated more and more near the projectile as the viscosity increases.

In Fig. 8 we show the two-particle correlations extracted from BAMPS calculations of the Mach Cones shown in Fig. 7. For the JET scenario (a) and sufficiently small $\eta/s = 0.005$ we observe only a peak in direction of the jet. The typical double peak structure, which has been proposed as a possible signature of the Mach cone in HIC, can only be observed for the PED scenario (b) and small $\eta/s$. However, the PED scenario has no correspondence in heavy-ion physics. We conclude that Mach cones can not be connected to double peak structures by any realistic picture of jets in HIC. In addition, finite values of the $\eta/s$ destroy any kind of Mach cone signatures.

**Acknowledgments** The authors are grateful to the Center for the Scientific Computing (CSC) at Frankfurt for the computing resources. This work was supported by the Helmholtz International Center for FAIR within the framework of the LOEWE program launched by the State of Hesse.

## References

1. S. Scott Adler et al., Elliptic flow of identified hadrons in Au+Au collisions at $\sqrt{s_{NN}} = 200$ GeV. Phys. Rev. Lett. **91**, 182301 (2003)
2. J. Adams et al., Particle dependence of azimuthal anisotropy and nuclear modification of particle production at moderate p(T) in Au + Au collisions at $\sqrt{s_{NN}} = 200$ GeV. Phys. Rev. Lett. **92**, 052302 (2004)
3. B.B. Back et al., Centrality and pseudorapidity dependence of elliptic flow for charged hadrons in Au+Au collisions at $\sqrt{s_{NN}} = 200$ GeV. Phys. Rev. C **72**, 051901 (2005)
4. J. Adams et al., Transverse momentum and collision energy dependence of high p(T) hadron suppression in Au+Au collisions at ultrarelativistic energies. Phys. Rev. Lett. **91**, 172302 (2003)
5. C. Adler et al., Centrality dependence of high p(T) hadron suppression in Au+Au collisions at $\sqrt{s_{NN}} = 130$ GeV. Phys. Rev. Lett. **89**, 202301 (2002)
6. K. Adcox et al., Suppression of hadrons with large transverse momentum in central Au+Au collisions at $\sqrt{s_{NN}} = 130$ GeV. Phys. Rev. Lett. **88**, 022301 (2002)
7. F. Wang, Measurement of jet modification at RHIC. J. Phys. G **30**, S1299–S1304 (2004)
8. H. Stöcker, Collective Flow signals the Quark Gluon Plasma. Nucl. Phys. A **750**, 121–147 (2005)
9. Z. Xu, C. Greiner, Thermalization of gluons in ultrarelativistic heavy ion collisions by including three-body interactions in a parton cascade. Phys. Rev. C **71**, 064901 (2005)
10. A. El, Z. Xu, C. Greiner, Thermalization of a color glass condensate and review of the 'Bottom-Up' scenario. Nucl. Phys. A **806**, 287–304 (2008)
11. Z. Xu, C. Greiner, Shear viscosity in a gluon gas. Phys. Rev. Lett. **100**, 172301 (2008)
12. A. El, A. Muronga, Z. Xu, C. Greiner, Shear viscosity and out of equilibrium dissipative hydrodynamics. Phys. Rev. C **79**, 044914 (2009)
13. Z. Xu, C. Greiner, H. Stöcker, PQCD calculations of elliptic flow and shear viscosity at RHIC. Phys. Rev. Lett. **101**, 082302 (2008)
14. Z. Xu, C. Greiner, Elliptic flow of gluon matter in ultrarelativistic heavy- ion collisions. Phys. Rev. C **79**, 014904 (2009)
15. O. Fochler, Z. Xu, C. Greiner, Towards a unified understanding of jet-quenching and elliptic flow within perturbative QCD parton transport. Phys. Rev. Lett. **102**, 202301 (2009)
16. K. Aamodt et al., Elliptic flow of charged particles in Pb+Pb collisions at $\sqrt{s_{NN}} = 2.76$ TeV. Phys. Rev. Lett. **105**, 252302 (2010)
17. J. Adams et al., Azimuthal anisotropy in Au + Au collisions at $\sqrt{s_{NN}} = 200$ GeV. Phys. Rev. C **72**, 014904 (2005)

18. K. Aamodt et al., Suppression of charged particle production at large transverse momentum in central Pb+Pb collisions at $\sqrt{s_{NN}} = 2.76$ TeV. Phys. Lett. B **696**, 30–39 (2011)
19. S. Albino, B.A. Kniehl, G. Kramer, AKK update: improvements from new theoretical input and experimental data. Nucl. Phys. B **803**, 42–104 (2008)
20. O. Fochler, Z. Xu, C. Greiner, Energy loss in a partonic transport model including bremsstrahlung processes. Phys. Rev. C **82**, 024907 (2010)
21. L.D. Landau, I. Pomeranchuk, Limits of applicability of the theory of bremsstrahlung electrons and pair production at high energies. Dokl. Akad. Nauk Ser. Fiz. **92**, 535–536 (1953)
22. A. Adare et al., Heavy quark production in $p+p$ and energy loss and flow of heavy quarks in au+au collisions at $\sqrt{s_{NN}} = 200$ GeV. Phys. Rev. C **84**, 044905 (2011)
23. J. Uphoff, O. Fochler, Z. Xu, C. Greiner, Heavy quark production at RHIC and LHC within a partonic transport model. Phys. Rev. **C 82**, 044906 (2010)
24. B. Abelev et al., Suppression of high transverse momentum D mesons in central Pb-Pb collisions at $\sqrt{s_{NN}} = 2.76$ TeV, J High Energy Phys. **09**, 112 (2012)
25. Chiara B, Measurement of $D^0$ $v_2$ in Pb-Pb collisions at $\sqrt{s_{NN}} = 2.76$ TeV with ALICE at the LHC, arXiv:1111.6886 (2011)
26. B.I. Abelev et al., Transverse momentum and centrality dependence of high-$p_t$ non-photonic electron suppression in Au+Au collisions at $\sqrt{s_{NN}} = 200$ GeV. Phys. Rev. Lett. **98**, 192301 (2007)
27. A. Adare et al., Energy loss and flow of heavy quarks in Au+Au collisions at $\sqrt{s_{NN}} = 200$ GeV. Phys. Rev. Lett. **98**, 172301 (2007)
28. S. Masciocchi, Investigation of charm and beauty production via semileptonic decays of heavy-flavour hadrons in pp at 7 TeV and Pb-Pb at 2.76 TeV with ALICE. J. Phys. G **38**, 124069 (2011)
29. L.Y. Dokshitzer, D.E. Kharzeev, Heavy quark colorimetry of QCD matter. Phys. Lett. B **519**, 199–206 (2001)
30. R. Abir, C. Greiner, M. Martinez, M.G. Mustafa, J. Uphoff, Soft gluon emission off a heavy quark revisited. Phys. Rev. D **85**, 054012 (2012)
31. P.B. Gossiaux, J. Aichelin, Towards an understanding of the RHIC single electron data. Phys. Rev. C **78**, 014904 (2008)
32. A. Peshier, Turning on the charm, Nucl. Phys. A **888**, 7-22 (2012)
33. J. Uphoff, O. Fochler, Z. Xu, C. Greiner, Elliptic flow and energy loss of heavy quarks in ultra-relativistic heavy ion collisions. Phys. Rev. C **84**, 024908 (2011)
34. A. Meistrenko, A. Peshier, J. Uphoff, C. Greiner, Collisional energy loss of heavy quarks, submitted to Physics Letters B, arXiv:1204.2397 (2012)
35. J. Uphoff, O. Fochler, Z. Xu, C. Greiner, Open heavy flavor at RHIC and LHC in a partonic transport model. Acta Phys. Pol. B Proc. Suppl. **5**, 555 (2012)
36. O. Fochler, J. Uphoff, Z. Xu, C. Greiner, Jet quenching and elliptic flow at RHIC and LHC within a pQCD-based partonic transport model. J. Phys. G **38**, 124152 (2011)
37. J. Uphoff, O. Fochler, Z. Xu, C. Greiner, Heavy quarks at RHIC and LHC within a partonic transport model. Nucl. Phys. A **855**, 444–447 (2011)
38. J. Uphoff, O. Fochler, Z. Xu, C. Greiner, Production, elliptic flow and energy loss of heavy quarks in the quark-gluon plasma. J. Phys. Conf. Ser. **270**, 012028 (2010)
39. V.N. Gribov, L.N. Lipatov, Deep inelastic e p scattering in perturbation theory. Sov. J. Nucl. Phys. **15**, 438–450 (1972)
40. Y.L. Dokshitzer, Calculation of the structure functions for deep inelastic scattering and e+ e- annihilation by perturbation theory in quantum chromodynamics. Sov. Phys. JETP **46**, 641–653 (1977)
41. G. Altarelli, G. Parisi, Asymptotic Freedom in Parton Language. Nucl. Phys. B **126**, 298 (1977)
42. G.P. Salam, Towards jetography. Eur. Phys. J. C **67**, 637–686 (2010)
43. M. Cacciari, G.P. Salam, G. Soyez, The anti-$k_t$ jet clustering algorithm. J. High Energy Phys. **04**, 063 (2008)
44. M. Cacciari, J. Rojo, G.P. Salam, G. Soyez, Jet reconstruction in heavy ion collisions. Eur. Phys. J. C **71**, 1539 (2011)

45. S. Chatrchyan, V. Khachatryan, A. Sirunyan, A. Tumasyan, W. Adam et al., Observation and studies of jet quenching in PbPb collisions at $\sqrt{s_{NN}} = 2.76$ TeV. Phys. Rev. C **84**, 024906 (2011)
46. G. Aad et al., Observation of a centrality-dependent dijet asymmetry in lead-lead collisions at $\sqrt{s_{NN}} = 2.76$ TeV with the ATLAS detector at the LHC. Phys. Rev. Lett. **105**, 252303 (2010)
47. CMS-Collaboration. Jet momentum dependence of jet quenching in PbPb collisions at $\sqrt{s_{NN}} =$ 2.76 TeV. Phys. Lett. B **712**, 176 (2012)
48. M. Glück, E. Reya, A. Vogt, Dynamical parton distributions of the proton and small x physics. Z. Phys. C **67**, 433–448 (1995)
49. T. Sjöstrand, S. Mrenna, P. Skands, PYTHIA 6.4 physics and manual. J. High Energy Phys. **05**, 26 (2006)
50. I. Bouras et al., Relativistic shock waves and mach cones in viscous gluon matter. J. Phys. Conf. Ser. **230**, 012045 (2010)
51. I. Bouras, A. El, O. Fochler, H. Niemi, Z. Xu et al., Transition from ideal to viscous mach cones in a kinetic transport approach. Phys. Lett. B **710**, 641–646 (2012)
52. I. Bouras et al., Relativistic shock waves in viscous gluon matter. Phys. Rev. Lett. **103**, 032301 (2009)
53. I. Bouras et al., Investigation of shock waves in the relativistic Riemann problem: a comparison of viscous fluid dynamics to kinetic theory. Phys. Rev. C **82**, 024910 (2010)
54. S.R. de Groot et al., *Relativistic Kinetic Theory: Principles and Applications* (North Holland, Amsterdam, 1980)
55. B. Betz, J. Noronha, G. Torrieri, M. Gyulassy, I. Mishustin et al., Universality of the diffusion wake from stopped and punch-through jets in heavy-ion collisions. Phys. Rev. C **79**, 034902 (2009)

# The QGP Phase in Relativistic Heavy-Ion Collisions

**E. L. Bratkovskaya, V. P. Konchakovski, O. Linnyk, W. Cassing, V. Voronyuk and V. D. Toneev**

**Abstract** The dynamics of partons, hadrons and strings in relativistic nucleus-nucleus collisions is analyzed within the novel Parton-Hadron-String Dynamics (PHSD) transport approach, which is based on a dynamical quasiparticle model for partons (DQPM) matched to reproduce recent lattice-QCD results—including the partonic equation of state—in thermodynamic equilibrium. The transition from partonic to hadronic degrees of freedom is described by covariant transition rates for the fusion of quark-antiquark pairs or three quarks (antiquarks), respectively, obeying flavor current-conservation, color neutrality as well as energy-momentum conservation. The PHSD approach is applied to nucleus-nucleus collisions from low SIS to RHIC energies. The traces of partonic interactions are found in particular in the elliptic flow of hadrons as well as in their transverse mass spectra.

## 1 Introduction

The 'Big Bang' scenario implies that in the first micro-seconds of the universe the entire state has emerged from a partonic system of quarks, antiquarks and gluons—a quark-gluon plasma (QGP)—to color neutral hadronic matter consisting

E. L. Bratkovskaya (✉)
Institute for Theoretical Physics, University of Frankfurt, Frankfurt, Germany
e-mail: Elena.Bratkovskaya@th.physik.uni-frankfurt.de

E. L. Bratkovskaya · V. Voronyuk · V. D. Toneev
Frankfurt Institute for Advanced Study, Frankfurt am Main, Germany

V. P. Konchakovski · O. Linnyk · W. Cassing
Institute for Theoretical Physics, University of Giessen, Giessen, Germany

V. Voronyuk
Bogolyubov Institute for Theoretical Physics, Kiev, Ukraine

V. Voronyuk · V. D. Toneev
Joint Institute for Nuclear Research, Dubna, Russia

W. Greiner (ed.), *Exciting Interdisciplinary Physics*,
FIAS Interdisciplinary Science Series, DOI: 10.1007/978-3-319-00047-3_19,

of interacting hadronic states (and resonances) in which the partonic degrees of freedom are confined. The nature of confinement and the dynamics of this phase transition has motivated a large community for several decades and is still an outstanding question of todays physics. Early concepts of the QGP were guided by the idea of a weakly interacting system of partons which might be described by perturbative QCD (pQCD). However, experimental observations at the Relativistic Heavy Ion Collider (RHIC) indicated that the new medium created in ultrarelativistic Au+Au collisions is interacting more strongly than hadronic matter and consequently this concept had to be severely questioned. Moreover, in line with theoretical studies in Refs. [1–4] the medium showed phenomena of an almost perfect liquid of partons [5–9] as extracted from the strong radial expansion and the scaling of elliptic flow $v_2(p_T)$ of mesons and baryons with the number of constituent quarks and antiquarks [5–8].

The question about the properties of this (nonperturbative) QGP liquid is discussed controversially in the literature and dynamical concepts describing the formation of color neutral hadrons from colored partons are scarce. A fundamental issue for hadronization models is the conservation of 4-momentum as well as the entropy problem, because by fusion/coalescence of massless (or low constituent mass) partons to color neutral bound states of low invariant mass (e.g. pions) the number of degrees of freedom and thus the total entropy is reduced in the hadronization process. This problem—a violation of the second law of thermodynamics as well as the conservation of four-momentum and flavor currents—has been addressed in Ref. [10] on the basis of the DQPM employing covariant transition rates for the fusion of 'massive' quarks and antiquarks to color neutral hadronic resonances or strings. In fact, the dynamical studies for an expanding partonic fireball in Ref. [10] suggest that the these problems have come to a practical solution.

A consistent dynamical approach—valid also for strongly interacting systems—can be formulated on the basis of Kadanoff-Baym (KB) equations [11, 12] or off-shell transport equations in phase-space representation, respectively [11, 12]. In the KB theory the field quanta are described in terms of dressed propagators with complex selfenergies. Whereas the real part of the selfenergies can be related to mean-field potentials (of Lorentz scalar, vector or tensor type), the imaginary parts provide information about the lifetime and/or reaction rates of time-like 'particles' [13]. Once the proper (complex) selfenergies of the degrees of freedom are known the time evolution of the system is fully governed by off-shell transport equations (as described in Refs. [11–13]). The determination/extraction of complex selfenergies for the partonic degrees of freedom has been performed before in Ref. [14] by fitting lattice QCD (lQCD) 'data' within the Dynamical QuasiParticle Model (DQPM). In fact, the DQPM allows for a simple and transparent interpretation of lattice QCD results for thermodynamic quantities as well as correlators and leads to effective strongly interacting partonic quasiparticles with broad spectral functions. For a review on off-shell transport theory and results from the DQPM in comparison to lQCD we refer the reader to Ref. [13].

The actual implementations in the PHSD transport approach have been presented in detail in Refs. [15, 16]. Here we present results for transverse mass spectra and elliptic flow of hadrons for heavy-ion collisions at relativistic energies in comparison to data from the experimental collaborations.

## 2 The PHSD Approach

The dynamics of partons, hadrons and strings in relativistic nucleus-nucleus collisions is analyzed here within the Parton-Hadron-String Dynamics approach [10, 15, 16]. In this transport approach the partonic dynamics is based on Kadanoff-Baym equations for Green functions with self-energies from the Dynamical Quasi-Particle Model (DQPM) [14] which describes QCD properties in terms of 'resummed' single-particle Green functions. In Ref. [16], the actual three DQPM parameters for the temperature-dependent effective coupling were fitted to the recent lattice QCD results of Ref. [17]. The latter lead to a critical temperature $T_c \approx 160$ MeV which corresponds to a critical energy density of $\varepsilon_c \approx 0.5$ GeV/fm$^3$. In PHSD the parton spectral functions $\rho_j$ ($j = q, \bar{q}, g$) are no longer $\delta$-functions in the invariant mass squared as in conventional cascade or transport models but depend on the parton mass and width parameters:

$$\rho_j(\omega, \mathbf{p}) = \frac{\gamma_j}{E_j}\left(\frac{1}{(\omega - E_j)^2 + \gamma_j^2} - \frac{1}{(\omega + E_j)^2 + \gamma_j^2}\right) \tag{1}$$

separately for quarks/antiquarks and gluons ($j = q, \bar{q}, g$). With the convention $E^2(\mathbf{p}^2) = \mathbf{p}^2 + M_j^2 - \gamma_j^2$, the parameters $M_j^2$ and $\gamma_j$ are directly related to the real and imaginary parts of the retarded self-energy, e.g. $\Pi_j = M_j^2 - 2i\gamma_j\omega$. The spectral function (1) is antisymmetric in $\omega$ and normalized as

$$\int_{-\infty}^{\infty} \frac{d\omega}{2\pi}\, \omega\, \rho_j(\omega, \mathbf{p}) = \int_{0}^{\infty} \frac{d\omega}{2\pi}\, 2\omega\, \rho_j(\omega, \mathbf{p}) = 1. \tag{2}$$

The actual parameters in Eq. (1), i.e. the gluon mass $M_g$ and width $\gamma_g$—employed as input in the PHSD calculations—as well as the quark mass $M_q$ and width $\gamma_q$, are depicted in Fig. 1 as a function of the scaled temperature $T/T_c$. As mentioned above these values for the masses and widths have been fixed by fitting the lattice QCD results from Ref. [17] in thermodynamic equilibrium.

One might worry whether the quasiparticle properties—fixed in thermal equilibrium—also should be appropriate for out-off equilibrium configurations. This question is nontrivial and can only be answered by detailed model investigations e.g. on the basis of Kadanoff-Baym equations. We recall that such studies have been summarized in Ref. [13] for strongly interacting scalar fields that initially are far off-equilibrium and simulate momentum distributions of colliding systems at high relative momentum. The results for the effective parameters $M$ and $\gamma$, which correspond to the time-dependent pole mass and width of the propagator, indicate that the

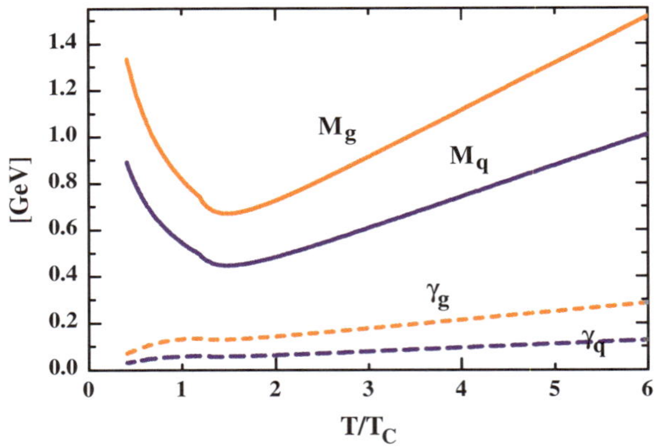


**Fig. 1** The effective gluon mass $M_g$ and width $\gamma_g$ as function of the scaled temperature $T/T_c$ (*red lines*). The *blue lines* show the corresponding quantities for quarks

quasiparticle properties—except for the very early off-equilibrium configuration—are close to the equilibrium mass and width even though the phase-space distribution of the particles is far from equilibrium (cf. Figs. 8 to 10 in Ref. [13]). Accordingly, we will adopt the equilibrium quasiparticle properties also for phase-space configurations out of equilibrium as appearing in relativistic heavy-ion collisions. The reader has to keep in mind that this approximation is far from being arbitrary, however, not fully equivalent to the exact solution.

We recall that the DQPM allows to extract a potential energy density $V_p$ from the space-like part of the energy-momentum tensor which can be tabulated e.g. as a function of the scalar parton density $\rho_s$. Derivatives of $V_p$ with respect to $\rho_s$ then define a scalar mean-field potential $U_s(\rho_s)$ which enters the equation of motion for the dynamical partonic quasiparticles. As one can see from Fig. 2, the scalar potential is rather large and nonlinearly increases with $\rho_s$. This implies that the repulsive force due to $U_s(\rho_s)$ will change in a non-monotonous way with the scalar density. The vector mean-field potential is not negligible, too, especially at high $\rho_s$ and induces a Lorentz force for the partons. Note that the vector mean-field vanishes with decreasing scalar density whereas the scalar mean-field approaches a constant value for $\rho_s \to 0$.

Furthermore, a two-body interaction strength can be extracted from the DQPM as well from the quasiparticle width in line with Ref. [4]. The transition from partonic to hadronic d.o.f. (and vice versa) is described by covariant transition rates for the fusion of quark-antiquark pairs or three quarks (antiquarks), respectively, obeying flavor current-conservation, color neutrality as well as energy-momentum conservation [15, 16]. Since the dynamical quarks and antiquarks become very massive close to the phase transition, the formed resonant 'prehadronic' color-dipole states ($q\bar{q}$ or $qqq$) are of high invariant mass, too, and sequentially decay to the groundstate meson and baryon octets increasing the total entropy.

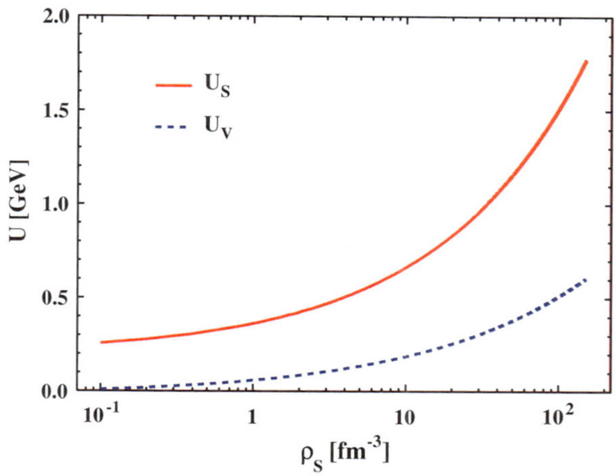


**Fig. 2** The scalar and vector mean-field potentials in the present PHSD model as a function of the scalar density $\rho_s$ of partons

On the hadronic side PHSD includes explicitly the baryon octet and decouplet, the $0^-$- and $1^-$-meson nonets as well as selected higher resonances as in the Hadron-String-Dynamics (HSD) approach [18, 19]. The color-neutral objects of higher masses (>1.5 GeV in case of baryonic states and >1.3 GeV in case of mesonic states) are treated as 'strings' (color-dipoles) that decay to the known (low-mass) hadrons according to the JETSET algorithm [20]. We discard an explicit recapitulation of the string formation and decay and refer the reader to the original work [20]. Note that PHSD and HSD (without explicit partonic degrees-of-freedom) merge at low energy density, in particular below the critical energy density $\varepsilon_c \approx 0.5$ GeV/fm$^3$.

The PHSD approach was applied to nucleus-nucleus collisions from $s_{NN}^{1/2} \sim 5$ to 200 GeV in Refs. [15, 16] in order to explore the space-time regions of 'partonic matter'. It was found that even central collisions at the top-SPS energy of $\sqrt{s_{NN}}$ =17.3 GeV show a large fraction of nonpartonic, i.e. hadronic or string-like matter, which can be viewed as a hadronic corona [21]. This finding implies that neither hadronic nor only partonic models can be employed to extract physical conclusions in comparing model results with data.

## 3 Application to Nucleus-Nucleus Collisions

In this Section we employ the PHSD approach to nucleus-nucleus collisions at moderate relativistic energies. It is of interest, how the PHSD approach compares to the HSD [19] model (without explicit partonic degrees-of-freedom) as well as to experimental data. In Fig. 3 we show the transverse mass spectra of $\pi^-$, $K^+$ and $K^-$ mesons for 7 % central Pb+Pb collisions at 40 and 80 A·GeV and 5 % central collisions at 158 A·GeV in comparison to the data of the NA49 Collaboration [22, 23]. Here the

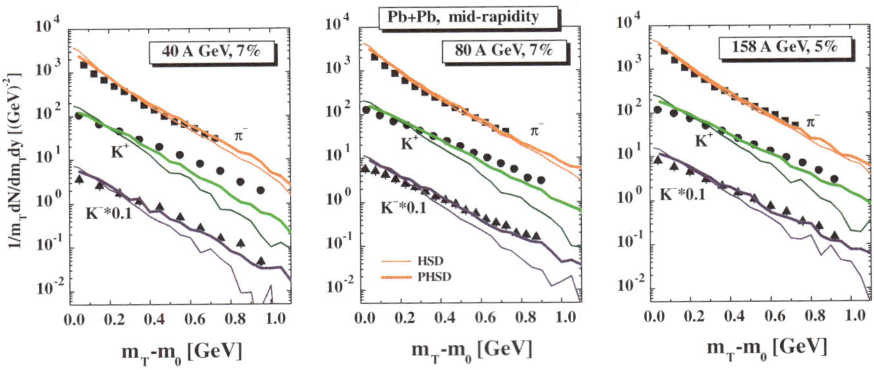


**Fig. 3** The $\pi^-$, $K^+$ and $K^-$ transverse mass spectra for central Pb+Pb collisions at 40, 80 and 158 A·GeV from PHSD (*thick solid lines*) in comparison to the distributions from HSD (*thin solid lines*) and the experimental data from the NA49 collaboration [22, 23]

**Fig. 4** The $\pi^-$, $K^+$ and $K^-$ transverse mass spectra for 5 % central Au+Au collisions at $\sqrt{s} = 200$ GeV from PHSD (*thick solid lines*) in comparison to the distributions from HSD (*thin solid lines*) and the experimental data from the BRAHMS, PHENIX and STAR collaborations [24–26] at midrapidity

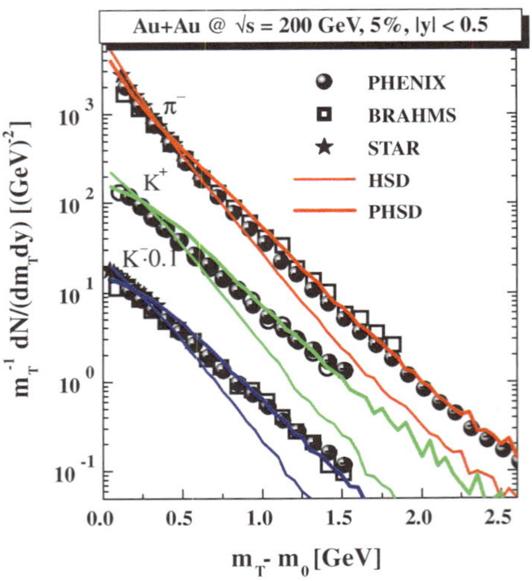


slope of the $\pi^-$ spectra is only slightly enhanced in PHSD relative to HSD which demonstrates that the pion transverse motion shows no sizeable sensitivity to the partonic phase. However, the $K^\pm$ transverse mass spectra are substantially hardened with respect to the HSD calculations at all bombarding energies—i.e. PHSD is more in line with the data—and thus suggests that partonic effects are better visible in the strangeness-degrees of freedom.

The PHSD calculations for RHIC energies show a very similar trend—the inverse slope increases by including the partonic phase—cf. Fig. 4 where we show the transverse mass spectra of $\pi^-$, $K^+$ and $K^-$ mesons for 5 % central Au+Au collisions at $\sqrt{s} = 200$ GeV in comparison to the data of the RHIC Collaborations [24–26].

The hardening of the kaon spectra can be traced back to parton-parton scattering as well as a larger collective acceleration of the partons in the transverse direction due to the presence of repulsive vector fields for the partons. The enhancement of the spectral slope for kaons and antikaons in PHSD due to collective partonic flow shows up much clearer for the kaons due to their significantly larger mass (relative to pions). We recall that in Refs. [27] the underestimation of the $K^{\pm}$ slope by HSD (and also UrQMD) had been suggested to be a signature for missing partonic degrees of freedom; the present PHSD calculations support this early suggestion.

The strange antibaryon sector is of further interest since here the HSD calculations have always underestimated the yield [28]. Our detailed studies in Ref. [15] show that the HSD and PHSD calculations both give a reasonable description of the $\Lambda + \Sigma^0$ yield of the NA49 Collaboration [29]; both models underestimate the NA57 data [30, 31] by about 30 %. An even larger discrepancy in the data from the NA49 and NA57 Collaborations is seen for $(\bar{\Lambda}+\bar{\Sigma}^0)/N_{\text{wound}}$; here the PHSD calculations give results which are in between the NA49 data and the NA57 data whereas HSD underestimates the $(\bar{\Lambda} + \bar{\Sigma}^0)$ midrapidity yield at all centralities.

The latter result suggests that the partonic phase does not show up explicitly in an enhanced production of strangeness (or in particular strange mesons and baryons) but leads to a different redistribution of antistrange quarks between mesons and antibaryons. In fact, as demonstrated in Ref. [15], we find no sizeable differences in the double strange baryons from HSD and PHSD—in a good agreement with the NA49 data—but observe a large enhancement in the double strange antibaryons for PHSD relative to HSD.

The anisotropy in the azimuthal angle $\psi$ is usually characterized by the even order Fourier coefficients $v_n = \langle \exp(i\, n(\psi - \Psi_{RP}))\rangle$, $n = 2, 4, \ldots$, since for a smooth angular profile the odd harmonics become equal to zero. As noted above, $\Psi_{RP}$ is the azimuth of the reaction plane and the brackets denote averaging over particles and events. In particular, for the widely used second order coefficient, denoted as an elliptic flow, we have

$$v_2 = \langle \cos(2\psi - 2\Psi_{RP})\rangle = \left\langle \frac{p_x^2 - p_y^2}{p_x^2 + p_y^2} \right\rangle, \tag{3}$$

where $p_x$ and $p_y$ are the $x$ and $y$ components of the particle momenta. This coefficient can be considered as a function of centrality, pseudo-rapidity $\eta$ and/or transverse momentum $p_T$. We note that the reaction plane in PHSD is given by the $(x - z)$ plane with the $z$-axis in the beam direction.

In Fig. 5 the experimental $v_2$ excitation function in the transient energy range is compared to the results from the PHSD calculations [38]; HSD model results are given as well for reference. We note that the centrality selection and acceptance are the same for the data and models.

We recall that the HSD model has been very successful in describing heavy-ion spectra and rapidity distributions from SIS to SPS energies. A detailed comparison of HSD results with respect to a large experimental data set was reported in Refs. [27, 34, 35] for central Au+Au (Pb+Pb) collisions from SIS to top SPS energies. Indeed, as shown in Fig. 5 (dashed lines), HSD is in good agreement with

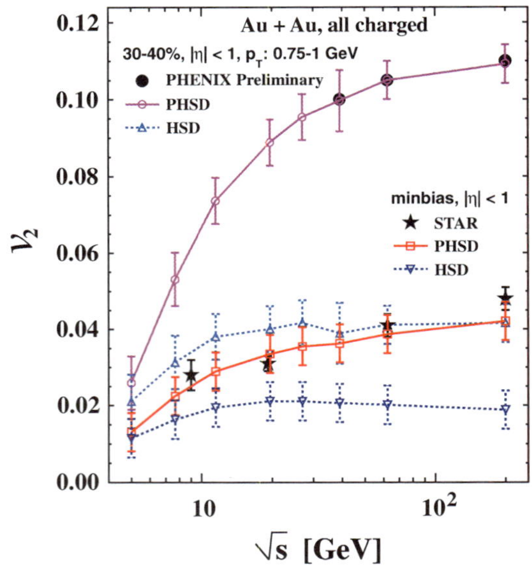


**Fig. 5** Average elliptic flow $v_2$ of charged particles at midrapidity for two centrality selections calculated within the PHSD (*solid curves*) and HSD (*dashed curves*). The $v_2$ STAR data compilation for minimal bias collisions are taken from [32] (*stars*) and the preliminary PHENIX data [33] are plotted by filled circles

experiment for both data sets at the lower edge ($\sqrt{s_{NN}} \sim 10$ GeV) but predicts an approximately energy-independent flow $v_2$ at larger energies and, therefore, does not match the experimental observations. This behavior is in quite close agreement with another independent hadronic model, the UrQMD (Ultra relativistic Quantum Molecular Dynamics) [36, 37] (cf. with Ref. [32]).

From the above comparison one may conclude that the rise of $v_2$ with bombarding energy is not due to hadronic interactions and models with partonic d.o.f. have to be addressed. Indeed, the PHSD approach incorporates the parton medium effects in line with a lQCD equation-of-state, as discussed above, and also includes a dynamic hadronization scheme based on covariant transition rates. It is seen from Fig. 5 that PHSD performs better: The elliptic flow $v_2$ from PHSD (solid curve) is fairly in line with the data from the STAR and PHENIX collaborations and clearly shows the growth of $v_2$ with the bombarding energy [38].

The $v_2$ increase is clarified in Fig. 6 where the partonic fraction of the energy density at mid-pseudorapidity with respect to the total energy density in the same pseudorapidity interval is shown. We recall that the repulsive scalar mean-field potential $U_s(\rho_s)$ for partons in the PHSD model leads to an increase of the flow $v_2$ as compared to that for HSD or PHSD calculations without partonic mean fields. As follows from Fig. 6, the energy fraction of the partons substantially grows with increasing bombarding energy while the duration of the partonic phase is roughly the same.

The $v_2$ coefficient measures the response of the heated and compressed matter to the spatial deformation in the overlap region of colliding nuclei, which is usually quantified by the eccentricity $\varepsilon_2 = \langle y^2 - x^2 \rangle / \langle x^2 + y^2 \rangle$. Since the flow response ($v_2$) is proportional to the driving force ($\varepsilon_2$), the ratio $v_2/\varepsilon_2$ is used to compare different impact parameters and nuclei.

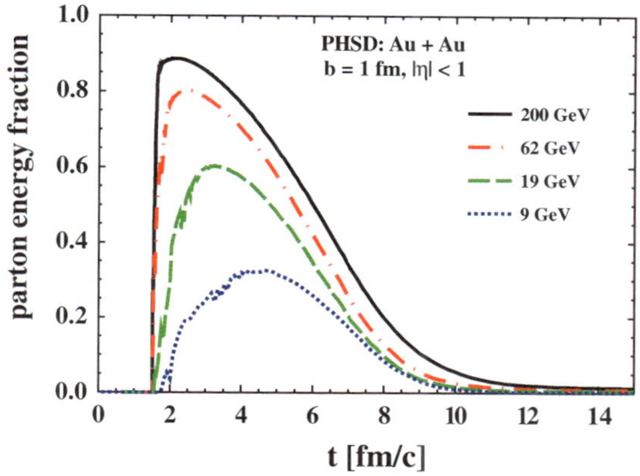


**Fig. 6** The evolution of the parton fraction of the total energy density at mid-pseudorapidity for different collision energies

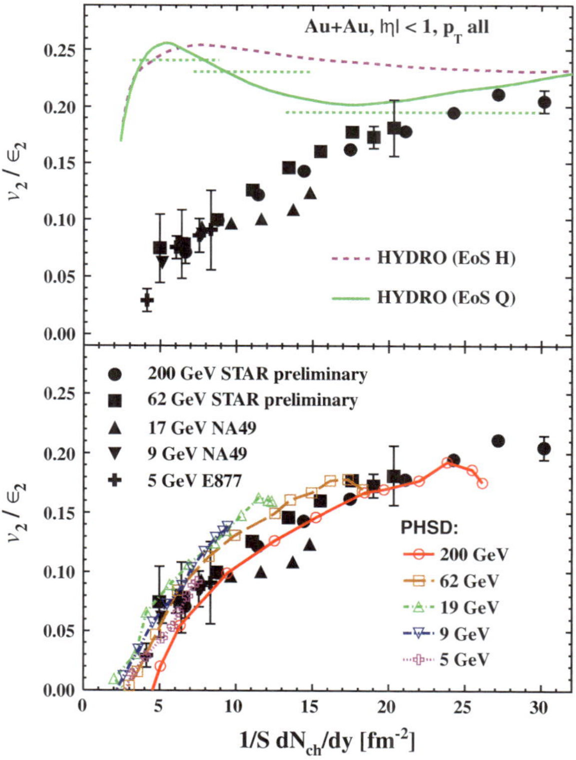


**Fig. 7** Scaling of $v_2/\varepsilon_2$ versus $(1/S)(dN_{ch}/dy)$. The PHSD results [40] are given by *lines* with open symbols. Predictions of ideal boost-invariant hydrodynamics are shown in the *top panel* (from [41]) and explained in the text. Our PHSD results are presented in the *bottom panel*. The experimental data points for Au+Au collisions at 200 GeV (*circles*) and 62 GeV (*squares*) are from Refs. [41, 42]

A remarkable property—*universal scaling*—has been proposed in Ref. [39] (see Fig. 7). It appears that $v_2/\varepsilon_2$ plotted versus $(1/S)dN_{ch}/dy$ falls on a 'universal' curve, which links very different regimes, ranging from AGS to RHIC energies. Here $S = \pi\sqrt{\langle x^2\rangle\langle y^2\rangle}$ is the overlap area of the collision system and $dN_{ch}/dy$ is the rapidity density of charged particles.

As can be seen from Fig. 7 (lower panel) the universal scaling of $v_2/\varepsilon_2$ versus $(1/S)dN_{ch}/dy$ is approximately reproduced by PHSD (see Ref. [40] for the details). This feature is not reproduced by hadronic transport models (such as HSD and UrQMD) and meets (severe) problems in the various hydrodynamic descriptions as demonstrated in the upper panel of Fig. 7 for a pure hadronic equation of state ('EoS H') as well as with a QGP phase transition ('EoS Q').

Thus, the experimentally observed scaling in Fig. 7 puts very strong constraints on the initial microscopic properties (entropy density, mean free path, etc.), as well as the global longitudinal structure [43].

**Acknowledgments** Work supported in part by the HIC for FAIR framework of the LOEWE program and by DFG.

## References

1. E. Shuryak, Prog. Part. Nucl. Phys. **53**, 273 (2004)
2. M.H. Thoma, J. Phys, G **31**, L7 (2005)
3. M.H. Thoma, Nucl. Phys. A **774**, 307 (2006)
4. A. Peshier, W. Cassing, Phys. Rev. Lett. **94**, 172301 (2005)
5. I. Arsene et al., Nucl. Phys. A **757**, 1 (2005)
6. B.B. Back et al., Nucl. Phys. A **757**, 28 (2005)
7. J. Adams et al., Nucl. Phys. A **757**, 102 (2005)
8. K. Adcox et al., Nucl. Phys. A **757**, 184 (2005)
9. T. Hirano, M. Gyulassy, Nucl. Phys. A **769**, 71 (2006)
10. W. Cassing, E.L. Bratkovskaya, Phys. Rev. C **78**, 034919 (2008)
11. S. Juchem, W. Cassing, C. Greiner, Phys. Rev. D **69**, 025006 (2004)
12. S. Juchem, W. Cassing, C. Greiner, Nucl. Phys. A **743**, 92 (2004)
13. W. Cassing, Eur. Phys. J. ST **168**, 3 (2009)
14. W. Cassing, Nucl. Phys. A **795**, 70 (2007)
15. W. Cassing, E.L. Bratkovskaya, Nucl. Phys. A **831**, 215 (2009)
16. E.L. Bratkovskaya, W. Cassing, V.P. Konchakovski, O. Linnyk, Nucl. Phys. A **856**, 162 (2011)
17. Y. Aoki et al., J. High Energy Phys. **0906**, 088 (2009)
18. W. Ehehalt, W. Cassing, Nucl. Phys. A **602**, 449 (1996)
19. W. Cassing, E.L. Bratkovskaya, Phys. Rep. **308**, 65 (1999)
20. H.-U. Bengtsson, T. Sjöstrand, Comp. Phys. Commun. **46**, 43 (1987)
21. J. Aichelin, K. Werner, Phys. Rev. C **79**, 064907 (2009)
22. C. Alt et al., NA49 Collaboration. Phys. Rev. C **66**, 054902 (2002)
23. C. Alt et al., Phys. Rev. C **77**, 024903 (2008)
24. S.S. Adler et al., Phys. Rev. C **69**, 034909 (2004)
25. J. Adams et al., Phys. Rev. Lett. **92**, 112301 (2004)
26. I.G. Bearden et al., Phys. Rev. Lett. **94**, 162301 (2005)
27. E.L. Bratkovskaya, S. Soff, H. Stöcker, M. van Leeuwen, W. Cassing, Phys. Rev. Lett. **92**, 032302 (2004)

28. J. Geiss, W. Cassing, C. Greiner, Nucl. Phys. A **644**, 107 (1998)
29. T. Anticic et al., Phys. Rev. C **80**, 034906 (2009)
30. F. Antinori et al., Phys. Lett. B **595**, 68 (2004)
31. F. Antinori et al., J. Phys. G: Nucl. Phys. **32**, 427 (2006)
32. M. Nasim, L. Kumar, P.K. Netrakanti, B. Mohanty, Phys. Rev. C **82**, 054908 (2010)
33. X. Gong et al., J. Phys. G **38**, 124146 (2011)
34. E.L. Bratkovskaya, W. Cassing, U. Mosel, Phys. Lett. B **424**, 244 (1998)
35. E.L. Bratkovskaya et al., Phys. Rev. C **69**, 054907 (2004)
36. S.A. Bass et al., Prog. Part. Nucl. Phys. **41**, 255 (1998)
37. M. Bleicher et al., J. Phys. G **25**, 1859 (1999)
38. V.P. Konchakovski et al., Phys. Rev. C **85**, 011902(R) (2012)
39. S.A. Voloshin, A.M. Poskanzer, Phys. Lett. B **474**, 27 (2000)
40. V.P. Konchakovski et al., arXiv:1201.3320 [nucl-th]
41. S.A. Voloshin, J. Phys. G **34**, S883 (2007)
42. M. Shimomura et al., PoS WPCF2011, 070 (2011)
43. G. Torrieri, Phys. Rev. C **76**, 024903 (2007)

# Recent HBT Results from a Hybrid Transport Approach to Heavy Ion Reactions

**Marcus Bleicher and Gunnar Graef**

**Abstract** We use the Ultrarelativistic Quantum Molecular Dynamics model to investigate heavy ion observables at various collision energies. The model is applied in cascade mode, hybrid mode (hydro+cascade) and also in using a core corona approach where only high density regions of the collision are treated hydrodynamically. Using these different modes we show the excitation function of strange particle ratios and the results of a HBT analysis from SPS up to LHC energies.

## 1 Introduction

All of everyday matter is made up of protons, neutrons and electrons. Nowadays a vast number of additional particles are known which are either leptons or composite of quarks and gluons, the fundamental particles of our universe. The interactions of quarks and gluons are described by the theory of strong interaction, namely Quantum-Chromo-Dynamics (QCD). Presently the full dynamical QCD can not be solved theoretically. That is the reason why one employs effective models to gain new insights out of the vast amount of data provided by the recent experiments at various accelerators such as the SPS, RHIC and LHC.

M. Bleicher
Institut für Theoretische Physik, Goethe Universität Frankfurt, Frankfurt, Germany

G. Graef (✉)
Frankfurt Institute for Advanced Studies, Ruth-Moufang-Str. 1, 60438 Frankfurt, Germany
e-mail: graef@th.physik.uni-frankfurt.de

W. Greiner (ed.), *Exciting Interdisciplinary Physics*,
FIAS Interdisciplinary Science Series, DOI: 10.1007/978-3-319-00047-3_20,

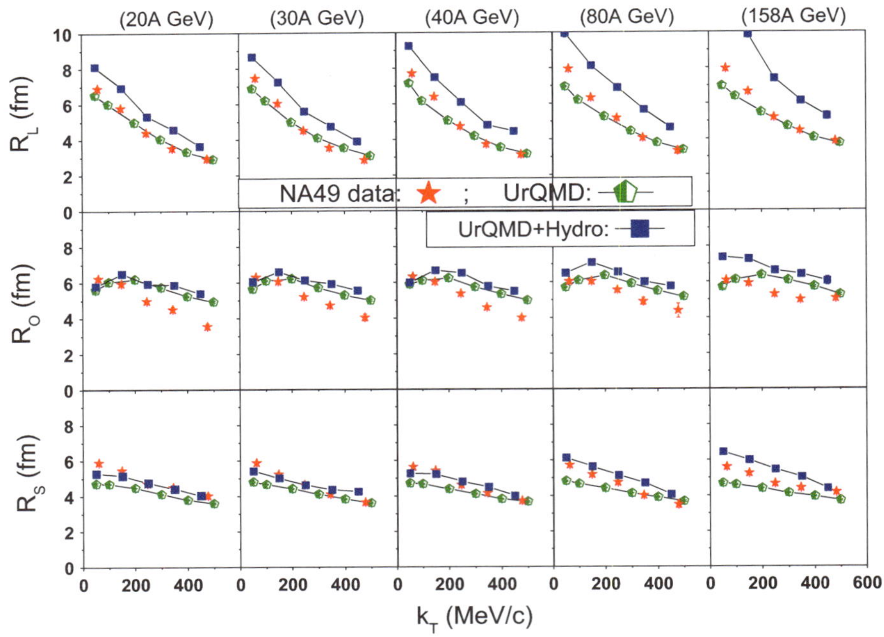


**Fig. 1** $k_T$ dependence of the HBT radii from UrQMD (*green pentagons*) and UrQMD+Hydro (*blue squares*) compared to NA49 data [3]. Figure taken from [4]

## 2 UrQMD+Hydro Approach

The Ultrarelativistic Quantum Molecular Dynamics (UrQMD) [1, 2] is a microscopic transport model with hadronic degrees of freedom. The particle production in the model is realized via string excitation and fragmentation and resonance decay and excitation.

The course of the simulated events is as follows. First the target and projectile nucleus are initialized. After that the initial scatterings and particle production take place between the colliding nucleons. Once the colliding nuclei have passed each other, namely after the time $t_{start} = 2R/\sqrt{\gamma^2 - 1}$ the particles are mapped onto a hydro grid using a Lorentz contracted gaussian with a width of 1 fm to gain an energy density. Then an ideal hydrodynamic expansion is performed using the SHASTA [5, 6] algorithm. As soon as the energy density in all cells of a transverse slice of thickness 0.2 fm is lower than five times the ground state energy this slice is particleized via the Cooper-Frye formalism.

For details of the up to date UrQMD version 3.3 and its hybrid results the reader is referred to [7, 8]. The model can be downloaded from [9].

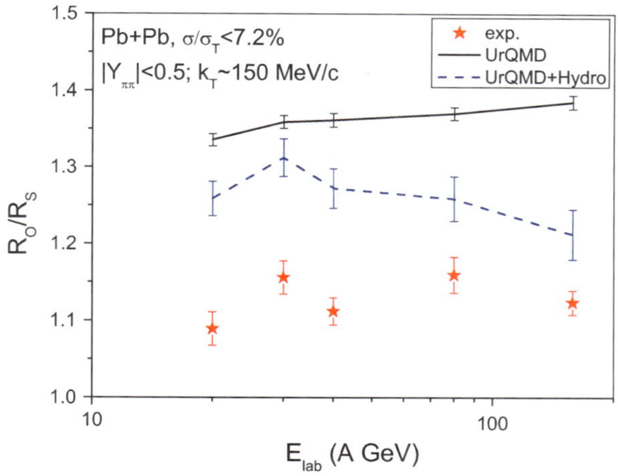


**Fig. 2** Excitation function of $R_O/R_S$ ratio for central PbPb collisions from UrQMD (*full black line*), UrQMD+Hydro (*dashed blue line*) and NA49 data [3]. Figure taken from [4]

## 3 Hanbury-Brown Twiss Radii

To investigate the influence of the intermediate hydro phase on the HBT radii we do a systematic comparison of UrQMD with and without intermediate hydro phase with data. Figure 1 shows the $k_T$ dependence of $R_O$, $R_S$ and $R_L$ for beam energies 20 to 158 AGeV. In general the experimental data [3] is well described by the UrQMD model while the additional hydro phase with a hadron gas equation of state produces bigger radii which leads to a slight over prediction of the data. However, the hydrodynamic phase leads to a better description of the $R_O/R_S$ ratio as presented in Fig. 2. We attribute the increase of the radii in the hybrid mode to the fact, that the freeze out of transverse slices as a whole is a simplistic description of the transverse dynamics. This delays the fluid to particle transition in cells and thus also the freeze out, leading to slightly bigger radii.

The results for higher energies e.g. $\sqrt{s} = 200$ GeV at RHIC and $\sqrt{s} = 2.76$ TeV at LHC are shown in Fig. 3. While there are only minor differences between the cascade calculation (black triangles) and data (blue diamonds) at $\sqrt{s} = 200$ GeV there is a major disagreement between the calculation (black crosses) and data (green stars) at $\sqrt{s} = 2.76$ TeV. We attribute this disagreement to the well known lack of pressure in the early phase of cascade calculations. A higher pressure would lead to a more explosive expansion of the particle emitting zone. This explosive expansion would lead to an faster decoupling and stronger flow, both reducing the HBT radii. To investigate this we repeated the calculation and artificially reduced the emission duration by a factor of two (red circles) and three (beige diamonds). For details of the procedure read [10]. This leads to a much better description of $R_{\text{out}}$ and $R_{\text{out}}/R_{\text{side}}$. The description of $R_{\text{long}}$ still fails, since this parameter is sensitive to the lifetime

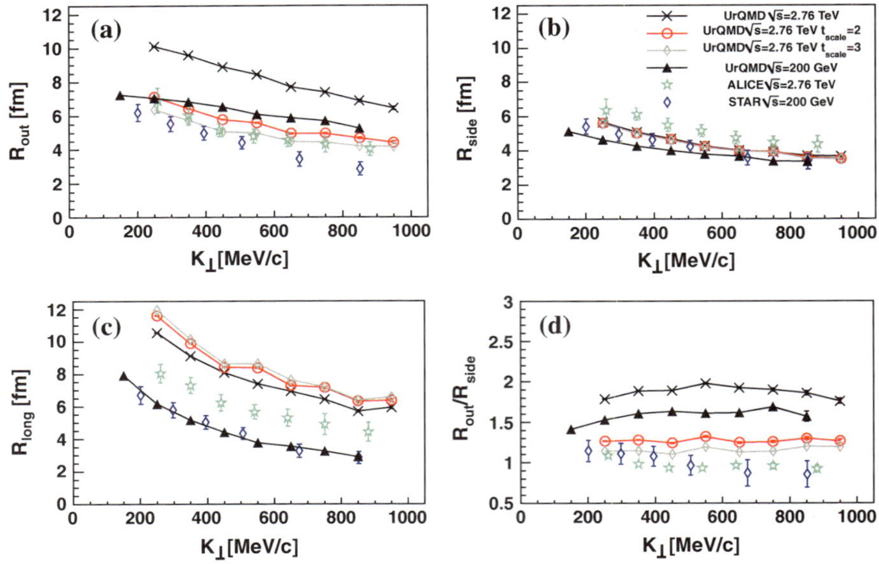


**Fig. 3** Experimental data for the $k_T$ dependence of the pion HBT radii and ratio for central PbPb collisions at LHC energy $\sqrt{s_{NN}} = 2.76$ TeV [13] (*green stars*) and AuAu collisions at RHIC energy $\sqrt{s_{NN}} = 200$ GeV [14] (*blue diamonds*) is shown. UrQMD results for $\sqrt{s_{NN}} = 200$ GeV (*black triangles*), $\sqrt{s_{NN}} = 2.76$ TeV (*black crosses*) and $\sqrt{s_{NN}} = 2.76$ TeV with artificially reduced source lifetimes by a factor of 2 (*red circles*) and 3 (*beige diamonds*) are also presented

of the system, that would be reduced by a more violent expansion, but not by the reduction of the emission duration time.

## 4 Core Corona

Since the hydrodynamic description is not valid for regions of low densities that are not properly thermalized we seek to improve the model. We do this by only describing high density regions hydrodynamically while performing a microscopic evolution of the lower energy regions simultaneously. For details of the approach see [11]. The left plot of Fig. 4 shows the excitation function for the fraction of energy that is transferred into the hydrodynamic phase for a cut at $n = 5n_{q0}$ (dashed red line) and $n = 4n_{q0}$ (full black line) where $n_{q0}$ is the ground state energy density. The right plot of Fig. 4 shows the $K^+/\pi^+$ ratio as a function of $\sqrt{s}$. It shows that the experimental data [12, 15–19] lies between the cascade and the complete hybrid calculation. The core-corona approach leads to a much better description of the experimental data improving on the magnitude of the ratio as well as on the position of the peak.

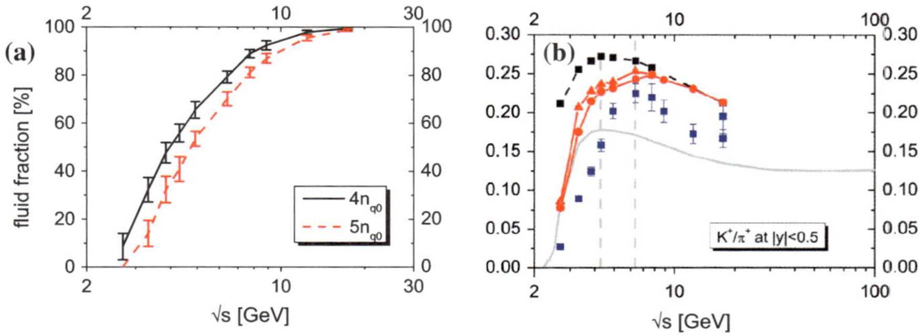


**Fig. 4** *Left* Total energy fraction which is transferred into the hydrodynamic phase as function of $\sqrt{s}$ for $b < 3.4$ fm collisions of AuAu/PbPb nuclei for a energy density cut of $n = 4n_{q0}$ (*black line*) or $n = 5n_{q0}$ (*dashed red line*). *Right* $K^+/\pi^+$ ratio as a function of $\sqrt{s}$ for UrQMD in cascade mode (*grey line*), energy cuts of $n = 4n_{q0}$ (*red circles*), $n = 5n_{q0}$ (*red triangles*) for the core-corona separation and a hybrid calculation with no core-corona separation (*black squares*). Experimental results are taken from [12, 15–19]

# 5 Summary

We presented HBT and $K^+/\pi^+$ results from the well known UrQMD over a broad energy range. For this we compare the cascade results with hybrid results combining the cascade calculation with an intermediate hydrodynamic phase. We found that the HBT data at SPS and RHIC energies are reasonable well described by the cascade model, while the intermediate hydro phase leads to an over prediction for all radii. At the same time the results of the $R_{\mathrm{out}}/R_{\mathrm{side}}$ get better. This is attributed to a simplified transverse particlization scheme. At LHC also the cascade only results fail to describe the data, we explained this with a lack of pressure in the early stage leading to an too slow expansion postponing the freeze out. To improve the transverse dynamics of the hybrid model we used a more realistic core-corona approach where only a part of the system is evolved hydrodynamically. This allowed improve the description of the $K^+/\pi^+$ ratio significantly.

**Acknowledgments** This work was supported by the Helmholtz International Center for FAIR within the framework of the LOEWE program launched by the State of Hesse, GSI, and BMBF. G.G. thanks the Helmholtz Research School for Quark Matter Studies.

# References

1. S.A. Bass, M. Belkacem, M. Bleicher, M. Brandstetter, L. Bravina, C. Ernst, L. Gerland, M. Hofmann et al., Prog. Part. Nucl. Phys. **41**, 255 (1998) [nucl-th/9803035]
2. M. Bleicher, E. Zabrodin, C. Spieles, S.A. Bass, C. Ernst, S. Soff, L. Bravina, M. Belkacem et al., J. Phys. G **5**, 1859 (1999) [hep-ph/9909407]
3. S. Kniege et al. [NA49 Collaboration], AIP Conf. Proc. **828**, 473 (2006) [nucl-ex/0601024]

4. H. Petersen, J. Steinheimer, Q. Li, G. Burau, M. Bleicher, arXiv:0806.1805 [hep-ph]
5. D.H. Rischke, Y. Pursun, J.A. Maruhn, Nucl. Phys. A **595**, 383 (1995) [Erratum-ibid. A 596, 717 (1996)] [nucl-th/9504021]
6. D.H. Rischke, S. Bernard, J.A. Maruhn, Nucl. Phys. A **595**, 346 (1995) [nucl-th/9504018]
7. H. Petersen, M. Bleicher, S.A. Bass, H. Stöcker, arXiv:0805.0567 [hep-ph]
8. H. Petersen, J. Steinheimer, G. Burau, M. Bleicher, H. Stöcker, Phys. Rev. C **78**, 044901 (2008)[arXiv:0806.1695 [nucl-th]]
9. Download the most recent UrQMD source code from http://urqmd.org/
10. Q. Li, G. Graef, M. Bleicher, Phys. Rev. C **85**, 034908 (2012) [arXiv:1203.4104 [nucl-th]]
11. J. Steinheimer, M. Bleicher, Phys. Rev. C **84**, 024905 (2011) [arXiv:1104.3981 [hep-ph]]
12. S.V. Afanasiev et al., The NA49 collaboration. Phys. Rev. C **66**, 054902 (2002)
13. K. Aamodt et al. [ALICE Collaboration], Phys. Lett. B **696**, 328 (2011) [arXiv:1012.4035 [nucl-ex]]
14. J. Adams et al. [STAR Collaboration], Phys. Rev. C **71**, 044906 (2005) [nucl-ex/0411036]
15. L. Ahle et al., E866 collaboration and E917 collaboration. Phys. Lett. B **476**, 1 (2000)
16. J. Adams et al., STAR collaboration. Phys. Rev. Lett. **92**, 112301 (2004)
17. K. Adcox et al., PHENIX collaboration. Phys. Rev. C **69**, 024904 (2004)
18. C. Alt et al., NA49 collaboration. Phys. Rev. C **77**, 024903 (2008)
19. M.K. Mitrovski et al., NA49 collaboration. J. Phys. G **32**, S43 (2006)

# Energy Loss of Heavy Quarks—A Signal of Plasma Properties

**J. Aichelin**

**Abstract** The possible observables for studying the properties of a plasma of quarks and gluons (QGP), which is presumably created in ultrarelativistic heavy-ion collisions, are discussed. While the light mesons do not contain the desired information about the QGP phase due to the strong final hadronic interaction, the 'heavy' mesons, i.e. those containing a c- or b- quark, are more useful. We demonstrate that our recent pQCD based approach for the energy loss of heavy quarks in a QGP combined with hydrodynamical model of Kolb and Heinz for the expansion of the plasma can successfully describe the variety of experimental data—as the transverse momentum spectra, $R_{AA}$, and the elliptic flow $v_2$ of heavy quarks—from RHIC to LHC energies.

## 1 Introduction

Hadrons have a finite radius. Therefore, if the density becomes too high one expects naively that the hadrons start to overlap and the constituents of the hadrons, the quarks and gluons, can move freely from one hadron to another. Thus hadrons loose their identity and form a new stage of matter—a plasma of unbound gluons and quarks which is in local thermal equilibrium. Because the density increases with temperature we expect such a transition at some given temperature in analogy with the boiling temperature of water when the phase transition from the liquid state to vapor occurs.

The phase transition from the hadronic to the partonic phase is also predicted by the fundamental theory for strongly interaction systems, the Quantum Chromo Dynamics (QCD). This follows from so called lattice gauge calculations [1], which are presently the only known way to predict expectation values of observables in strongly interacting systems. In this approach one creates by a Monte Carlo procedure

J. Aichelin (✉)
SUBATECH, University of Nantes, EMN, IN2P3/CNRS, 4 rue Alfred Kastler, Nantes cedex 344307, France
e-mail: aichelin@subatech.in2p3.fr

W. Greiner (ed.), *Exciting Interdisciplinary Physics*,
FIAS Interdisciplinary Science Series, DOI: 10.1007/978-3-319-00047-3_21,

field configurations on a finite size lattice according the the QCD Lagrangian which are subsequently used to calculate expectation values of operators. They have then to be extrapolated to the continuum limit. This is a very subtle method and therefore only recently the different groups found agreement on the transition temperature from the QGP to hadronic phase ($\approx$ 170 MeV or $1.5 \cdot 10^6$ K at zero baryon chemical potential) . These calculations are done assuming an infinite system in global thermal equilibrium.

A QGP has been most probably created in the early universe—few microseconds after the big bang. Direct evidence for this is, however, not available. It may also exist in the interior of neutron stars [2] at a very low temperature and a high density, so one talks about quark matter and not about a QGP, but this is presently only speculation because the interior may also consist of neutron matter. Radius and mass measurements of neutron stars can provide information on the their interior. Both are related by the equation of state [3], which is different for quark and nuclear matter. The present available data do not allow for a conclusion about the phase of the interior of these stars.

If one wants to do systematic studies of the QGP and its properties there is only one possibility, to create a QPG in heavy ion reactions. There the situation differs substantially from the conditions used for the lattice QCD calculations, i.e. the infinite partonic matter in equilibrium: In heavy ion reactions such a state can be only created for a very short time (of the order of $10^{-23}$ s) and has an extension of a couple of $10^{-15}$ m. Then the system, which expands with almost the speed of light, forms hadrons which are finally observed in the detector. Thus, it is quite difficult to prove—from the theoretical as well as from the experimental side—that in a such small system in a very short time a partonic plasma has be formed.

The problem is therefore to reconstruct from the observed hadrons the existence and the properties of such a QGP. In this situation a physicists feel like a fire expert who has to determine the cause of the fire from the left overs long after the fire is extinguished. The vast majority of the observed hadrons are formed from light (u,d,s) quarks which, unfortunately, tell us little about the formation of a plasma. It turned out that the multiplicity of light hadrons is very well described by statistical models [4]. This means that the system of light quarks or hadrons (made of light quarks) has reached a state of equilibrium (and the temperature is close to that predicted by lattice gauge calculations for the transition temperature). From statistical mechanics we know that then all information about the properties of the system prior to the equilibrium is lost. As we cannot conclude from the presence of water at 0° that at a lower temperature ice is formed, we cannot conclude from the existence of an equilibrated hadron gas close to the transition temperature that at higher temperatures it goes over to a QGP. One may argue that the spectra of light hadrons may give additional information. Unfortunately hadronic final state interactions are too strong and contain too many experimentally unknown cross sections in order to infer from the measured spectra the spectra at the moment when hadrons move out of equilibrium.

There are some correlations like the ridge which could only be formed very early in the reaction and the mass, momentum and centrality dependence of collective variables, like the elliptic flow $v_2$, gives a glimpse on what happens early during the reaction but their interpretation is not unique.

The light particle spectra are well described by so called event generators [5] which model the whole reaction on a computer. They assume that after a violent initial phase hadrons are formed which, if the energy density exceeds a given value, form a QGP. The expansion of the QGP is modelled either in viscous or in non viscous hydrodynamics. At a given energy density which is around 1 GeV/fm$^3$ [6] a sudden transition to the hadronic world takes place followed by a final state interactions among the hadrons. The problem is that the physics of the initial state is little known and leaves a lot of room for different assumptions. As a consequence, some model assume that the measured centrality dependence of the elliptic flow presents evidence for the need of viscous hydrodynamics [7] whereas other models describe the data equally well in an ideal hydrodynamical approach [8] assuming that not all particles take part in the hydrodynamical expansion.

Also the fast equilibration is not yet understood. Therefore other models [9] do not assume that a thermal equilibrium is established and describe the strongly interacting quark-gluon plasma by relativistic transport equations, derived from many-body Kadanoff–Baym equations, with a dynamical hadronization of partons to hadrons. This concept leads as well to a very satisfying description of a variety of data.

In a situation like this it is evident that one looks for possible observables which do not suffer from this memory erasing equilibrium phase. There are essentially two: High $p_t$ hadrons which originate from jets as well as the $p_t$ and $v_2$ distribution of heavy mesons which contain either a c or a b quark because neither jets nor heavy quarks come to an equilibrium with the plasma. Jets have the problem that the leading particle, i.e. that with the highest momentum, may change by interactions with the plasma. This makes the understanding of jets difficult.

Heavy quarks are produced in hard binary initial collisions between the incoming protons. Their production cross sections are known from pp collisions and can as well be calculated in pQCD calculations. Therefore the initial transverse momentum distribution of the heavy quarks is known. Comparing this distribution with that measured in heavy ion collisions allows defining $R_{AA} = (d\sigma_{AA}/dp_t^2)/(N_c\, d\sigma_{pp}/dp_t^2)$, where $N_c$ is the number of the initial binary collisions between projectile and target. The deviation of $R_{AA}$ from one measures the interaction of the heavy quark with the plasma because the hadron cross sections of heavy mesons are small. The heavy quark does not come to thermal equilibrium with the QGP therefore $R_{AA}$ contains the information on the interaction of the heavy quark while it traverses the plasma. In addition, the distribution of heavy quarks at the moment of their creation is isotropic in azimuthal direction, therefore the elliptic flow $v_2 = \langle \cos 2(\phi - \phi_R)\rangle$, where $\phi$ ($\phi_R$) is the azimuthal angle of the emitted particle (reaction plane) is 0. The observed finite $v_2$ value of the observed heavy meson can only originate from interactions between light QGP constituents and the heavy quarks. The simultaneous description of $R_{AA}$ and $v_2$ and their centrality dependence, presently the only observables for which data exist, give then the possibilities to understand the interactions inside the QGP.

Unfortunately the experimental results depend not only on the elementary interaction but also on the description of the expansion of the QGP [10]. Therefore the ultimate aim is to control the expansion by results on the light meson sector. This has not been achieved yet for the LHC and therefore it is difficult to asses the influence of the expansion on the observables. We use here the approach from Kolb and Heinz which has reasonably well described the midrapidity light mesons at RHIC [11]. We adjust only the charged particle multiplicity to the value measured at LHC.

The $R_{AA}$ of 0.2 values observed for large $p_t$ heavy mesons are much smaller than originally expected. Early theoretical approaches based on perturbative QCD (pQCD) calculation gave much larger values and it has been doubted, whether pQCD is the right tool to describe this interaction. This early calculation, however, used ad hoc assumptions on the coupling constant $\alpha_s$ and the infrared regulator $\mu$. With a standard choice $\mu$ and $\alpha_s$ an artificial K factor, an overall multiplication factor of the elastic cross section of around 10 [12, 13] had to be introduced to match the experimental data.

A while ago we advanced an approach for the collisional energy loss of heavy quarks in the QGP [14–16] in which (a) $\mu$ has been fixed by the demand that more realistic calculations using the hard thermal loop approach give the same energy loss as our Born type pQCD calculation and (b) the coupling constant is running and fixed by the sum rule advanced by Dokshitzer and later used by Peshier. Both these improvements increased the cross section especially for small momentum transfers and reduced therefore the necessary K factor to 2. Here we include in addition the radiative energy loss [17, 18].

## 2 Model

Our approach extended by including radiative energy loss has been well described the heavy quark data at RHIC. Therefore it is worthwhile to calculate what we expect for LHC energies if we modify the model only in a minimal way by adjusting the initial condition do $dN/dy = 1600$, as observed at RHIC. To include radiation we have to consider the following 5 matrix elements, displayed in Fig. 1, which contributes to radiation. The commutation relation

$$T^b T^a = T^a T^b - i f_{abc} T^c \tag{1}$$

allows us to regroup the 5 matrix elements into 3 combinations, each of them being independently gauge invariant:

$$\begin{aligned} iM^{QED}_{h.q.} &= C_a i(M_1 + M_2) \\ iM^{QED}_{l.q.} &= C'_a i(M_3 + M_4) \\ iM^{QCD} &= C_c i(M_1 + M_3 + M_5). \end{aligned} \tag{2}$$

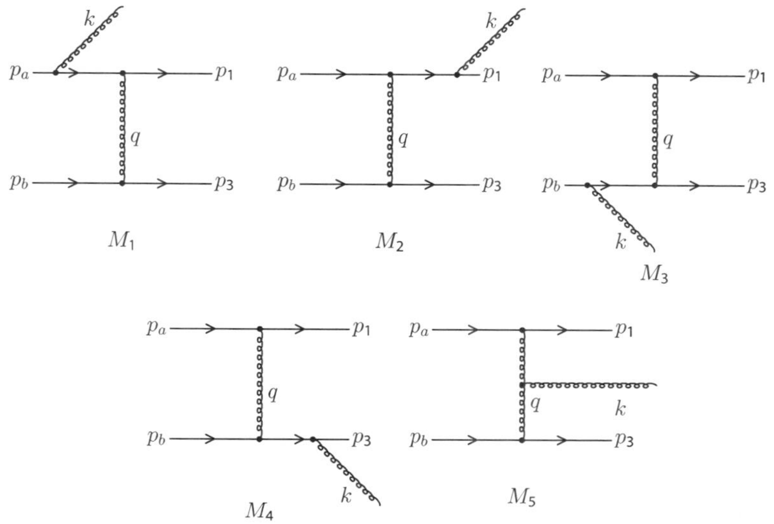


**Fig. 1** The five matrix elements which contribute to the gluon bremsstrahlung

h.q. (l.q.) mark the emission of the gluon from the heavy (light quark) line. $C_a, C_a'$ and $C_c$ are the color algebra matrix elements. The matrix elements labeled as QED are the bremsstrahlung diagrams already observed in Quantum Electrodynamics (QED), whereas that labeled QCD is the genuine diagram of Quantum Chromodynamics (QCD). The QCD diagram is the main objet of interest here because it dominates the energy loss of heavy quarks.

We evaluate the matrix elements in scalar QCD (see Ref. [19]). They are given by

$$\begin{aligned}
iM_1^{SQCD} &= C_A(ig)^3 \frac{(p_b+p_3)^\mu}{(p_3-p_b)^2} D_{\mu\nu}[p_3-p_b]\left(\frac{(p_a+p_1-k)^\nu(2p_a-k)\epsilon}{(p_a-k)^2-m^2} - \varepsilon^\nu\right)\\
iM_5^{SQCD} &= C_c(ig)^3 D^{\mu\mu'}[p_3-p_b]D^{\nu\nu'}[p_1-p_a]\left[g_{\mu'\nu'}(p_a-p_1+p_3-p_b)_\sigma\right.\\
&\quad \left.+\, g_{\nu'\sigma}(p_1-p_a-k)_{\mu'} + g_{\sigma\mu'}(p_b-p_3+k)_{\nu'}\right]\varepsilon^\sigma\\
&\quad \cdot \frac{(p_3+p_b)^\mu(p_a+p_1)^\nu}{(p_3-p_b)^2(p_1-p_a)^2}
\end{aligned} \tag{3}$$

$M_3$ is obtained by replacing $p_a \to p_b$ and $p_1 \to p_3$ in $M_1$. Using light cone gauge and keeping only the leading term in $\sqrt{s}$ we find that the square of the matrix element factorizes

$$|M|^2 = |M_{elast}(s,t)|^2 P_g(m,t,\mathbf{k_t},x) \tag{4}$$

with $|M_{elast}(s,t)|^2 = g^4 \frac{4s^2}{t^2}$ being the matrix element squared for the elastic cross section in a coulomb-like interaction between the heavy quark and a light quark (gluon). $P_g(m,t,s,\mathbf{k_t})$ describes the distribution function of the produced gluons.

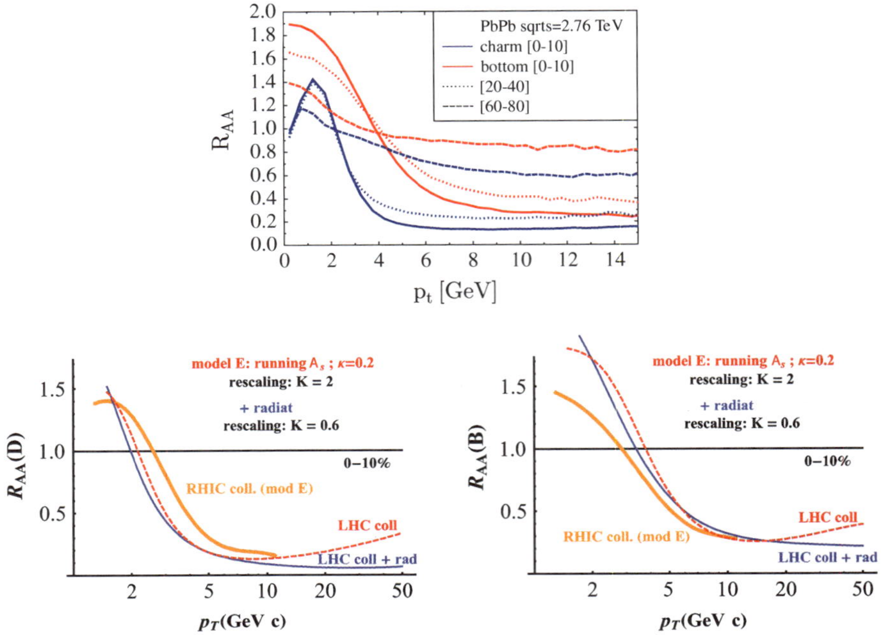


**Fig. 2** The transverse momentum distribution of $R_{AA}$ at midrapidity for different centralities and for bottom (*blue*) and charm (*red*) quarks. *Top* The details at small $p_t$, *bottom left* (*right*) $R_{AA}$ at large $p_t$ for D-mesons (B-mesons)

To discuss the physics we adopt the following light cone vectors

$$
\begin{aligned}
p_a &= \left\{\sqrt{s-m^2}, \frac{m^2}{\sqrt{s-m^2}}, 0, 0\right\} \\
p_b &= \{0, \sqrt{s-m^2}, 0, 0\} \\
k &= \{x\sqrt{s-m^2}, 0, \mathbf{k_t}\} \\
p_1 &= p_a + q - k = \left\{p_a^+(1-x) - \frac{q_t^2}{p_b^-}, \frac{(\mathbf{k_t}-\mathbf{q_t})^2+m^2}{(1-x)p_a^+}, \mathbf{q_t}-\mathbf{k_t}\right\} \\
p_3 &= p_b - q = \left\{\frac{q_t^2}{p_b^-}, p_b^- - \frac{(1-x)k_t^2 - x(\mathbf{k_t}-\mathbf{q_t})^2 + m^2x^2}{p_a^+(1-x)x}, -\mathbf{q_t}\right\}
\end{aligned}
\tag{5}
$$

The scalar product is defined as $p_a p_b = \frac{p_a^+ p_b^- + p_a^- p_b^+}{2} - p_{at} p_{bt}$ and $q^2 = t \approx q_t^2$. In this coordinate system x is given by $k^+ = x p_a^+$ and represents the relative longitudinal momentum fraction of the gluon with respect to the incoming heavy quark. In this coordinate system $|M_{SQCD}|^2$ has a very simple form:

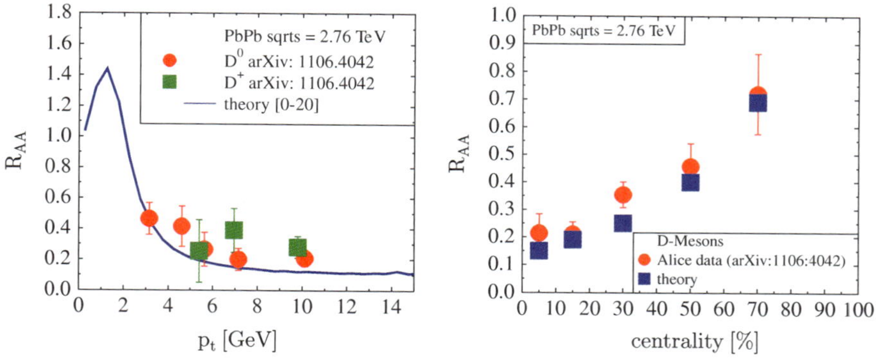


**Fig. 3** *Left* $R_{AA}$ as a function of $p_t$ for 0–20 % centrality, *right* centrality dependence of $R_{AA}$. We compare data from the ALICE collaboration [21] with our prediction

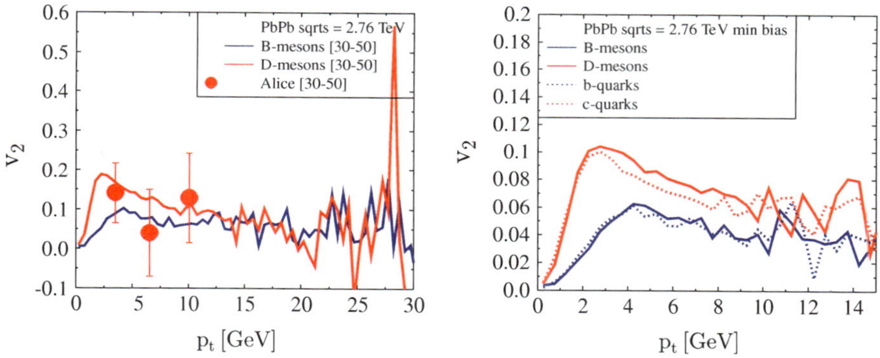


**Fig. 4** $p_t$ dependence of $v_2$. On the left hand side we compare our calculations for D and B meson for [30–50 %] centrality with the experimental data shown as this conference, on the right hand side we display $v_2$ for minimum bias separately for c-quarks and D-mesons and b-quarks and B-mesons, respectively

$$|M_{SQCD}|^2 = g^2 D^{QCD} 4(1-x)^2 |M_{elast}|^2 \left( \frac{\mathbf{k}_t}{k_t^2 + x^2 m^2} - \frac{\mathbf{k}_t - \mathbf{q}_t}{(\mathbf{q}_t - \mathbf{k}_t)^2 + x^2 m^2} \right)^2 \tag{6}$$

with the color factor $D^{QCD} = C_A * C_{el}^{qq} = \frac{2}{3}$. The first term in the bracket describes the emission from the incoming heavy quark line, the second term the emission from the gluon. This shows that in light cone gauge and in this coordinate system in leading order of $\sqrt{s}$ the matrix element for the emission from the light quark do not contribute. In the case of massless quarks we recover the squared matrix element of Gunion and Bertsch (GB) of Ref. [20].

## 3 Results

Having the matrix elements we can calculate the cross section of the elastic and radiative collisions of the heavy quarks with the plasma particles. At RHIC we found that the agreement is best when we multiply all cross section with a constant K factor of 0.6. A K factor of one is also compatible with the data but at the limits of the error bars. These cross sections are embedded in the plasma expansion as described in Refs. [14–16]. Figure 2 displays the $p_t$ dependence of $R_{AA}$ at midrapidity for different centrality bins and for c and b quarks separately. Charm quarks, being lighter, suffer a larger energy loss than bottom quarks and are therefore pushed more toward low $p_t$. Below a centrality of 40 % $R_{AA}$ does not change substantially. At small momenta we see an enhancement. There the energy loss accumulates the heavy quarks. For large $p_t$, shown in the bottom part of Fig. 2, radiative collisions act differently than elastic collisions. If we employ only elastic collisions (model E, with a K factor of 2) we see an increase of $R_{AA}$ with $p_T$ whereas for elastic and radiative collisions (with a K-factor of 0.6) $R_{AA}$ remains almost flat. If we include the Landau Pomeranchuck Migdal effect which suppresses radiation we would expect a moderate increase of $R_{AA}$ with increasing pt. For comparison we display as well the calculation for the RHIC data which matched the experimental results.

Figure 3 shows the comparison of our calculations with $R_{AA}$ ALICE data [21]. On the left hand side we display $R_{AA}$ as a function of $p_t$ of [0–20 %] centrality. The calculations follow closely the experimental data. On the right hand side we see $R_{AA}$ for mesons with $p_t > 6$ GeV as a function for the centrality. Also here we see a good agreement between theory and experiment

Figure 4 show the comparison of our calculations with recent $v_2$ ALICE data [22]. We see that at low $p_t$ $v_2$ for B-mesons is substantially smaller than for D-mesons. This is again the consequence of the smaller mass of the c-quarks which can more easily absorb the $v_2$ of the light plasma particles with whom they collide during the expansion. We see that the prediction of our model (the data have been presented for the first time at this conference when the calculations have been already performed) agrees with the experimental value in between the error bars. The right hand side highlights the difference of $v_2$ between b and c quarks at intermediate $p_t$. This difference is inherent in the model and may therefore serve as a verification if perturbative QCD is the right theory to describe the data. Whereas the $v_2$ of D-mesons is slightly higher than that of the c-quarks due to the hadronisation, the heavy B-meson has practically the same $v_2$ as the b-quark before hadronisation.

In conclusions we have shown that pQCD like models which include a running coupling constant as well as a infrared regulator of the gluon propagator in the elastic cross section which is based on hard thermal loop calculations reproduce the LHC data as they reproduced the RHIC data. The model predicts different $v_2$ values for D- and B-mesons as well as an increase of $R_{AA}$ in central collisions with $p_t$ for $p_t$ larger than 15 GeV. The model can therefore be verified by future experimental data. The results show that collisional as well as radiative energy loss is necessary to describe the data. Both contribute to $R_{AA}$ in a comparable way. In this analysis we used the hydrodynamical model of Heinz and Kolb. It remains to be seen how other models for the expansion of the plasma change the numerical values of $R_{AA}$ and $v_2$. Studies of different expansion scenarios as well as of the influence of the Landau Pomernschuk Migdal effect are under way.

## References


1. S. Borsanyi et al., J. Phys. G **38**, 124101 (2011)
2. S. Weissenborn, I. Sagert, G. Pagliara, M. Hempel, J. Schaffner-Bielich, Astrophys. J. **740**, L14 (2011) [arXiv:1102.2869 [astro-ph.HE]]
3. J. Aichelin, J. Schaffner-Bielich, arXiv:0812.1341 [nucl-th]
4. A. Andronic, P. Braun-Munzinger, K. Redlich, J. Stachel, J. Phys. G **38**, 124081 (2011). [arXiv:1106.6321 [nucl-h]]
5. K. Werner, I. Karpenko, T. Pierog, M. Bleicher, K. Mikhailov, Phys. Rev. C **82**, 044904 (2010). [arXiv:1004.0805 nucl-th]]
6. J. Steinheimer, J. Aichelin, M. Bleicher, arXiv:1203.5302 [nucl-th]
7. M. Luzum, P. Romatschke, Phys. Rev. C **78**, 034915 (2008) [Erratum-ibid. C 79 (2009) 039903] [arXiv:0804.4015 [nucl-th]]
8. J. Aichelin, K. Werner, J. Phys. G **37**, 094006 (2010). [arXiv:1008.5351 [nucl-th]]
9. W. Cassing, E.L. Bratkovskaya, Nucl. Phys. **A831**, 215 (2009) arXiv:0907.5331 [nucl-th]
10. P.B. Gossiaux, S. Vogel, H. van Hees, J. Aichelin, R. Rapp, M. He, M. Bluhm, arXiv:1102.1114 [hep-ph]
11. P.F. Kolb, U.W. Heinz, arXiv:nucl-th/0305084. Published in Qurak Gluon Plasma, World Scientific
12. G.D. Moore, D. Teaney, Phys. Rev. C **71**, 064904 (2005)
13. B. Svetitsky, Phys. Rev. D **37**, 2484 (1988)
14. P.B. Gossiaux, J. Aichelin, Phys. Rev. **78**, 014904 (2008) arXiv:0802.2525 [hep-ph]
15. P.B. Gossiaux, R. Bierkandt, J. Aichelin, arXiv:0901.0946 [hep-ph]
16. P. B. Gossiaux, J. Aichelin, arXiv:0901.2462 [nucl-th]
17. P.B. Gossiaux, J. Aichelin, T. Gousset, V. Guiho, J. Phys. G **37**, 094019 (2010) [arXiv:1001.4166 [hep-ph]]
18. J. Aichelin, P.B. Gossiaux, T. Gousset, to be published
19. E. Meggiolaro, Phys. Rev. D **53**, 3835 (1996). [arXiv:hep-th/9506043]
20. J.F. Gunion, G. Bertsch, Phys. Rev. D **25**, 746 (1982)
21. A. Dainese [ALICE Collaboration], arXiv:1203.2160 [nucl-ex]
22. F. Prino, C. Bianchin, *Contributions to the Conference "Strangeness in Quark Matter"*, Cracow, Sept 2011

# The Thermal Model and the Tsallis Distribution at the Large Hadron Collider

**J. Cleymans**

**Abstract** An analysis is presented of identified particles at the Large Hadron Collider. Possible deviations from standard statistical distributions are investigated by considering in detail the Tsallis distribution. Matter-antimatter production is discussed within the framework of chemical equilibrium in p-p and heavy ion collisions.

## 1 The Hadronic World

The available energy for heavy ions at the Large Hadron Collider (LHC) is $\sqrt{s} =$ 2760 AGeV yet the observed temperature is only of the order of $T \approx 0.160$ GeV. To understand this enormous change from the initial state to the final state we first clarify how this temperature is determined. There are several independent ways of doing this.

1. From the number of hadronic resonances listed in the particle data booklet [1]. This method was first proposed by Hagedorn [2]. Note that this involves no transverse momentum spectrum, no energy distribution, only the number of particles listed in the PDG [1] table. A recent updated version of this determination is shown in Fig. 1 where the logarithm of the number of resonances below a certain mass [3] is plotted versus the mass. The fitted line corresponds to a Hagedorn temperature of

$$T_H = 174 \pm 11 \quad \text{MeV}. \tag{1}$$

   Other recent determinations are consistent with this value [4–7]. At masses above 3 GeV the increase stops due to the difficulty in identifying heavy hadronic resonances, a situation which will probably never be resolved experimentally.

J. Cleymans (✉)
UCT-CERN Research Centre and Department of Physics, University of Cape Town, Cape Town, South Africa
e-mail: jean.cleymans@uct.ac.za

W. Greiner (ed.), *Exciting Interdisciplinary Physics*, FIAS Interdisciplinary Science Series, DOI: 10.1007/978-3-319-00047-3_22,

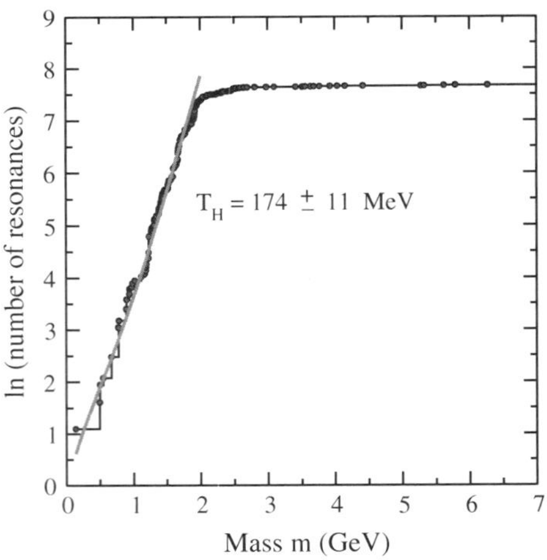


**Fig. 1** Cumulative number of hadronic resonances as a function of $m$ [3]. The hadronic data include baryons, mesons and also heavy resonances made up of charm and *bottom* quarks

2. The multiplicity of particles in the final state. This has been an ongoing effort for the past two decades [8–10]. Again this involves no transverse momentum or energy distribution. In this case it is only the number of identified particles in the final state. The temperature at $\mu_B = 0$ is remarkably close to the original Hagedorn temperature [2] obtained by summing the number of hadronic resonances.
3. The critical temperature determined from Lattice QCD is again remarkably close to the Hagedorn temperature and to the chemical freeze-out temperature at $\mu_B = 0$ [11, 12].
4. The temperature can also be determined from the slope of the transverse momentum spectrum. This leads to a lower temperature, at least in p-p collisions and will be discussed below.

## 2 Transverse Momentum Distribution

The Tsallis distribution has gained prominence recently in high energy physics with high quality fits of the transverse momentum distributions made by the STAR [13] and PHENIX [14] collaborations at the Relativistic Heavy Ion Collider and by the ALICE [15], CMS [16] and ATLAS [17] collaborations at the Large Hadron Collider.

In the literature there exists more than one version of the Tsallis distribution [18, 19]. In this paper we investigate a version which we consider suited for describing results in high energy particle physics. Our main guiding criterium will be thermodynamic consistency which has not always been implemented correctly. The explicit form which we use is [20]:

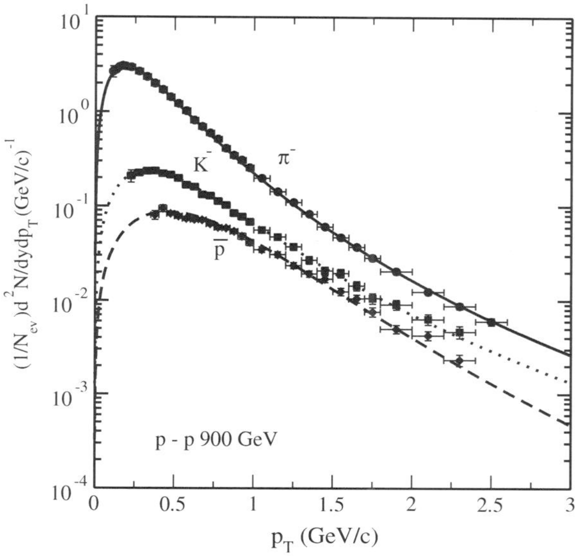


**Fig. 2** Fit to the data [15] using the Tsallis distribution [20]

$$\frac{d^2N}{dp_T dy} = gV\frac{p_T m_T \cosh y}{(2\pi)^2}\left[1+(q-1)\frac{m_T\cosh y-\mu}{T}\right]^{q/(1-q)}, \quad (2)$$

where $p_T$ and $m_T$ are the transverse momentum and mass respectively, $y$ is the rapidity, $T$ and $\mu$ are the temperature and the chemical potential, $V$ is the volume, $g$ is the degeneracy factor.

The motivation for preferring this form is presented in detail in [20]. The parameterization given in Eq. (2) is close (but different) from the one used by STAR, PHENIX, ALICE, CMS and ATLAS [13–17]:

$$\frac{d^2N}{dp_T dy} = p_T\frac{dN}{dy}\frac{(n-1)(n-2)}{nC(nC+m_0(n-2))}\left(1+\frac{m_T-m_0}{nC}\right)^{-n}, \quad (3)$$

where $n$, $C$ and $m_0$ are fit parameters. The analytic expression used in Refs. [13–16] corresponds to identifying

$$n \to \frac{q}{q-1} \quad (4)$$

and

$$nC \to \frac{T+(q-1)m}{q-1}. \quad (5)$$

But differences do not allow for the above identification to be made complete due to an additional factor of the transverse mass on the right-hand side. In particular,

**Table 1** Fitted values of the $T$ and $q$ parameters for strange particles measured by the ALICE [15] and CMS collaborations [16] using the Tsallis-B form for the momentum distribution

| Particle | $q$ | $T$ (GeV) |
|---|---|---|
| $\pi^+$ | $1.154 \pm 0.036$ | $0.0682 \pm 0.0026$ |
| $\pi^-$ | $1.146 \pm 0.036$ | $0.0704 \pm 0.0027$ |
| $K^+$ | $1.158 \pm 0.142$ | $0.0690 \pm 0.0223$ |
| $K^-$ | $1.157 \pm 0.139$ | $0.0681 \pm 0.0217$ |
| $K_S^0$ | $1.134 \pm 0.079$ | $0.0923 \pm 0.0139$ |
| $p$ | $1.107 \pm 0.147$ | $0.0730 \pm 0.0425$ |
| $\bar{p}$ | $1.106 \pm 0.158$ | $0.0764 \pm 0.0464$ |
| $\Lambda$ | $1.114 \pm 0.047$ | $0.0698 \pm 0.0148$ |
| $\Xi^-$ | $1.110 \pm 0.218$ | $0.0440 \pm 0.0752$ |

no clear pattern emerges for the values of $n$ and $C$ while an interesting regularity is obtained for $q$ and $T$ as seen in Table 1. The striking feature is that the values of $q$ are consistently between 1.1 and 1.2 for all species of hadrons. The fit to negatively charged particles in p-p collisions measured by the ALICE collaboration is shown in Fig. 2. An interpretation of the parameter $q$ in terms of fluctuations around a Boltzmann distribution has been given in [21, 22].

## 3 Antimatter

One of the striking features of particle production at high energies is the near equal abundance of matter and antimatter in the central rapidity region [23, 24]. As is well known, a similar symmetry existed in the initial stage of the early universe and it still remains a mystery as to how this got lost in the subsequent evolution of the universe reaching a stage with no visible amounts of antimatter being present.

Closely related to this matter/antimatter symmetry is the production of light antinuclei, hypernuclei, and antihypernuclei at high energies. Since the first observation of hypernuclei in 1952 [25], there has been a steady interest in searching for new hypernuclei, exploring the hyperon-nucleon interaction which is relevant (see, e.g., [26, 27]) for nuclear physics. Hypernuclei decay with a lifetime which depends on the strength of the hyperon-nucleon interaction. While several hypernuclei have been discovered since the first observations in 1952, no antihypernucleus has ever been observed until the recent discovery of the antihypertriton in Au+Au collisions at $\sqrt{s_{NN}} = 200$ GeV by the STAR collaboration at RHIC [28]. The yield of (anti)hypernuclei measured by STAR is very large, in particular, they seem to be produced with a similar yield as other (anti)nuclei, in particular (anti)helium-3. This abundance is much higher than measured for hypernuclei and nuclei at lower energies [29]. It is of interest to understand the nature of this enhancement, and for this the mechanism of production of (anti)hypernuclei should be investigated.

The thermalization assumption applies successfully to hadrons produced in a large number of particle and nuclear reactions at different energies (see, e.g., [30–34]). This fact allows us to estimate thermal parameters characterizing the particle source for each colliding system, relevant for the understanding of the thermal properties of dense and hot matter, and in particular for studies of QCD phase transitions.

Using the parameterizations of thermal parameters found in the THERMUS model [35, 36], estimates have been made of the yields of (anti)hypernuclei, that can be directly compared to the recently measured yields at RHIC, as well as of (anti)matter and (anti)hypernuclei production at the Large Hadron Collider (LHC) [37]. A similar analysis, not including p-p results, has been presented recently in [38] where it was shown that ratios of hypernuclei to nuclei show an energy dependence similar to the $K^+/\pi^+$ one with a clear maximum at lower energies.

A quantitative study as to how the matter/antimatter symmetry is reached as the beam energy is increased has been presented in [37]; estimates of ratios of hypernuclei and antihypernuclei yields in Au+Au collisions at RHIC using the above mentioned parameterizations of thermal parameters that best fit hadron production at RHIC have also been presented [37]. The analysis uses a thermal model and aims to elucidate the production mechanism of hypernuclei and antihypernuclei in heavy ion collisions at RHIC and LHC energies, thus providing insight in the surprising increase of (anti)hypernuclei production at high energies.

In heavy-ion collisions the increase in the antimatter to matter ratio with the center-of-mass energy of the system has been observed earlier by the NA49 [39, 40] and the STAR [13] collaborations. The trend of the $\overline{\mathrm{p}}$/p ratio increase with the energy towards unity is shown in Fig. 3, where the open squares refer to heavy ion collisions and the solid circles refer to p-p collisions. It includes results from the NA49 [39], STAR [13] and the new results from the ALICE collaboration [24]. The two input parameters, the chemical freeze-out temperature $T$ and the baryon chemical potential $\mu_B$ as a function of $\sqrt{s}$ are taken from Ref. [10]. The solid circles represent $\mu_B$, obtained after fitting experimental data with the THERMUS model [35, 36]. The solid line is a new parameterization adjusted for p-p collisions [37]. In view of the fact that peripheral and central collisions show no noticeable change in the temperature, the same $T$ dependence for p-p as in heavy ion collisions was used [37]. It is important to note that $\mu_B$ is always lower in p-p collisions than in heavy ion collisions, e.g., the freeze-out chemical potential follows a different pattern, due to the lower stopping power in p-p collisions.

The relation between the $\overline{\mathrm{p}}$/p ratio and $\mu_B$ can be shown easily within the statistical concept using the Boltzmann statistics. The production of light nuclei including hypertritons ($^3_\Lambda$H) and antihypertritons ($^3_{\overline{\Lambda}}\overline{\mathrm{H}}$) was recently observed by the STAR collaboration [28]. The abundances of such light nuclei and antinuclei follow a consistent pattern in the thermal model. The temperature remains the same as before but an extra factor of $\mu_B$ is picked up each time the baryon number is increased. Each proton or neutron thus simply adds a factor of $\mu_B$ to the Boltzmann factor. The production of nuclear fragments is therefore very sensitive to the precise value of the baryon chemical potential and could thus lead to a precise determination of $\mu_B$. Deuterium has an additional neutron and the antideuterium to deuterium ratio is given by the square of the antiproton to proton ratio:

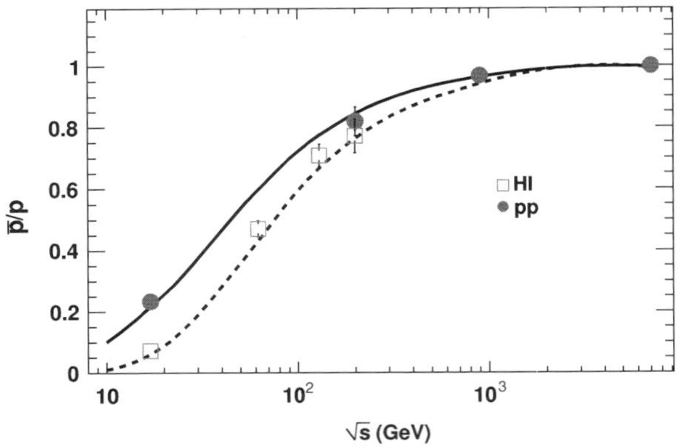


**Fig. 3** The p̄/p ratio as function of $\sqrt{s}$. The *solid circles* are results from p-p collisions and the *open squares* are results from HI collisions as a function of the invariant beam energy [13, 23, 24, 39, 40]

$$\frac{n_{\overline{\mathrm{d}}}}{n_{\mathrm{d}}} = e^{-(4\mu_B)/T}. \tag{6}$$

Helium-3 has three nucleons and the corresponding antihelium-3 to helium-3 ratio is given by

$$\frac{n_{3\overline{\mathrm{He}}}}{n_{3\mathrm{He}}} = e^{-(6\mu_B)/T}. \tag{7}$$

If the nucleus carries strangeness, this leads to an extra factor of $\mu_S$

$$\frac{n_{{}^3_{\overline{\Lambda}}\overline{\mathrm{H}}}}{n_{{}^3_{\Lambda}\mathrm{H}}} = e^{-(6\mu_B-2\mu_S)/T}. \tag{8}$$

In mixed ratios, the different degeneracy factors are also taken into account, e.g., 6 for ${}^3_{\Lambda}H$ and 2 for ${}^3_{\Lambda}H$

$$\frac{n_{{}^3_{\Lambda}\mathrm{H}}}{n_{3\mathrm{He}}} = 3e^{-(6\mu_B-\mu_S)/T}. \tag{9}$$

A detailed description of the results can be found in [37, 38, 41].

## 4 Conclusions

The Tsallis distribution gives a very good description of the transverse momentum spectrum, the parameter $q$ which is a measure for the deviation from a standard Boltzmann distribution is found to be around 1.1. The thermal model provides valuable insights in the composition of the final state produced in heavy ion and in p-p collisions. It shows a clear systematic way of interpreting results concerning identified particles. The production of antimatter like antinuclei, hypernuclei and antihypernuclei shows a new region of applications for the thermal model which promises to be very useful.

**Acknowledgments** Numerous discussions with S. Kabana, A. Kalweit, I. Kraus, K. Redlich, H. Oeschler, N. Sharma, A. Sorin and D. Worku are at the basis of the results presented here.

## References

1. C. Caso et al. (Particle Data Group), Eur. Phys. J. **3**, 1 (2008)
2. R. Hagedorn, Supp. Nuovo Cimento **III**, 147 (1965)
3. J. Cleymans, D. Worku, Mod. Phys. Lett. A **26**, 1197 (2011)
4. S. Chatterjee, S. Gupta, R.M. Godbole, Phys. Rev. C **81**, 044907 (2010)
5. W. Broniowski, W. Florkowski, Phys. Lett. B **490**, 223 (2000)
6. W. Broniowski, contribution to the Proceedings of the Mini-Workshop "Few-Quark Problems", Bled, Slovenia, 8–15 July 2000, arXiv:hep-ph/0008112
7. W. Broniowski, W. Florkowski, L.Y. Glozman, Phys. Rev. D **70**, 117503 (2004)
8. A. Andronic, P. Braun-Munzinger, J. Stachel, Nucl. Phys. A **834**, 237C (2010)
9. F. Becattini, J. Mannninen, M. Gazdzicki, Phys. Rev. C **73**, 044905 (2006)
10. J. Cleymans, H. Oeschler, K. Redlich, S. Wheaton, Phys. Rev. C **73**, 034905 (2006)
11. S. Borsanyi et al. (Wuppertal-Budapest Collaboration), JHEP **1009**, 073 (2010)
12. M. Cheng, S. Ejiri, P. Hegde, F. Karsch, O. Kaczmarek, E. Laermann, R.D. Mawhinney, C. Miao et al., Phys. Rev. D **81**, 054504 (2010)
13. B.I. Abelev et al. (STAR), Phys. Rev. C **75**, 064901 (2007)
14. A. Adare et al. (PHENIX), Phys. Rev. C **83**, 064903 (2011)
15. K. Aamodt et al. (ALICE Collaboration), Eur. Phys. J. C **71**, 1655 (2011)
16. V. Khachatryan et al. (CMS), JHEP **05**, 064 (2011)
17. G. Aad et al. (ATLAS Collaboration), New J. Phys. **13**, 053033 (2011)
18. C. Tsallis, J. Stat. Phys. **52**, 479 (1988)
19. C. Tsallis, R.S. Mendes, A.R. Plastino, Phys. A **261**, 534 (1998)
20. J. Cleymans, D. Worku, J. Phys. G **39**, 025006 (2012)
21. G. Wilk, Z. Wlodarczyk, Phys. Rev. Lett. **84**, 2770 (2000)
22. G. Wilk, Z. Wlodarczyk, Eur. Phys. J. A **40**, 299 (2009)
23. B.I. Abelev et al. (STAR Collaboration), Phys. Rev. C **79**, 034909 (2009)
24. K. Aamodt et al. (ALICE Collaboration), Phys. Rev. Lett. **106**, 072002 (2010)
25. M. Danysz, J. Pniewski, Phil. Mag. **44**, 348 (1953)
26. D. Hahn, H. Stöcker, Nucl. Phys. A **476** (1988)
27. H. Stöcker, W. Greiner, Phys. Rep. **137**, 277 (1986)
28. B.I. Abelev et al. (STAR Collaboration), Science 328, **58** (2010)
29. R. Rapp, E.V. Shuryak, Phys. Rev. Lett. **86**, 2980 (2001)
30. F. Becattini, Z. Phys. C **76**, 269 (1997)
31. F. Becattini, U. Heinz, Z. Phys. C **69**, 485 (1996)
32. K. Redlich, J. Cleymans, H. Oeschler, A. Tounsi, Acta Phys. Pol. B **33**, 1609 (2002)
33. P. Braun-Munzinger, K. Redlich, J. Stachel, nucl-th/0304013, *Invited Review in Quark Gluon Plasma 3*, ed. by R.C. Hwa, X.N. Wang, (World Scientific Publishing, 2004)
34. P. Braun-Munzinger, J. Cleymans, H. Oeschler, K. Redlich, Nucl. Phys. A **697**, 902 (2002)
35. S. Wheaton, J. Cleymans, J. Phys. G **31**, S1069 (2005)

36. S. Wheaton, J. Cleymans, M. Hauer, Comput. Phys. Commn. **180**, 84 (2009)
37. J. Cleymans, S. Kabana, I. Kraus, H. Oeschler, K. Redlich, N. Sharma, Phys. Rev. C **84**, 054916 (2011)
38. A. Andronic, P. Braun-Munzinger, J. Stachel, H. Stöcker, S., Phys. Lett. B **697**, 203 (2011)
39. C. Alt et al. (NA49 Collaboration), Phys. Rev. C **73**, 044910 (2005)
40. C. Alt et al. (NA49 Collaboration), Phys. Rev. C **77**, 024903 (2008)
41. A. Andronic, P. Braun-Munzinger, J. Stachel, Nucl. Phys. A **772**, 167 (2006)

# The Mini Bang and the Big Bang: From Collider to Cosmology

**Bikash Sinha**

## 1 Introduction

Collisions of two nuclei at ultra-relativistic energy such as in RHIC and/or LHC are expected to lead to a new state of matter, usually referred to as Quark Gluon Plasma (QGP). Although there is a fair amount of controversy about the exact nature of the phase transition from hadrons to QGP, there is little doubt that QGP has been found.

Lattice calculations [1–3] tend to indicate that at small hadronic chemical potential but high temperature the transition is like a crossover, such that after the transition no memory is left over of the matter before the transition. This scenario is applicable, it is argued, both at LHC and in the very early universe, about a microsecond after the Big Bang.

Recent discoveries at RHIC clearly establish that QGP behaves as a perfect fluid with $\eta/s$ satisfying the AdS/CFT limit of 0.08, almost zero [4, 5]. This phenomenon almost uniquely comes from the elliptic flow characterised by $v_2$ the flow velocity. Higher order flow components $v_n$ $(n = 3\ldots)$ are extremely important as amply demonstrated by [6, 7] PHENIX and STAR at RHIC and ALICE, ATLAS and CMS at LHC [8].

It has been pointed out by the present author and his collaborators that the thermometric signals such as $\gamma$, $\mu^+\mu^-$ are rather efficient signals of QGP, especially of the early times, almost immediately after QGP is formed. It was further pointed out [9] that the ratio $\gamma/\mu^+\mu^-$ will have the added advantage of universality in the sense that being a ratio, the dependence on boundary conditions, such as model dependent initial temperature, initial time etc. get cancelled out, thus effectively turning the ratio more sensitive to the actual state of the matter.

Finally the scenario of quark hadron phase transition around a microsecond after the Big Bang—what possible relics can be left over today? Recent work of Boeckel

B. Sinha (✉)
Variable Energy Cyclotron Centre, 1/AF, Bidhan Nagar, Kolkata 700064, India
e-mail: bikash@vecc.gov.in

W. Greiner (ed.), *Exciting Interdisciplinary Physics*,
FIAS Interdisciplinary Science Series, DOI: 10.1007/978-3-319-00047-3_23,

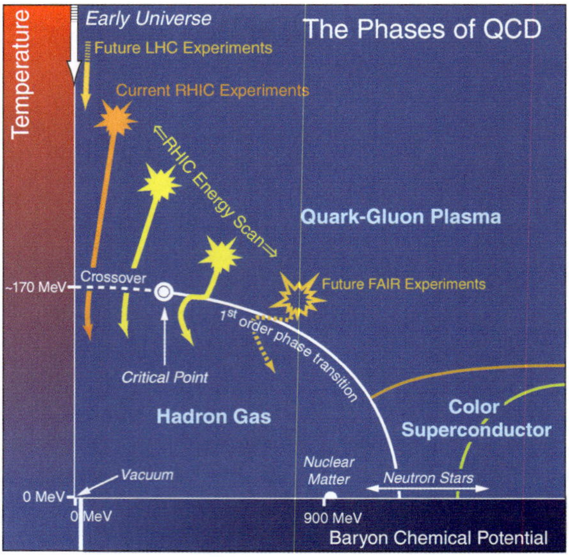


**Fig. 1** The phases of QCD

and Schaffner-Bielich [10] has demonstrated that a little inflation (7e-foldings) will force $\mu/T$ to be of the order of unity, precipitating a first-order phase transition from quark to hadrons at this primordial epoch.

Our previous work [11, 12] and this work [10] predict a large number of very interesting relics; I argue that if the relics are found that will corroborate a first order phase transition scenario.

## 2 Dileptons and Photons

In Fig. 1 the QCD phase diagram is shown in all its glory. Over the years this so called "cartoon" has evolved to a level of sophistication or complication, depending on one's taste, that it is no longer funny, at all!

We see at the top extreme left, the microsecond old universe "crossing over" to the hadronic world, a little on the right (along the baryonic chemical potential axis) LHC making the reverse journey from hadrons to quarks along the crossover critical temperature. RHIC energy regime takes us to the edge of the crossover and then, along the phase boundary "rainbow" we meet up with the future FAIR experiment and then at very low temperatures but very high baryonic chemical potential we encounter the neutron star regime even the land of Quarkonium, colour superconductivity and all kinds of exotica!

It has been amply demonstrated by lattice calculations [1–3] that the energy density scaled by $T^4$, $\varepsilon/T^4$ as a function of the temperature turns invariant of the temperature beyond some critical temperature $T_c$, Fig. 2, either the standard critical

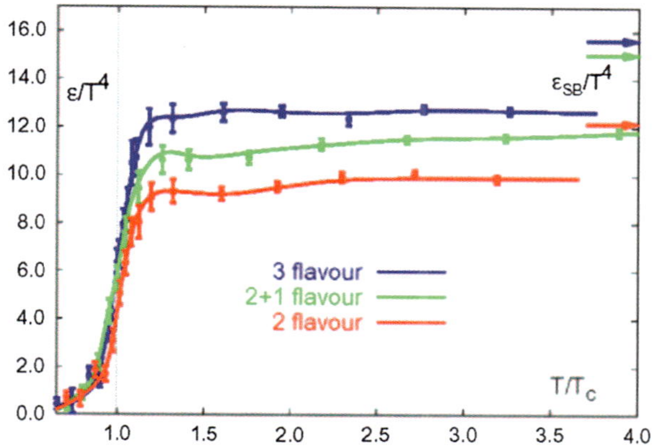


**Fig. 2** Temperature dependence of the scaled energy density as predicted by lattice QCD

temperature of a first order phase transition or the so called crossover critical temperature, just discussed. What is referred to as (2+1) flavour implies that unlike the other two lines which deals with zero quark mass, one flavour, strange quark in this case has finite mass, leading to crossover.

In Fig. 3 the diagrams in lowest order involved in photon and dimuon productions from QGP are shown. It was shown by the present author [9] that the cross section of production of the just mentioned thermometric signals vary as $T^4$ in the QGP sector. Thus the ratio $R(\gamma/\mu^+\mu)$ will saturate to a constant value signaling *the onset of QGP*. For the simplest case, with some approximation

$$R(\gamma/\mu^+\mu) \approx 2/\alpha \propto s\ln(1/\alpha s) \tag{1}$$

The ratio is modified in form [13] with the addition of more sophistication but essentially, the message remains the same.

In Fig. 4 even at SPS, we present the data from the Pb+Pb run of WA98 at 17.3 GeV/A; the fits to the data using the Parton Cascade model for hard scattering leading to hard photons and the standard [14, 15] hydrodynamical model leading to thermal photons are shown. It is already clear even at SPS energies that neither the thermal alone nor just hard scattering will fit the data, rather a combination of the two does the job well. We have mentioned many times [14, 15] that the typical window of $p_T$ for thermal photons lie within $1.5 \leq p_T \leq 3.5$ GeV. Clearly, even at SPS, for relatively low $p_T \simeq 1.5$ GeV, as expected, hard photons have virtually no role, but still thermal photons fall short of the data, whereas for high $p_T$ regime $\approx 3.5$ GeV, hard photons are crucial. It tells us that even at SPS energies we have to have thermal photons and hard photons cannot be neglected.

Thus, we already see the tantalising hint of QGP photons at SPS; this point was emphasized by Srivastava and Sinha [16].

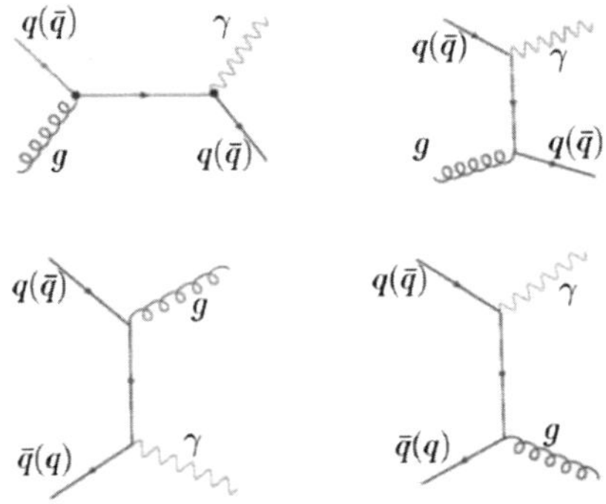


**Fig. 3** Light from QGP

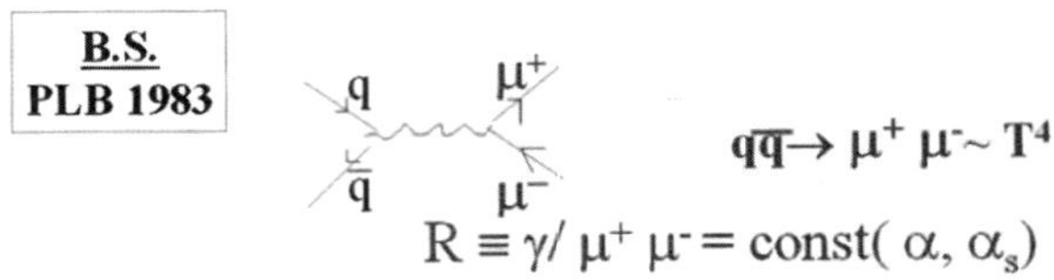


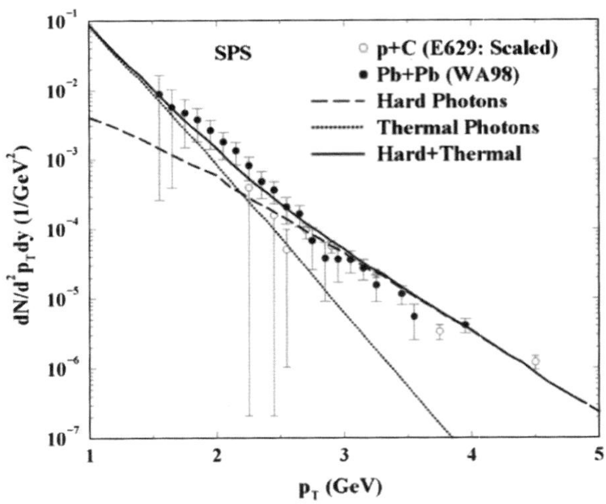


**Fig. 4** Photons at SPS

In Fig. 4 data from PHENIX at $\sqrt{s} = 200$ GeV/A (RHIC energies) is shown, emphasizing the above points more substantially: The ratio

$$R_{\mathrm{em}} = \frac{d^2 N_\gamma}{d^2 p_T dy} \Big/ \frac{d^2 N_\gamma^*}{d^2 p_T dy} \tag{2}$$

$\gamma/\mu^+\mu^-$, as has already been mentioned, is an excellent handle for determining the initial temperature.

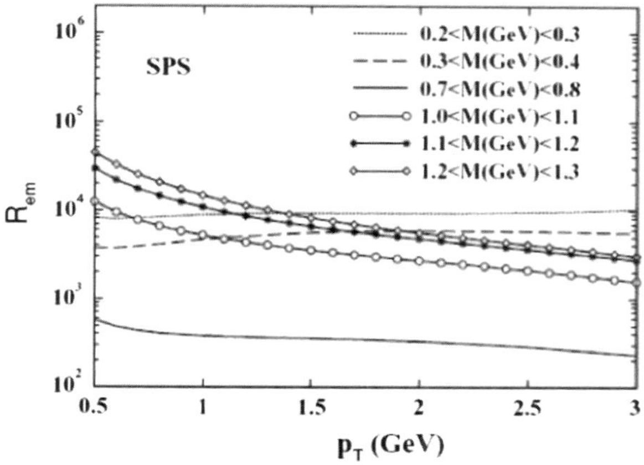


**Fig. 5** The thermal photon to dilepton ratio $R_{\mathrm{em}}$ as a function of transverse momentum $p_T$ for various invariant mass windows

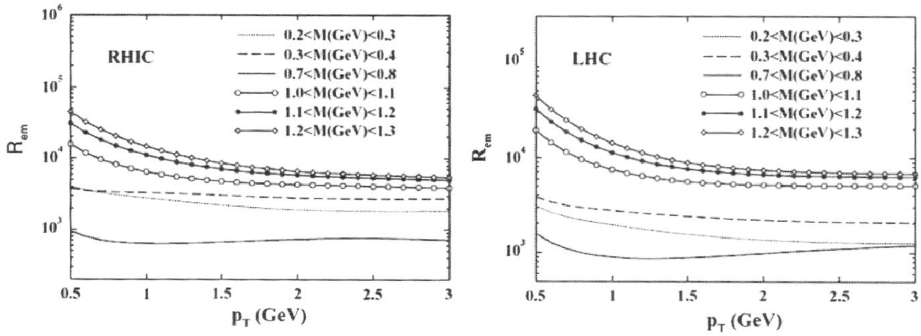


**Fig. 6** The thermal photon to dilepton ratio $R_{\mathrm{em}}$ at RHIC (*left*) and LHC (*right*)

It is clear (Figs. 5, 6) that the quantiy $R_{\mathrm{em}}$ reaches a plateau beyond $p_T \approx 1.5$ GeV for all three cases of SPS, RHIC and LHC. Interestingly enough the degree of flatness goes up with energy, as we go from SPS to RHIC to LHC. The difference in $R_{\mathrm{em}}$ in the plateau originates from different values of the initial temperature, thus, it can be a measure of $T_i$. However, the ratio is largely independent of $T_c$, transverse velocity $v_0$ even being independent of the equation of state. It is further observed [17] that $R_{\mathrm{em}}$ is a sensitive measure of flow as well.

## 3 Universality of $\eta/s$

By 2004, RHIC experiments determined and reported several key properties of the hot dense matter. Its opacity to energetic quarks and gluon indicates extremely high

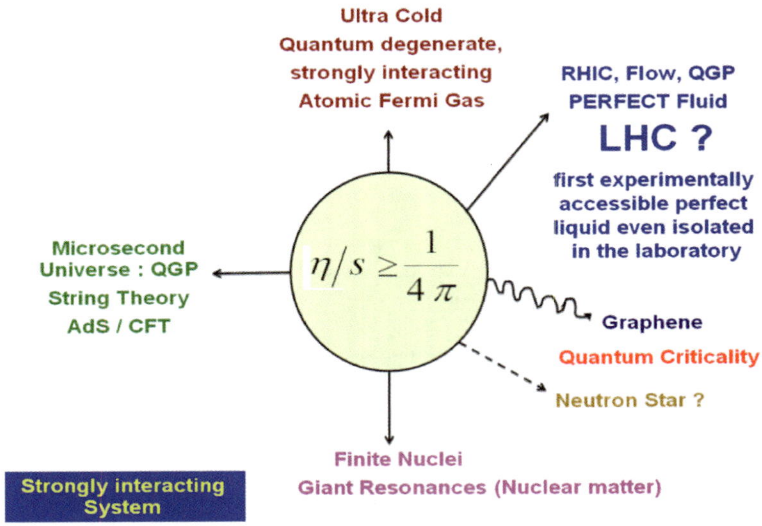


**Fig. 7** The $\eta/s$ saga

density. Hydrodynamic descriptions reproduce the data in general from the early times, through expansion, cooling and hadrons formation. One has to take into account the elliptic flow, and there comes viscosity per entropy $\eta/s$. The above, just mentioned is only possible if $\eta/s$ is close to zero.

Thus the significant discovery is that the QGP at RHIC is not the weakly interacting gas of almost free moving quarks and gluons one would have naively expected from the characteristic of asymptotic freedom. Instead, it is strongly coupled.

The strong coupling [18] further implies that some correlation among the quarks and gluons may survive within the plasma phase near $T_c$ and produce multiparticle interaction with near neighbours. Indeed, lattice QCD studies of energy density correlations in a QGP at temperatures $1 \ldots 2T_c$ show correlation peaks. This kind of correlation is rather similar to short range order observed in ordinary liquids near the liquid gas phase transition.

On the other hand, ultracold quantum degenerate, strongly interacting atomic Fermi gases also give rise to a very small value of $\eta/s$.

The so called Kovtun, Son and Starinets (KSS bound) for strongly coupling behavior of conformal gauge theories has a bound for $\eta/s$, the so called AdS/CFT limit $\eta/s \geq 1/4\pi = 0.08$. In Fig. 7 we display a large number of systems, very different from each other manifest this universal property of $\eta/s \approx 0.08$, even graphene.

It is not obviously clear as to the fundamental source of this universality, I only suspect, as was just pointed out, it has to do with the strong correlation around some critical point, somewhat independent of the very nature of interaction, as long as it is strongly correlated. This was most brilliantly pointed out by Efimov [19].

In Figs. 8, 9, 10, 11 the significance of $\eta/s$, especially with a value near the AdS/CFT level ($\eta/s \approx 0.08$) for understanding elliptic flow is displayed.

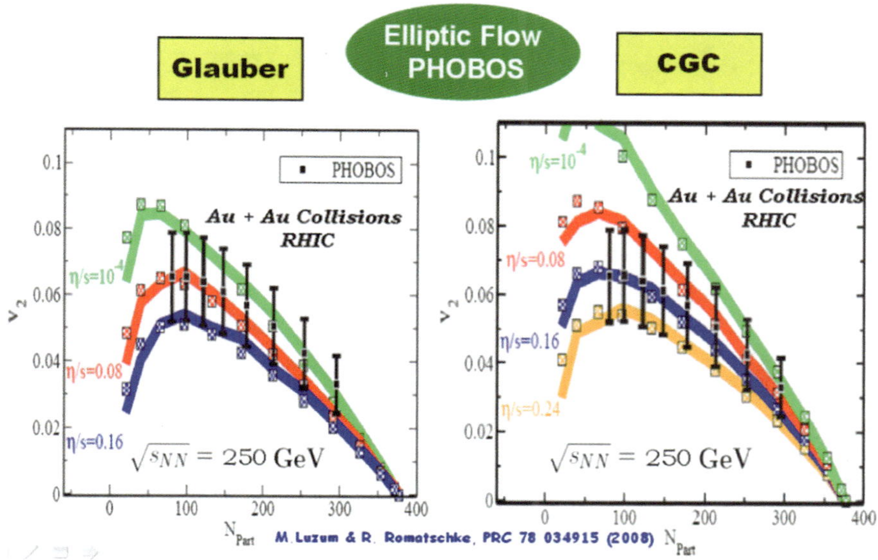


**Fig. 8** Hydrodynamic models to experimental data on charged hadron integrated elliptic flow by PHOBOS

The ratio $\eta/s$ is uniquely suited to determine how strongly the excitations in a quantum fluid interact. We determine $\eta/s$ in clean undoped graphene using quantum kinetic theory. It is remarkable that $\eta/s$ in this case comes close to a lower bound conjectured in the context of the quark gluon plasma [20].

It is thus remarkable that both the coldest and hottest matter on earth exhibit very similar elliptic flow patterns with $\eta/s$ near the conjectured lower bound of AdS/ CFT. It will be most interesting to find out how the "flow" at LHC is different from RHIC [21]. In particular, does $\eta/s$ change with energy?

Recently, Chaudhuri and Sinha [22] investigated the effect of viscous drag on photons and dileptons [23]. The space-time evolution of the fluid was obtained by solving Israel-Stewart's second-order hydrodynamics; the details of which are given in Ref. [22]. The equations are solved with the code AZHYDRO Kolkata developed at the Cyclotron Centre, Kolkata. The results are shown in Figs. 11, 12. It is quite remarkable that $\eta/s$ for QGP at LHC energies is remarkably close to strongly coupled QGP produced in RHIC collisions; though the initial temperature is much higher at LHC. This is a unique property of $\eta/s$ which need to be investigated at a fundamental level [24].

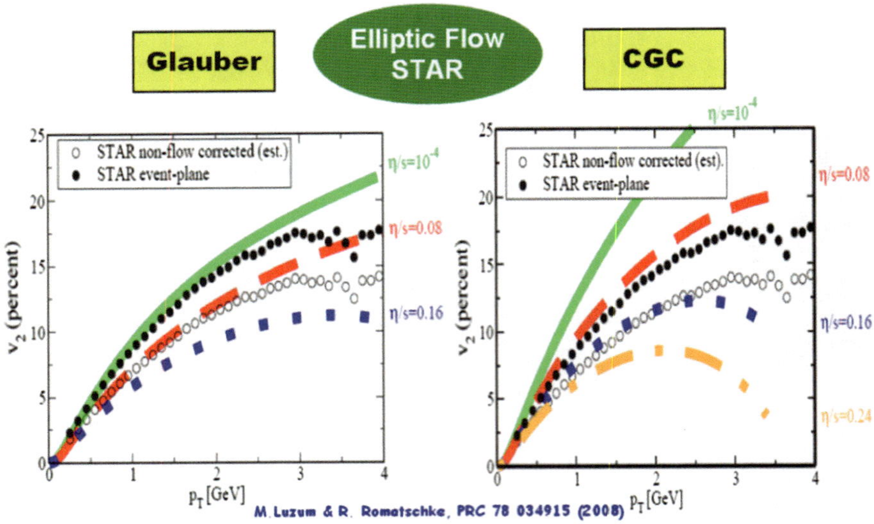


**Fig. 9** Hydrodynamic models to experimental data on charged hadron minimum bias elliptic flow by STAR

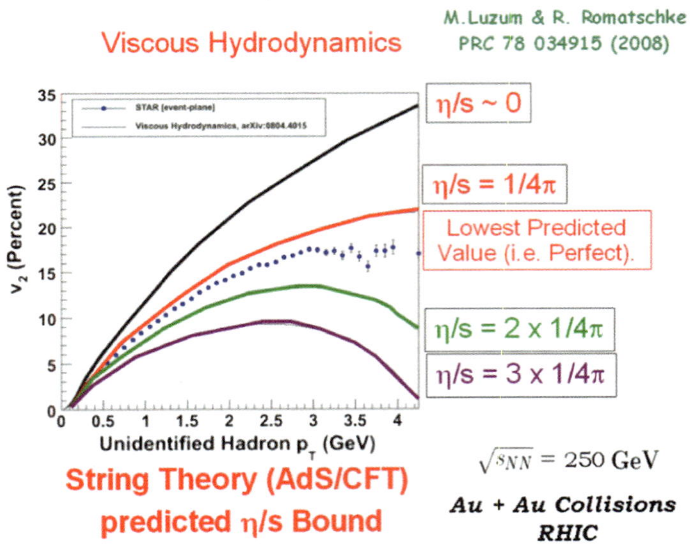


**Fig. 10** Quark-Gluon Plasma: a perfect fluid

# 4 From the Terrestrial Light to the Cosmic Light

The evolution of the universe during the QCD phase transition is governed by Einstein's equations

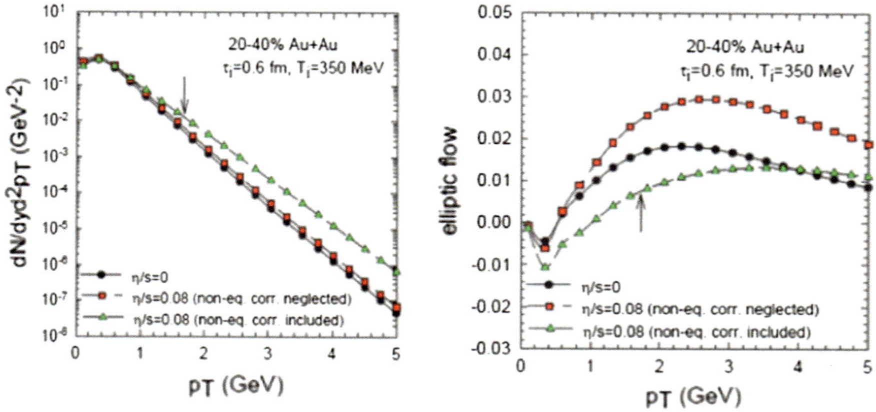


**Fig. 11** Effect of viscosity on photon spectra and elliptic flow. Validity of hydrodynamics requires that non-equilibrium contribution to photon spectra is smaller than the equilibrium contribution. For AdS/CFT limit of viscosity to entropy ratio, $\eta/r = 1/4\pi$, hydrodynamics is applicable only in a limited $p_T$ region (marked by *arrow*). It is important to have a consistent model, e.g., neglect of non-equilibrium correction to distribution function can lead to increased elliptic flow

Production rate of dileptons of in variant mass M from the QGP phase can be approximated as

$$E\frac{dN}{dM^2d^3p} \approx \frac{\alpha^2}{8\pi^4}\sum e_q^2 f_{\mathrm{neq}}(E)$$

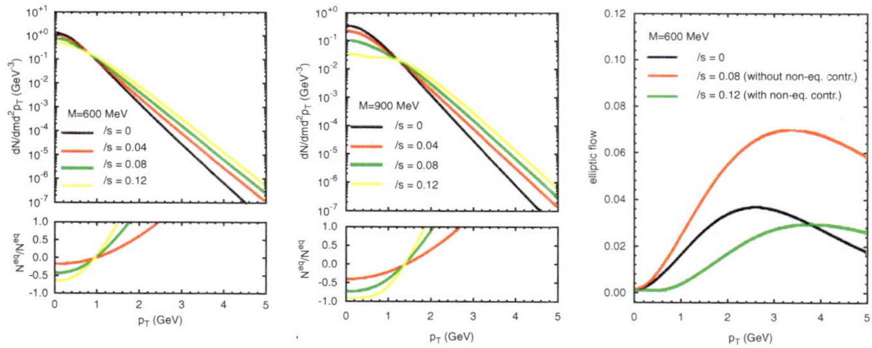


**Fig. 12** Viscous effects on dilepton production. Similar to photons, viscous effects on dilepton production is also large. $p_T$ spectra are hardened, elliptic flow is reuduced. Also, viscous hydrodynamics remains applicable only in a limited $p_T$ range. The applicability range increases with invariant mass

$$\left(\frac{\dot{R}}{R}\right)^2 = \frac{8\pi\rho}{3m_{\mathrm{Pl}}^2}$$

$$\frac{d(\rho R^3)}{dt} + P\frac{dR^3}{dt} = 0 \qquad (3)$$

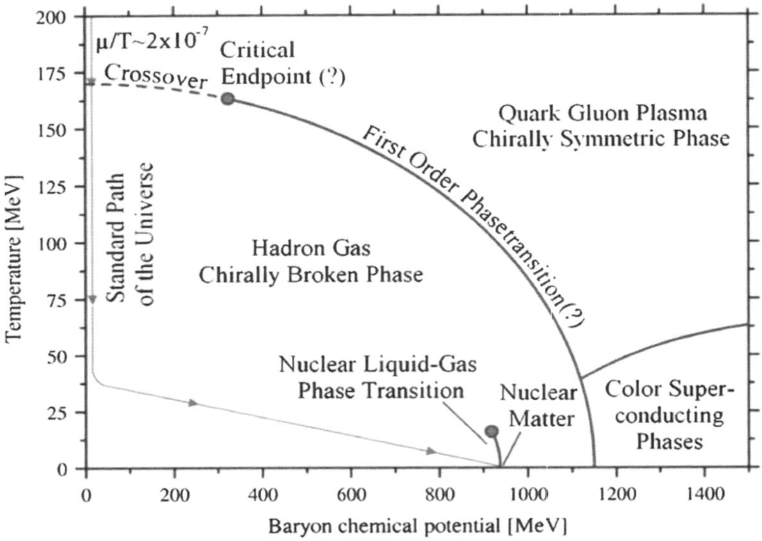


**Fig. 13** Sketch of a possible QCD phase diagram with the commonly accepted standard evolution path of the universe as calculated, e.g., in [25] depicted by the green path. *Source* Refs. [25, 26]

with $\rho$, the energy density, $P$, the pressure and $m_{\rm Pl}$ is the Planck mass. A simple equation of state with a bag constant $B$ is used.

$$\begin{aligned} \varepsilon &= 3aT^4 + B \\ P &= aT^4 - B \qquad\qquad t = 0.74/T^2 \end{aligned} \tag{4}$$

In Fig. 13, the standard scenario of the universe, crossing over at a temperature beyond the critical end point is shown. The universe as it goes over to the hadronic world in a crossover scenario erases its immediate past memory and thus, no relics of this transition remains with the world.

Recently however, Boeckel and Schaffner [26] have demonstrated that introducing a little inflation, $N = 7$ e-foldings, we can resurrect a first-order phase transition of the quark hadron phase transition, see Fig. 14.

It is shown by them and by us [11, 12] that the relics of this first-order phase transition are the best witness of the order of the phase transition. Crossover will lead to no surviving relics from the quark hadron phase transition in today's universe! First-order phase transition on the other hand will necessarily leave a rich harvest of relics.

We give a few examples.

Some time ago Bhattacharya et al. [28, 29], using chromo electric flux-tube examined the survivability of quark blobs or quark nuggets with time. We wanted to examine, in particular, the survivability of these nuggets even in the present day universe. The guiding equations are essentially of the structure of the famous Saha equations

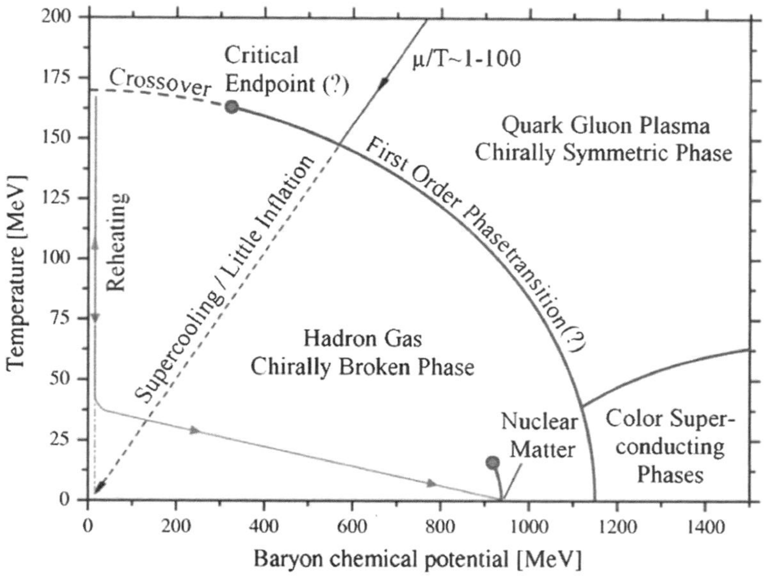


**Fig. 14** Sketch of a possible QCD phase diagram with the evolution path of the universe in the little inflation scenario. *Source* Ref. [26]

of thermalisation:

$$\frac{dN_B}{dt} = \left(\frac{dN_B}{dt}\right)_{\text{ev}} + \left(\frac{dN_B}{dt}\right)_{\text{abs}}$$

For further details, I refer to [27, 28].

In Fig. 15 the survivability is clearly demonstrated for $N_B > 10^{43}$; $N_B$ is the baryon number of the quark nugget; (please recall a quark has the baryon number of 1/3).

In Ref. [26] the relevance of dark energy and a first-order phase transition is pointed out. Boeckel and Schaffner have discussed this issue of dark energy in great detail.

Banerjee et al. [28], on the other hand, have demonstrated the most natural explanation of these surviving nuggets as the evolutionary product of metastable false vacuum domains, the so called strong quark nuggets. Indeed these nuggets are in fact the Massive Compact Halo Objects, MACHO [29, 30], very identifiable relics of the cosmic quark-hadron phase transition. These MACHO objects, it was argued [29, 30], can comfortably be candidate for Cold Dark Matter and that the total number of MACHOs is about $10^{23-24}$.

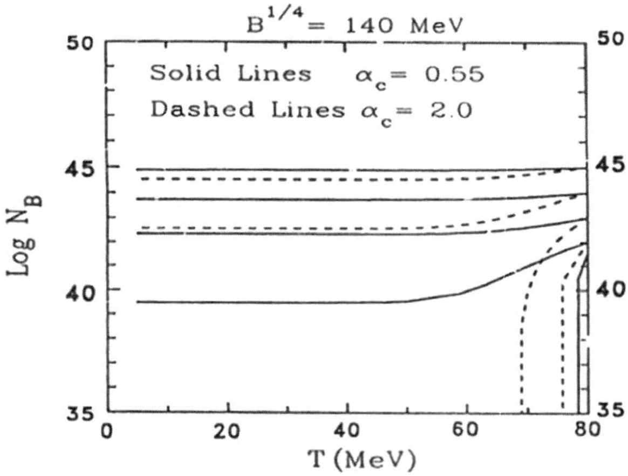


**Fig. 15** Baryon number of quark nuggets in the present universe

## 5 Epilogue

Our epilogue is through the looking glass (Lewis Carroll):
"ALICE IN THE QUARK LAND"

"The time has come", the Walrus said,
"To talk of many things
Of shoes—and ships—and sealing wax—
Of cabbages—and kings—
And why the sea is boiling hot—
And whether pigs have wings."

**Acknowledgments** I thank Jane Alam, Sibaji Raha, Asis Chaudhuri and many of my colleagues.

## References

1. Y. Aoki, Z. Fodor, S.D. Katz, K.K. Szabo, Phys. Lett. B **643**, 46 (2006)
2. F. Karsch, RBC-Bielefeld collaboration. J. Phys. G **34**, 627 (2007)
3. Z. Fodor, S.D. Katz, J. High Energy Phys. **4**, 50 (2004)
4. M. Luzum, P. Romatschke, Phys. Rev. C **78**, 034915 (2008)
5. Special issue "First Three years of Operation of RHIC", Nucl. Phys. A **757**, (2005)
6. C. Adler et al. (STAR Collaboration), Phys. Rev. C **66**, 034904 (2002)
7. A. Adare et al. (PHENIX Collaboration), Phys. Rev. Lett. **107**, 252301 (2011)
8. K. Aamodt et al. (ALICE Collaboration), Phys. Rev. Lett. **105**, 252302 (2011)
9. B. Sinha, Phys. Lett. B **128**, 91 (1983)
10. T. Boeckel, J. Schaffner-Bielich, Phys. Rev. Lett. **105**, 041301 (2010)
11. P. Bhattacharya, J. Alam, B. Sinha, S. Raha, Phys. Rev. D **48**, 4630 (1993)
12. J. Alam, B. Sinha, S. Raha, Nucl. Phys. A **638**, 523 (1998)

13. J. Nayak, J. Alam, S. Sarkar, Phys. Rev. C **78**, 034903 (2008)
14. J. Alam, B. Sinha, S. Raha, Phys. Rep. **273**, 243 (1996)
15. J. Alam, D.K. Srivastava, B. Sinha, S. Chakrabarty, S. Raha, Nucl. Phys. A **554**, 493C (1992)
16. D.K. Srivastava, B. Sinha, Phys. Rev. Lett. **73**, 2421 (1994)
17. J. Nayak, J. Alam, S. Sarkar, B. Sinha, J. Phys. G **35**, 104161 (2008)
18. B. Jacak, P. Steinberg, Phys. Today **63**(5), 39 (2010)
19. V. Efimov, Phys. Lett. B **33**, 563 (1970)
20. M. Mueller, J. Schmalian, L. Fritz, Phys. Rev. Lett. **103**, 025301 (2009)
21. H. Niemi, G.S. Denicol, P. Huovinen, E. Molnar, D.H. Rischke, J. Phys. G **38**, 124050 (2011)
22. A.K. Chaudhuri, B. Sinha, Phys. Rev. C **83**, 034905 (2011)
23. A.K. Chaudhuri, B. Sinha, arxiv:1201.0569 (2012)
24. C. Greene, Phys. Today **63**(3), 40 (2010)
25. M. Fromerth, J. Rafelski, arXiv:astro-ph/0211346 (2002)
26. T. Boeckel, J. Schaffner-Bielich, Phys. Rev. D **85**, 103506 (2012)
27. S. Banerjee, A. Bhattacharyya, S.K. Ghosh, E.-M. Ilgenfritz, S. Raha, B. Sinha, E. Takasugi, H. Toki, Phys. Lett. B **611**, 27 (2005)
28. S. Banerjee, A. Bhattacharyya, S.K. Ghosh, E.-M. Ilgenfritz, S. Raha, B. Sinha, E. Takasugi, H. Toki, Nucl. Phys. A **774**, 769 (2006)
29. S. Raha, S. Banerjee, A. Bhattacharyya, S.K. Ghosh, E.-M. Ilgenfritz, B. Sinha, E. Takasugi, H. Toki, J. Phys. G **31**, S857 (2005)
30. S. Banerjee, A. Bhattacharyya, S.K. Ghosh, S. Raha, B. Sinha, H. Toki, Mon. Not. R. Astron. Soc. **340**, 284 (2003)

# From d-Bars to Antimatter- and Hyperclusters

**J. Steinheimer, Zhangbu Xu, P. Rau, C. Sturm and H. Stöcker**

**Abstract** The Facility for Antiproton and Ion Research (FAIR) is going to be constructed within the next six years adjacent to the existing accelerator complex of the GSI Helmholtz Centre for Heavy Ion Research at Darmstadt/Germany, expanding the research goals and technical possibilities substantially. Providing a broad spectrum of unprecedented fore-front research at worldwide unique accelerator and experimental facilities, FAIR will open the way for a large variety of experiments in hadron, nuclear, atomic and plasma physics as well as applied sciences which will be briefly described in this article. As an example the article presents research efforts on strangeness at FAIR using heavy ion collisions, exotic nuclei from fragmentation and antiprotons to tackle various topics in this area. In particular the creation of hypernuclei as well as metastable exotic multi-hypernuclear objects (MEMOs) and anti-matter will be investigated.

J. Steinheimer (✉)
Lawrence Berkeley National Laboratory, Berkeley, CA 94720, USA
e-mail: JSFroschauer@lbl.gov

J. Steinheimer · P. Rau · H. Stöcker
Frankfurt Institute for Advanced Studies (FIAS), Ruth-Moufang-Str 1, 60438 Frankfurt am Main, Germany

Z. Xu
Physics Department, Brookhaven National Laboratory, Upton, NY 11973, USA

P. Rau · H. Stöcker
Institut für Theoretische Physik, Goethe-Universität, Max-von-Laue-Str. 1, 60438 Frankfurt am Main, Germany

C. Sturm · H. Stöcker
GSI Helmholtzzentrum für Schwerionenforschung GmbH, Planckstr. 1, 64291 Darmstadt, Germany

C. Sturm
Institut für Kernphysik, Goethe Universität Frankfurt, Max-von-Laue Str.1, 60438 Frankfurt Main, Germany

W. Greiner (ed.), *Exciting Interdisciplinary Physics*,
FIAS Interdisciplinary Science Series, DOI: 10.1007/978-3-319-00047-3_24,

# 1 The FAIR Project

The Facility for Antiproton and Ion Research, FAIR [1–3], will provide an extensive range of particle beams from protons and their antimatter partners, antiprotons, to ion beams of all chemical elements up to the heaviest one, uranium, with in many respects world record intensities. As a joint effort of 16 countries the new facility builds, and substantially expands, on the present accelerator system at GSI, both in its research goals and its technical possibilities. Compared to the present GSI facility, an increase of a factor of 100 in primary beam intensities, and up to a factor of 10000 in secondary radioactive beam intensities, will be a technical property of the new facility.

After the official launch of the project on November 7th, 2007, on October 4th, 2010, nine countries[1] signed the international agreement on the construction of FAIR. Civil work for the first buildings of FAIR will start during this year and first beams will be delivered in 2018. The start version of FAIR, the so-called *Modularized Start Version* [4, 5], includes a basic accelerator SIS100 (module 0) as well as three experimental modules (module 1–3) as it is illustrated in Fig. 1. The superconducting synchrotron SIS100 with a circumference of 1100 m and a magnetic rigidity of 100 Tm is at the heart of the FAIR accelerator facility. Following an upgrade for high intensities, the existing GSI accelerators UNILAC and SIS18 will serve as an injector. Adjacent to the SIS100 synchrotron are two storage-cooler rings and experiment stations, including a superconducting nuclear fragment separator (Super-FRS) and an antiproton production target. The Modularized Start Version secures a swift start of FAIR with outstanding science potential for all scientific pillars of FAIR within the current funding commitments. Moreover, after the start phase and as additional funds become available the facility will be upgraded by experimental storage rings enhancing capabilities of secondary beams and upgraded by SIS300 providing particle energies 20-fold higher compared to those achieved so far at GSI.

# 2 The Experimental Programme of FAIR

The main thrust of FAIR research focuses on the structure and evolution of matter on both a microscopic and on a cosmic scale. The approved FAIR research programme embraces 14 experiments, which form the four scientific pillars of FAIR and offers a large variety of unprecedented forefront research in hadron, nuclear, atomic and plasma physics as well as applied sciences. Already today, over two 2500 scientists and engineers are involved in the design and preparation of the FAIR experiments. They are organized in the experimental collaborations APPA, CBM, NuSTAR, and PANDA.

---

[1] In alphabetical order: Finland, France, Germany, India, Poland, Romania, Russia, Slovenia and Sweden

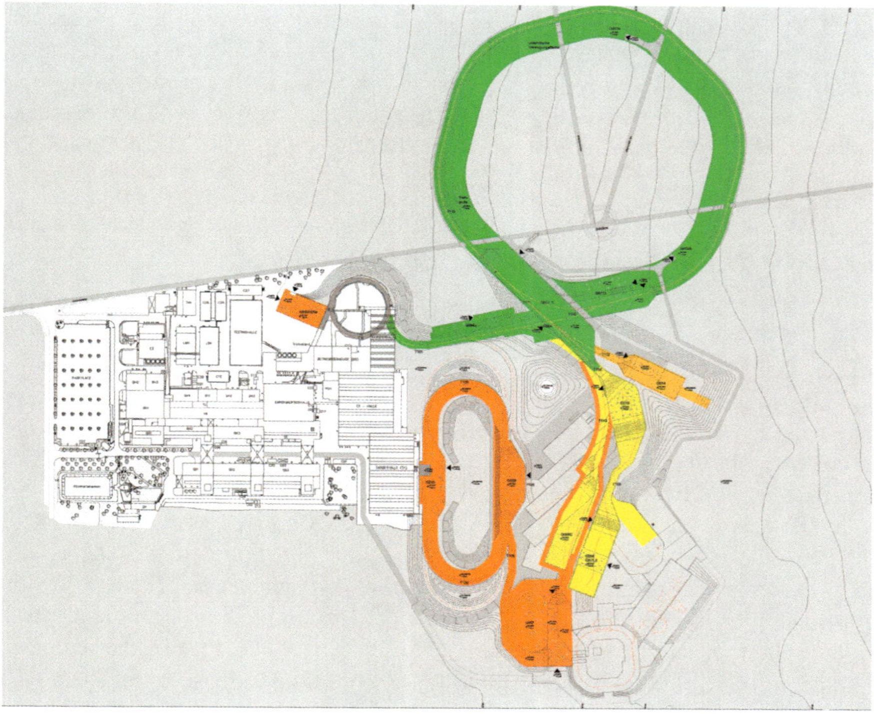

**Fig. 1** On the *left* the existing GSI facility is shown. Displayed in color is the so called Modularized Start Version of FAIR including module 0, 1, 2 and 3. Coloring: the 100 Tm super conducting synchrotron SIS100 (module 0)—*green*; the experimental area for CBM/HADES (module 1)—*red*; the NuSTAR facility including the Super-FRS (module 2)—*yellow*; The Antiproton facility including the PANDA experiment (module 3)—*orange*. Not shown is the additional experimental area above ground for the APPA community (module 1)

## *2.1 APPA—Atomic Physics, Plasma Physics and Applications*

Atomic physics with highly charged ions [6] will concentrate on two central research themes: (a) the correlated electron dynamics in strong, ultra-short electromagnetic fields including the production of electron-positron pairs and (b) fundamental interactions between electrons and heavy nuclei—in particular the interactions described by Quantum Electrodynamics, QED. Here bound-state QED in critical and supercritical fields is the focus of the research programme. In addition, atomic physics techniques will be used to determine properties of stable and unstable nuclei and to perform tests of predictions of fundamental theories besides QED.

For Plasma physics the availability of high-energy, high-intensity ion-beams enables the investigation of High Energy Density Matter in regimes of temperature, density and pressure not accessible so far [7]. It will allow probing new areas in the phase diagram and long-standing open questions of basic equation of state

(EoS) research can be addressed. The biological effectiveness of high energy and high intensity beams was never studied in the past. It will afford to investigate the radiation damage induced by cosmic rays and protection issues for the Moon and Mars missions. Furthermore, the intense ion-matter interactions with projectiles of energies above 1 GeV/u will endorse systematic studies of material modifications.

## 2.2 CBM/HADES—Compressed Baryonic Matter

Violent collisions between heavy nuclei promise insight into an unusual state in nature, that of highly compressed nuclear matter. In addition to its relevance for understanding fundamental aspects of the strong interaction, this form of matter may exist in various so far unexplored phases in the interior of neutron stars and in the core of supernovae. The mission of high-energy nucleus-nucleus collision experiments worldwide is to investigate the properties of strongly interacting matter under these extreme conditions. At very high collision energies, as available at RHIC and LHC, the measurements concentrate on the study of the properties of deconfined QCD matter at very high temperatures and almost zero net baryon densities. Results from lattice QCD indicate that the transition from confined to deconfined matter at vanishing net baryon density is a smooth crossover, whereas in the region of high baryon densities, accessible with heavy-ion reactions at lower beam energies, a first-order phase transition is expected [8]. Its experimental confirmation would be a substantial progress in the understanding of the properties of strongly interacting matter.

Complementarily to high-energy nucleus-nucleus collision experiments at RHIC and LHC, the CBM experiment [9, 10] as well as HADES [11, 12] at SIS100/300 will explore the QCD phase diagram in the region of very high baryon densities and moderate temperatures by investigating heavy-ion collision in the beam energy range 2–35 AGeV. This approach includes the study of the nuclear matter equation-of-state, the search for new forms of matter, the search for the predicted first order phase transition to the deconfinement phase at high baryon densities, the QCD critical endpoint, and the chiral phase transition, which is related to the origin of hadron masses. In the case of the predicted first order phase transition, basically one has to search for non-monotonic behaviour of observables as function of collision energy and system size. The CBM experiment at FAIR is being designed to perform this search with a large range of observables, including very rare probes like charmed hadrons. Produced near threshold, their measurement is well suited to discriminate hadronic from partonic production scenarios. The former requires pairwise creation of charmed hadrons, the latter the recombination of c-quarks created in first chance collisions of the nucleus-nucleus reaction. Ratios of hadrons containing charm quarks as a function of the available energy may provide direct evidence for a deconfinement phase.

The properties of hadrons are expected to be modified in a dense hadronic environment which is eventually linked to the onset of chiral symmetry restoration at high baryon densities and/or high temperatures. The experimental verification of this theoretical prediction is one of the most challenging questions in modern strongly interacting matter physics. The dileptonic decays of the light vector mesons ($\rho, \omega, \phi$) provide the tool to study such modifications since the lepton daughters do not undergo strong interactions and can therefore leave the dense hadronic medium essentially undistorted by final-state interaction. For these investigations the $\rho$ meson plays an important role since it has a short lifetime and through this a large probability to decay inside the reaction zone when created in a nucleus-nucleus collision. As a detector system dedicated to high-precision di-electron spectroscopy at beam energies of 1–2 AGeV, the modified HADES detector at SIS100 will measure $e^+e^-$ decay channels as well as hadrons [13, 14] up to 10 AGeV beam energy. Complementarily, the CBM experiment will cover the complete FAIR energy range by measuring both the $e^+e^-$ and the $\mu^+\mu^-$ decay channels.

Most of the rare probes like lepton pairs, multi-strange hyperons and charm will be measured for the first time in the FAIR energy range. The goal of the CBM experiment as well as HADES is to study rare and bulk particles including their phase-space distributions, correlations and fluctuations with unprecedented precision and statistics. These measurements will be performed in nucleus–nucleus, proton–nucleus, and proton–proton collisions at various beam energies. The unprecedented beam intensities will allow studying extremely rare probes with high precision which have not been accessible by previous heavy-ion experiments at the AGS and the SPS.

## 2.3 NuSTAR—Nuclear Structure, Astrophysics and Reactions

The main scientific thrusts in the study of nuclei far from stability are aimed at three areas of research: (i) the structure of nuclei, the quantal many-body systems built by protons and neutrons and governed by the strong force, towards the limits of stability, where nuclei become unbound, (ii) nuclear astrophysics delineating the detailed paths of element formation in stars and explosive nucleosynthesis that involve short-lived nuclei, (iii) and the study of fundamental interactions and symmetries exploiting the properties of specific radioactive nuclei.

The central part of the NuSTAR programme at FAIR [15, 16] is the high acceptance Super-FRS with its multi-stage separation that will provide high intensity mono-isotopic radioactive ion beams of bare and highly-ionized exotic nuclei at and close to the driplines. This separator, in conjunction with high intensity primary beams with energies up to 1.5 AGeV, is the keystone for a competitive NuSTAR physics programme. This opens the unique opportunity to study the evolution of nuclear structure into the yet unexplored territory of the nuclear chart and to determine the properties of many short-lived nuclei which are produced in explosive astrophysical events and crucially influence their dynamics and associated nucleosynthesis processes.

### *2.4 PANDA—AntiProton ANnihilation in Darmstadt*

The big challenge in hadron physics is to achieve a quantitative understanding of strongly interacting complex systems at the level of quarks and gluons. In $p\bar{p}$-annihilation, particles with gluonic degrees of freedom as well as particle-antiparticle pairs are copiously produced, allowing spectroscopic studies with unprecedented statistics and precision. The PANDA experiment at FAIR [17–19] will bring new fundamental knowledge in hadron physics by pushing the precision barrier towards new limits. The charmonium ($c\bar{c}$) spectroscopy will take advantage by precision measurements of mass, width, decay branches of all charmonium states. Particular emphasis is placed on mesons with open and hidden charm, which extends ongoing studies in the light quark sector to heavy quarks, and adds information on contributions of the gluon dynamics to hadron masses. The search for exotic hadronic matter such as hybrid mesons or heavy glueballs gains enormously by precise scanning of resonance curves of narrow states as well. Recently, this field has attracted much attention with the surprise observation at electron-positron colliders of the new X, Y and Z states with masses around 4 GeV. These heavy particles show very unusual properties, whose theoretical interpretation is entirely open. Additionally the precision gamma-ray spectroscopy of single and double hypernuclei will allow extracting information on their structure and on the hyperon-nucleon and hyperon-hyperon interaction.

## 3 The Creation of Anti-Matter

The history of antimatter is a brief and fascinating history of scientific discoveries. In 1928, Dirac predicted the existence of negative energy states of electrons based on the application of symmetry principles to quantum mechanics. The states were recognized as antimatter partner of electrons (positrons) discovered by Anderson in the cosmic rays in 1932. The constructions of accelerators have provided the necessary energy and luminosity for the discoveries of heavier antimatters. The extension of Dirac's theory implied the existence of antimatter protons and neutrons, and both particles were discovered at Bevatron in 1955. The scientific investigation of antimatter has three major focuses since then:

(a) Antiparticles are produced as by-products of high-energy particle collisions. Many particle and antiparticle pairs are created in such collisions through strong or electromagnetic processes. Antiparticles are merely part of the energy and chemical (baryon, isospin or lepton) conservation laws.
(b) Precise measurements of particle and antiparticle properties, which can provide insights into the fundamental CPT conservation and baryon asymmetry in the Universe.
(c) Constructing more complex system of antimatter.

It is clear that all these three topics are related to each other with a different emphasis for each subject. Although it may sound trivial to define what antimatter is, its definition is not without controversy. There are particles and antiparticles (such as $\mu^+$ and $\mu^-$), which annihilate when put together. However, neither of them annihilates the ordinary matter. Antonino Zichichi (2008) argues that there is a basic difference between antiparticle and antimatter, and even anti-hydrogen is not antimatter. In this proceeding, we mainly focus on constructing more complex systems of antimatter: antinuclei and antihypernuclei.

After the discoveries of antiprotons and antineutrons, one of the important questions was whether the building blocks in the antimatter world have the same force to glue together the antinucleons into nuclei and eventually anti-atoms by adding positrons. Figure 2 depicts the history of the discoveries of antimatter. We note that the antimatter project span eight decades with four decades per step in our discoveries. There are effectively three periods in these 80 years. The first discovery was made in the cosmic ray in 1932. The second period of discovery was between 1955 and 1975 when the fixed target accelerators provided increasing intensity and energy for producing heavier and heavier antimatter. The third period was made possible with high energy relativistic heavy-ion collider at RHIC and at the LHC. At the same time, the technology advance also enables us to decelerate antiproton beams and trap antimatter hydrogen. The necessity of the long term commitment was expressed by Walter Greiner (2001) in 'Fundamental Issues in the Physics of Elementary Matter': "The extension of the periodic system into the sectors of hypermatter (strangeness) and antimatter is of general and astrophysical importance. [...] The ideas proposed here, the verification of which will need the *commitment for 2–4 decades of research, could be such a vision* with considerable attraction for the best young physicists." [20].

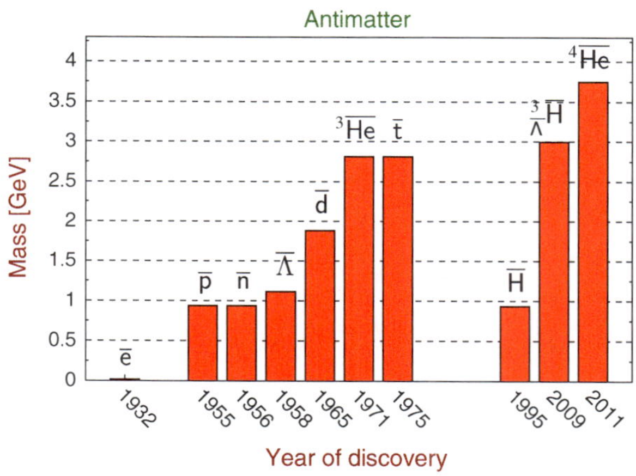


**Fig. 2** Discovery year of the antimatter and its associated mass

Understanding the asymmetry of antimatter and matter is one of the frontiers of modern physics. Nuclei are abundant in the universe, but anti-nuclei with $|A| \geq 2$ have not been found in nature. Relativistic heavy ion collisions, simulating the condition at the early universe, provide an environment with abundant antinucleons and antihyperons and produce antinuclei and antihypernuclei by coalescing them together [21]. This offers the first opportunity for discovery of antihypernuclei [22] and heavier antinuclei [23] having atomic mass numbers (or baryon numbers) $|A| > 2$. The production of antimatter nuclei can be explained by coalescence of antiprotons and antineutrons close in position and momentum. Figure 3 compiles all the antideuteron production in $e^+e^-$, $\gamma p$, $pp$, $pA$ and $AA$ collisions [24]. The results are shown for $\bar{d}/\bar{p}$ ratio as a function of beam energy. One can see that this ratio increases from $10^{-5}$ at low energy to $10^{-3}$ at high energy. Each additional antinucleon into the heavier antimatter decreases its production rate by that same penalty factor. At a center of mass energy of 100 GeV and above, this factor is relatively flat at slightly below $10^{-3}$. It is interesting to note that this effective measure of antibaryon density shows no difference among $pp$, $pA$ and $AA$ collisions. In heavy ion collisions, more antiprotons are produced in each collision than in $pp$ collisions. However, if more $pp$ collisions are collected to match the amount of antiproton yields in heavy-ion collisions, one can essentially produce the same amount of heavy antimatter in $pp$ and heavy-ion collisions. Now we understand that there are two deciding facts that RHIC discovered the last two heavy antimatters: sufficient energy to provide the highest antibaryon density for antinuclear production, and high luminosity heavy-ion collisions for effective data collection and particle identification.

Figure 3 shows the matter and antimatter yields as a function of baryon numbers as measured by the STAR Collaboration at RHIC [23]. The fit lines yield the production reduction rate by a factor of $1.6 \times 10^{-3}$ ($1.1 \times 10^{-3}$) for matter (antimatter) for each additional nucleon (antinucleon). The sensitivity of current and planned space based charged particle detectors is below what would be needed to observe antihelium

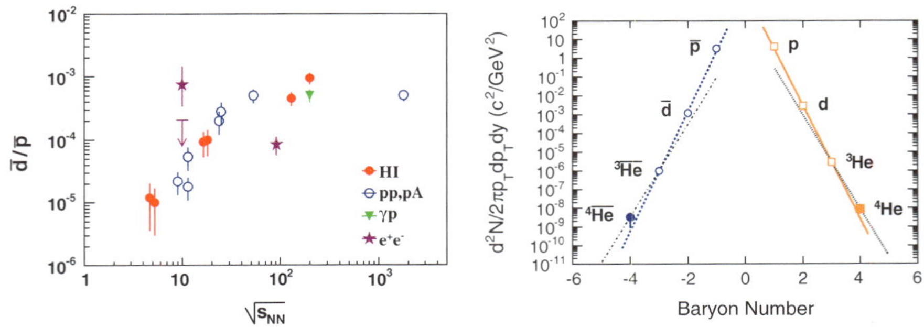


**Fig. 3** *Left* antideuteron and antiproton yield ratio as a function of the center of mass energy of the colliding target and projectile particles. *Right* matter and antimatter invariant differential yields as a function of the baryon number. The *solid* and *dashed lines* are a fit to all the experimental data. The *dotted* and *dotted dash lines* are extrapolations from the yields of $|A| = 3$ and $|A| = 4$.

produced by nuclear interactions in the cosmos. This implies that any observation of antihelium or even heavier antinuclei in space would indicate the existence of a large amount of antimatter elsewhere in the universe. In particular, finding antimatter $^4$He in the cosmos is one of the major motivations for space detectors such as the Alpha Magnetic Spectrometer [25]. We have shown that antimatter $^4$He exists and provided a measure of the background rate in nuclear collisions for possible future observations in cosmic radiation.

The next stable antimatter nucleus would be $A = 6$ ($^6$He; $^6$Li). However, the penalty factor on the production rate for an additional antinucleon is about 1500 as shown in Fig. 3. This means that the $A = 6$ antinuclei are produced at a rate 2 × 106 lower than that of an $A = 4$ antialpha particle. Unless production mechanisms or collider technology change dramatically, it is unlikely that $A = 6$ antinuclei can be produced in collider or fixed-target experiments (STAR 2011). On the other hand, the ratio of the $^4\mathrm{He}/^3\mathrm{He} = 3.1 \times 10^{-3}$ and $^4\overline{\mathrm{He}}/^3\overline{\mathrm{He}} = 2.4 \times 10^{-3}$. There is a factor of 2 higher yield of $|A| = 4$ over $|A| = 3$ than the extrapolation from the fit. The excess is visible even in a log-scale plot of 13 orders of magnitude. This ratio is also much higher than that shown in Fig. 3 for the $|A| = 2$ over $|A| = 1$. It has been argued that a more economic way of producing heavier antimatter and/or nuclear matter containing large amount of strange quark contents is through excitation of complex nuclear structure from the vacuum or through strangeness distillation from a QGP. Is this enhanced yield an indicative of a new production mechanism or a minor deviation due to trivial configuration of nuclear binding? Where do we go from here into the future in search and construction of heavier and more exotic antimatter? The indicative enhancement of higher antialpha yields suggests that even higher enhanced yields of heavier antimatter. Besides the possible high yields of $|A| = 6$ antimatter, the heaviest antimatter that can be produced and detected with a tracking detector in high-energy accelerators are likely to be $A = 4$ or 5 unstable antinuclei: $^4\mathrm{He}^* \rightarrow \mathrm{t} + \mathrm{p}$, $^4\mathrm{Li} \rightarrow {}^3\mathrm{He} + \mathrm{p}$, and $^5\mathrm{Li} \rightarrow {}^4\mathrm{He} + \mathrm{p}$. New trigger scheme and high data acquisition rate have been proposed to improve the effective data taking rate by two orders of magnitude in STAR during the heavy ion collisions [26]. This should confirm if the enhancement indeed exists and provide a possible path for discovering even heavier antimatter. In addition, as mentioned in the previous section, the antimatter yield reduction factor is similar in $p+p$ and $AA$ collisions. One expects that the penalty factor to persist for antimatter heavier than antideuteron in $p+p$ collisions. A comparison between the antimatter yields as shown in Fig. 3 in $p+p$ and $A+A$ collisions will provide a reference for whether the enhancement seen in antialpha production in $AA$ collisions is due to new production mechanism. Both RHIC and LHC have sufficient luminosity in $p+p$ collisions to produce antialpha. The only experimental issue is how to trigger and identify those particles. STAR has proposed a new trigger and TPC readout schemes for heavy antimatter search by using the Electromagnetic Calorimeter (EMC) for charged hadrons and only readout small sector of TPC associated with that struck EMC.

# 4 Hypermatter

Relativistic heavy ion collisions are an abundant source of strangeness. As strange quarks have to be newly produced during the hot and dense stage of the collision, they are thought of carrying information on the properties of the matter that was created [27]. Exotic forms of deeply bound objects with strangeness have been proposed [28] as states of matter, either consisting of baryons or quarks. The H di-baryon was predicted by Jaffe [29] and later, many more bound di-baryon states with strangeness were proposed using quark potentials [30, 31] or the Skyrme model [32]. However, the non-observation of multi-quark bags, e.g. strangelets is still one of the open problems of intermediate and high energy physics. Lattice calculations suggest that the H-dibaryon is a weakly unbound system [33], while recent lattice studies report that there could be strange di-baryon systems including $\Xi$'s that can be bound [34]. Because of the size of these clusters lattice studies are usually very demanding on computational resources and have large lattice artifacts, therefore an experimental confirmation of such a state would be an enormous advance in the understanding of the hyperon interaction.

Hypernuclei are known to exist and be produced in heavy ion collisions already for a long time [35–38]. The recent discoveries of the first anti-hypertriton [39] and anti-$\alpha$ [40] (the largest anti-particle cluster ever reported) has fuelled the interest in the field of hypernuclear physics. Metastable exotic multi-hypernuclear objects (MEMOs) as well as purely hyperonic systems of $\Lambda$'s and $\Xi$'s were introduced in [41, 42] as the hadronic counterparts to multi-strange quark bags [43, 44].

In the work presented in this section we will focus on the production of hypernuclei in high energy collisions of Au+Au ions. In such systems strangeness is produced abundantly and is likely to form clusters of different sizes. Hypernuclear clusters can emerge from the hot and dense fireball region of the reaction. In this scenario the cluster is formed at, or shortly after, the (chemical-)freeze out of the system. A general assumption is, that these clusters are then formed through coalescence of different newly produced hadrons. To estimate the production yield we can employ thermal production of clusters froma fluid dynamical description to heavy ion collisions. Though thermal production differs significantly in its assumptions from a coalescence approach one would expect to obtain different results, depending on the method used. However it can be shown that both approaches can lead to very similar results [49]. More detailed information on the calculations performed for the results in this section can be found in [50].

Figure 4 shows our results for the mid rapidity yields ($|y| < 0.5$) of di-baryons and hypernuclei as a function of the beam energy $E_{\mathrm{lab}}$. In our calculations we considered most central ($b < 3.4$ fm) Pb+Pb/Au+Au collisions at $E_{\mathrm{lab}} = 1$–$160$ AGeV. In addition, Fig. 4 shows the $\Lambda$ yield (black lines and squares) for the model compared to data [46–48]. In these figures, the UrQMD hybrid model calculations are shown as lines. At lower energies the cluster production should be suppressed additionally due to the non-equilibrium of strangeness. In the thermal calculations restrictions of energy and momentum conservation, resulting in a phase space reduction for produced strange particles, strongly decreases strange particle yields [51–53]. This behavior was also observed in a core-corona implementation in the hybrid model [54].

Di-baryon production rates have been calculated in a coalescence approach using the RQMD model for $\sqrt{s_{NN}} = 200$ GeV collisions of Au nuclei [45]. To relate our calculations to these results, they are indicated as the colored bars on the right axis of Fig. 4. The RQMD model used was in particular tuned to reproduce multi strange particle yields (such as the $\Xi$) and the results are therefore close to the ones obtained with our thermal/hydrodynamic approach. When the beam energy of the collisions is increased, the system created becomes almost net-baryon free. This means that the probability to create an anti-particle cluster approaches that of the particle cluster. Figure 5 shows the results for anti-particle cluster production at mid-rapidity ($|y| < 0.5$) in collisions of Pb+Pb/Au+Au at center of mass energies of $\sqrt{s_{NN}} = 3$–200 GeV. The yields of the anti-particle clusters show a monotonous increase with beam energy. They show that, at the highest RHIC energy (and at the LHC) the reconstruction of ${}^{4}_{\Lambda}$He might be a feasible task.

As another promising experimental tool for the production of antimatter clusters and hypernuclei we propose collisions of asymmetric sized nuclei, e.g. Ne+U, Ca+U.

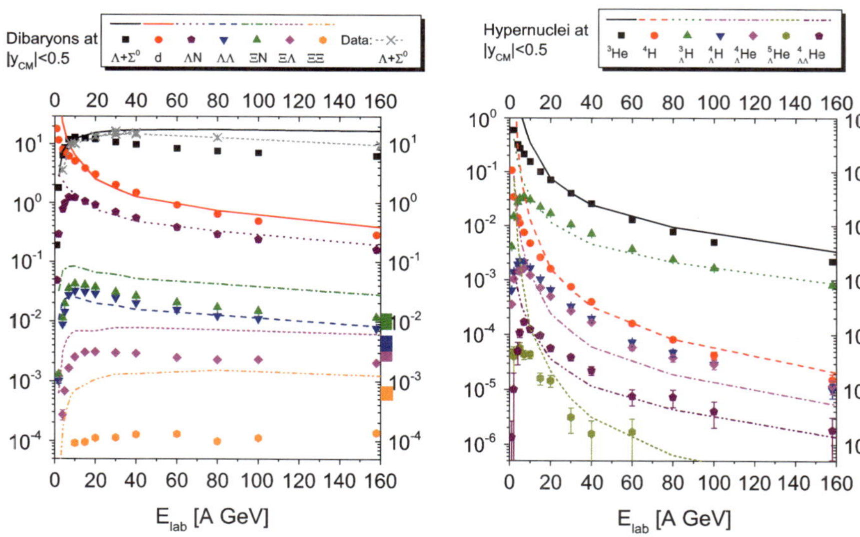


**Fig. 4** *Left* yields of different di-baryons in the mid rapidity region ($|y| < 0.5$) of most central collisions of Pb+Pb/Au+Au. Shown are the results from the thermal production in the UrQMD hybrid model (*lines*) as compared to coalescence results with the DCM model (*symbols*). The *small bars* on the right hand axis denote results on di-baryon yields from a previous RQMD calculation at $\sqrt{s_{NN}} = 200$ GeV [45]. In addition, the black lines and symbols depict results for the production rate of $\Lambda$'s from both models, compared to data (*grey crosses*) from [46–48]. *Right* yields of different (hyper-)nuclei in the mid rapidity region ($|y| < 0.5$) of most central collisions of Pb+Pb/Au+Au. Shown are the results from the thermal production in the UrQMD hybrid model (*lines*) as compared to coalescence results with the DCM model (*symbols*)

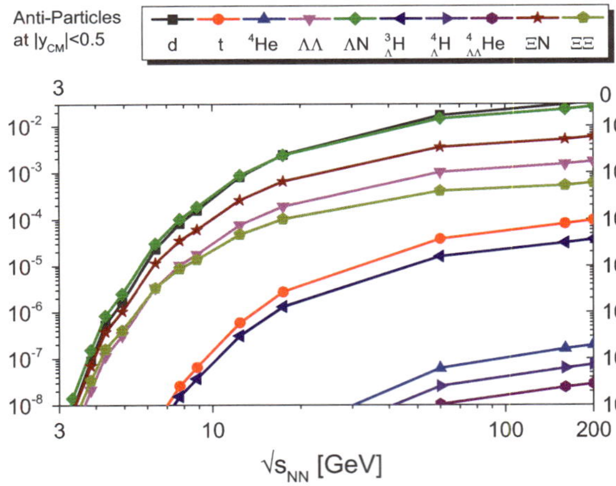


**Fig. 5** Yields of anti-particle clusters in the mid rapidity region ($|y| < 0.5$) of most central collisions of Pb+Pb/Au+Au as a function of $\sqrt{s_{NN}}$. Shown are only the results from the thermal production in the UrQMD hybrid model (*lines* with *symbols*)

In Fig. 6 we show the energy (a) and baryon number (b) density distributions in the reaction plane at $t = 16$ fm of a Ne+U collision at $E_{\text{lab}} = 15$ AGeV with an impact parameter $b = 4$ fm as calculated with a hydrodynamic model. At the collision zone

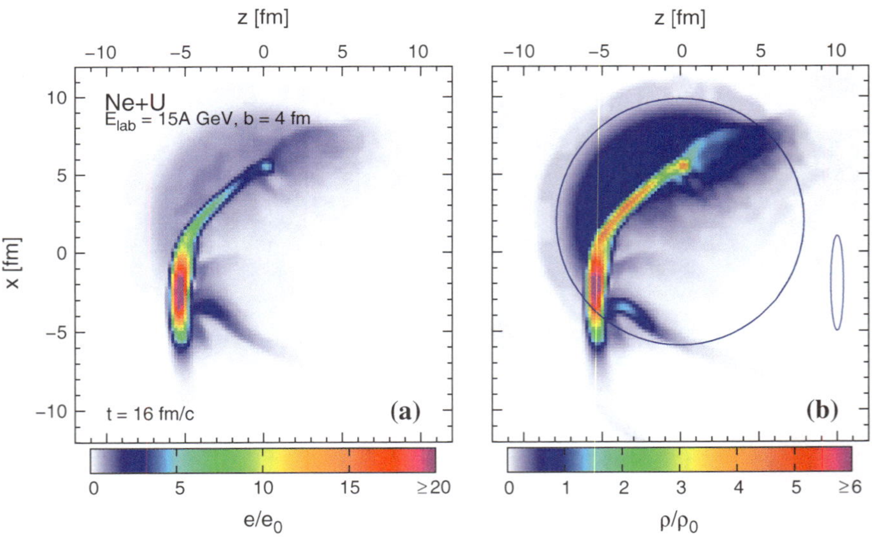


**Fig. 6** Density distributions of the energy (**a**) and baryon number (**b**) in units of the nuclear ground state values $e_0$, $\rho_0$ in the reaction plane of the asymmetric collision Ne+U with an impact parameter $b = 4$ fm and $E_{\text{lab}} = 15$ AGeV. The *blue lines* in (**b**) show the initial setup of the colliding Ne and U nuclei for the calculation with a hydrodynamic model

a highly compressed Mach shock wave is created and propagates through the target nucleus causing a directed cone like emission of particles [55]. Depending on the underlying EoS [56] high densities above the phase transition are reached already at rather small beam energies. In this very dense and hot shock zone, antibaryons are abundantly produced. Due to the directed, correlated emission through the surface, many of those $\bar{p}$ and $\bar{n}$ are able to escape annihilation and coalesce to antimatter clusters. Similarly, multi hyperon clusters form and can be detected downstream.

## 5 Summary

On October 4th, 2010, and after about ten years of negotiations, R&D, and writing reports, nine countries signed the international agreement on the construction of the Facility for Antiproton and Ion Research FAIR. Construction of the first FAIR buildings will start in 2012, so that the first beams will be delivered in 2018. The start version of FAIR, the so-called *Modularized Start Version*, includes the superconducting synchrotron SIS100 as well as three experimental modules to perform experiments for all research pillars. It will allow to carry out an outstanding and world-leading research programme in hadron, nuclear, atomic and plasma physics as well as applied sciences. Due to the high luminosity which exceeds current facilities by orders of magnitude, experiments will be feasible that could not be done elsewhere. FAIR will expand the knowledge in various scientific fields beyond current frontiers. Moreover, the exploitation of exiting strong cross-topical synergies promise novel insights.

**Acknowledgments** This work was supported by HGS-HIRe and the Hessian LOEWE initiative through the Helmholtz International center for FAIR (HIC for FAIR). The computational resources were provided by the Frankfurt Center for Scientific Computing (CSC). J. S. acknowledges a Feodor Lynen fellowship of the Alexander von Humboldt foundation.

## References

1. H.H. Gutbrod et al. (eds.), FAIR Baseline Technical Report. ISBN 3-9811298-0-6, Nov 2006
2. W.F. Henning, FAIR: recent developments and status. Nucl. Phys. A **805**, 502 (2008)
3. H. Stöcker, FAIR: challenges overcome and still to be met, in *Proceedings of EPAC08*, MOY-CGM01, Genoa, Italy, 2008
4. C. Sturm, B. Sharkov, H. Stöcker, 1,2,3. FAIR !. Nucl. Phys. A **834**, 682c (2010)
5. Green Paper of FAIR: The Modularized Start Version, www.gsi.de/documents/DOC-2009-Nov-124-1.pdf. Oct 2009
6. Th. Stöhlker et al., Atomic physics with highly-charged ions at the future FAIR facility: a status report. Nucl. Inst. Meth. B **261**, 234 (2007)
7. I.V. Lomonosov, N.A. Tahir, Prospects of high-energy density matter research at the future FAIR facility at Darmstadt. Nucl. Phys. News **16**(1), 29 (2006)
8. B. Friman, C. Höhne, J. Knoll, S. Leupold, J. Randrup, R. Rapp, P. Senger (eds.), in *The CBM Physics Book*. Lecture notes in physics, vol. 814, 2011

9. P. Senger et al., [CBM Collaboration], Compressed baryonic matter: experiments at GSI and FAIR. Phys. Part. Nucl. **39**, 1055 (2008)
10. P. Senger et al., [CBM Collaboration], Probing dense baryonic matter. Prog. Part. Nucl. Phys. **62**, 375 (2009)
11. G. Agakishiev et al., [HADES Collaboration], The high-acceptance dielectron spectrometer HADES. Eur. Phys. J. A **41**, 243 (2009)
12. J. Stroth et al., [HADES Collaboration], Di-electron emission from resonance matter. Prog. Part. Nucl. Phys. **62**, 481 (2009)
13. I. Fröhlich for the HADES Collaboration, Future perspectives at SIS-100 with HADES-at-FAIR, arXiv:0906.0091 [nucl-ex]
14. P. Tlusty et al., [HADES Collaboration], Future experiments with HADES at FAIR. AIP Conf. Proc. **1322**, 116 (2010)
15. R. Krücken et al., [NuSTAR Collaboration], Then nustar facility at FAIR. J. Phys. G **31**, S1807 (2005)
16. B. Rubio, T. Nilsson, NuSTAR. Nucl. Phys. News **16**, 9 (2006)
17. Physics Performance Report for PANDA: Strong Interaction Studies with Antiprotons. arXiv:0903.3905v1 [hep-ex]
18. K. Fohl et al., [PANDA Collaboration], The PANDA detector at the future FAIR laboratory. Eur. Phys. J. ST **162**, 213 (2008)
19. J.S. Lange for the PANDA Collaboration, The PANDA experiment: hadron physics with antiprotons at FAIR. Int. J. Mod. Phys. A **24**, 369 (2009)
20. W. Greiner, Int. J. Mod. Phys. E **5**, 1 (1996)
21. J. Adams et al., STAR Collaboration, Nucl. Phys. A **757**, 102 (2005)
22. B. I. Abelev [STAR Collaboration], Science **328**, 58 (2010)
23. H. Agakishiev et al. [STAR Collaboration], Nature **473**, 353 (2011), (Erratum ibid.**475**, 412 (2011))
24. H. Liu, Z. Xu, arXiv:nucl-ex/0610035
25. S.P. Ahlen, V.M. Balebanov et al., Nucl. Instrum. Meth. A **350**, 351 (1994)
26. STAR decadal plan 2011, www.bnl.gov/npp/docs/STAR_decadal_Plan_Final.pdf
27. P. Koch, B. Müller, J. Rafelski, Phys. Rep. **142**, 167 (1986)
28. A.R. Bodmer, Phys. Rev. D **4**, 1601 (1971)
29. R.L. Jaffe, Phys. Rev. Lett. **38**, 195 (1977) (Erratum ibid. **38**, 617 (1977))
30. T. Goldman, K. Maltman, G.J. Stephenson Jr, K.E. Schmidt, F. Wang, Phys. Rev. Lett. **59**, 627 (1987)
31. J.T. Goldman et al., Mod. Phys. Lett. A **13**, 59 (1998)
32. B. Schwesinger, F.G. Scholtz, H.B. Geyer, Phys. Rev. D **51**, 1228 (1995)
33. I. Wetzorke, F. Karsch, Nucl. Phys. Proc. Suppl. **119**, 278 (2003)
34. S.R. Beane et al. [NPLQCD Collaboration], Phys. Rev. D **85**, 0544511 (2012)
35. P. Braun-Munzinger, J. Stachel, J. Phys. G **21**, L17 (1995)
36. J.K. Ahn et al., Phys. Rev. Lett. **87**, 132504 (2001)
37. H. Takahashi et al., Phys. Rev. Lett. **87**, 212502 (2001)
38. A. Andronic, P. Braun-Munzinger, J. Stachel, H. Stöcker, Phys. Lett. B **697**, 203 (2011)
39. S.T.A.R. Collaboration, Science **328**, 58 (2010)
40. S.T.A.R. Collaboration, Nature **473**, 353 (2011)
41. J. Schaffner, H. Stöcker, C. Greiner, Phys. Rev. C **46**, 322 (1992)
42. J. Schaffner, C.B. Dover, A. Gal, C. Greiner, H. Stöcker, Phys. Rev. Lett. **71**, 1328 (1993)
43. E.P. Gilson, R.L. Jaffe, Phys. Rev. Lett. **71**, 332 (1993)
44. J. Schaffner-Bielich, C. Greiner, A. Diener, H. Stöcker, Phys. Rev. C **55**, 3038 (1997)
45. J. Schaffner-Bielich, R. Mattiello, H. Sorge, Phys. Rev. Lett. **84**, 4305 (2000)
46. S. Ahmad et al., Phys. Lett. B **382**, 35 (1996)
47. A. Mischke et al., [NA49 Collaboration], J. Phys. G **28**, 1761 (2002)
48. C. Alt et al., [NA49 Collaboration], Phys. Rev. C **78**, 034918 (2008)
49. J. Steinheimer et al., to be published

50. J. Steinheimer, M. Mitrovskia, T. Schuster, H. Petersen, M. Bleicher, H. Stöcker, Phys. Lett. B **676**, 126 (2009)
51. F. Becattini, U.W. Heinz, Z. Phys. C **76**, 269 (1997) (Erratum ibid. C **76**, 578 (1997))
52. J. Cleymans, K. Redlich, E. Suhonen, Z. Phys. C **51**, 137 (1991)
53. A. Andronic, P. Braun-Munzinger, J. Stachel, Nucl. Phys. A **772**, 167 (2006)
54. J. Steinheimer, M. Bleicher, Phys. Rev. C **84**, 024905 (2011)
55. P. Rau, J. Steinheimer, B. Betz, H. Petersen, M. Bleicher, H. Stöcker, arXiv:1003.1232 [nucl-th]
56. P. Rau, J. Steinheimer, S. Schramm, H. Stöcker, Phys. Rev. C **85**, 025204 (2012)

# Part IV
# Astrophysics, Particle Physics

# Astronomical Tests of General Relativity and the Pseudo-Complex Theory

**Thomas Boller and Andreas Müller**

**Abstract** Gravitation is very well described by Einstein's General Relativity. However, several theoretical predictions like the existence of curvature singularities and event horizons are under debate. This motivated to modify the standard theory of gravity. Here, we contrast predictions made by General Relativity with the pseudo-complex field theory proposed recently. Among them we study the gravitational redshift effect, perihelion shift, orbital motion, timing measurements and spectral lines. We consider supermassive black holes as ideal testbeds to test the theoretical predictions in the regime of strong gravity. In particular, we investigate the innermost centers of active galaxies and the Galactic Centre. This involves high-performance astronomical instruments of the next generation. We present feasibility studies with the proposed Athena X-ray experiment and with the upcoming GRAVITY near-infrared instrument to be mounted at the Very Large Telescope.

## 1 Introduction

Gravitation is successfully described by Einstein's General Relativity (GR) invented 100 years ago. The success of GR consists in an impressive number of experimental tests and by now GR passes them all. We use GR to describe the spacetime of our Earth and meanwhile GR effects are daily business in navigation systems. Einstein's theory also adequately describes massive bodies like our Sun or even more massive and compact objects like stellar mass black holes and neutron stars which are endpoints of stellar evolution as well as supermassive black holes. A breakthrough was certainly

T. Boller (✉)
Max Planck-Institute for extraterrestrial physics, PSF 1312, 85741 Garching, Germany
e-mail: bol@mpe.mpg.de

A. Müller
Technische Universiät München, Excellence Cluster Universe, 85741 Garching, Germany
e-mail: andreas.mueller@universe-cluster.de

W. Greiner (ed.), *Exciting Interdisciplinary Physics*,
FIAS Interdisciplinary Science Series, DOI: 10.1007/978-3-319-00047-3_25,

the description of our whole Universe by GR. The Friedmann cosmology successfully describes the dynamics of our Universe.

However, some mysteries remain in the framework of GR, such as the appearance of curvature singularities and event horizons. This motivated some scientists to go beyond GR. Since the advent of GR many other, alternative gravitational theories have been developed. Early after Einstein's publication of General Relativity in 1916, alternative theories of gravitation entered the stage. Some of them involved at least one more spatial extra dimension like the Kaluza-Klein theory or string theory. Others involve a different ansatz for the Einstein tensor, the left-hand side of Einstein's field equation, like $f(\mathrm{R})$ gravity which assumes a more general curvature expression for the Einstein tensor. A new ansatz is called pseudo-complex field theory [1] which goes in a similar direction as the latter one.

Interestingly, it is astronomy which offers a zoo of cosmic objects to test the strong gravity effects predicted by GR. Here, we confront these predictions of the standard GR picture with the predictions given by the pseudo-complex theory. We are lucky enough to find significant differences which allow to discriminate between the two theories. As we will see, these tests involve especially cosmic black holes. If we want to probe the strong gravity of black holes we have to get very close to these beasts. Therefore, these studies naturally involve (but not only) X-ray astronomy because X-rays are the signals coming from the immediate black hole surroundings. We will show what will be observational signatures accessible by X-ray telescopes and also by infrared instruments of the next generation of modern astronomical instruments.

## 2 Gravitational Theories

In this section we first sketch two gravitational theories, General Relativity in Sect. 2.1, and the pseudo-complex field theory proposed recently in Sect. 2.2.

### *2.1 Einstein's General Relativity*

Albert Einstein published a new theory for gravity in 1915. It is called General Relativity (GR) and is a completely different ansatz to describe gravity than Newtonian gravity. The Newtonian forces are not longer existent. Gravity is described by a four-manifold, a four dimensional continuum of space and time: spacetime. This is the dynamical stage for matter and for light. In Special Relativity, spacetime is flat and is described by the Minkowski metric. In GR, the spacetime is generally curved. The sources for gravity are any type of energy E and according to $E = mc^2$ also by mass m. The speed of light, c, is a fundamental constant in this framework. Its concrete value is not given by theory and has to be measured by experiments.

The fundamental field equation in GR involves more complicated mathematical objects called tensors. The essential statement of GR is that matter and energy curves spacetime, and spacetime dictates matter and light where to move. The tracks along which particles and light move are called geodesics and GR allows to extract a geodesics equations for each given metric. The field equation of GR holds simply G = T (ignoring any constants and indices) whereas G is the Einstein tensor containing curvature and T is the stress-energy tensor containing energy and mass. In more detail the field equation of GR looks like this

$$R^{\mu\nu} - \frac{1}{2} g^{\mu\nu} R = \frac{8\pi\kappa}{c^2} T^{\mu\nu}, \tag{1}$$

where $R^{\mu\nu}$ denotes the Ricci tensor (a contraction of the Riemann tensor) and $R$ is the Ricci scalar. $\kappa$ is the coupling constant of gravity which according to the correspondence principle is proportional to Newton's constant. This equation is very powerful, but also very complicated. Written in its whole beauty it is a coupled system of ten partial and non-linear differential equations. Mathematics cannot provide a full set of solutions for equations of this kind. Therefore, theorists found again and again new special solutions for the field equations of GR.

The two most important solutions in the context of this work are the Schwarzschild solution and the Kerr solution. The (outer) Schwarzschild solution was found by Karl Schwarzschild in 1916 [2] and describes a point mass in GR. The Schwarzschild spacetime is spherically symmetric and static. The Kerr solution was found in 1963 (significantly later!) by Roy Kerr [3] and describes a rotating mass in GR. The Kerr spacetime is axially symmetric and stationary. Both spacetimes describe cosmic black holes. They only have very few parameters ("No-hair theorem"). A Schwarzschild black hole has only mass $M$ and a Kerr black hole has mass $M$ and angular momentum $J$. Usually theorists use the specific angular momentum, $a = J/M$.

It is now the task of astronomers to find cosmic sources where black holes could be present. Then, it would be interesting to develop methods to measure the black hole parameters $M$ and $a$ by observations.

## 2.2 The Pseudo-Complex Theory

A new formulation of a field theory for gravity, based on a pseudo-complex description has been first published by [1]. An update of the pseudo-complex theory is given by [5]. Here, we only sketch the ansatz. A pseudo-complex number $X$ can be written as:

$$X = X_R + I \times X_I\ , \quad \text{with} \quad I^2 = +1. \tag{2}$$

From this a new Einstein equation follows and can be formulated as

$$R^{\mu\nu} - \frac{1}{2} g^{\mu\nu} R = \frac{8\pi\kappa}{c^2} T^{\mu\nu} \sigma_- \quad \text{with} \quad \sigma_- = \frac{1}{2}(1-I), \quad \sigma_-\sigma_+ = 0 \quad \sigma_-^2 = \sigma - \tag{3}$$

where the energy-stress tensor represents a field with repulsive properties (c.f. [5] for a detailed description). A comparison of this new field equation with Eq. 1 shows just one more additional quantity on the right. $\sigma_-$ is called zero divisor basis. The authors mentioned above deduced a new metric tensor from the new Einstein equations. Its 00 component satisfies:

$$g_{00} = \frac{r^2 - 2Mr + a^2\cos^2(\theta) + \frac{B}{2r}}{r^2 + a^2\cos^2(\theta)} \tag{4}$$

with $a$ as introduced before and an integration constant B.

Here, we use the gravitational radius $R_G = GM/c^2$ with black hole mass $M$, Newton's constant G and vacuum speed of light c. An interesting new feature in pseudo-complex field theory is that it removes the coordinate singularity at the Schwarzschild radius $r = 2\ R_G \equiv R_S$, which is a prediction of GR. Interestingly, there is therefore also no event horizon. This means that a classical black hole is absolutely dark at the horizon whereas a pseudo-complex black hole is rather gray, i.e. light originating at this region might escape to an external observer. We will return to this aspect in Sect. 3.

# 3 Predictions and Tests of General Relativity Versus the Pseudo-Complex Field Theory

In this section we work out several tests for General Relativity and confront them with those of the pseudo-complex field theory. We propose to test the innermost stable circular orbit (Sect. 3.1), the gravitational redshift (Sect. 3.2), perihelion shift (Sect. 3.3), timing studies (Sect. 3.4) and, the profile of relativistic emission lines (Sect. 3.5).

We note that experimental tests of the pseudo-complex theory have also been published by [4].

## *3.1 Comparison of the Effective Potentials and Innermost Stable Orbits*

In classical mechanics potentials are tools to investigate the motion of point masses. In celestrial mechanics, astronomers use the gravitational potential in Newtonian gravity to study e.g. the motion of planets around the Sun. Effective potentials are suitable approaches in GR to study the orbital motion of test particles.

The effective potential obtained from the pseudo-complex theory differs from that obtained in the standard GR as we show in Fig. 1 taken from [5]. The effective potential of classical black holes can be found in the literature, e.g. [6].

In both plots the curves are parametrized by the angular momentum, $L$, of the orbiting test particle. The relative minima of the curves correspond to stable Kepler

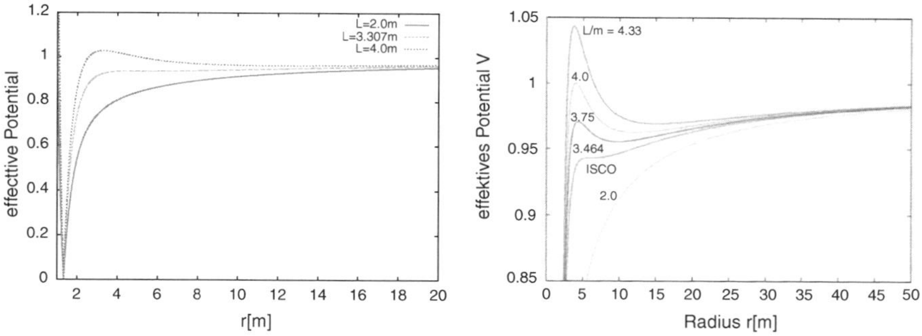


**Fig. 1** *Left panel*: effective potential as a function of radius in the pseudo-complex theory for a black hole spin of $a = 0$ (plot adopted from [5]). The *green curve* shows a reversal point at 5.3 $R_G$ which defines the ISCO. The ISCO is different from the standard theory. Below 4/3 $R_G$ the effective potential is increasing, corresponding to repulsion as described in Sect. 2.2. *Right panel*: effective potential in the standard GR for $a = 0$ (plot taken from [7]). The ISCO is located at 6 $R_G$

orbits. The absolute minima indicate the infall into the central mass concentration. The innermost stable circular orbit (ISCO) is defined by the reversal point, i.e. the second derivative is equal to zero.

A common method in X-ray astronomy is to determine the ISCO from observations and to link it to a spin value as explained by theory. GR essentially says: The closer the stable orbit, the higher the spin. The ISCO varies from 9 $R_G$ ($a = -M$, retrograde) over 6 $R_G$ ($a = 0$) to 1 $R_G$ ($a = M$, prograde).

Now, the left panel of Fig. 1 shows the effective potential for the pseudo-complex theory. The ISCO is at 5.3 $R_G$ in the pseudo-complex theory, which is below 6.0 $R_G$ where the ISCO of a non-spinning black hole with $a = 0$ is located. Here we see that, as a consequence, the spin determination is different between the two theories. This allows to discriminate between the two gravitational theories.

Pseudo-complex field theory exhibits another interesting feature. This is shown in the left panel of Fig. 1. Moving further inwards, we see that the effective potential is increasing again very steeply at a distance less than 4/3 $R_G$. It looks like a repulsion which is similar to the Yukawa potential in nuclear physics. This region sits deep in the black hole, too close to be feasible with current observational techniques. But maybe this might be tested in the future.

The orbital motion around black holes will be tested to unprecedented accuracy with a new infrared instrument at ESO's Very Large Telescope (VLT) in Chile. This detector of the forthcoming second generation of Very Large Telescope Interferometry (VLTI) instrumentation is called GRAVITY. GRAVITY will provide astrometric measurements with a precision of the order of one Schwarzschild radius, $R_S = 2\,GM/c^2$, of the black hole Sgr A* in the centre of the Milky Way [8]. The GRAVITY project will allow to probe physics in the strong field limit (c.f. Sect. 3.3 and Sect. 3.4) and will revolutionize measurements of motions of stellar orbits in the Galactic Centre. A summary of the whole science cases and the instrument capabilities are given in [9].

## 3.2 Gravitational Redshift Near Black Holes

Gravitational redshift characterizes the effect that light is trapped by a gravitational source, i.e. a mass. The influence on radiation is twofold: first, the energy of the photons is shifted towards lower energies, i.e. to the red end of the spectrum. Hence, it is a redshift. Second, the effect influences the spectral flux, i.e. it lowers the observed intensity. Gravitational redshift dims the light. The gravitational redshift effect is omnipresent for any mass and a prediction by any metric gravitational theory, therefore also for GR. For black holes this effect is extraordinarily strong. A black hole is in a sense defined by this effect because it exhibits an event horizon which marks the region where any local emission is reduced to zero as observed externally.

The Schwarzschild radius $\mathrm{R_S} = 2\,\mathrm{G}M/\mathrm{c}^2$ only depends on the mass parameter, i.e. the more massive the black hole the larger it will appear from the outside. However, the distance plays a role. The more distant the black hole, the smaller its event horizon region will appear. This apparent size, $\theta_{\mathrm{BH}}$, can be easily computed from black hole mass $M$ and its distance $d$ and satifies

$$\theta_{\mathrm{BH}} = 39.4 \times \frac{M}{10^6\, M_\odot} \times \frac{1\ \mathrm{kpc}}{d}\ \mu\mathrm{arcsec}, \tag{5}$$

where $M_\odot = 2 \times 10^{33}$ g denotes the mass of the Sun, 1 kpc = 3260 light years is a common distance unit, and 1 $\mu$arcsec $= 10^{-6}$ arcsec. Plugging in the values for the Galactic Centre black hole ($M_{\mathrm{GC}} = 4 \times 10^6\, M_\odot$, $d_{\mathrm{GC}} = 8$ kpc) [10] and M87, the massive elliptical galaxy in the Virgo Cluster ($M_{\mathrm{M87}} = 6 \times 10^9\, M_\odot$, $d_{\mathrm{M87}} = 16\,\mathrm{Mpc}$) delivers $\theta_{\mathrm{BH}}, GC = 20\,\mu$arcsec and $\theta_{\mathrm{BH}}, M87 = 15\,\mu$arcsec which is remarkably similar. The M87 supermassive black hole is significantly more massive but also significantly more distant.

So, these are the apparent sizes of the two black holes at the sky. They are very tiny, compared e.g. to the apparent size of the full moon, $\theta_{\mathrm{Moon}} = 0.5^0 =$ 1800 arcsec $= 1.8 \times 10^9\,\mu$arcsec. However, modern interferometric techniques are capable to resolve such tiny regions. Among them are Very Long Baseline Interferometry (VLBI) in radio astronomy (e.g. [11–13]) and the GRAVITY instrument [14, 15]. GRAVITY has sufficient resolution to test the gravitational redshift effect in the Galactic Centre as well as in the extragalactic source M87. However, the darkening towards the black hole is in GR different from the pseudo-complex theory as shown in Fig. 2. We expect that the observations with GRAVITY will be good enough to discriminate between GR and the pseudo-complex field theory at the innermost few gravitational radii.

Not only the Galactic Centre host a supermassive black hole, they are also present in other galaxies, most probably in all galaxy centres—sometimes even more than one massive black hole. One particular class is different from the centre of the Milky Way because the luminosity is very high. Astrophysicists call them active galactic nuclei (AGN), i.e. the luminous cores of galaxies powered by an accreting supermassive black hole. There are various AGN families, e.g. Seyfert galaxies which are rather

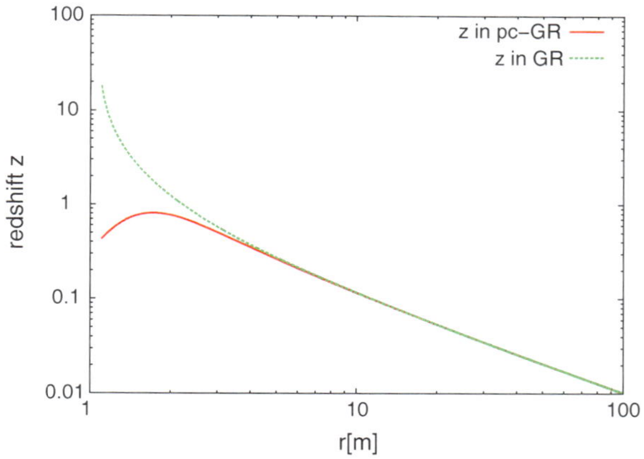


**Fig. 2** Gravitational redshift as a function of $R_G$ for the pseudo-complex theory (*green curve*) and the standard theory (*red curve*) for a black hole spin of $a = 0.998$. The figure has been adopted from [5]

low in luminosity and quasars which are very luminous. According to a common geometrical model there are two AGN types. AGN type-1 are seen face-on, i.e. the observer on Earth looks into the AGN core. AGN type-2 are rather seen edge-on, i.e. the observer cannot look into the AGN core. The view is blocked by a giant and massive dust torus sitting at large radii. We are dealing here with a special AGN type called Narrow-line Seyfert 1 (NLS1) which belongs to the AGN type-1.

In X-ray bright AGN, it is assumed that most of the X-ray emission is arsing within only a few $R_G$ and that the central emission is highly peaked with emissivity indices of about 6, e.g. [16, 17]. The future Athena mission will provide even more precise measurements in many other AGNs [18]. Athena is a large mission candidate of ESA's Cosmic Vision program 2015-2025. The main science objectives of Athena are to (i) explore the extreme physical conditions around black holes, (ii) map the large scale structure of the universe, and (iii) study the physics of feedback on all astrophysical scales. Significant differences between the ratio of the observed intensities arise from both theories. Assuming that most of the emission is arising at 1.2 $R_G$, the gravitational redshift $z_G$ is about 0.7 in the pseudo-complex theory and about 6 in the standard theory for a Kerr black hole (c.f. Fig. 2). This translates into generalized Doppler factors [19] $g = 1/(1 + z_G)$ of 0.6 and 0.14 for the pseudo-complex and the standard theory, respectively. As the observed and rest frame intensities scale according to the Liouville theorem with

$$I_\nu^{\text{obs}} = g^3 \times I_\nu^{\text{rest}}. \tag{6}$$

the ratio of the observed intensities between the pseudo-complex and the standard theory is about 70 assuming a Kerr black hole. As a consequence pseudo-complex black holes are brighter than standard GR black holes. With the Athena X-ray satellite measurements in the immediate vicinity around many black holes will become much more precise and this will allow to test different gravitational theories.

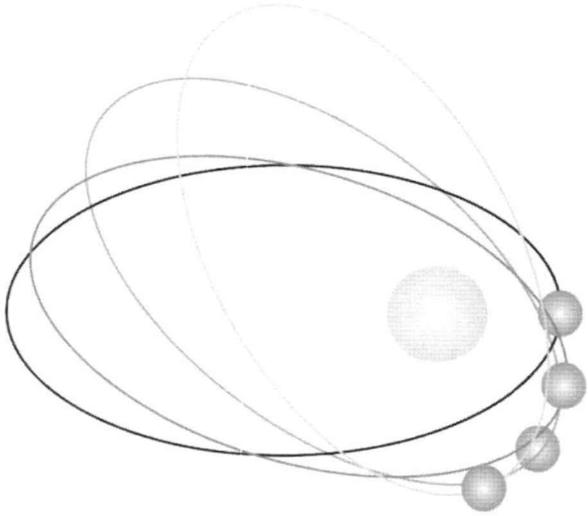

**Fig. 3** Sketch which illustrates the perihelion shift, i.e. the motion of the complete orbit of a celestial body. This was observed for the innermost planet Mercury orbiting the Sun

## *3.3 Perihelion Shifts*

With the advent of GR, first tests were proposed already in the second decade of the 20th Century. One observational fact remained unexplained so far, namely the the motion of Mercury's orbit. Since the the 19th Century, Mercury's orbit has been known with much more accuracy. At that time, the French astronomer Urbain Le Verrier used Mercury's transits to track the orbits very precisely. In particular, the so-called perihelion shift[1] represented a mystery. This phenomenon is shown in Fig. 3. Mercury's as the closest planet to the Sun shows a remarkably large perihelion shift amount. This phenomenon is also present in the classical Newtonian gravitational theory. However, the observations did not match the Newtonian prediction. A discrepancy of 43 arcsec remained unexplained. Einstein's General Relativity did this perfect match and hence the triumphal procession of GR started one hundred years ago.

Meanwhile, the performance of astronomical instruments increased significantly. Today, it is possible to track perihelion shift beyond the solar system, i.e. in the centre of the Milky Way. Here, the GRAVITY experiment comes into play again. The GRAVITY experiment in its astrometric mode will allow to precisely track the motion of stars around the supermassive black hole in Sgr A* with an accuracy of 10 μarcsec (c.f. [14, 15]). Based on simulations of the stellar orbits, the authors show in their Fig. 2 that GRAVITY will probe radial precession and even the Lense-Thirring effect ("frame-dragging"). This brings us to another test already present in GR: frame-dragging. This is an effect where the rotating spacetime drags any test particle and also light. However, this effect decays steeply as moving away from the gravitational rotating source. For a rotating black hole classically described by the Kerr metric, the rotation of spacetime (i.e the gradient of frame dragging frequency) decays with the third power of the distance. Therefore, astronomers have to get close

[1] Perihelion denotes the point on the orbit closest to the Sun.

to the gravitational rotating source, which means that the spatial resolution has to be high. This is what GRAVITY can perform.

Recently, the frame-dragging effect for the rotating Earth was observed with the experiments LAGEOS and Gravity Probe B. Here, the positions of satellites were accurately under control by the use of lasers. The gyroscopes onboard these missions were sensitive enough to test the rotation of Earth's spacetime. Perihelion and frame dragging effects are the classical tests for the standard theory and GRAVITY will allow test field theories in the strong gravity limit. Currently, the concrete results in the pseudo-complex field theory are work in progress. As soon as they are available they could be tested against the standard GR picture.

## 3.4 Keplerian Motion and Timing Analyses

The planets in the solar system move on Keplerian orbits. The classical Keplerian laws can be proven by using Newtonian gravity. With the advent of GR, these laws have to be modified to apply them to relativistic bodies such as black holes. However, in a moderate distance to the black hole the good old Keplerian laws apply. This is the case for one of the innermost stars orbiting the central massive black hole in the Milky Way, close to the radio source Sgr A* in the constellation Sagittarius (Sgr). Astronomers were able to track the complete orbit for the star S2 [20]. The 3rd Keplerian law states

$$\frac{\tau^2}{a^3} = \frac{4\pi^2}{\mathrm{GM}} = \text{const}, \tag{7}$$

with the orbital time $\tau$, the length of the semi-major axis $a$, Newton's constant G and the central mass $M$.

In fact, $\tau$ and $a$ are observables at Sgr A* by means of infrared observations, e.g. with observations at the VLT or with Keck on Hawaii. This delivered the high mass concentration in the heart of the Milky Way. Approximately 4 million solar masses in a region comparable with the solar system in size. The best interpretation for the compact object sitting there is the one of a classical massive black hole described by GR.

Quasi-periodic frequencies of infrared and X-ray emission have been detected in the Galactic Centre as well a in a few active galaxies. Usually, this is interpreted as modulated emission coming from orbiting hot spots [15] report on quasi-periodic frequencies of this kind in the Galactic Centre which exhibit a time scale of about 20–22 min (c.f. their Fig. 2). Such quasi-periodic frequencies can be compared with the prediction of Keplerian frequencies from the standard and the pseudo-complex theory. In Fig. 4 we show the Keplerian frequencies as a function of the distance to the black hole for both theories, calculated for a mass of $4.3 \times 10^6$ solar masses and a black hole spin of a = 0.995. The lower limit for the black hole spin in the Galactic Center is 0.52 [10]. Assuming that the black hole spin of the supermassive black hole in the Galactic Centre is determined with forthcoming GRAVITY and Athena observations, the Keplerian frequencies shown in Fig. 4 for the standard and the

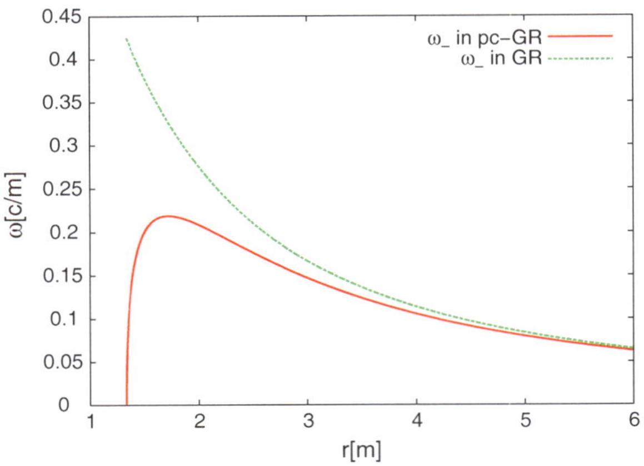


**Fig. 4** Keplerian frequencies as a function of the distance to a black hole with a mass of $M = 4.3 \times 10^6$ solar masses and a black hole spin of $a = 0.995$. The plot has been adopted from [5]

pseudo-complex theory can be directly compared. We note that at distances smaller than about 2 $R_G$ there occur significant differences between both theories. The same holds for the quasi-periodic oscillations detected in AGNs (c.f. Fig. 5).

Matter which is infalling into a black hole emits significant amounts of high-energetic X-rays. Typically, bright and hot emission features form, so-called hot spots. They orbit the black hole a few times and finally the black hole swallows the clumb. Temperature inhomogeneities in the accretion disc, often referred to as X-ray hot spots, are expected to produce a special signature of the Fe K$\alpha$ line emission in the energy-time plane. Figure 5 taken from [21] shows the smoothed theoretical time-energy map of emission features from an orbiting flare observed in the Seyfert galaxy NGC 3516. It could be shown that the feature varies systematically in flux at intervals of 25 ks. The peak moves in energy between 5.7 and 6.5 keV. The spectral evolution of the feature agrees with Fe K emission arising from a spot on the accretion disc, illuminated by a co-rotating flare located at a radius of (7-16) $R_G$, modulated by Doppler and gravitational effects as the flare orbits around the black hole.

Astronomers who would like to observe this phenomena caught-in-the-act need X-ray telescopes with a high time-resolution. Time-resolved X-ray spectroscopy with Athena is a technique to follow X-ray emission features of this kind. Such observations will allow to test different theories of the strong gravity limit. While approaching the black hole the emission is characteristically influenced by the dynamics, but also by the curved spacetime of the black hole. A first effect is the relativistic version of the Doppler effect. Relativistic Doppler boosting beams the emission towards the observer while the orbiter is approaching along the line of sight. As a consequence, the emission is shifted to higher energies and is brighter than in the rest frame (Doppler blueshift). On the receding side of the orbital track the emission is beamed away from the observer. Hence, the emission is shifted to lower energies

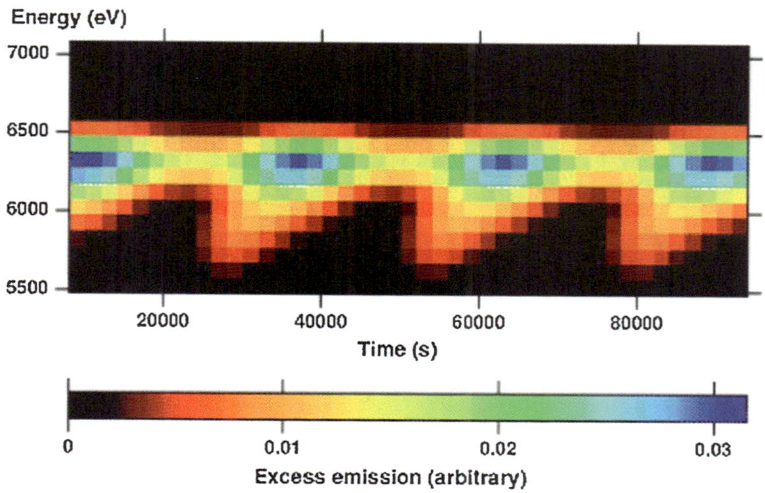


**Fig. 5** Smoothed theoretical time-energy map of emission features from an orbiting flare in NGC 3516. The Doppler effect produces the characteristic sinusoidal variations. The period is determined by the orbital time scale

and is dimmer than in the rest frame (Doppler redshift). If the orbital plane should be oriented in a face-on manner then only the quadratic Doppler effect survives.

In addition to this dynamical effect, we have a black hole sitting close by the orbiter. Its highly curved space-time drags the light and causes the relativistic gravitational redshift effect. The presence of the black hole's deep gravitational potential hinders the light from escaping—at least the light which comes to close to the hole. At some critical surface called the event horizon nothing can escape the black hole. This is where the emission dies out and the black hole itself becomes visible as an (like GR says) absolutely dark spheroidal zone. This stands in contrast to the background which has some brightness, e.g. from the surrounding accretion flow or, if there is no accretion, at least from the ambient cosmic microwave background radiation.

## *3.5 Relativistic Emission Line Studies*

One prominent X-ray feature is the iron K$\alpha$ line at 6.4 keV rest frame energy. This spectral line is produced by a fluorescence process. Electrons are excited into a higher state on the L shell and decay either by emitting an Auger electron (66 % probability), or by the emission of a fluorescence photon with 6.4 keV while the electron drops from the L to K shell (33 % probability). Typically a spectral line is rather sharp in the rest frame and can be sufficiently modeled by a narrow Gaussian profile. However, in the observer's frame the spectral line is distorted by the aforementioned relativistic Doppler and gravitational redshift effects. In the astronomical context

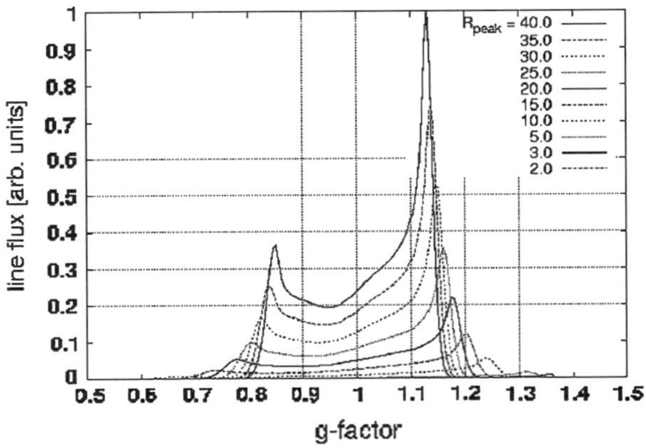


**Fig. 6** Decay of a relativistic line profile as a function of an emission region approaching the black hole. Due to the gravitation redshift effect the line profile becomes broader and gets dimmer (Image taken from [23])

the iron K line is produced in hot accretion disks around stellar and supermassive black hole. Astronomers observe a typical line profile which is coming from low to intermediately inclined accretion disks, e.g. active galaxies of type-1. In total, the observed fluorescent iron K lines for these sources are typically broad, skew with a long smeared red tail on the red line wing due to gravitational redshift and a peaked beamed blue wing due to Doppler blueshift.

The relativistic ray tracing technique is a standard method in relativistic astrophysics to visualize GR effects and to simulate relativistically broadened line profiles, see e.g. [19, 22]. Figure 6 [23] illustrates how the line profile decays as the emission region approaches the black hole. $R_{Peak}$ is the radius where the emission of the ring is maximal, given in units of gravitational radii $R_G$. Due to the gravitation redshift effect the line profiles broadens and gets dimmer and vanishes finally.

Relativistic line profiles from infalling hot spots.

The analysis is based on the theoretical model presented by [24]. The authors assume that the 6 keV line features are due to localized spots which occur on the surface of an accretion disk around a Schwarzschild black hole. They presented simulated line profiles as a function of orbital phase of the spot and its radial distance to the black hole. The models predict a specific behavior of the light curves and of the variability in the energy-time plane. In the model the hot spot starts at 5.6 $R_G$, slightly below the marginally stable orbit (at 6 $R_G$ for a Schwarzschild black hole) and disappears at the horizon at 2 $R_G$, i.e the Schwarzschild radius. The infall time corresponds to roughly 1.6 orbits which correspond to 30 ks for a black hole with 50 million solar masses. The size of the spot is 0.25 $R_G$. The trajectory of the hot spot, the spectrum in the energy-time plane, and the unfolded spectra for several infall segments as shown in Fig. 7.

Feasibility studies of infall motion for the Athena instrument.

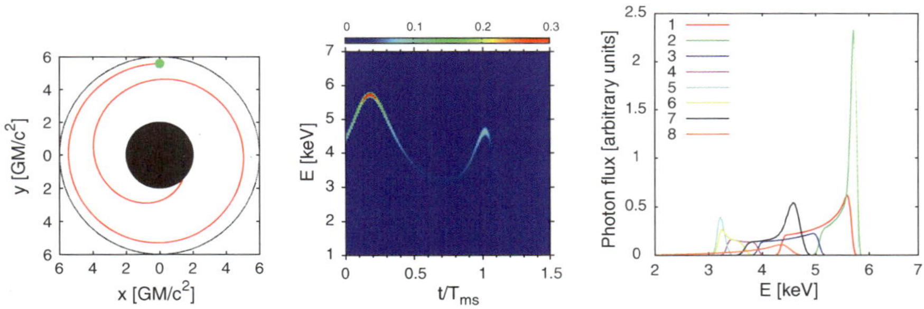


**Fig. 7** *Left* The trajectory of the hot spot is depicted by the *red*, the spot size is colored in *green*. The central black hole is the black hole region surrounded by the marginally stable. *Middle* The dynamics of the Fe K$\alpha$ emission line in the energy-time plane. The time is given in units of Keplerian orbital time at the ISCO. The mid panel shows the subsequent die out of the line due to gravitational redshift of an orbiter approaching more and more to the black hole. *Right* The model spectra for time bins 1 to 8 while the infall time is divided into 10 bins. The photon flux for time bins 9 and 10 is very low, i.e. invisible in the plot

Now, we link the infall model to concrete X-ray observations with Athena [18]. We have folded the 5 time segments of the infalling hot spot with the model parameters described above with the Athena Wide Field Imager [25] response. In Fig. 8 we show, that the signature of the infalling hot spot shows off and on phases so that even the duration of the non-detections give important information about the kinematics and timescale of the infalling material. As the gravitational redshift and the Kepler frequencies are different at small distances to the black hole for standard and the pseudo-complex theory (c.f. Figs. 2 and 4), significant differences are expected for the Fe K line profiles and the infall frequencies, which can be measured and tested with Athena.

Here we have shown the expected effects based on Einstein's GR theory. The theoretical calculation of the relativistic line profiles and infall times for the pseudo-complex theory is currently under investigation. Quantitative results for relativistic line studies for the the pseudo-complex theory will be reported elsewhere. The theoretical and consequently the observed line profiles and infall times will significantly differn from Einsteins GR theory and will provide another important test for both theories.

## 4 Astronomical Observations of Active Galaxies

In this section we summarize astronomical observations which serve as the basis for the proposed tests of General Relativity and its pseudo-complex extension.

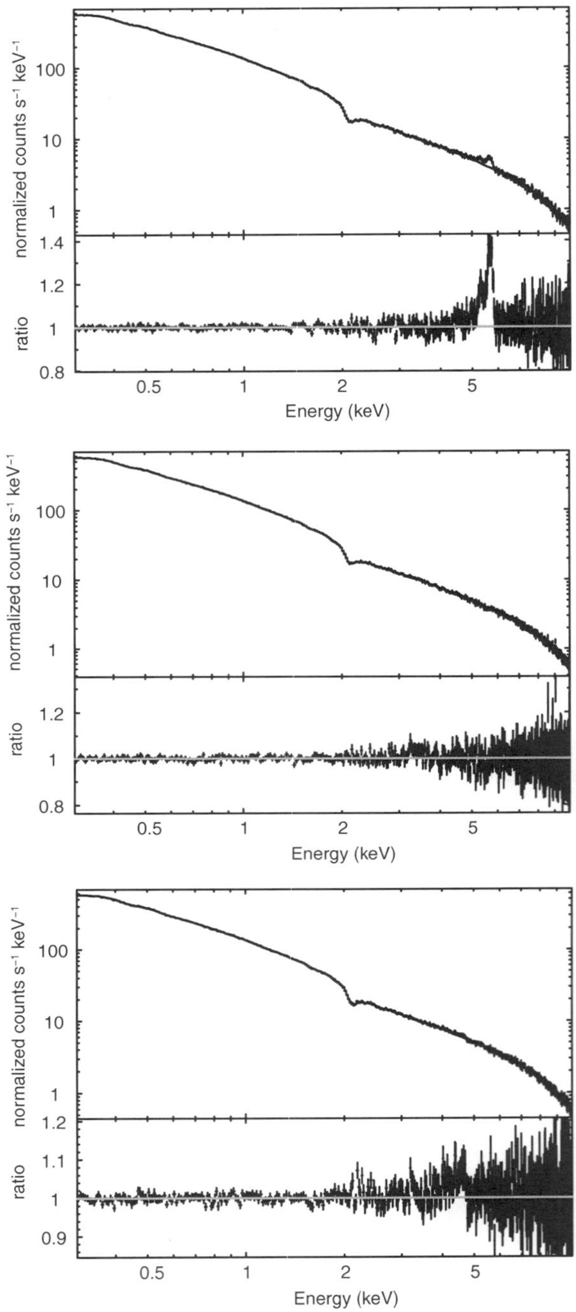


**Fig. 8** Simulation of Athena observations. *Left* first infall segment: the residua are clearly visible due to the Doppler boosting. This can be considered as the Athena on-phase of the detection of the infalling spot. The second infall segment is not plotted, as the line profile becomes undectable due to smearing in a large range of energies and so buried in the continuum. *Middle* Third infall segment: At this stage, the spot is receding from the observer. Doppler boosting creates a red peak around 3 keV which is not visible in the data because the relative contribution of the line is smaller whereas more flux is coming from the continuum. The statistics is too poor to detect relativistic line emission. *Right* fourth infall segment: interestingly, the relativistic line profile becomes visible again, at late infall times and at distances very close to the black hole. This is due to Doppler boosting (beaming). However, the line core is now shifted to lower energies around 4.5 keV, because gravitational redshift is getting stronger. For the final stages of the infall the gravitational redshift effect is too strong to reveal relativistic signatures

## 4.1 Previous Research Work

The fact that accretion onto black holes powers the most luminous sources in the Universe is known for decades ([26–28]). Pioneering work on flaring emission near black holes was performed by [29]. First work on relativistic light curves of a star orbiting a black hole was done by [30]. One breakthrough was the detection of relativistic broad emission lines which are emitted on an accretion disk by fluorescence of hard X-ray radiation. In this way, radiation from a nearby primary hard X-ray source, called corona, is reprocessed. The dominant line feature is produced by iron which has the largest fluorescence yield among all elements. The core of the broad iron K$\alpha$ line can be found at a rest frame energy of 6.4 keV and line fits revealed that it originates only a few gravitational radii away from the black hole. This feature is visible in several AGN (e.g. [31–33]) and galactic black hole candidates (e.g. [34]). Reverberation mapping studies exploit the physics between the first power-law continuum flare emission and the lagging emission line response. In this way, one can constrain the position of the flare emitter and the spacetime (black hole spin), see e.g. [35, 36]. The flare could be linked to the disk and therefore orbit with the disk or it could be a stationary emitter on the disk rotational axis, e.g. the jet base. A detailed understanding of this geometry is still lacking.

## 4.2 Present X-ray Observations Near Black Holes

### 4.2.1 Spectral Analysis

Observations with XMM-Newton, Chandra and Suzaku revealed that the inner accretion flows around black holes emit significant amounts of X-rays. Both X-ray spectra and the time variability of this X-ray emission contain a wealth of information about the innermost matter flow and the black hole itself. Over the last decades, many X-ray observations of stellar black holes in X-ray binaries and supermassive black holes in active galactic nuclei delivered insights into the black hole-accretion flow system. It possible to fit, e.g. temperature of the accretion flow, inclination angle of the disk towards the observer, disk emissivities as well as mass and spin of the black hole. The radiation originates so close to the black hole that it allows for probing the dynamics of matter and the interactions between matter and radiation in the strong gravity limit. X-ray astronomers have found that the primary X-ray emission is concentrated solely to the central part of the accretion disc and must lie within 1 gravitational radius of the event horizon of the black hole. This was convincingly shown by [16] for X-ray observations in the narrow-line Seyfert 1 galaxy 1H 0707-495 in its low flux state.

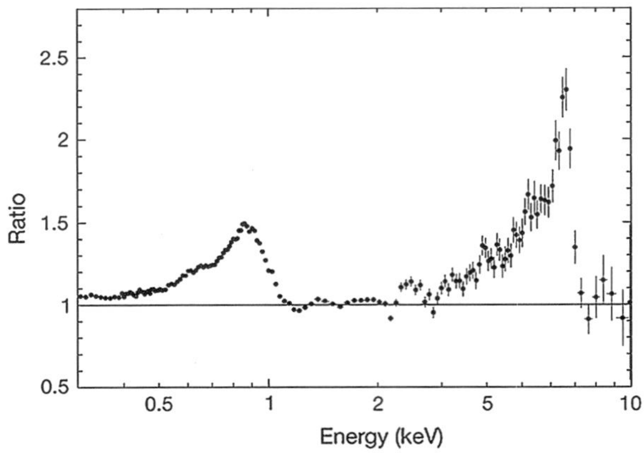


**Fig. 9** The first detection of both Fe K and Fe L line emission in the NLS1 1H 0707-495. The ratio of the normalization of the Fe K to the Fe L line in photon flux is 20 to 1, in agreement to atomic physics. The Fe L line becomes most probably detectable due to the high Fe abundance which is about 9 times higher than in the solar environment ([37])

### 4.2.2 Timing Analysis

Signatures of X-ray hot spots orbiting a supermassive black hole have been detected in a few AGN (c.f. Sect. 3.4 and Fig. 5). These observations are based on the observed Fe K$\alpha$ emission as a function of time and allows to constrain the distance of the X-ray hot spots to the central black hole.

Cross-correlation analyses of time series from different spectral bands yield further insight into the intertwined physical processes of the accretion disc. They certainly have a great discovery potential — as it was impressively demonstrated by the detection of the reverberation signature, namely the detection of time lags between the Fe K and L line (c.f. Figs. 9, 10 and its physical interpretation [37]). The Fe K and L line emission is caused by X-ray fluorescence at the return of an electron of iron of the L to the K-shell after excitation of the iron with an X-ray photon, i.e. the Fe K line emission is caused by the return of an electron to the K shell, and the Fe L emission is due to the return of an electron to the L shell. Relativistic distortion of the line makes it sensitive to the strong gravity and spin of the black hole. The normalization of the Fe K and Fe L lines in photon spectra are in the ratio 20 to 1—in agreement to atomic physics. The bright iron L emission allows the detection of a reverberation lag of about 30s between the direct X-ray continuum and its reflection from matter falling into the black hole.

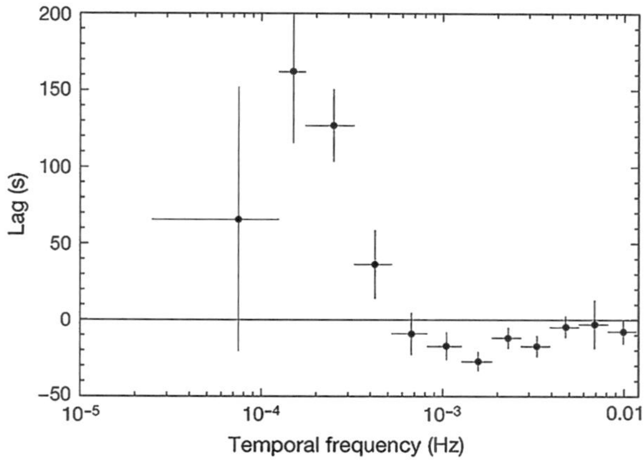


**Fig. 10** Detection of a frequency dependent time lag of 1H 0707-495. The soft band is defined in the energy range between 0.3 and 1.0 keV, the hard band between 1.0 and 4.0 keV. A positive time lag indicates that the hard band follows the soft band, opposite to negative time lags. At the lowest frequencies positive time lags are detected. These are interpreted as accretion rate fluctuations at larger distances to the black hole. A significant negative time lag lag has been detected at 30 s. This is interpreted as the detection of a reverberation lag

### *4.3 Future X-ray Observations*

We have shown that present X-ray missions are very successful in delivering data from inner accretion flows near black holes. So far, the accretion onto black holes is a paradigm to power luminous X-ray sources, however detailed timing studies of the infalling matter are still lacking. Furthermore, the event horizon of a black hole was not proven observationally so far. One may put doubts whether or not an observational proof by means of electromagnetic waves is possible. So far, there are a few standard methods for determining black hole spin. However, various techniques contradict to each other which might be a hint that either the model or even the theory is wrong. The proposed Athena satellite mission will bring important new results from the innermost matter flow around black holes and we expect essential new insight into the nature of gravity.

## 5 Summary

Einstein's General Relativity is the best theory we currently have to describe gravitational effects. However, it is a non-quantized theory and the interesting question is whether or not there are regimes where we have to go beyond Einstein's Relativity. Do curvature singularities in black holes really exist in nature, or do they signal a problem? And what about event horizons? There are also a product of GR but never have been proven so far. The regime of strong gravity is very fascinating in this context because it offers an opportunity to get answers to this interesting questions.

Here, we confront standard GR with a new suggestion for a gravitational theory which is called pseudo-complex field theory. We elaborated test examples which enable us to test strong gravity and to discriminate between GR and the alternative pseudo-complex theory. So far, the tests involve in particular gravitational redshift and Keplerian motion. We also presented an outlook on how further tests could look like. They involve perihelion shifts of close orbiting particles, orbital motions of matter around supermassive black holes as well as X-ray timing and spectral analyses. This is work in progress and we have just shown a preview. The detailed results will be discussed elsewhere.

The ideal experimental objects to test the theoretical predictions are places where black holes harbor. These is the Galactic Centre and some very suited active galaxies. These astronomical testbeds will be our targets to test strong gravity. These tests involve instruments which are not yet available. We need high-performance detectors to come as close as possible to the black holes. We selected two proposals of forthcoming instruments of the next generation, namely Athena, an ESA X-ray mission, and GRAVITY, an instrument mounted at ESO's Very Large Telescope in Chile. With both high-performance instruments we expect to be able to learn more about the nature of gravity in its strong regime.

**Acknowledgments** TB and AM greatly acknowledge extensive scientific discussions with P. Hess, W. Greiner, T. Schönenbach and G. Caspar on the topics presented in this paper. TB is much obliged to W. Greiner for his longstanding scientific support. We thank M. Dovciak and J. Svoboda from the Astronomical Institute of the Academy of Science in Prag, Czech Republic, who significantly contributed in writing up Sect. 3.5. TB is greateful to S. Gillessen for intense discussion on the Galactic Center research.

## References

1. P.O. Hess, W. Greiner, Pseudo-complex general relativity. Int. J. Mod. Phys. E **18**, 51–77 (2009)
2. K. Schwarzschild, Über das Gravitationsfeld eines Massenpunktes nach der Einsteinschen Theorie. Sitzungsber. Preuss. Akad. Wiss. Berlin (Math.Phys.), 189–196 (1916)
3. R.P. Kerr, Gravitational field of a spinning mass as an example of algebraically special metrics. Phys. Rev. Lett. **11**, 237 (1963)
4. G. Caspar et al., Pseudo-complex general relativity: Schwarzschild, Reissner-Nordström and Kerr solutions. Int. J. Mod. Phys. E **21**(1250015), 1–39 (2012)
5. T. Schönenbach, G. Caspar, P.O. Hess, T. Boller, A. Müller, M. Schäfer, W. Greiner, Experimental tests of pseudo-complex General relativity, arXiv:1209.2815 [gr-qc] (2012)
6. M. Camenzind, *Compact Objects in Astrophysics: White Dwarfs Neutron Stars and Black Holes* (Springer, New York, 2007)
7. A. Mueller, www.astronomiewissen.de.
8. F.H. Vincent et al., in *The Galactic Center: A Window to the Nuclear Environment of Disk Galaxies*, ed. by M.R. Morris, Q.D. Wang, F. Yuan. Proceedings of a workshop held at Shanghai, China, 19–23 Oct 2009. (Astronomical Society of the Pacific, San Francisco, 2011), p. 275
9. F. Eisenhauer et al., GRAVITY: observing the universe in motion. The Messenger **143**, 16–24 (2011)

10. R. Genzel et al., Near-infrared flares from accreting gas around the supermassive black hole at the Galactic Centre. Nature **425**, 934 (2003)
11. V.L. Fish et al., 1.3 mm wavelength VLBI of Sagittarius A*: detection of time-variable emission on event horizon scales. ApJ **727**, 36 (2011)
12. R.S. Lu et al., Multiwavelength VLBI observations of Sagittarius A*. A&A **525**, 76 (2011)
13. S. Doeleman et al., Imaging an event horizon: submm-VLBI of a super massive black hole (2012), http://de.arxiv.org/abs/0906.3899
14. F.H. Vincent et al., Performance of astrometric detection of a hotspot orbiting on the innermost stable circular orbit of the Galactic Centre black hole. MNRAS **412**, 653 (2011)
15. F. Eisenhauer et al., GRAVITY: getting to the event horizon of Sgr A*. SPIE **7013**, 69 (2008)
16. A.C. Fabian et al., 1H 0707–495 in 2011: an X-ray source within a gravitational radius of the event horizon. MNRAS **419**, 116 (2012)
17. Th. Boller, in *Soft X-ray reflection and strong and weak field limit determination in Narrow-Line Seyfert 1 Galaxies, Exploring Fundamental Issues in Nuclear Physics*, ed. by D. Bandyopadhyay. Proceedings of the Symposium on Advances in Nuclear Physics in Our Time, (World Scientific Publishing Co. Pte. Ltd., 2012) ISBN-13 978–981-4355-72-8
18. K. Nandra, ATHENA: The advanced telescope for high energy astrophysics, the X-ray Universe 2011, Presentations of the conference held in Berlin, Germany, 27–30 June 2011. Available online at: http://xmm.esac.esa.int/external/, articel id.022, 2011
19. A. Müller, M. Camenzind, Relativistic emission lines from accreting black holes. The effect of disk truncation on line profiles. A&A **413**, 861 (2004)
20. S. Gillessen et al., The orbit of the star S2 around SGR A* from very large telescope and keck data. ApJ **707**, L114 (2009)
21. K. Iwasawa et al., Flux and energy modulation of redshifted iron emission in NGC 3516: implications for the black hole mass. MNRAS **355**, 1073 (2004)
22. A.C. Fabian et al., X-ray reflection in the narrow-line Seyfert 1 galaxy 1H 0707–495. MNRAS **353**, 1071 (2004)
23. A. Müller, M. Wold, On the signatures of gravitational redshift: the onset of relativistic emission lines. A&A **457**, 485 (2006)
24. M. Dovciak et al., Relativistic spectral features from X-ray-illuminated spots and the measure of the black hole mass in active galactic nuclei. MNRAS **350**, 745 (2004)
25. L. Strüder et al., The wide-field imager for IXO: status and future activities. SPIE **7732**, 46 (2010)
26. Y.B. Zel'dovich, I.D. Novikov, The radiation of gravity waves by bodies moving in the field of a collapsing star. Sov. Phys. Dokl. **9**, 246 (1964)
27. E.E. Salpeter, Accretion of interstellar matter by massive objects. ApJ **140**, 796 (1964)
28. D. Lynden-Bell, Galactic nuclei as collapsed old quasars. Nature **223**, 690 (1969)
29. J.M. Bardeen et al., Rotating black holes: locally nonrotating frames, energy extraction, and scalar synchrotron radiation. ApJ **178**, 347 (1972)
30. C.T. Cunningham, J.M. Bardeen, The Optical Appearance of a Star Orbiting an Extreme Kerr Black Hole. ApJ **183**, 237 (1973)
31. Y. Tanaka et al., Gravitationally redshifted emission implying an accretion disk and massive black hole in the active galaxy MCG-6-30-15. Nature **375**, 659 (1995)
32. A.C. Fabian, K. Iwasawa, Broad Fe-K lines from Seyfert galaxies. Adv. Space Res. **25**, 471–480 (2000)
33. K. Nandra et al., An XMM-Newton survey of broad iron lines in Seyfert galaxies. MNRAS **382**, 194 (2007)
34. J.M. Miller et al., in *Relativistic Iron Lines in Galactic Black Holes: Recent Results and Lines in the ASCA Archive, The Tenth Marcel Grossmann Meeting* ed. by Mário Novello; Santiago Perez Bergliaffa; Remo Ruffini. Proceedings of the MG10 Meeting held at Brazilian Center for Research in Physics (CBPF), Rio de Janeiro, Brazil, 20–26 -July 2003. Singapore: World Scientific Publishing, in 3 volumes, ISBN 981-256-6678 (set), ISBN 981-256-980-4 (Part A), ISBN 981-256-979-0 (Part B), ISBN 981-256-978-2 (Part C), 2005, XLVIII + 2492 p.: 2005, p. 1296, 2005

35. C.S. Reynolds, in *Compton Reflection and Iron Fluorescence in Active Galactic Nuclei and Galactic Black Hole Candidates, High Energy Processes in Accreting Black Holes*, ed. by J. Poutanen, R. Svensson. ASP Conference Series 161, ISBN 1-886733-81-3 (1999), p. 178 (1999)
36. A.J. Young, C.S. Reynolds, Iron line reverberation mapping with constellation-X. ApJ **529**, 101 (2000)
37. A.C. Fabian et al., Broad line emission from iron K- and L-shell transitions in the active galaxy 1H0707-495. Nature **459**, 540 (2009)

# Black Holes or Gray Stars? That's the Question: Pseudo-Complex General Relativity

**Peter O. Hess, W. Greiner, T. Schönenbach and G. Caspar**

**Abstract** After a short review on attempts to extend General Relativity, pseudo-complex variables are introduced. We restate the main properties of these variables. The variational principle has to be modified in order to obtain a new theory. An additional contribution appears, whose origin is a repulsive, dark energy. The general formalism is presented. As examples, the Schwarzschild and the Kerr solutions are discussed. It is shown that a collapsing mass inceasingly accumulates dark energy until the collapse is stopped. Rather than a black hole, a gray star is formed. We discuss a possible experimental verification, investigating the orbital frequency of a particle in a circular orbit.

## 1 Introduction

General Relativity (GR) is a well accepted theory which has been verified by many experimental measurements. One prediction of this theory is the existence of *black holes*, which are formed once a very large mass suffers a gravitational collapse. Astronomical observations seem to confirm this prediction, finding large mass concentrations in the center of most galaxies. These masses vary from several million solar masses to up to several billion solar masses. However, a black hole implies the appearance of an event horizon, below which an external observer cannot penetrate, thus, excluding a part of space from observation. A black hole also implies a singularity at its center. Both consequences from GR may be, from a philosophical point of view, unacceptable and one would like to find a possibility to avoid them. A black hole is an extreme object and one would not be surprised that GR has to be modified

Peter O. Hess (✉)
Instituto de Ciencias Nucleares, UNAM, C.U., A.P. 70-543, 04510 México D.F., Mexico
e-mail: hess@nucleares.unam.mx

W. Greiner, G. Caspar and T. Schönenbach
FIAS, J. W.-Goethe University, Ruth-Moufang-Str. 1, 60438 Frankfurt am Main, Germany

W. Greiner (ed.), *Exciting Interdisciplinary Physics*,
FIAS Interdisciplinary Science Series, DOI: 10.1007/978-3-319-00047-3_26,

for such situations. For example, the singularity could be avoided, considering a quantized GR, not yet available.

There have been several attempts to generalize GR. Einstein [1, 2] introduced complex variables in order to unify GR with Electrodynamics. Later on, other groups continued this research (see for example [3, 4] and references therein) calling the new theory *complexified GR*. The real component of the complex variable is given by $x^\mu$ while the imaginary component is given by $l\frac{p^\mu}{m}$, where $p^\mu$ is the momentum of a particle and $m$ is its mass. As a by-product a minimal length parameter $l$ appears, for dimensional reasons. One of the motivations to continue the investigation of the complex GR is the *Born's equivalence principle*. Born noted [5, 6] that in GR there is an asymmetry between the coordinates and momenta, while in Quantum Mechanics they occur in a symmetric manner. In order to recuperate the symmetry he proposed a modified length element, adding to $ds^2$, the length square element, an additional term $l^2 g_{\mu\nu} du^\mu du^\nu$, with $u^\mu$ as the four velocity and $g_{\mu\nu}$ the metric (Born used instead of $u^\mu$ the $p^\mu/m$). Again the minimal length parameter appears due to dimensional reasons. In [7] it was recognized that the new length element is related to a *maximal acceleration*, $a \leq 1/l$. Many other groups joined in this investigation [8–14] and we will show that it is automatically contained in the proposed pseudo-complex extension of GR (which we will call from here on pc-GR). In [15, 16] a non-symmetric metric is considered and we will also show that it is contained within a pseudo-complex (pc) description.

In Sect. 2 we will introduce the pc-variables and mention some important properties. In the same section the formulation of the pc-GR is resumed. In Sect. 3 we present the results of the pc-Schwarzschild and pc-Kerr solution. It will be shown that in the pc-GR dark energy accumulates around a large mass concentration, which will finally stop the gravitational collapse, forming rather a gray star than a black hole. There will be no event horizon, thus allowing an external observer to access all region of space. Also in this section, the circular motion of a particle around a gray star is considered, with possible experimental verification. In Sect. 4 the conclusions will be drawn.

## 2 Formulation of the Pseudo-Complex General Relativity

First we resume some basic properties of pc-variables: A pseudo-complex variable is given by $X = X_R + IX_I$, with $X_R$ as the pseudo-real and $X_I$ the pseudo-imaginary component. It is of great advantage to write it in terms of the *zero divisor basis* (the notation becomes obvious further below) $X = X_+\sigma_+ + X_-\sigma_-$, with $\sigma_\pm = \frac{1}{2}(1 \pm I)$. The $\sigma_\pm$ obey the relations $\sigma_\pm^2 = \sigma_\pm$ and $\sigma_+\sigma_- = 0$. The last property is the definition of a zero divisor. When one defines as the complex conjugate $X^* = X_R - IX_I$, which implies $\sigma_\pm^* = \sigma_\mp$, then for elements in the zero divisor basis ($X = \lambda\sigma_\pm$) the norm squared $\mid X \mid^2 = XX^*$ is zero. One can look at it as a "generalized" zero. Calculations in the zero divisor basis are particularly simple. For example, products and division of functions can be done independently in each zero divisor component. Also differentiation and integration can be defined, similar to complex analysis (with some slight changes). For more details, please consult [17, 18]. In the literature there exist several names for the pc-variables. Sometimes they are called hyper-complex, hyperbolic or semi-complex.

The consequences of using pc-variables for the Lorentz transformation are as follows: A finite Lorentz transformation is given by

$$
\begin{aligned}
e^{i\omega_{\mu\nu}\Lambda_{\mu\nu}} &= e^{i\omega^+_{\mu\nu}\Lambda^+_{\mu\nu}}\sigma_+ + e^{i\omega^-_{\mu\nu}\Lambda^-_{\mu\nu}}\sigma_- \\
\Lambda_{\mu\nu} &= X_\mu P_\nu - X_\nu P_\mu \\
\Lambda^\pm_{\mu\nu} &= X^\pm_\mu P^\pm_\nu - X^\pm_\nu P^\pm_\mu \\
\omega_{\mu\nu} &= \omega^+_{\mu\nu}\sigma_+ + \omega^-_{\mu\nu}\sigma_- \quad . \qquad (1)
\end{aligned}
$$

It divides into a Lorentz transformation in each zero-divisor component. The generators look the same, except now the variables are pseudo-complex. In the zero-divisor component the coordinates are given by $X^\pm_\mu$ and the momenta by $P^\pm_\nu$. Because $\sigma_+\sigma_- = 0$, the two Lorentz transformations commute, thus we have

$$
SO_+(3,1) \otimes SO_-(3,1) \supset SO(3,1). \qquad (2)
$$

The standard Lorentz group is contained in the direct product and is reached by projecting the pseudo-complex parameters, coordinates and momenta to their real parts, i.e.,

$$
\begin{aligned}
\omega_{\mu\nu} \rightarrow \omega^R_{\mu\nu} &= \frac{1}{2}\left(\omega^+_{\mu\nu} + \omega^-_{\mu\nu}\right) \\
X_\mu \rightarrow x_\mu & \\
P_\nu \rightarrow p_\nu & \quad . \qquad (3)
\end{aligned}
$$

This projection method has to be applied also to the metric components.

That pseudo-complex variables also proved to be very useful was demonstrated in [19]: As shown in [19], the field equation for a scalar boson field is obtained from the Lagrangian density $\frac{1}{2}\left(D_\mu \Phi D^\mu \Phi - M^2\Phi^2\right)$, where $\Phi$ is the pc-boson field, $M = M_+\sigma_+ + M_-\sigma_-$ is a pc-mass and $D_\mu$ a pc-derivative. The propagators of this theory are the ones of Pauli-Villars, which already are regularized. One obtains the same propagator in the standard theory, with a non-pc scalar field, using the Lagrange density $-\frac{1}{(M_+^2 - M_-^2)}\phi\left(\partial_\mu\partial^\mu + M_+^2\right)\left(\partial_\mu\partial^\mu + M_-^2\right)\phi$, where $\phi$ is now a real valued function, $M_+$ is identified with the physical mass $m$ and $M_- >> M_+$ with the regularizing mass. Note, that this theory is highly non-linear while the pc-description is linear. This indicates that a pc-description can substantially simplify the structure of the theory and we can expect something similar in the pc-formulation of GR.

Let us now return to the pc-GR: The pc-extension of GR is quite direct within the zero divisor components. The first attempts are published in [20, 21] and in a more recent article [22] which includes modifications. Here we will present a short review. We introduce the pc-metric via

$$g_{\mu\nu}(X,A) = g^{+}_{\mu\nu}(X_+,A_+)\sigma_+ + g^{-}_{\mu\nu}(X_-,A_-)\sigma_-, \tag{4}$$

were the metric is assumed to be symmetric (in Moffat's theory of a non-symmetric metric [15, 16], the $\sigma_+$ component is the metric $g_{\mu\nu}$, while the $\sigma_-$ component is its transposed, so in principle Moffat's theory is contained in our theory, if we skip the restriction to a symmetric metric). The metric components depend on the variables $X^{\mu}_{\pm}$ and parameters, denoted shortly as $A_{\pm}$. In each zero-divisor component a GR is constructed in the same manner as in standard GR. The pc-coordinates have the structure

$$X^{\mu} = x^{\mu} + Ilu^{\mu} \quad . \tag{5}$$

Again, due to dimensional reasons, a minimal length parameter has to be introduced. Because it is just a parameter, it is not affected by any relativistic transformation, contrary to the believe that a minimal length is related to the breaking of Lorentz symmetry. The error made is to relate a minimal length to a physical length, which is affected by a Lorentz transformation. Here, the minimal length is a parameter and thus cannot be affected by such a transformation. The consequences are very important. For example, in [19] a pc-Field Theory was developed, demonstrating that a minimal length parameter does not affect the known symmetries, thus the calculations of Feynman diagrams remain very simple and that the propagators of the theory are automatically regularized.

In mathematical terms we can explain the pc-extension of GR in terms of the following chain

$$G_+ \otimes G_+ \supset G. \tag{6}$$

In each component a standard GR is formulated. The base manifold is given by $X^{\mu}_{\pm}$ and the tangent spaces are given by $U^{\mu}_{\pm}$. Note, that $U^{\mu}$ includes the acceleration. Excluding the acceleration leads to $G$.

The pc-length square element is given by

$$d\omega^2 = g_{\mu\nu}(X,A)DX^{\mu}DX^{\nu} \quad , \tag{7}$$

where $D$ refers to a pc-differential [19, 20].

One may ask, what are the corrections due to the minimal length $l$? This will lead to the conclusion that all other theories, mentioned in the introduction, are a consequence of a pc-description. An expansion up to $lu^{\mu}$ is given by

$$g_{\mu\nu}(X) \approx g_{\mu\nu}(x) + lu_{\lambda}F^{\lambda}_{\mu\nu}(x). \tag{8}$$

The norm of the four-velocity can not be larger than 1. Considering that the minimal length is probably very small (Planck length), one can safely take into account only the first term. Thus, in the metric tensor $g_{\mu\nu}(X)$ the pc-coordinates $X^{\mu}$ are substituted by $x^\mu$. With this and expressing the pc-coordinates explicitly in terms of their pseudo-real and pseudo-imaginary components, the $d\omega^2$ acquires the form

$$d\omega^2 \approx g_{\mu\nu}(x)\left(dx^\mu dx^\nu + l^2 du^\mu du^\nu\right) + 2Ilg_{\mu\nu}(x)dx^\mu du^\nu \quad . \tag{9}$$

The terms in $du^\mu$ can not be neglected when effects near the maximal accelerations are considered. The $du^\mu$ are *differentials* of velocities, thus accelerations, and can reach values of the order of $1/l$. When the motion of a particle is considered, the $d\omega^2$ has to be real. This provides the condition

$$g_{\mu\nu}(x)dx^\mu du^\nu = 0 \quad , \tag{10}$$

which is nothing but the dispersion relation. With (10) the length square element acquires the form as used in the theories mentioned in the introduction. There, the dispersion relation is introduced by hand while here it appears as a logical consequence.

When maximal acceleration effects are of no importance, one can also neglect the terms proportional to $l$ and $l^2$ in (9).

All properties of tensors, four derivatives, Christoffel symbols, etc. can be directly extended from standard GR, defining them in each zero-divisor component as done in standard GR [20, 22, 23]. The only concept which has to be modified is the variational principle. If one uses ($S$ denotes the action) $\delta S = \delta S_+ \sigma_+ + \delta S_- \sigma_- = 0$, then we would obtain $\delta S_\pm = 0$, which correspond to two separated theories. In order to get a new theory, in [24, 25] a modified variational principle was proposed, namely that the variation has to be within the zero divisor (it can be interpreted as a "generalized zero"). This leads to field equations which on their right hand side are not zero but proportional to an element in the zero divisor. Our convention is to set it proportional to $\sigma_-$. Thus the Einstein equations read ($c = 1$)

$$G_{\mu\nu} = R_{\mu\nu} - \frac{1}{2}g_{\mu\nu}R \;=\; -8\pi\kappa T_{\mu\nu}\sigma_- \quad . \tag{11}$$

The $R_{\mu\nu}$ are the components of the pc-Ricci tensor, while $R$ is the Ricci scalar. On the right hand side appears an energy-momentum tensor which describes the presence of an additional field which is always there in a pc-description. This field will turn out to have the properties of a dark energy and it will introduce a repulsion against gravitational collapse.

## 3 pc-Schwarzschild and pc-Kerr solutions

In [22] we presented the pc-Schwarzschild and pc-Kerr solutions. Of interest here is the $g_{00}$ component, namely

$$g_{00} = \left(1 - \frac{2m}{r} + \frac{\Omega(r)}{r}\right) \quad . \tag{12}$$

(Here, we neglect for the moment a possible factor $e^f$ [22], which we set to 1.) We already restricted to the first term in the expansion in $lu^\mu$. The $\Omega(r)$ is a not yet known function in the radial distance $r$. We model it by $\Omega = \frac{B}{2r^2}$. This leads to a correction in the metric of $\frac{B}{2r^3}$. The correction to the metric components have to depend at least on $1/r^3$, because a dependence on $1/r^2$ with a large $B$ is excluded by experiments in the solar system [26].

One may speculate about the origin of the dark energy. One possibility are the vaccum fluctuations (Casimir effect): In [27] the Casimir effect in a gravitational backgound is investigated, within the Hartle-Hawking vacuum. No recoupling of the vacuum fluctuations with the gravitational field is considered. Thus, there is still the Schwarzschild metric present with an event horizon at the Schwarzschild radius. As a result, the expectation value of the trace of the energy-momentum tensor, due to the vacuum fluctuations, falls off proportional to $1/r^6$. This would mean that the mass, represented by the energy density, falls of proportional to $1/r^3$. Because no recoupling with the gravitational field is considered, the calculation has to stop at the Schwarzschild radius. Below that, no time can be defined in the same way as outside. In the pc-GR the recoupling of the dark energy energy-momentum tensor with the gravitational field is automatically included in (11). This leads to the correction in (12). Using the result in [27] literally, would imply a correction to the metric proportional to $1/r^4$. We will assume that the correction to the metric falls off like $1/r^3$ instead. This is the minimal correction which can be implemented not yet in conflict with current astronomical observations [26]. We expect to change the $r$-dependence, when the recoupling to the gravitational field is included in the calculation of the Casimir effect. Therefore, the model assumption that the corrections to the metric behave as $1/r^3$ is a rather good one. Proposing $1/r^4$ does not change our results significantly!

After this consideration, we return to the discussion of the pc-GR: In order to have the same interpretation of time in all regions of space, the $g_{00}$ component has to be larger than zero. This introduces a minimal value of $B$.

Note, that the $\sqrt{g_{00}}$ component is proportional to an effective potential, with angular momentum zero [28]. With this, the effective potential is proportional to

$$\sqrt{1 - \frac{2m}{r} + \frac{B}{2r^3}} \quad . \tag{13}$$

For large distance, the potential is similar to the standard Schwarzschild solution. The differences start to appear near the Schwarzschild radius. The event horizon vanishes, because $g_{00}$ never becomes zero. At smaller radial distances, the potential becomes repulsive, which is the consequence of the accumulation of dark energy. This changes the picture of a gravitational collapse: When a large mass is contracted due its gravitational influence, dark energy starts to accumulate and increases when the collapse advances. The collapse is finally stopped when enough dark energy accumulates and acts against the gravitational attraction. Thus, instead of a black hole the result is rather a *gray star*, though the gray star resembles pretty much a black hole seen from far apart. Therefore, from now on we will always refer to a gray star.

Today we know that the gray stars in the centers of galaxies rotate nearly at maximum speed. Thus, instead of the Schwarzschild solution one has to take the Kerr solution, which describes stars in rotation. The pc-Kerr solution was obtained in [22, 29]. Please look there for details.

In order to relate the theory to experiment, we investigated the motion of a particle in a circular geodesic orbit around a gray star. This may be related to the possible observation of a plasma cloud orbiting such a star [30]. In Fig. 1 the orbital frequency is plotted versus the radial distance. As can be seen, the orbital frequency differs little from the standard Kerr solution until $r$ is of the order of the Schwarzschild radius. Towards smaller radial distances, the orbital frequency is smaller in the pc-description, showing a maximum value, after which it diminishes. The maximum is a result of the structure of $g_{00}$ which has a global minimum at about two-thirds of the Schwarzschild radius. For radii below that value the expression for the orbital frequency gets imaginary and we do not expect to observe circular geodesic orbits

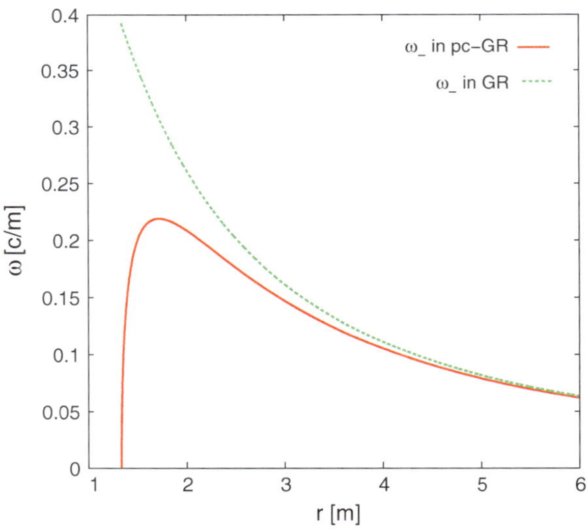


**Fig. 1** The orbital frequency of a particle in a circular orbit around a gray star, as a function on the radial distance $r$. The units of $\omega$ are in $\frac{m}{c}$ while the radial distance is in units of half the Schwarzschild radius. $r = 2$ corresponds to the Schwarzschild radius and $\omega = 0.22$ is equivalent to about 0.11/min (For this computation we took the mass of Sagittarius A, the center of our galaxy, which is of the order $3 \times 10^7 M_{\mathrm{sun}}$). The standard Kerr solution is given by the upper line, while the pc-solution is given by the lower line

anymore. The curve for the pc-Kerr solution stops at this value. The curve for the standard GR stops at the point of the last stable orbit.

The result was obtained assuming that $\Omega = \frac{B}{2r^2}$. If it goes with a larger power in $r$, the pc-solution approaches the standard Kerr solution, but will always show a maximum and the last stable orbit will be further out, i.e., the basic results will be the same.

This result has important consequences in the experimental verification of pc-GR and we refer to the talk given by T. Boller [30].

## 4 Conclusions

In this contribution we reviewed the pseudo-complex General Relativity. The extension of the standard GR to pc-variables is direct due to the property that the zero-divisor components commute. In each component a standard GR is constructed. In order to obtain a new theory, the variational principle has to be changed. The variation of the action has now to be within the zero-divisor, i.e., it has to be a "generalized zero". This introduces a new energy-momentum tensor in the Einstein equations, describing a dark energy field.

As a consequence of this dark energy-field, the gravitational collapse of a large mass is halted as soon as enough dark energy has accumulated. Due to this, no event horizon is formed and no singularity either. Instead of a black hole rather a *gray star* is formed. This answers the question in the title!

A possible experimental verification is proposed, determining the orbital frequency of a particle around a gray star.

**Acknowledgments** Financial support from the Frankfurt Institute for Advanced Studies (FIAS), "Stiftung Polytechnische Gesellschaft Frankfurt am Main" (SPTG) and from CONACyT are acknowledged.

## References

1. A. Einstein, Ann. Math. **578**, 46 (1945)
2. A. Einstein, Rev. Mod. Phys. **35**, 20 (1948)
3. C. Mantz, T. Prokopec: arXiv:gr-qcl0804.0213v1, 2008
4. D. Lovelook, Annali di Matematica **83**(1), 43 (1969)
5. M. Born, Proc. Roy. Soc. A **165**, 291 (1938)
6. M. Born, Rev. Mod. Phys. **21**, 463 (1949)
7. E.R. Caianiello, Nuovo Cim. Lett. **32**, 65 (1981)
8. H.E. Brandt, Found. Phys. Lett. **2**, 39 (1989)
9. H.E. Brandt, Found. Phys. Lett. **4**, 523 (1989)
10. H.E. Brandt, Found. Phys. Lett. **6**, 245 (1993)
11. R.G. Beil, Found. Phys. **33**, 1107 (2003)
12. R.G. Beil, Int. J. Theor. Phys. **26**, 189 (1987)

13. R.G. Beil, Int. J. Theor. Phys. **28**, 659 (1989)
14. R.G. Beil, Int. J. Theor. Phys. **31**, 1025 (1992)
15. J.W. Moffat, Phys. Rev. D **19**, 3554 (1979)
16. G. Kunstatter, J.W. Moffat, J. Malzan, J. Math. Phys. **24**, 886 (1983)
17. F. Antonuccio, Semi-Complex Analysis and Mathematical Physics, gr-qc/9311032 (1993)
18. I.L. Kantor, A.S. Solodovnikov, *Hypercomplex Numbers* (An Elementary Introduction to Algebra, Springer, Heidelberg, 1989)
19. P.O. Hess, W. Greiner, Int. J. Mod. Phys. E **16**, 1643 (2007)
20. P.O. Hess, W. Greiner, Int. J. Mod. Phys. E **18**, 51 (2009)
21. P.O. Hess, L. Maghlaoui, W. Greiner, Int. J. Mod. Phys. E **19**, 1315 (2010)
22. G. Caspar, T. Schönenbach, P.O. Hess, W. Greiner, Int. J. Mod. Phys. E **21**, 250015 (2012)
23. R. Adler, M. Bazin, M. Schiffer, *Introduction to General Relativity* (McGraw Hill, New York, 1975)
24. F. P. Schuller, Dirac-Born-Infeld Kinematics, Maximal Acceleration and Almost Product Manifolds, Ph.D. Thesis, Cambridge, 2003
25. F.P. Schuller, M.N.R. Wohlfarth, T.W. Grimm, Class. Quantum Grav. **20**, 4269 (2003)
26. C.M. Will, Living Rev. Relativ. **9**, 3 (2006)
27. M. Visser, Phys. Rev. D **54**, 5103 (1996)
28. C.W. Misner, K.S. Thorne, J.A. Wheeler, *Gravitation* (W. H. Freeman Company, San Francisco, 1973)
29. T. Schönenbach, arXiv:1106.1316v1 [gr-qc], 2011
30. T. Boller, A. Müller, contribution to this conference, p. 293

# Structure and Cooling of Neutron and Hybrid Stars

**S. Schramm, V. Dexheimer, R. Negreiros, T. Schürhoff and J. Steinheimer**

**Abstract** The study of neutron stars is a topic of central interest in the investigation of the properties of strongly compressed hadronic matter. Whereas in heavy-ion collisions the fireball, created in the collision zone, contains very hot matter, with varying density depending on the beam energy, neutron stars largely sample the region of cold and dense matter with the exception of the very short time period of the existence of the proto-neutron star. Therefore, neutron star physics, in addition to its general importance in astrophysics, is a crucial complement to heavy-ion physics in the study of strongly interacting matter. In the following, model approaches will be introduced to calculate properties of neutron stars that incorporate baryons and quarks. These approaches are also able to describe the state of matter over a wide range of temperatures and densities, which is essential if one wants to connect and correlate star observables and results from heavy-ion collisions. The effect of exotic particles and quark cores on neutron star properties will be considered. In addition to the gross properties of the stars like their masses and radii their expected inner composition is quite sensitive to the models used. The effect of the composition can be studied through the analysis of the cooling curve of the star. In addition, we consider the effect of rotation, as in this case the particle composition of the star can be modified quite drastically.

## 1 Introduction

A large number of experimental programs and theoretical efforts is devoted to the study of strong interaction physics under extreme conditions. These conditions comprise large temperatures, densities, as well as extreme values of nuclear isospin.

---

S. Schramm (✉) · V. Dexheimer · R. Negreiros · T. Schürhoff · J. Steinheimer
Frankfurt Institute for Advanced Studies, J.W. Goethe-Universität, Ruth Moufang-Str. 1, 60438 Frankfurt am Main, Germany
e-mail: schramm@fias.uni-frankfurt.de

W. Greiner (ed.), *Exciting Interdisciplinary Physics*,
FIAS Interdisciplinary Science Series, DOI: 10.1007/978-3-319-00047-3_27,

In ultra-relativistic heavy-ion collisions a very hot fireball is created in the collision zone where hadronic matter is assumed to have melted into its constituents, quarks and gluons. The net density in such reactions can be affected by the beam energy. At energies aimed for in the upcoming FAIR facility at GSI a hot and also relatively dense system will be produced. On the other hand, in order to reach densities several times nuclear ground state density at relatively low temperatures, a study of the properties of neutron stars is essential. In the following we will discuss a theoretical approach that is able to describe the conditions found in compact stars as well as those created in heavy-ion collisions.

## 2 Recent Observations

The main observational information about neutron stars is still the stellar mass that can help to constrain model descriptions of compact stars. Here, a new benchmark has been set with the accurate measurement of the mass of pulsar PSR J1614-2230 of $M = 1.97 \pm 0.04\, M_{\odot}$ [1] with several, much less certain, potential higher-mass candidates. A statistical analysis of measured masses suggests that there is no sign of a cut-off behavior of the mass distribution at the upper end, implying that higher values than 2 solar masses are plausible [2].

This new value serves to exclude a number of models or specific parameter sets, especially including hyperons, that have been in use before (see e.g. [3–6]).

In the case of hybrid stars with a quark core in the center of the star, a quark phase based on a simple non-interacting quark model like the MIT bag model also tends to reduce the maximum mass significantly (see the discussion in [7]).

A quark phase that includes strong repulsive interactions, however, can have an equation of state quite similar to a nucleonic one, which avoids the softening of the matter and therefore the drop in maximum mass [8–10]. Note, however, a potential problem with reproducing lattice QCD susceptibilities at small chemical potential in models with a strong repulsive quark-quark interaction [11].

Another important measurement is the first observation of the real-time cooling behavior of a neutron star in the supernova remnant Cassiopeia A, where a rather steep drop of the surface temperature in the last 10 years has been recorded [12]. This result has significant impact on cooling studies of compact stars and might help to constrain the properties of matter in the interior of the star.

## 3 The Quark-Hadron Model

A long-lasting problem in modeling strong interaction physics originates from the fact that, depending on density $\rho_B$ and temperature $T$, the effective degrees of freedom of QCD are completely different. Whereas at low values of $\rho_B$ and $T$ the world is hadronic, there is a transition to a deconfined and chirally symmetric system at some values of density and temperature. Whether this transition is a first- or higher-order phase transition or a smooth crossover is not known. The only relatively certain information on this point comes from lattice QCD simulations at zero chemical potential, indicating a cross-over transition. Although far from certain, at high densities and low temperatures one generally expects a first-order phase transition. If this is the case, somewhere in-between there should be at least one critical end-point of second order, the location of which is a prominent topic in heavy-ion research. A model description that should be valid over a large range of $T$ and $\rho_B$ (or, alternatively, the chemical potential $\mu_B$), should be able to describe also a cross-over transition. Gluing together a hadronic and a quark equation of state (without extreme fine-tuning of parameters) necessarily leads to a first-order transition over the whole range of $T$ and $\mu_B$. In order to avoid this and related problems we developed a unified model of hadrons and quarks with the correct asymptotic degrees of freedom in the different regions of thermodynamic parameters.

The hadronic part of the model is based on an effective chiral flavor-SU(3) model that includes the lowest SU(3) multiplets for baryons and mesons. A detailed description of this general ansatz can be found in [13, 14]. Restricting the discussion to the time-independent mean-field approximation the interaction of the baryons with the scalar and vector mesons reads

$$L_{Int} = -\sum_i \bar{\psi}_i [\gamma_0 (g_{i\omega}\omega + g_{i\phi}\phi + g_{i\rho}\tau_3\rho) + m_i^*]\psi_i, \tag{1}$$

summing over the baryon species $i$. $\omega$ and $\rho$ are the non-strange isovector 0 and 1 vector mesons, whereas $\phi$ denotes the vector meson consisting of an $s\bar{s}$ quark pair. The coupling to the scalar mesons is contained in the effective baryon masses $m_i^*$:

$$m_i^* = g_{i\sigma}\sigma + g_{i\zeta}\zeta + g_{i\delta}\delta + \delta m_i. \tag{2}$$

The terms include the coupling of the baryons to the non-strange scalar isoscalar $\sigma$, isovector $\delta$ and strange fields $\zeta$. In addition, there is a small explicit mass term $\delta m_i$. Baryonic vacuum masses are generated by non-vanishing vacuum expectation values of the scalar mesons. These are the result of the structure of the SU(3)-invariant scalar self-interactions, given by

$$L_{Self} = k_0(\sigma^2 + \zeta^2 + \delta^2) + k_1(\sigma^2 + \zeta^2 + \delta^2)^2 + k_2\left(\frac{\sigma^4}{2} + \frac{\delta^4}{2} + 3\sigma^2\delta^2 + \zeta^4\right)$$
$$+ k_3(\sigma^2 - \delta^2)\zeta + k_4 \ln\frac{(\sigma^2 - \delta^2)\zeta}{\sigma_0^2\zeta_0}. \tag{3}$$

In addition, the full Lagrangian includes self-interactions of the vector mesons and a further term that breaks chiral symmetry explicitly generating the masses of the pseudo-scalar mesons. A detailed discussion of the model and specific values of the parameters can be found in [15].

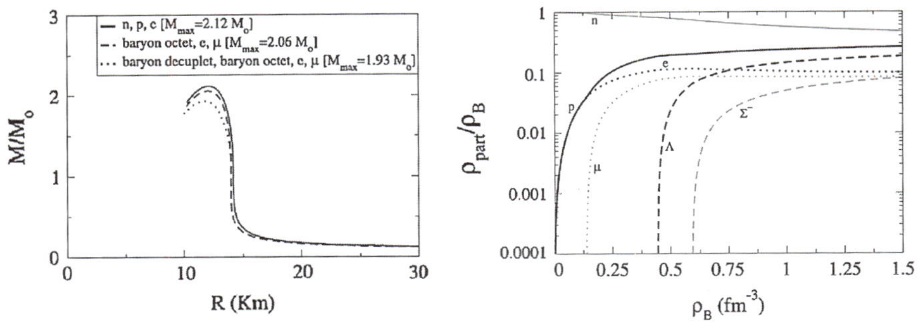


**Fig. 1** *Left* Star masses as function of radii including nucleons, hyperons, and the baryonic spin 3/2 decuplet [15]. *Right* Relative particle populations as function of baryon density including the lowest baryon octet

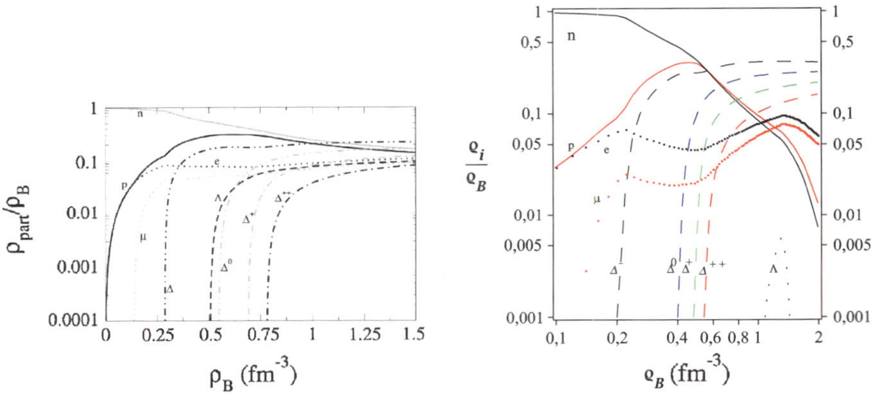


**Fig. 2** *Left* Relative particle densities as function of baryon density including the spin-3/2 decuplet. *Right* same as the left panel for a relative strength of the $\Delta - \omega$ coupling $r_V = g_{N\omega}/g_{\Delta\omega} = 0.9$

In order to study compact stars we determine the equation of state within this model and solve the Tolman-Oppenheimer-Volkoff equations for static stars [16, 17]. The resulting star masses and radii of stars are shown in Fig. 1. As could be expected, including hyperons in addition to nucleons, and further taking into account the baryonic spin 3/2 decuplet (mainly, the $\Delta$ resonances) reduces the maximum mass of the compact stars from 2.12 to 1.93 solar masses as more degrees of freedom translate to a softer equation of state. However, the influence of the hyperons is rather weak with a value of $f_s$, the amount of strangeness per baryon, of about 0.1 in the core of the heaviest star. Note, that the internal structure of the stars can change quite considerably, although the maximum masses drop by only 10%. This can be seen by comparing Fig. 1 (right panel) and Fig. 2 (left panel). Including the $\Delta$ resonances changes the population of particles substantially, removing the $\Sigma^-$ and suppressing the $\Lambda$ hyperon, replacing them by the corresponding $\Delta$ states $\Delta^-$ and $\Delta^0$. In these calculations it was assumed that the baryon decuplet has the same vector meson

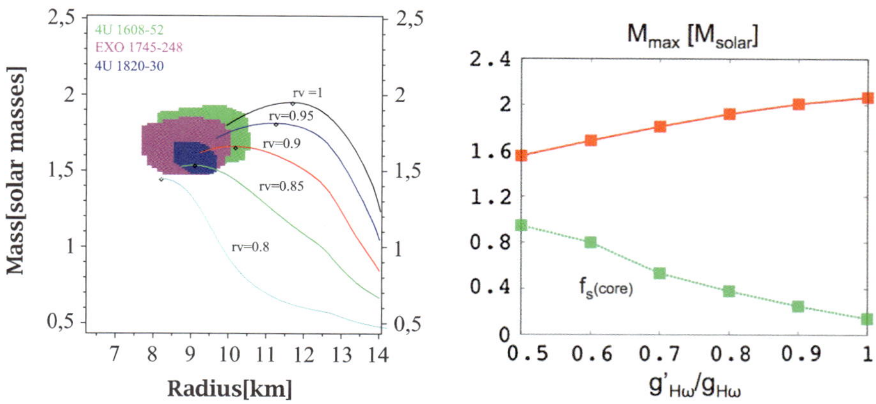


**Fig. 3** *Left* mass-radius diagrams for different values of $r_V$. Also shown are the results of an analysis of the probable value range of mass and radius for the cases of three low-mass X-ray binaries [19]. *Right* Maximum star mass and strangeness content $f_s$ of the star as function of the strength of the vector repulsion of the hyperons

coupling strengths as the octet, that is, for instance in the case of the $\Delta$, $g_{N\omega} = g_{\Delta\omega}$. Changing this value moderately, by reducing the $\Delta$ coupling, can alter the results significantly as it was investigated in [18]. The results are shown in Fig. 2. In the right panel of the figure the particle densities are shown for a ten percent reduction of $g_{\Delta\omega}$. It is interesting to note that for densities beyond four times saturation density the $\Delta$'s dominate the system compared to nucleons. This parameter adjustment, however, has a sizable effect on masses and radii of the stars as can be seen in Fig. 3 (left panel). Also shown are possible ranges for masses and radii of three X-ray binaries as analyzed in [19]. By lowering the coupling one can reproduce the results of the experimental analysis, but the maximum masses are below the observed value of two solar masses.

In a similar way one can investigated the influence of the hyperons on the results. The rather small influence of the hyperons on the star properties is largely due to the meson interactions that keep the strange scalar field relatively large at high densities. Thus hyperons stay heavy and are not very strongly populated [15]. Fig. 3 (right panel) shows the result of a calculation using the same model but reducing by hand the vector coupling constant of the hyperons at densities beyond nuclear matter densities (thus without changing the reasonably well-known optical potential depths in normal nuclear matter that are reproduced in the model). One can see that a reduction of the coupling strength by 50% increases the strangeness fraction in the core of the star ten-fold from 0.1 to 1, which corresponds to an average of one strange quark per baryon. This softens the equation of state significantly, and the maximum mass is reduced by half a solar mass.

Following the ideas of the PNJL approach [20, 21] we include quark degrees of freedom and an effective field $\Phi$, in analogy to the Polyakov loop, describing the deconfinement phase transition. Here, the quark fields couple linearly to the scalar

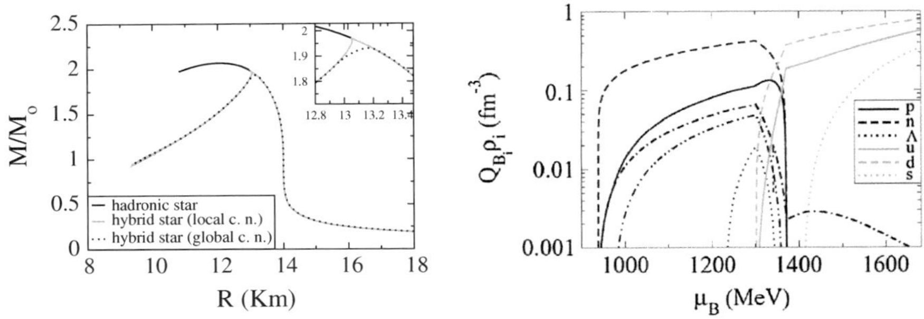


**Fig. 4** *Left* mass-radius diagram of the quark-hadron model compared to the purely baryonic case. The inset shows the effect of introducing a Gibbs mixed phase. *Right* Particle population in the star as function of baryon density. Note that there are practically no strange particles in the interior of the star

and vector condensates as in Eq. (1). $\Phi$ couples to the hadron and quark masses such that quarks attain a high mass in the confined phase at low values of $\Phi$ and correspondingly hadrons obtain a large mass for large values of the field, removing the baryons as degrees of freedom. The values of the parameters are quoted in [22, 23]. In addition to the usual structure of the effective potential of the field $\Phi$ we add chemical-potential dependent terms. They can be chosen to reproduce the position of the critical end point of a first-order phase transition line as suggested by lattice calculations [24]. The star masses and radii get modified due the quark contributions. The results are shown in the left panel of Fig. 4. The quarks largely cut off the branch of stable masses with a maximum value of $M = 1.93\,M_\odot$. A Gibbs phase mixture in the core of the star leads to a 2 km mixed region of quarks and baryons.

In a different theoretical approach to chirally symmetric models we studied the chiral symmetry restoration in the so-called parity-doublet model [25–27]. Here one extends the baryonic states by the (hypothetical) partner states with opposite parity. In the case of the nucleon one candidate is the N(1535) resonance. In this formulation the signal for chiral symmetry restoration is given by the degeneracy of the parity partners, in analogy to the scalar and pseudo-scalar mesons in the linear $\sigma$ model, which obtain equal masses at high temperature. Extending this approach to the whole SU(3) octet [28] the equations look similar to the ones above with the exception of the effective baryon masses that read:

$$m^*_{i\pm} = \sqrt{\left[(g^{(1)}_{\sigma i}\sigma + g^{(1)}_{\zeta i}\zeta)^2 + (m_0 + n_s m_s)^2\right]} \pm g^{(2)}_{\sigma i}\sigma \pm g^{(1)}_{\zeta i}\zeta, \tag{4}$$

where $\pm$ denote the parity partners, $m_s$ is the strange quark mass and $n_s$ is given by the number of strange quarks in the corresponding baryon. Due to the doublet structure there are now twice as many scalar coupling constants. From this expression one can see that in the case of vanishing fields $\sigma$ and $\zeta$, the particle masses of the parity partners are degenerate with a mass of the parameter $m_0$, in the case of nucleons.

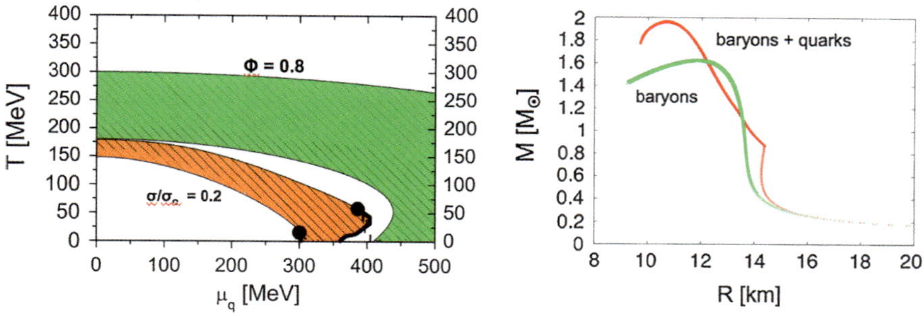


**Fig. 5** *Left panel* Phase diagram of the quark-hadron model as function of temperature and quark chemical potential. First-order liquid-gas and chiral transitions are marked by full lines. The lower shaded area represents the range, wherein the scalar field drops from 80 to 20% of its vacuum value. The upper band corresponds to values of the Polyakov loop field $\Phi$ between 0.2 and 0.8. *Right panel* Star masses and radii for the quark-hadron parity model and for the purely hadronic parity model. The latter does not take into account excluded volume corrections

A full discussion of the model is given in [28]. The quarks are again coupled to the meson fields. Here we take into account an excluded volume correction for the hadrons in a thermodynamically consistent formulation, which automatically leads to a switch to quarks at high temperature and/or densities.

The phase diagram of the model is shown in the left panel of Fig. 5. There are two first-order phase transition lines both ending in a critical end point, corresponding to the usual liquid-gas transition and to the chiral phase transition. Separately, at higher temperatures and densities the deconfinement transition occurs as symbolized by the upper shaded area. At vanishing chemical potential the model describes the temperature dependence of the thermodynamical quantities and order parameters as seen in lattice calculations very well [28]. Calculating stars with this approach, one obtains a mass-radius diagram as shown in the right panel of Fig. 5 . The figure contains curves for the purely hadronic and the quark-hadron model. In the case of the hadronic stars no excluded volume correction was taken into account, as such a term in purely hadronic models inevitably generates acausal speeds of sound larger than the speed of light. This is not the case in the QH model as at high densities the system switches to quarks (that do not have an excluded volume) before such conditions can be reached. In the quark-hadron case two-solar mass stars are possible as can be inferred from the figure.

## 4 Effects of Rotation and the Relevance of Cooling

Neutron stars can rotate at very high frequencies. Currently the fastest known rotator is the pulsar PSR J1748-2446ad with a rotational frequency of 716 Hz [29]. As rotation leads to a decrease of the central density of the star, rotating stars can support larger masses than in the static case as seen in Fig. 6 (left panel). Following the nearly horizontal lines in the figure corresponds to a spinning down from the maximum

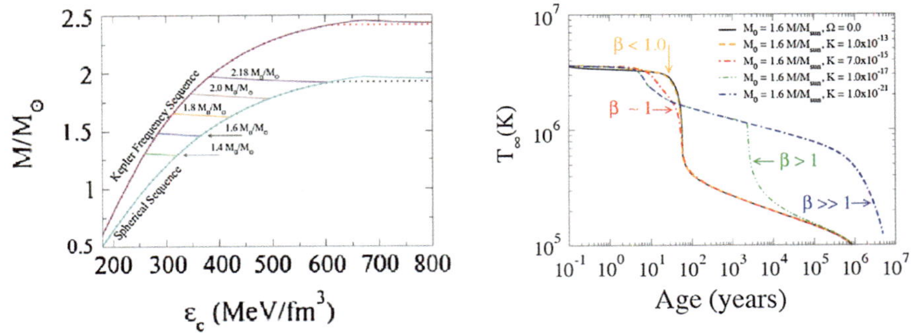


**Fig. 6** *Left* Star masses as function of central energy density. Results for the static solutions and for stars rotating at their Kepler frequency are shown. The nearly horizontal lines indicate the gravitational mass of slowing-down stars at a fixed baryon number. *Right* Cooling curves of neutron stars depending on the ratio $\beta$ of the time scale of the thermal core-crust coupling and spin down

(Kepler) frequency to zero rotation at a constant baryon number. If there is no sizable accretion or mass loss these lines are the evolutionary paths of a slowing-down pulsar. It is very interesting to note that although the gravitational masses change very little, the central energy density increases by typically 50%. This has a big impact on the structure of the star and can lead to a late appearance of hyperons or quarks in the star while it is spinning down. Furthermore, combining the effects of rotation and the cooling of the star can lead to intriguing results. This can be seen in the right panel of Fig. 6, where the temperature evolution of the neutron star with time is shown. Depending on the mass of the star a rapid drop of the temperature occurs after about 100 years. The reason for this is that the core of the star cools mainly by neutrino emission via direct Urca processes like $n \rightarrow p + e^- + \bar{\nu}_e$, which are only possible at higher densities of about 2 times saturation density or more (depending on the specific equation of state). Therefore, cooling occurs mainly in the core of the star. The cooling wave travels outward and reaches the surface after about 100 years.This is true for the static star. In the case of a rotating star this drop can be delayed as the star first has to slow down in order to reach inner densities such that the direct Urca process can take place. This is parametrized by the quantity $\beta$, the ratio of the spin-down to the core-crust coupling time (in the static case). As the figure shows, for $\beta$ larger than 1 a delayed temperature drop can be observed [30]. This effect might help to explain the observed cooling evolution of the neutron star in Cas A, that shows a rather steep drop of temperature 300 years after the supernova explosion. This could be understood as the result of the onset of the direct Urca process that was delayed due to rotation as outlined before (for details, see [30]).

## 5 Conclusions, Outlook

We studied the properties of compact stars and the phase diagram of strongly interacting matter. To that end we extended our hadronic chiral SU(3) model to include quarks. In addition to study compact stars, in such an approach one can also obtain a very reasonable description of hot, low-density matter as it is produced in heavy-ion collisions.

The resulting star masses are in agreement with recent observations. However, as we have discussed, depending on the degrees of freedom taken into account the inner composition can vary quite substantially without modifying the masses too severely. Therefore additional observables that are sensitive to the inner structure of the star are important. Here the study of neutron star cooling is important, as for at least the first thousand years the cooling is neutrino-dominated. The neutrinos originate mainly in the dense core of the star. Therefore the cooling behavior directly depends on the properties of the constituents in the core of the star. Especially in conjunction with rotational effects this can lead to new effects as discussed above, that might be tested against observational data. Here, it is important to perform a full two-dimensional simulation of the cooling of the star. Work along this line is in progress [31].

**Acknowledgments** We acknowledge the use of the CSC computer facilities at Frankfurt university for our work. R. N. acknowledges financial support from the LOEWE program HIC for FAIR. T. S. is supported by the Nuclear Astrophysics Virtual Institute (NAVI).

## References

1. P.B. Demorest, T. Pennucci, S.M. Ransom, M.S.E. Roberts, J.W.T. Hessels, Nature **467**, 1081 (2010)
2. B. Kiziltan, A. Kottas, S.E. Thorsett, arXiv:1011.4291v1 [astro-ph.GA], (2010)
3. I. Vidana, D. Logoteta, C. Providencia, A. Polls, I. Bombaci, Europhys. Lett. **94**, 11002 (2011)
4. H. Chen, M. Baldo, G.F. Burgio, H.-J. Schulze, Phys. Rev. D **84**, 105023 (2011)
5. S. Weissenborn, D. Chatterjee, J. Schaffner-Bielich, Phys. Rev. C **85**, 065802 (2012)
6. H. Djapo, B.-J. Schaefer, J. Wambach, Phys. Rev. C **81**, 035803 (2010)
7. R.P. Negreiros, V. Dexheimer, S. Schramm, Phys. Rev. C **85**, 035805 (2012) [arXiv:1011.2233 [astro-ph.HE]]
8. M. Alford, M. Braby, M.W. Paris, S. Reddy, Astrophys. J. **629**, 969 (2005)
9. S. Weissenborn, I. Sagert, G. Pagliara, M. Hempel, J. Schaffner-Bielich, Astrophys. J. **740**, L14 (2011)
10. L. Bonanno, A. Sedrakian, Astron. Astrophys. **539**, A16 (2012) [arXiv:1108.0559 [astro-ph.SR] (2011)]
11. J. Steinheimer, S. Schramm, Phys. Lett. B **696**, 257 (2011)
12. C.O. Heinke, W.C.G. Ho, Astrophys. J. **719**, L167 (2010)
13. P. Papazoglou, S. Schramm, J. Schaffner-Bielich, H. Stöcker, W. Greiner, Phys. Rev. C **57**, 2576 (1998)
14. P. Papazoglou, D. Zschiesche, S. Schramm, J. Schaffner-Bielich, H. Stöcker, W. Greiner, Phys. Rev. C **59**, 411 (1999)
15. V. Dexheimer, S. Schramm, Astrophys. J. **683**, 943 (2008)
16. R.C. Tolman, Phys. Rev. **55**, 364 (1939)
17. J.R. Oppenheimer, G.M. Volkoff, Phys. Rev. **55**, 374 (1939)
18. T. Schürhoff, S. Schramm, V. Dexheimer, Astrophys. J. **724**, L74 (2010)
19. F. Ozel, G. Baym, T. Guver, Phys. Rev. D **82**, 101301 (2010)
20. K. Fukushima, Phys. Lett. B **591**, 277 (2004)
21. C. Ratti, M.A. Thaler, W. Weise, Phys. Rev. D **73**, 014019 (2006)

22. V.A. Dexheimer, S. Schramm, Phys. Rev. C **81**, 045201 (2010)
23. R. Negreiros, V.A. Dexheimer, S. Schramm, Phys. Rev. C **82**, 035803 (2010)
24. Z. Fodor, S.D. Katz, JHEP **0404**, 050 (2004)
25. C.E. Detar, T. Kunihiro, Phys. Rev. D **39**, 2805 (1989)
26. V. Dexheimer, S. Schramm, D. Zschiesche, Phys. Rev. C **77**, 025803 (2008)
27. V. Dexheimer, G. Pagliara, L. Tolos, J. Schaffner-Bielich, S. Schramm, Eur. Phys. J. A **38**, 105 (2008)
28. J. Steinheimer, S. Schramm, H. Stöcker, Phys. Rev. C **84**, 045208 (2011) [arXiv:1108.2596 [hep-ph]]
29. J.W.T. Hessels, S.M. Ransom, I.H. Stairs, P.C.C. Freire, V.M. Kaspi, F. Camilo, Science **311**, 1901 (2006)
30. R. Negreiros, S. Schramm, F. Weber, Phys. Lett. B **718**, 1176 (2013) [arXiv:1103.3870 [astro-ph]]
31. R. Negreiros, S. Schramm, F. Weber, Phys. Rev. D **85**, 104019 (2012)

# Nuclei in Strongly Magnetised Neutron Star Crusts

**Rana Nandi and Debades Bandyopadhyay**

**Abstract** We discuss the ground state properties of matter in outer and inner crusts of neutron stars under the influence of strong magnetic fields. In particular, we demonstrate the effects of Landau quantization of electrons on compositions of neutron star crusts. First we revisit the sequence of nuclei and the equation of state of the outer crust adopting the Baym, Pethick and Sutherland (BPS) model in the presence of strong magnetic fields and most recent versions of the theoretical and experimental nuclear mass tables. Next we deal with nuclei in the inner crust. Nuclei which are arranged in a lattice, are immersed in a nucleonic gas as well as a uniform background of electrons in the inner crust. The Wigner-Seitz approximation is adopted in this calculation and each lattice volume is replaced by a spherical cell. The coexistence of two phases of nuclear matter—liquid and gas, is considered in this case. We obtain the equilibrium nucleus corresponding to each baryon density by minimizing the free energy of the cell. We perform this calculation using Skyrme nucleon-nucleon interaction with different parameter sets. We find nuclei with larger mass and charge numbers in the inner crust in the presence of strong magnetic fields than those of the zero field case for all nucleon-nucleon interactions considered here. However, SLy4 interaction has dramatic effects on the proton fraction as well as masses and charges of nuclei. This may be attributed to the behaviour of symmetry energy with density in the sub-saturation density regime. Further we discuss the implications of our results to shear mode oscillations of magnetars.

R. Nandi
Astroparticle Physics and Cosmology Division, Saha Institute of Nuclear Physics, 1/AF Bidhannagar, Kolkata 700064, India
e-mail: rana.nandi@saha.ac.in

D. Bandyopadhyay (✉)
Astroparticle Physics and Cosmology Division and Centre for Astroparticle Physics, Saha Institute of Nuclear Physics, 1/AF Bidhannagar, Kolkata 700064 , India
e-mail: debades.bandyopadhyay@saha.ac.in

W. Greiner (ed.), *Exciting Interdisciplinary Physics*,
FIAS Interdisciplinary Science Series, DOI: 10.1007/978-3-319-00047-3_28,

## 1 Introduction

Neutron star crust is a possible site where neutron rich heavy nuclei might reside. Extreme physical conditions exist at the crust of a neutron star. The temperature is $\sim 10^{10}$ K and the density varies from $10^4$–$10^{14}$ g/cm$^3$ there. Recently, it was observed that certain neutron stars called magnetars had surface magnetic fields $\sim 10^{15}$ G. The internal fields could be several times higher than the surface fields of magnetars. Soft gamma repeaters (SGRs) are suitable candidates for magnetars [1–3]. Giant flares were observed from SGRs in several cases. Those giant flare events are thought to be the results of star quakes in magnetars. This might be attributed to the tremendous magnetic stress due to the evolving magnetic field leading to cracks in the crust. Quasi-periodic oscillations discovered in three giant flares are the evidences of torsional shear mode oscillations in magnetar crusts.

Such strong magnetic fields in magnetars are expected to influence charged particles such as electrons in the crust through Landau quantization. The effects of strong magnetic fields on dense matter in neutron star interior were studied earlier [4, 5]. It was also noted that atoms, molecules became more bound in a magnetic field [6]. In this article, we discuss the effects of strongly quantising magnetic fields on compositions and equation of state of the ground state matter in neutron star crusts and its connection to torsional shear mode oscillations.

We organise the article in the following way. Neutron star crusts in strong magnetic fields are described in Sects. 2, 3 and 4. Torsional shear mode oscillations of magnetars are discussed in Sect. 5. Finally, we summarise in Sect. 6.

## 2 Crusts in Strong Magnetic Fields

We investigate compositions and equations of state (EoS) of outer and inner crusts in strong magnetic fields. Nucleons are bound in nuclei in the outer crust. Nuclei are immersed in a uniform background of electron gas which becomes relativistic beyond $10^7$ g/cm$^3$. Neutrons start to drip out of nuclei at higher densities. This is the beginning of the inner crust. In this case, nuclei are embedded both in electron and neutron gases. Magnetic fields may influence the ground state properties of crusts either through magnetic field and nuclear magnetic moment interaction or through Landau quantisation of electrons. In a magnetic field $\sim 10^{17}$ G, magnetic field and nuclear magnetic moment interaction would not produce any significant change. However such a strong magnetic field is expected to influence charged particles such as electrons in the crust through Landau quantization. Our main focus is to study the effects of Landau quantisation on the ground state properties of neutron star crusts. Later we discuss shear mode frequencies using our results of magnetised neutron star crusts.

## 2.1 Landau Quantisation of Electrons

We consider electrons are noninteracting and placed under strongly quantising magnetic fields. In the presence of a magnetic field, the motion of electrons is quantized in the plane perpendicular to the field. We do not consider Landau quantisation of protons because magnetic fields in question in this calculation are below the critical field for protons. However, protons in nuclei would be influenced by a magnetic field through the charge neutrality condition. We take the magnetic field (**B**) along Z-direction and assume that it is uniform throughout the inner crust. If the field strength exceeds a critical value $B_c = m_e^2/e \simeq 4.414 \times 10^{13}$ G, then electrons become relativistic [6]. The energy eigenvalue of relativistic electrons in a quantizing magnetic field is given by

$$E_e(\nu, p_z) = \left[p_z^2 + m_e^2 + 2eB\nu\right]^{1/2} , \tag{1}$$

where $p_z$ is the Z-component of momentum, $\nu$ is the Landau quantum number. The Fermi momentum of electrons, $p_{F_{e,\nu}}$, is obtained from the electron chemical potential in a magnetic field

$$p_{F_{e,\nu}} = \left[{\mu_e}^2 - m_e^2 - 2eB\nu\right]^{1/2} . \tag{2}$$

The number density of electrons in a magnetic field is calculated as

$$n_e = \frac{eB}{2\pi^2} \sum_{\nu=0}^{\nu_{max}} g_\nu p_{F_{e,\nu}} , \tag{3}$$

where the spin degeneracy is $g_\nu = 1$ for the lowest Landau level ($\nu = 0$) and $g_\nu = 2$ for all other levels.

The maximum Landau quantum number ($\nu_{max}$) is obtained from

$$\nu_{max} = \frac{{\mu_e}^2 - m_e^2}{2eB} . \tag{4}$$

The energy density of electrons is,

$$\varepsilon_e = \frac{eB}{4\pi^2} \sum_{\nu=0}^{\nu_{max}} g_\nu \left( p_{F_{e,\nu}}\mu_e + (m_e^2 + 2eB\nu) \ln \frac{p_{F_{e,\nu}} + \mu_e}{\sqrt{(m_e^2 + 2eB\nu)}} \right) . \tag{5}$$

Similarly the pressure of the electron gas is determined by

$$P_e = \frac{eB}{4\pi^2} \sum_{\nu=0}^{\nu_{max}} g_\nu \left( p_{F_{e,\nu}}\mu_e - (m_e^2 + 2eB\nu) \ln \frac{p_{F_{e,\nu}} + \mu_e}{\sqrt{(m_e^2 + 2eB\nu)}} \right) . \tag{6}$$

## 3 Magnetic BPS Model of Outer Crust Revisited

We describe the BPS model in the presence of strong magnetic fields $B \sim 10^{16}$G to determine the sequence of equilibrium nuclei and the equation state of the outer crust [7, 8]. Nuclei are arranged in a bcc lattice in the outer crust. Here we adopt the Wigner-Seitz (WS) approximation and replace each lattice volume by a spherical cell which contains one nucleus at the center. Further each cell is to be charge neutral such that equal numbers of protons and electrons are present there. The Coulomb interaction among cells is neglected. An equilibrium nucleus (A, Z) at a given pressure P is obtained by minimising the Gibbs free energy per nucleon with respect to A and Z. In this calculation, we modify the magnetic BPS model including the finite size effect in the lattice energy and adopting recent experimental and theoretical mass tables. The total energy density of the system is given by

$$E_{tot} = n_N(W_N + W_L) + \varepsilon_e \, . \tag{7}$$

The energy of the nucleus (including rest mass energy of nucleons) is

$$W_N = m_n(A - Z) + m_p Z - bA \, , \tag{8}$$

where $n_N$ is the number density of nuclei, $b$ is the binding energy per nucleon. Experimental nuclear masses are obtained from the atomic mass table compiled by Audi, Wapstra and Thibault [9]. For the rest of nuclei we use the theoretical extrapolation of Möller et al. [10]. $W_L$ is the lattice energy of the cell and is given by

$$W_L = -\frac{9}{10}\frac{Z^2e^2}{r_C}\left(1 - \frac{5}{9}\left(\frac{r_N}{r_C}\right)^2\right) . \tag{9}$$

Here $r_C$ is the cell radius and $r_N \simeq r_0 A^{1/3}$ ($r_0 \simeq 1.16$ fm) is the nuclear radius. The first term in $W_L$ is the lattice energy for point nuclei and the second term is the correction due to the finite size of the nucleus (assuming a uniform proton charge distribution in the nucleus). Further $\varepsilon_e$ is the electron energy density as given by Eq. (5) and P is the total pressure of the system given by

$$P = P_e + \frac{1}{3} W_L n_N \, , \tag{10}$$

where $P_e$ is the pressure of electron gas in a magnetic field as given by Eq. (6).

The Gibbs free energy per nucleon is

$$g = \frac{E_{tot} + P}{n} = \frac{W_N + 4/3 W_L + Z\mu_e}{A} \, , \tag{11}$$

where $n$ is the total baryon number density.

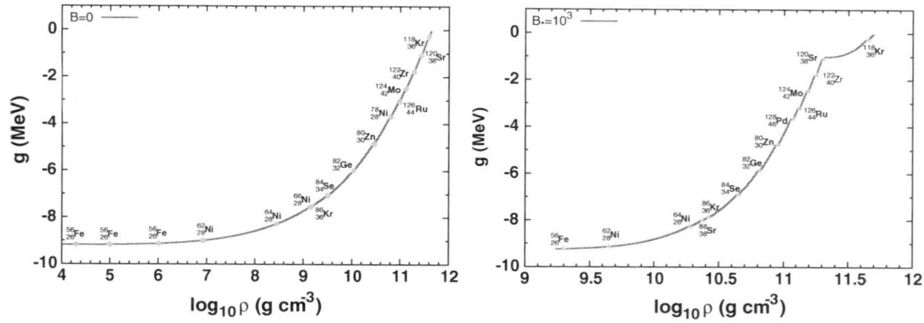


**Fig. 1** Gibbs free energy per nucleon is plotted with mass density for zero magnetic field (*left panel*) and $B_* = 10^3$ (*right panel*). Equilibrium nuclei are shown with solid symbols in both panels

At a fixed pressure $P$, we minimise $g$ varying $A$ and $Z$ of a nucleus. The sequence of equilibrium nuclei and their corresponding free energies are shown in Fig. 1. Here we define $B_* = B/B_c$. The left panel shows results for $B = 0$ and the right panel corresponds to $B_* = 10^3$. It is evident from the figure that some nuclei disappear and new nuclei appear under the influence of strong magnetic fields. It is attributed to the phase space modification of electrons due to Landau quantisation which enhances the electron number density [8].

# 4 Inner Crust in Quantizing Magnetic Fields

Now we describe the ground properties of matter of inner crusts in presence of strong magnetic fields using the Thomas-Fermi (TF) model at zero temperature. Inner crust nuclei are immersed in a nucleonic gas as well as a uniform background of electrons. Furthermore, nuclei are arranged in a bcc lattice. As in the case of outer crust, we again adopt the Wigner-Seitz (WS) approximation in this calculation. Here each cell is taken to be charge neutral and the Coulomb interaction between cells is neglected. Electrons are uniformly distributed within a cell. The system is in $\beta$-equilibrium. We assume that the system is placed in a uniform magnetic field. Though electrons are directly affected by strongly quantizing magnetic fields, protons in the cell are influenced through the charge neutrality condition [11]. The interaction of nuclear magnetic moment with the field is not considered because it is negligible in a magnetic field below $10^{18}$ G [12].

The spherical cell in the WS approximation does not define a nucleus. We exploit the prescription of Bonche, Levit and Vautherin [13, 14] to subtract the gas part from the cell and obtain the nucleus. It was shown that the TF formalism at finite temperature generated two solutions [15]—one for the nucleus plus neutron gas and the other representing the neutron gas. The nucleus is obtained as the difference of two solutions. This formalism is adopted in our calculation at zero temperature as described below.

The thermodynamic potentials for nucleus plus gas (NG) and only gas (G) phases are defined as [13, 14]

$$\Omega = \mathcal{F} - \sum_{q=n,\,p} \mu_q n_q \,, \tag{12}$$

where $\mathcal{F}$, $\mu_q$ and $n_q$ are the free energy density, baryon chemical potential and number density, respectively. The nucleus plus gas solution coincides with the gas solution at large distance i.e. $\Omega_{NG} = \Omega_G$. The free energy which is a function of baryon number density and proton fraction ($Y_p$), is defined as [11]

$$\mathcal{F}(n_q, Y_p) = \int [\mathcal{H} + \varepsilon_c + \varepsilon_e] d\mathbf{r} \,. \tag{13}$$

The nuclear energy density is calculated using the Skyrme nucleon-nucleon interaction and it is given by [16–18]

$$\begin{aligned} \mathcal{H}(r) = {} & \frac{\hbar^2}{2m_n^*}\tau_n + \frac{\hbar^2}{2m_p^*}\tau_p + \frac{1}{2}t_0\left[\left(1+\frac{x_0}{2}\right)n^2 - \left(x_0+\frac{1}{2}\right)\left(n_n^2+n_p^2\right)\right] \\ & - \frac{1}{16}\left[t_2\left(1+\frac{x_2}{2}\right) - 3t_1\left(1+\frac{x_1}{2}\right)\right](\nabla n)^2 \\ & - \frac{1}{16}\left[3t_1\left(x_1+\frac{1}{2}\right) + t_2\left(x_2+\frac{1}{2}\right)\right]\left[(\nabla n_n)^2 + (\nabla n_p)^2\right] \\ & + \frac{1}{12}t_3 n^\alpha\left[\left(1+\frac{x_3}{2}\right)n^2 - \left(x_3+\frac{1}{2}\right)\left(n_n^2+n_p^2\right)\right] , \end{aligned} \tag{14}$$

and the effective nucleon mass

$$\begin{aligned} \frac{m}{m_q^*(r)} = {} & 1 + \frac{m}{2\hbar^2}\left\{\left[t_1\left(1+\frac{x_1}{2}\right) + t_2\left(1+\frac{x_2}{2}\right)\right]n \right. \\ & \left. + \left[t_2\left(x_2+\frac{1}{2}\right) - t_1\left(x_1+\frac{1}{2}\right)\right]n_q\right\} , \end{aligned} \tag{15}$$

where total baryon density is $n = n_n + n_p$.

The direct parts of Coulomb energy densities for the nucleus plus gas and gas phases follow from [11, 19]

$$\begin{aligned} \varepsilon_c^{NG}(r) &= \frac{1}{2}(n_p^{NG}(r) - n_e)\int \frac{e^2}{|\mathbf{r}-\mathbf{r}'|}(n_p^{NG}(r') - n_e)d\mathbf{r}' \\ \varepsilon_c^{G}(r) &= \frac{1}{2}(n_p^{G}(r) - n_e)\int \frac{e^2}{|\mathbf{r}-\mathbf{r}'|}(n_p^{G}(r') - n_e)d\mathbf{r}' \\ & \quad + n_p^N(r)\int \frac{e^2}{|\mathbf{r}-\mathbf{r}'|}(n_p^{G}(r') - n_e)d\mathbf{r}' , \end{aligned} \tag{16}$$

where $n_p^{NG}$ and $n_p^G$ are proton densities in two respective phases. The exchange parts of coulomb energy densities are small and neglected in this calculation.

The average electron chemical potential in a magnetic field given by Eq. (2) is modified to [11]

$$\mu_e = \left[p_{F_{e,\nu}}(\nu)^2 + m_e^2 + 2eB\nu\right]^{1/2} - \langle V^c(r)\rangle \,, \tag{17}$$

where $\langle V^c(r)\rangle$ denotes the average single particle Coulomb potential and for both phases it is given by

$$V^c(r) = \int \left[n_p^{NG}(r') - n_e\right] \frac{e^2}{\mid \mathbf{r} - \mathbf{r}' \mid} d\mathbf{r}' \,. \tag{18}$$

The density profiles of neutrons and protons with or without magnetic fields are obtained by minimising the thermodynamic potential in the TF approximation

$$\frac{\delta\Omega_{NG}}{\delta n_q^{NG}} = 0 \,, \qquad \frac{\delta\Omega_G}{\delta n_q^G} = 0 \,, \tag{19}$$

with the condition of number conservation of each species from

$$Z_{cell} = \int n_p^{NG}(r) d\mathbf{r} \,, \qquad N_{cell} = \int n_n^{NG}(r) d\mathbf{r} \,, \tag{20}$$

where $N_{cell}$ and $Z_{cell}$ are neutron and proton numbers in the cell, respectively.

We obtain the mass number $A = N + Z$ and atomic number using the subtraction procedure as

$$Z = \int \left[n_p^{NG}(r) - n_p^G(r)\right] d\mathbf{r} \,, \qquad N = \int \left[n_n^{NG}(r) - n_n^G(r)\right] d\mathbf{r} \,. \tag{21}$$

Here we again obtain the equilibrium nucleus at each density by minimising the free energy of the nuclear cluster in the cell along with charge neutrality and $\beta$-equilibrium conditions [11]. In the left panel of Fig. 2, proton fraction is shown for $B = 0$ and $B_* = 10^4$. Protons are influenced by the Landau quantisation of electrons through charge neutrality condition. At lower densities, only the zeroth Landau level is populated by electrons whereas a few Landau levels are populated above density 0.005 fm$^{-3}$ for $B_* = 10^4$ i.e. $4.414\times10^{17}$ G. This is reflected in the

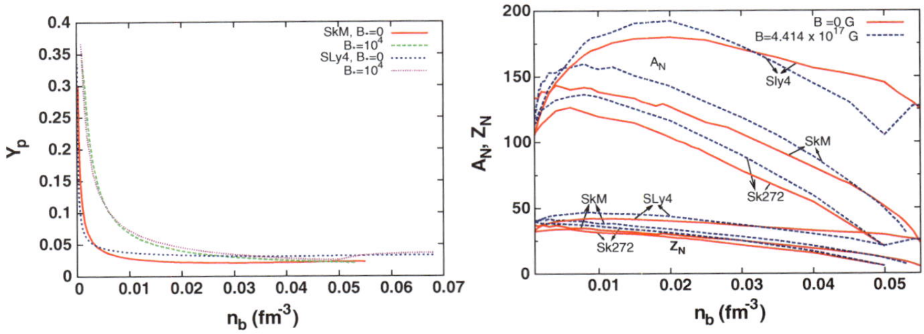


**Fig. 2** Proton fraction (*left panel*) and mass and atomic numbers of equilibrium nuclei (*right panel*) are plotted with average baryon density for different magnetic field strengths and Skyrme interaction parameter sets

proton fraction which rises hugely at lower densities and approaches to the zero field case at higher densities. Further we estimate the effects of different parameter sets of Skyrme interaction on the proton fraction. It is noted that the SLy4 set [20] results in higher proton fraction due to the stiffer density dependence of the symmetry energy at sub-saturation densities than that of the SkM set.

We exhibit mass and atomic numbers of equilibrium nuclei after subtraction of free neutrons as a function of average baryon density in the right panel of Fig. 2. Results are obtained for $B = 0$ and $B_* = 10^4$. Besides SkM and SLy4 parameter sets, we also exploit Sk272 [21] parameter set for this calculation. In all three cases, mass and atomic numbers are higher than zero field cases as long as only the zeroth Landau level is populated. However, the situation is changed at higher densities when electrons jump from the zeroth Landau level to the first level. This leads to jumps in mass and atomic numbers in nuclei as noted for the SLy4 set. Further, the variation of parameters for nucleon-nucleon interaction affects mass and atomic numbers of nuclei as it is evident from the figure. We also note that the free energy of the ground state matter in strong magnetic fields is reduced and becomes more bound compared with the field free case.

## 5 Shear Mode Oscillations in Magnetars

Giant x-ray flares caused by the tremendous magnetic stress on the crust of magnetars were observed in several cases. Star quakes associated with these giant flares excite seismic oscillations. Quasi-periodic oscillations (QPOs) were found in the decaying tail of giant flares from SGR 1900+14 and SGR 1806-20. Those QPOs were identified as shear mode oscillations of magnetar crusts [3, 22]. Frequencies of the observed QPOs ranged from 18 to 1800 Hz.

Shear mode frequencies are sensitive to the shear modulus of neutron star crust. The shear modulus is again strongly dependent on the composition of neutron star crust. It might be possible to constrain the properties of neutron star crusts by studying the observed frequencies of QPOs. Torsional shear mode oscillations were investigated both in Newtonian gravity [23, 24] and general relativity [25–27]. In both cases, it was assumed that the magnetised crust was decoupled from the fluid core.

Here we describe the calculation of shear mode frequencies adopting the model of Sotani et al. [26]. In this case, we study torsional shear oscillations of spherical and non-rotating relativistic stellar models. The metric used here has the form,

$$ds^2 = -e^{2\Phi}dt^2 + e^{2\Lambda}dr^2 + r^2\left(d\theta^2 + \sin^2\theta d\phi^2\right) . \tag{22}$$

The equilibrium models are obtained by solving Tolman-Oppenheimer-Volkoff equation. Next the equilibrium star is assumed to be endowed with a strong dipole magnetic field [26]. The deformation in the equilibrium star for magnetic fields $\sim 10^{16}$ G is neglected. Torsional shear modes are the results of material velocity oscillations. These modes are incompressible and do not result in density perturbation in equilibrium stars. Consequently, this leads to negligible metric perturbations and justifies the use of the relativistic Cowling approximation [26]. The relevant perturbed matter quantity for shear modes is the $\phi$-component of the perturbed four velocity $\partial u^\phi$ [26]

$$\partial u^\phi = e^{-\phi}\partial_t \mathcal{Y}(t,r)\frac{1}{\sin\theta}\partial_\theta P_l(\cos\theta) , \tag{23}$$

where $\partial_t$ and $\partial_\theta$ correspond to partial derivatives with respect to time and $\theta$, respectively, $P_l(\cos\theta)$ is the Legendre polynomial of order $\ell$ and $\mathcal{Y}(t,r)$ is the angular displacement of the matter. The perturbation equation is obtained from the linearised equation of motion. Finally, we estimate eigenfrequencies by solving two first order differential equations Eqs. (69) and (70) of Sotani et al. [26].

Now we study the dependence of shear mode frequencies on the compositions of magnetised crusts which are already described in Sects. 3 and 4. Earlier calculations were performed with non-magnetised crusts [3, 26–28]. One important input for the shear mode calculation is the knowledge of shear modulus of the magnetised crust. Here we adopt the expression of shear modulus as given by [29, 30]

$$\mu = 0.1194\frac{n_i(Ze)^2}{a} , \tag{24}$$

where $a = 3/(4\pi n_i)$, $Z$ is the atomic number of a nucleus and $n_i$ is the ion density. This zero temperature form of the shear modulus was obtained by assuming a bcc lattice and performing directional averages [31]. Later the dependence of the shear modulus on temperature was investigated with Monte Carlo sampling technique by Strohmayer et al. [30]. However we use the zero temperature shear modulus of Eq. (24) in this calculation.

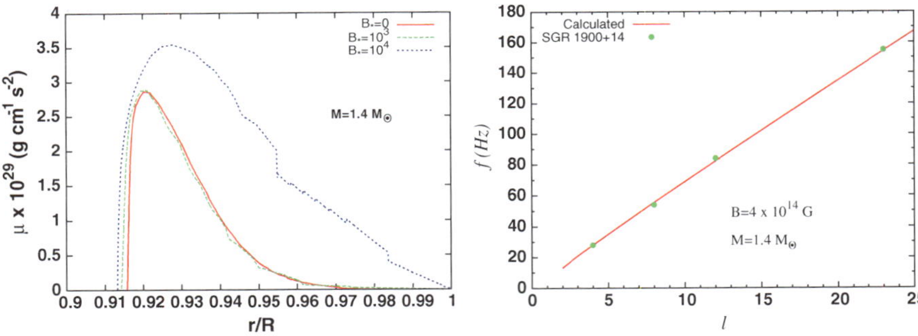


**Fig. 3** Shear modulus is plotted as a function of normalised distance for different magnetic field strengths (*left panel*) and shear mode frequencies are plotted with different $\ell$ values for a neutron star mass of 1.4 $M_\odot$ and $B = 4 \times 10^{14}$ G (*right panel*)

We calculate the shear modulus using Eq. (24) and the compositions and equations of state of magnetised crusts obtained in Sects. 3 and 4. This is shown as a function of normalised distance with respect to radius ($R$) of the star for different field strengths $B = 0$, $B_* = 10^3$ and $B_* = 10^4$ and a neutron star mass of 1.4 $M_\odot$ in the left panel of Fig. 3. Shear modulus increases initially with decreasing distance and drops to zero at the crust-core boundary. For $B_* = 10^4$ i.e. $4.414 \times 10^{17}$ G or more, the shear modulus is enhanced appreciably compared with the zero field case.

It was argued that shear mode frequencies are sensitive to shear modulus [3, 28]. We perform our calculation for shear mode frequencies using the model of Sotani et al. [26] and the shear modulus of magnetised crusts as described above. We calculate fundamental shear mode frequencies for a neutron star mass of 1.4 $M_\odot$ as well as magnetic fields as high as $4.414 \times 10^{17}$ G. When we compare those frequencies involving magnetised crust with those of the non-magnetised crust, we do not find any noticeable change between two cases. For SGR 1900+14 having $B = 4 \times 10^{14}$ G and a neutron star mass of 1.4 $M_\odot$, we show in the right panel of Fig. 3 that the observed QPO frequencies match nicely with frequencies estimated using our magnetised crust model. Further we observe that the first radial overtones calculated with our magnetised crust model have higher frequencies than those calculated with the non-magnetised crust model. This is in agreement with the prediction that the radial overtones are susceptible to magnetic effects [23].

## 6 Summary

We have constructed the model of magnetised neutron star crusts and applied it to shear mode oscillations of magnetars. In particular, we highlighted the effects of strongly quantising magnetic fields on the properties of ground state matter of outer and inner crusts in this article. It is noted that compositions and equations of state of neutron star crusts are significantly altered in strong magnetic fields. Consequently, shear modulus of the crust which is sensitive to the compositions of crusts, is enhanced. We have observed that our model of the magnetised crust might explain the observed shear mode frequencies quite well.

**Acknowledgments** We thank S. K. Samaddar, J. N. De, B. Agrawal, D. Chatterjee, I. N. Mishustin and W. Greiner for many fruitful discussions. We also acknowledge the support under the Research Group Linkage Programme of Alexander von Humboldt Foundation.

## References

1. R.C. Duncan, C. Thompson, Astrophys. J. **392**, L9 (1992)
2. C. Thompson, R.C. Duncan, Mon. Not. Roy. Astron. Soc. **275**, 255 (1995)
3. A.L. Watts, arXiv:1111.0514
4. S. Chakrabarty, D. Bandyopadhyay, S. Pal, Phys. Rev. Lett. **78**, 2898 (1997)
5. D. Bandyopadhyay, S. Chakrabarty, S. Pal, Phys. Rev. Lett. **79**, 2176 (1997)
6. D. Lai, Rev. Mod. Phys. **62**, 629 (2001)
7. D. Lai, S.L. Shapiro, Astrophys. J. **383**, 745 (1991)
8. R. Nandi, D. Bandyopadhyay, J. Phys. Conf. Ser. **312**, 042016 (2011)
9. G. Audi, A.H. Wapstra, C. Thibault, Nucl. Phys. A **729**, 337 (2003)
10. P. Moller, J.R. Nix, W.D. Myers, W.J. Swiatecki, At. Data Nucl. Data Tables **59**, 185 (1995)
11. R. Nandi, D. Bandyopadhyay, I.N. Mishustin, W. Greiner, Astrophys. J. **736**, 156 (2011)
12. A. Broderick, M. Prakash, J.M. Lattimer, Astrophys. J. **537**, 351 (2000)
13. P. Bonche, S. Levit, D. Vautherin, Nucl. Phys. A **427**, 278 (1984)
14. P. Bonche, S. Levit, D. Vautherin, Nucl. Phys. A **436**, 265 (1985)
15. E. Suraud, Nucl. Phys. A **462**, 109 (1987)
16. H. Krivine, J. Treiner, O. Bohigas, Nucl. Phys. A **336**, 115 (1980)
17. M. Brack, C. Guet, H.B. Hâkansson, Phys. Rep. **123**, 275 (1985)
18. J.R. Stone, J.C. Miller, R. Koncewicz, P.D. Stevenson, M.R. Strayer, Phys. Rev. C **68**, 034324 (2003)
19. T. Sil, J.N. De, S.K. Samaddar, X. Vinas, M. Centelles, B.K. Agrawal, S.K. Patra, Phys. Rev. C **66**, 045803 (2002)
20. E. Chabanat et al., Nucl. Phys. A **635**, 231 (1998)
21. B.K. Agrawal, S. Shlomo, V. Kim Au, Phys. Rev. C **68**, 031304 (2003)
22. A.L. Watts, T.E. Strohmayer, Adv. Space. Res. **40**, 1446 (2007)
23. A.L. Piro, Astrophys. J. **634**, L153 (2005)
24. P.N. McDermott, H.M. van Horn, C.J. Hansen, Astrophys. J. **325**, 725 (1988)
25. B.L. Schumaker, K.S. Thorne, Mon. Not. R. Astron. Soc. **203**, 457 (1983)
26. H. Sotani, K.D. Kokkotas, N. Stergioulas, Mon. Not. Astron. Soc. **375**, 261 (2007)
27. H. Sotani, Mon. Not. Astron. Soc. **417**, L70 (2011)
28. A.W. Steiner, A.L. Watts, Phys. Rev. Lett. **103**, 181101 (2009)
29. S. Ogata, S. Ichimaru, Phys. Rev. A **42**, 4867 (1990)
30. T. Strohmayer, H.M. van Horn, S. Ogata, H. Iyetomi, S. Ichimaru, Astrophys. J. **375**, 679 (1991)
31. P. Haensel, "Neutron Star crusts" in *Lecture Notes in Physics: Physics of Neutron Star Interiors*, ed. by D. Blaschke, N.K. Glendenning, A. Sedrakian, vol. 578 (Springer, Heidelberg, 2001), p. 127

# Generation Model of Particle Physics and the Parity of the Neutral Pion

**Brian Robson**

**Abstract** The chapter emphasizes that the Generation Model is obtained from the Standard Model of particle physics essentially by interchanging the roles of the mass eigenstate and weak eigenstate quarks. In the Generation Model the mass eigenstate quarks of the same generation form weak isospin doublets analogous to the mass eigenstate leptons of the same generation while the weak eigenstate quarks form the constituents of hadrons. This allows a simpler and unified classification scheme in terms of only three conserved additive quantum numbers for both leptons and quarks. This unified classification scheme of the Generation Model makes feasible a composite model of the leptons and quarks, which predicts that the weak eigenstate quarks are mixed-parity states. In the Standard Model pions have parity $P = -1$ and the chapter describes that this value of the parity of pions led to the overthrow of both parity conservation and CP conservation in weak interactions. In the Generation Model pions exist in mixed-parity states leading to an understanding of the apparent CP violation observed by Christenson et al. in the decay of the long-lived neutral kaon.

## 1 Introduction

This chapter introduces a new model of particle physics, which has been developed over the last decade, called the Generation Model (GM) [1–3]. Basically, the GM is obtained from the Standard Model (SM) [4] by making two postulates, which together maintain the same transition probabilities for both leptonic and hadronic processes as the SM so that agreement with experiment is preserved.

The differences between the GM and the SM lead to several new paradigms in particle physics: strange quarks in nucleons [3]; origin of mass [5, 6]; origin of gravity

B. Robson (✉)
Department of Theoretical Physics, Research School of Physics and Engineering, The Australian National University, Canberra ACT 0200, Australia

W. Greiner (ed.), *Exciting Interdisciplinary Physics*,
FIAS Interdisciplinary Science Series, DOI: 10.1007/978-3-319-00047-3_29,

[5, 6] and mixed-parity quark states in hadrons [3, 7–9]. However, the discussion in this chapter is restricted to showing that the GM provides an understanding of charge-conjugation–parity (CP) symmetry in the decay of the long-lived neutral kaon, which is absent in the SM.

Section 2 demonstrates that the intrinsic parity of pions played a decisive role in the overthrow of both parity conservation in 1957 and CP conservation in 1964 within the framework of the SM. Section 3 presents the classification of the elementary matter particles of the SM in terms of additive quantum numbers and discusses the main problems arising from this classification. Section 4 introduces the GM and indicates how the simpler and unified GM classification scheme overcomes the problems inherent in the SM. Finally Sect. 5 states the conclusions.

## 2 Parity of Pions

In the SM the pions are assumed to have parity $P = -1$. This value was first obtained in 1954 by Chinowsky and Steinberger [10] using the capture of negatively charged pions by deuterium to form two neutrons: $\pi^- + D \rightarrow 2n$. This was prior to the quark model so that in the analysis of the experiment, the pion, the proton and the neutron were each assumed to be elementary particles with no substructure.

Following the adoption of the quark model [11, 12] as part of the SM, the parity of the pions remained accepted as $P = -1$. In the quark model, the pions were proposed to be combinations of up ($u$) and down ($d$) quarks and their antiparticles $\bar{u}$ and $\bar{d}$:

$$\pi^+ \equiv [u\bar{d}] \; , \; \pi^0 \equiv ([u\bar{u}] - [d\bar{d}])/\sqrt{2} \; , \; \pi^- \equiv [d\bar{u}] \; . \tag{1}$$

Assuming the quarks have intrinsic parity $P_q = +1$ while the corresponding antiquarks have intrinsic parity $P_{\bar{q}} = -1$, all the three pions have parity $P = -1$. This follows since parity is a multiplicative operator in quantum mechanics and consequently e.g. $P\pi^- \equiv P[d\bar{u}] = P_d P_{\bar{u}}[d\bar{u}] = -[d\bar{u}] \equiv -\pi^-$.

This value of the parity of pions led to the overthrow of both parity conservation and CP conservation in weak interactions. We shall now describe briefly how this came about.

### *2.1 Tau-Theta Puzzle*

In the period 1947–1953 several new particles were discovered. In particular one charged meson, named the tau meson [13], decayed to three charged pions, while another charged meson, named the theta meson [14], decayed to two pions:

$$\tau^+ \rightarrow \pi^+ + \pi^+ + \pi^- \; , \; \theta^+ \rightarrow \pi^+ + \pi^0 \; . \tag{2}$$

Analysis of the decays of both these particles indicated that they had closely similar lifetimes and masses. These properties suggested that the tau and theta mesons were simply two different decay modes of the *same* particle.

In 1953–1954 Dalitz [15] suggested that a study of the energy distribution of the three pions in tau meson decays would provide information about the spin and parity of the tau meson. Such analyses, assuming $P = -1$ for pions, led to the conclusion that the tau meson had $J^P = 0^-$ or $2^-$. On the other hand, the two-body decay of the theta meson gave the opposite parity for the two even spin values, i.e. $J^P = 0^+$ or $2^+$. This indicated that the tau and theta mesons were *different* particles.

This tau-theta puzzle was resolved in 1956 by Lee and Yang [16], who suggested that parity may be violated in weak interactions. This suggestion was rapidly confirmed in 1957 in three independent experiments [17–19]. These experiments indicated that the tau and theta mesons were indeed the same meson (now called the $K^+$) and also showed that charge-conjugation (C) was violated in weak interactions.

## 2.2 Neutral Kaon Mixing

In 1955 Gell-Mann and Pais [20] considered the behavior of neutral particles under the charge-conjugation operator. In particular they considered the $K^0$ meson and realized that unlike the photon and the neutral pion, which transform into themselves under the C operator so that they are their own antiparticles, the antiparticle of the $K^0$ meson (strangeness $S = +1$), $\bar{K}^0$, was a distinct particle, since it had a different strangeness quantum number ($S = -1$). They concluded that the two neutral mesons, $K^0$ and $\bar{K}^0$, are degenerate particles that exhibit unusual properties, since they can transform into each other via weak interactions such as

$$K^0 \rightleftharpoons \pi^+\pi^- \rightleftharpoons \bar{K}^0. \tag{3}$$

In order to treat this novel situation, Gell-Mann and Pais suggested that it was more convenient to employ different particle states, rather than $K^0$ and $\bar{K}^0$, to describe neutral kaon decay. They suggested the following representative states:

$$K_1^0 = (K^0 + \bar{K}^0)/\sqrt{2}\,, \quad K_2^0 = (K^0 - \bar{K}^0)/\sqrt{2}\,, \tag{4}$$

and concluded that these particle states must have different decay modes and lifetimes. In particular they concluded that $K_1^0$ could decay to two charged pions, while $K_2^0$ would have a longer lifetime and more complex decay modes. This conclusion was based upon the conservation of C in the weak interaction processes: both $K_1^0$ and the $\pi^+\pi^-$ system are even (i.e. $C = +1$) under the C operation. On the other hand $CK_2^0 = -K_2^0$.

**Table 1** SM additive quantum numbers for leptons

| Particle | $Q$ | $L$ | $L_\mu$ | $L_\tau$ |
|---|---|---|---|---|
| $\nu_e$ | 0 | 1 | 0 | 0 |
| $e^-$ | $-1$ | 1 | 0 | 0 |
| $\nu_\mu$ | 0 | 1 | 1 | 0 |
| $\mu^-$ | $-1$ | 1 | 1 | 0 |
| $\nu_\tau$ | 0 | 1 | 0 | 1 |
| $\tau^-$ | $-1$ | 1 | 0 | 1 |

The particle-mixing theory of Gell-Mann and Pais was confirmed [21] in 1957 by experiment, in spite of the incorrect assumption of C-invariance in weak interaction processes.

This led to a suggestion by Landau [22] that the weak interactions may be invariant under the combined operation CP, although both C and P are individually maximally violated.

Landau's suggestion implied that the Gell-Mann–Pais model of neutral kaons would still apply if the states, $K_1^0$ and $K_2^0$, were eigenstates of CP with eigenvalues $+1$ and $-1$, respectively. Since the charged pions were considered to have intrinsic parity $P = -1$, it was clear that only the $K_1^0$ state could decay to two charged pions, if CP was conserved.

The suggestion of Landau was accepted for several years since it nicely restored some degree of symmetry in weak interaction processes. However, in 1964 the surprising discovery was made by Christenson et al. [23] that the long-lived neutral kaon could decay to two charged pions. The observed violation of CP conservation turned out to be very small ($\approx 0.2$ %) compared with the maximal violations ($\approx 100$ %) of both C and P conservation separately. Indeed the very smallness of the apparent CP violation led to a variety of suggestions explaining it in a CP-conserving way [24, 25]. However, these efforts were unsuccessful and CP violation in weak interactions was accepted.

## 3 Standard Model (SM)

Let us now discuss briefly the Standard Model (SM) [4] of particle physics. Tables 1 and 2 show the additive quantum numbers allotted to the six leptons (electron ($e^-$), muon ($\mu^-$), tau particle ($\tau^-$) and their three associated neutrinos, $\nu_e$, $\nu_\mu$, $\nu_\tau$) and the six quarks (up ($u$), down ($d$), charmed ($c$), strange ($s$), top ($t$) and bottom ($b$)), respectively, which constitute the elementary matter particles of the SM. For the leptons we have: charge $Q$, lepton number $L$, muon lepton number $L_\mu$ and tau lepton number $L_\tau$. For the quarks we have: charge $Q$, baryon number $A$, strangeness $S$, charm $C$, bottomness $B$ and topness $T$. For each particle additive quantum number $N$, the corresponding antiparticle has the additive quantum number - N.

**Table 2** SM additive quantum numbers for quarks

| Particle | $Q$ | $A$ | $S$ | $C$ | $B$ | $T$ |
|---|---|---|---|---|---|---|
| $u$ | $+\frac{2}{3}$ | $\frac{1}{3}$ | 0 | 0 | 0 | 0 |
| $d$ | $-\frac{1}{3}$ | $\frac{1}{3}$ | 0 | 0 | 0 | 0 |
| $c$ | $+\frac{2}{3}$ | $\frac{1}{3}$ | 0 | 1 | 0 | 0 |
| $s$ | $-\frac{1}{3}$ | $\frac{1}{3}$ | $-1$ | 0 | 0 | 0 |
| $t$ | $+\frac{2}{3}$ | $\frac{1}{3}$ | 0 | 0 | 0 | 1 |
| $b$ | $-\frac{1}{3}$ | $\frac{1}{3}$ | 0 | 0 | $-1$ | 0 |

It should be noted that, except for charge, leptons and quarks are allotted different kinds of additive quantum numbers so that this classification is *non-unified*. Each of the additive quantum numbers is conserved in any interaction, except for $S$, $C$, $B$ and $T$, which can undergo a change of one unit in weak interactions.

The six leptons and six quarks are all spin-$\frac{1}{2}$ particles and fall naturally into three families or generations: (i) $\nu_e$, $e^-$, $u$, $d$ ; (ii) $\nu_\mu$, $\mu^-$, $c$, $s$ ; (iii) $\nu_\tau$, $\tau^-$, $t$, $b$ . Each generation consists of particles which have similar properties apart from mass: two leptons with charges $Q = 0$ and $Q = -1$ and two quarks with charges $Q = +\frac{2}{3}$ and $Q = -\frac{1}{3}$. The masses of the particles increase significantly with each generation with the possible exception of the neutrinos, whose very small masses have yet to be determined.

### 3.1 Basic Problem Inherent in SM

The basic problem with the SM is the classification of its elementary particles employing a diverse complicated scheme of additive quantum numbers (Tables 1 and 2), some of which are not conserved in weak interaction processes; and at the same time failing to provide any physical basis for this scheme.

A good analogy of the SM situation is the Ptolemaic model of the universe, based upon a stationary Earth at the center surrounded by a rotating system of crystal spheres refined by the addition of epicycles (small circular orbits) to describe the peculiar movements of the planets around the Earth. Although the Ptolemaic model yielded an excellent description of the observed movements of the constituents of the universe, it is a complicated diverse scheme for predicting the movements of the Sun, Moon, planets and the stars around a stationary Earth and unfortunately provides no understanding of these complicated movements.

### *3.2 Universality of Charge-Changing (CC) Weak Interactions*

Another problem with the SM concerns the method it employs to accommodate the *universality* of the CC weak interactions [26, 27]. The CC weak interactions are mediated by the $W^+$ and $W^-$ vector bosons, which have zero additive quantum numbers apart from charge.

In the SM, the observed universality of the CC weak interactions in the lepton sector is described by assuming that each mass eigenstate charged lepton forms a weak isospin doublet ($i = \frac{1}{2}$) with its respective neutrino, i.e. $(\nu_e, e^-)$, $(\nu_\mu, \mu^-)$, $(\nu_\tau, \tau^-)$, with each doublet having the third component of weak isospin $i_3 = (+\frac{1}{2}, -\frac{1}{2})$. In addition each doublet is associated with a different lepton number so that there are no CC weak interaction transitions between generations.

Restricting the discussion to only the first two generations for simplicity, this means that $\nu_e$ and $\nu_\mu$ interact with $e^-$ and $\mu^-$, respectively, with the *full* strength of the CC weak interaction but $\nu_e$ and $\nu_\mu$ do *not* interact at all with $\mu^-$ and $e^-$, respectively. This is guaranteed by the conservation of lepton numbers.

On the other hand the universality of the CC weak interactions in the quark sector is treated differently in the SM. Again for simplicity, restricting the discussion to only the first two generations, it is assumed that the $u$ and $c$ quarks form weak isospin doublets with so-called weak eigenstate quarks $d'$ and $s'$, respectively. These weak eigenstate quarks are linear superpositions of the mass eigenstate quarks ($d$ and $s$):

$$d' = d \cos\theta_c + s \sin\theta_c \ , \quad s' = -d \sin\theta_c + s \cos\theta_c \ , \tag{5}$$

where $\theta_c$ is a mixing angle introduced by Cabibbo [28] in 1963 into the transition amplitudes prior to the development of the quark model in 1964.

The SM assumes that $u$ and $c$ interact with $d'$ and $s'$, respectively, with the *full* strength of the CC weak interaction and that $u$ and $c$ do *not* interact at all with $s'$ and $d'$, respectively. However, this latter assumption is dubious since, unlike the lepton sector, there are no conserved quantum numbers to guarantee this.

It should be noted that the extension of the above discussion to all three generations is straightforward [29]. In this case, the quark-mixing parameters correspond to the Cabibbo-Kobayashi-Maskawa (CKM) matrix elements [30], which indicate that inclusion of the third generation would have minimal effect on the above discussion.

## 4 Generation Model (GM)

The Generation Model (GM) of particle physics has been developed over the last decade. In the initial chapter [1] a new classification of the elementary particles, the six leptons and the six quarks, of the SM was proposed. This classification was based upon the use of only three additive quantum numbers: charge ($Q$), particle number ($p$) and generation quantum number ($g$), rather than the nine additive quantum numbers

**Table 3** GM additive quantum numbers for leptons and quarks

| Particle | $Q$ | $p$ | $g$ | Particle | $Q$ | $p$ | $g$ |
|---|---|---|---|---|---|---|---|
| $\nu_e$ | 0 | $-1$ | 0 | $u$ | $+\frac{2}{3}$ | $\frac{1}{3}$ | 0 |
| $e^-$ | $-1$ | $-1$ | 0 | $d$ | $-\frac{1}{3}$ | $\frac{1}{3}$ | 0 |
| $\nu_\mu$ | 0 | $-1$ | $\pm 1$ | $c$ | $+\frac{2}{3}$ | $\frac{1}{3}$ | $\pm 1$ |
| $\mu^-$ | $-1$ | $-1$ | $\pm 1$ | $s$ | $-\frac{1}{3}$ | $\frac{1}{3}$ | $\pm 1$ |
| $\nu_\tau$ | 0 | $-1$ | $0, \pm 2$ | $t$ | $+\frac{2}{3}$ | $\frac{1}{3}$ | $0, \pm 2$ |
| $\tau^-$ | $-1$ | $-1$ | $0, \pm 2$ | $b$ | $-\frac{1}{3}$ | $\frac{1}{3}$ | $0, \pm 2$ |

(see Tables 1 and 2) of the SM. Thus the new classification is both simpler and *unified* in that leptons and quarks are assigned the same kind of additive quantum numbers unlike those of the SM.

Another feature of the new classification scheme is that all three additive quantum numbers, $Q$, $p$ and $g$, are required to be conserved in all leptonic and hadronic processes. In particular the generation quantum number $g$ is strictly conserved in weak interactions unlike some of the quantum numbers, e.g. strangeness $S$, of the SM. This latter requirement led to a new treatment of quark mixing in hadronic processes [1, 3], which will be discussed in Sect. 4.1.

The development of the GM unified classification scheme indicated that leptons and quarks are intimately related and led to the development of composite versions of the GM [6, 31].

Table 3 displays a set of three additive quantum numbers for the unified classification of the leptons and quarks corresponding to the current composite GM [6]. As for Tables 1 and 2, the corresponding antiparticles have the opposite sign for each particle additive quantum number. Each generation of leptons and quarks has the same set of values for the additive quantum numbers $Q$ and $p$. The generations are differentiated by the generation quantum number $g$, which in general can have multiple values. The latter possibilities arise from the composite nature of the leptons and quarks in the composite GM.

## 4.1 Conservation of Generation Quantum Number

The conservation of the generation quantum number in weak interactions was only achieved by making two postulates, which requires the GM to differ fundamentally from the SM in two ways. Again, for simplicity, the discussion is restricted to the first two generations.

Firstly, the GM postulates that it is the *mass* eigenstate quarks of the same generation, which form weak isospin doublets: $(u, d)$ and $(c, s)$. Thus the GM assumes that the $u$ and $c$ quarks interact with the $d$ and $s$ quarks, respectively, with the *full* strength of the CC weak interaction and do *not* interact at all with the $s$ and $d$ quarks, respectively. These properties are guaranteed by the conservation of generation quantum number.

Secondly, the GM postulates that hadrons are composed of *weak* eigenstate quarks such as $d'$ and $s'$ rather than the corresponding *mass* eigenstate quarks, $d$ and $s$, as in the SM. Essentially, in the GM the roles of the mass eigenstate quarks and the weak eigenstate quarks are *interchanged* from that in the SM.

## 4.2 Composite Generation Model

The unified classification scheme of the GM makes feasible a composite version of the GM [6, 31]. This is not possible in terms of the non-unified classification scheme of the SM, involving different additive quantum numbers for leptons than for quarks and the non-conservation of some additive quantum numbers, such as strangeness, in the case of quarks.

In the composite GM, the leptons and quarks are not elementary particles as in the SM but are composed of rishons and/or antirishons [6]. For the present purposes, it should be noted that this composite model predicts that the down and strange quarks have *opposite intrinsic parities*: the $d$ quark consists of two rishons and one antirishon while the $s$ quark consists of three rishons and two antirishons and it is assumed that rishons and antirishons have opposite intrinsic parities. This is important because it implies that the weak eigenstate quarks, $d'$ and $s'$, are *mixed-parity* states so that pions exist in mixed-parity states.

## 4.3 Pions in GM

In the GM the pions consist of weak eigenstate quarks:

$$\pi^+ \equiv [u\bar{d}'] \,,\; \pi^0 \equiv ([u\bar{u}] - [d'\bar{d}'])/\sqrt{2} \,,\; \pi^- \equiv [d'\bar{u}] \,. \tag{6}$$

rather than mass eigenstate quarks as in the SM (Eq. (1)).

Recently it has been demonstrated [8] that the early experiment of Chinowsky and Steinberger [10] is *indeterminate* with respect to the determination of the intrinsic parity of the negatively charged pion, if the pion, neutron and proton have a complex substructure as in the composite GM: the experiment is also compatible with the mixed-parity nature of the charged pions.

Similarly, it has been shown [9] that the recent determination [32] of the parity of the neutral pion, using the double Dalitz decay $\pi^0 \to e^+e^-e^+e^-$, is also compatible with the mixed-parity nature of the neutral pion predicted by the composite GM.

Analysis of this experiment placed a limit on scalar contributions to the decay amplitude of the $\pi^0$ of less than 3.3 %. The GM predicts a scalar contribution to the decay amplitude of about 2.5 %, which is of the same order of magnitude as the experimental result, suggesting that further experimentation may determine a non-zero scalar contribution to the decay amplitude.

Thus the mixed-parity natures of both charged and neutral pions are compatible with experiment.

### *4.4 Conservation of CP in Neutral Kaon System*

Recently, Morrison and Robson [7] have demonstrated that the indirect CP violation observed by Christenson et al. [23] for the $K^0 - \bar{K}^0$ system can be described in terms of mixed-quark states in hadrons. In the GM, within the two generation approximation, the long-lived neutral kaon exists in a $CP = -1$ eigenstate and the mixed-parity of the charged pions provides the two-charged pion system $(\pi^+\pi^-)$ with a small component of a $CP = -1$ eigenstate so that the long-lived neutral kaon can decay to two charged pions without CP violation. Thus the GM predicts that essentially the decay of the long-lived neutral kaon into two charged pions arises from the mixed-parity of the charged pions and *not* from CP violation.

## 5 Conclusion

The GM has been developed from the SM essentially by *interchanging the roles* of the mass eigenstate and weak eigenstate quarks from that in the SM. In the GM the mass eigenstate quarks of the same generation form weak isospin doublets analogous to the mass eigenstate leptons of the same generation while the weak eigenstate quarks form the constituents of hadrons. This allows a simpler and unified classification scheme in terms of only three conserved additive quantum numbers for both leptons and quarks.

This unified classification scheme of the GM makes feasible a composite model of the leptons and quarks, which predicts that the weak eigenstate quarks of the GM are mixed-parity states. Consequently, in the GM, pions exist in mixed-parity states, which to date are compatible with experiment. The mixed-parity nature of the neutral pion predicted by the GM suggests the existence of a scalar contribution of about 2.5 % to the double Dalitz decay amplitude for $\pi^0 \rightarrow e^+e^-e^+e^-$, which further experimentation may be able to detect. Finally, the mixed-parity nature of the charged pions allows the observed decay of the long-lived neutral kaon to be described with the *conservation* of CP symmetry, indicating that the observed decay of the long-lived neutral kaon into two charged pions by Christenson et al. [23] in 1964 does not correspond to indirect CP violation.

## References

1. B.A. Robson, A generation model of the fundamental particles. Int. J. Mod. Phys. E **11**, 555–566 (2002)
2. B.A. Robson, Relation between strong and weak isospin. Int. J Mod. Phys. E **13**, 999–1016 (2004)
3. P.W. Evans, B.A. Robson, Comparison of quark mixing in the standard and generation models. Int J. Mod. Phys. E **15**, 617–625 (2006)
4. E. Gottfried, V.F. Weisskopf, *Concepts of Particle Physics*, vol. 1 (Oxford University Press, New York, 1984)
5. B.A. Robson, The generation model and the origin of mass. Int. J. Mod. Phys. E **18**, 1773–1780 (2009)
6. B.A. Robson, A quantum theory of gravity based on a composite model of leptons and quarks. Int. J. Mod. Phys. E **20**, 733–745 (2011)
7. A.D. Morrison, B.A. Robson, $2\pi$ decay of the $K_L^0$ meson without CP violation. Int. J. Mod. Phys. E **18**, 1825–1830 (2009)
8. B.A. Robson, Parity of charged pions. Int. J. Mod. Phys. E **20**, 1677–1686 (2011)
9. B.A. Robson, Parity of neutral pion. Int. J. Mod. Phys. E **20**, 1961–1965 (2011)
10. W. Chinowsky, J. Steinberger, Absorption of negative pions in deuterium: parity of pion. Phys. Rev. **95**, 1561–1564 (1954)
11. M. Gell-Mann, A schematic model of baryons and mesons. Phys. Lett. **8**, 214–215 (1964)
12. G. Zweig, CERN Reports 8182/TH 401 and 8419/TH 412 (1964), unpublished, but reproduced in D.B. Lichtenberg, S.P. Rosen, Developments in the Quark Theories of Hadrons, Vol. 1. (Hadronic Press, Nonamtum Mass, 1980)
13. R. Brown et al., Observations with electron-sensitive plates exposed to cosmic radiation. Nature **163**, 82–87 (1949)
14. G.G. Rochester, C.C. Butler, Evidence for the existence of new unstable elementary particles. Nature **160**, 855–857 (1947)
15. R.H. Dalitz, Decay of $\tau$ mesons of known charge. Phys. Rev. **94**, 1046–1051 (1954)
16. T.D. Lee, C.N. Yang, Question of parity conservation in weak interactions. Phys. Rev. **104**, 254–258 (1956)
17. C.S. Wu, E. Ambler, R.W. Hayward, D.D. Hoppes, R.P. Hudson, Experimental test of parity conservation in beta decay. Phys. Rev. **105**, 1413–1415 (1957)
18. R.L. Garwin, L.M. Lederman, M. Weinrich, Observations of the failure of conservation of parity and charge conjugation in meson decays: the magnetic moment of the free muon. Phys. Rev. **105**, 1415–1417 (1957)
19. J.I. Friedman, V.L. Telegdi, Nuclear emulsion evidence for parity nonconservation in the decay chain $\pi^+ - \mu^+ - e^+$. Phys. Rev. **105**, 1681–1682 (1957)
20. M. Gell-Mann, A. Pais, Behavior of neutral particles under charge conjugation. Phys. Rev. **97**, 1387–1389 (1955)
21. F. Eisler, R. Plano, N. Samios, M. Schwartz, J. Steinberger, Systematics of $\Lambda^0$ and $\theta^0$ decay. Nuovo Cimento **5**, 1700–1710 (1957)
22. L.D. Landau, On the conservation laws for weak interactions. Nucl. Phys. **3**, 127–131 (1957)
23. J.H. Christenson, J.W. Cronin, V.L. Fitch, R. Turlay, Evidence for the $2\pi$ decay of the $K_2^0$ meson. Phys. Rev. Lett. **13**, 138–140 (1964)
24. P.K. Kabir, *The CP Puzzle: Strange Decays of the Neutral Kaon* (Academic Press, London, 1968)
25. A. Franklin, *The Neglect of Experiment* (Cambridge University Press, Cambridge UK, 1986)
26. R.P. Feynman, M. Gell-Mann, Theory of the Fermi interaction. Phys. Rev. **109**, 193–198 (1958)
27. E.C.G. Sudarshan, R.E. Marshak, Chirality invariance and the universal Fermi interaction. Phys. Rev. **109**, 1860–1862 (1958)
28. N. Cabibbo, Unitary symmetry and leptonic decays. Phys. Rev. Lett. **10**, 531–533 (1963)
29. M. Kobayashi, T. Maskawa, CP-violation in renormalizable theory of weak interaction. Prog. Theor. Phys. **49**, 652–657 (1973)

30. D.H. Perkins, *Introduction to High Energy Physics*, 4th edn. (Cambridge University Press, Cambridge UK, 2000)
31. B.A. Robson, A generation model of composite leptons and quarks. Int. J. Mod. Phys. E **14**, 1151–1169 (2005)
32. E. Abouzaid et al., Determination of the parity of the neutral pion via its four-electron decay. Phys. Rev. Lett. **100**, 182001 (2008)

# Fundamental Neutrinos Properties

**Fedor Šimkovic**

**Abstract** After about six decades since the discovery of the neutrino, we have started to understand the role of neutrinos in our world. The discoveries of oscillations of atmospheric, solar, accelerator and reactor neutrinos have opened a new excited era in neutrino physics and represents a big step forward in our knowledge of neutrino properties. The observed small neutrino masses have profound implications for our understanding of the Universe and are now a major focus in astro, particle and nuclear physics and in cosmology. The physics community worldwide is embarking on the challenging problem, finding whether neutrinos are indeed Majorana particles (i.e., identical to its own antiparticle) as many particle models suggest or Dirac particles (i.e., is different from its antiparticle). The search for the $0\nu\beta\beta$-decay represents the new frontiers of neutrino physics, allowing to determine the Majorana nature of neutrinos and to fix the neutrino mass scale and possible CP violation effects, which could explain the matter-antimatter asymmetry in the Universe.

## 1 Introduction

Until the early part of the twentieth century neutrinos were unknown. On 4 December 1930 Wolfgang Pauli postulated a new particle in an attempt not to abandon the energy law and angular momentum conservation in the nuclear beta decay. Enrico Fermi called this new particle the neutrino and incorporated it in his theory of weak interaction.

F. Šimkovic (✉)
Department of Nuclear Physics and Biophysics, Comenius University, Bratislava, Slovakia
e-mail: simkovic@fmph.uniba.sk

F. Šimkovic
Laboratory of Theoretical Physics, JINR, Dubna, Russia
e-mail: simkovic@theor.jinr.ru

W. Greiner (ed.), *Exciting Interdisciplinary Physics*,
FIAS Interdisciplinary Science Series, DOI: 10.1007/978-3-319-00047-3_30,

Three types or flavor of neutrinos are known. There is strong evidence that no additional neutrinos exist, unless their properties are unexpectedly very different from the known types. According to the charged leptons produced with the neutrinos they are referred to as electron neutrino $\nu_e$, muon neutrino $\nu_\mu$, or tau neutrino $\nu_\tau$. In the Standard Model of Particle Physics they belong to the family of leptons. The three different neutrinos are complemented by anti-neutrinos, which have the same mass as neutrinos but inverse characteristics. As neutrinos are neutral and carry no charge they can be their own anti-particles (Majorana neutrinos).

Neutrinos are very special particles, which allow unique insights into fundamental questions in particle physics. The symmetries of Standard model of particle physics are associated with total baryon and lepton number and lepton flavor conservation. But, there is no known fundamental principle, which would require this. While massless neutrinos are part of the Standard model, from the observation of neutrino oscillations we now know that at least two neutrino species are massive, i.e. the lepton flavor is violated. Thus, neutrinos constitute direct evidence for physics beyond the Standard model. The smallness of neutrino masses can be explained by the interplay of weak-scale Dirac masses with much larger Majorana masses within the so-called sees-saw mechanism. In this way neutrinos are potentially a window to very high mass scales, directly inaccessible to foreseeable colliders.

There are many open questions about neutrinos that need both theoretical and experimental exploration. The subject of interest is the absolute mass scale, mixing, the Majorana or the Dirac nature of neutrinos, their electromagnetic properties and the possible existence of CP violation in the leptonic sector. A large enough CP violation is necessary to create the asymmetry between matter and anti-matter in the early Universe, and a large CP violation discovery in neutrino oscillations or neutrinoless double beta decay would support the evidence for the role of neutrinos in this mechanism.

The puzzle of absolute mass scale of neutrinos can be solved by the tritium end-point distortion measurement, from the evaluation of large scale structure in the Universe and with next generation of the neutrinoless double beta decay experiments. The question, whether neutrinos and anti-neutrinos are distinct (Dirac particle), or in fact the same particle (Majorana particle) is of great importance. Neutrinoless double beta-decay is the most important source of information about this problem. There is a chance that this process could be observed in the foreseeable future.

From the radioactive decay of atoms to the evolution of the Universe, neutrinos are central to our understanding of particle physics and astrophysics.

## 2 Sources of Neutrinos

In number, neutrinos exceed the constituents of ordinary matter (electrons, protons, neutrons) by a factor of ten billion. There are about $10^{87}$ neutrinos per flavor in the visible Universe, which corresponds to the number density per flavor of relic (anti-)neutrinos in average over the Universe about $56\,\mathrm{cm}^{-3}$. They were first created in the Big Bang, in the beginning of the Universe, and continue to be created in nuclear reactions and particle interactions. They travel through the Universe with close to the speed of light.

Relic neutrinos has not yet been confirmed by direct observations due to their small energy. This represents a challenging problem of modern cosmology and experimental astroparticle physics.

The Sun, where intense nuclear reactions are constantly occurring, emits $2 \times 10^{38}$ electron-neutrinos every second. They have an energy between 0 and 20 MeV, depending of the type of solar nuclear reaction they come. There are about 70 billion neutrinos per square centimeter and second streaming through the Earth from the Sun.

Neutrinos govern the dynamics of supernovae, and hence the production of heavy elements in the Universe. When the mass of the inert core exceeds the Chandrasekhar limit of about 1.4 solar masses, electron degeneracy alone is no longer sufficient to counter gravity. When a massive star at the end of its life collapses to a neutron star, it radiates almost all of its binding energy in the form of neutrinos, most of which have energies in the range 10–30 MeV. These neutrinos come in all flavors, and are emitted over a timescale of several tens of seconds. The total number of neutrinos spewed out by the supernova SN 1987A at Large Magellanic Cloud (distance of 168000 light-years) exceeded $10^{58}$ while over a period of ten seconds $5 \times 10^{28}$ of them passed right through the Earth and 24 of them were detected at underground laboratories.

Neutrinos are created in the collision of primary cosmic rays (typically protons) with nuclei in the upper atmosphere of Earth. This creates a shower of hadrons, mostly pions. The pions decay to a muon and a muon neutrino. The muons decay to an electron, another muon neutrino, and an electron neutrino. The averaged energy of atmospheric neutrinos is about few GeV. The first detections of atmospheric neutrinos were made in the early sixties in deep mines by Reines et al. in South Africa. Currently, atmospheric neutrinos with large energy range from hundreds of MeV up to TeV region are detected in various underground (SuperKamiokande), underwater (Baikal experiment, ANTARES, NEMO) experiments and other (IceCube) experiments.

Neutrinos are produced inside nuclear power plants as by-product of nuclear fissions. Nuclear reactors emit about $5 \times 10^{20}$ antineutrinos per second and per GW of thermal power with a mean energy of 4 MeV. Man-made neutrinos are produced also by using beams of high energy protons generated by large accelerators like the ones at CERN, Fermilab and KEK.

The interior of the Earth is a source of antineutrinos due to natural radioactivity. Radiogenic heat arises mainly from the decay (chains) of $^{238}U$, $^{232}Th$ and $^{40}K$. All these elements produce heat together with antineutrinos, with well fixed ratios heat/neutrinos. The geo-neutrinos has been successfully detected for the first by the KamLAND and Borexino experiments. The direct information of the heat generation mechanism obtained by neutrinos means the opening of a new field of science called "Neutrino Geophysics".

Depending on the source and the environment in which neutrinos are produced, different types of neutrinos are created, and their energies range from 0.0004 eV (Big Bang neutrinos) to $30 \times 10^9$ eV, or higher (at accelerators). Because of their weak interaction with other particles, matter is almost completely transparent to neutrinos and large sensitive detectors are needed to capture them.

## 3 Neutrino Oscillations

Neutrino oscillation is a quantum mechanical phenomenon predicted by Bruno Pontecorvo [1, 2]. He thought that there is an analogy between leptons and hadrons and believed that in the lepton world a phenomenon does exist, which is analogue to the well-known $K^0\,\overline{K}^0$ oscillations. In the neutrino oscillation a neutrino created with a specific lepton flavor (electron, muon or tau) can later be measured to have a different flavor.

Neutrino flavor states $|\nu_\alpha\rangle$ are related to the mass-eigenstates $|\nu_j\rangle$ with masses $m_j$ (j =1, 2, 3) and vice versa by the linear combinations

$$|\nu_\alpha\rangle = \sum_{j=1,2,3} U_{\alpha j}|\nu_j\rangle, \quad |\nu_j\rangle = \sum_{j=1,2,3} U^*_{\alpha j}|\nu_\alpha\rangle, \tag{1}$$

where $U$ is the Pontecorvo-Maki-Nakagawa-Sakata unitary mixing matrix, which can be written as

$$U = \begin{pmatrix} c_{12}c_{13} & c_{13}s_{12} & e^{-i\delta}s_{13} \\ -c_{23}s_{12} - e^{i\delta}c_{12}s_{13}s_{23} & c_{12}c_{23} - e^{i\delta}s_{12}s_{13}s_{23} & c_{13}s_{23} \\ s_{12}s_{23} - e^{i\delta}c_{12}c_{23}s_{13} & -e^{i\delta}c_{23}s_{12}s_{13} - c_{12}s_{23} & c_{13}c_{23} \end{pmatrix} \tag{2}$$

with $c_{ij} \equiv \cos(\theta_{ij})$, $s_{ij} \equiv \sin(\theta_{ij})$. $\theta_{12}, \theta_{13}$ and $\theta_{23}$ and three mixing angles and $\delta$ is the CP-violating phase. If neutrinos are Majorana particles $U$ in Eq. (2) is multiplied by a diagonal phase matrix $P = \mathrm{diag}(e^{i\alpha_1}, e^{i\alpha_2}, e^{i\delta})$, which contains two additional CP-violating Majorana phases $\alpha_1$ and $\alpha_2$:

The transition probability of different neutrino species into each other is

$$P_{\alpha\to\beta}(E, L) = \sum_{j=1,2,3}\sum_{k=1,2,3} U_{\alpha j}U^*_{\beta j}U^*_{\alpha k}U_{\beta k}e^{i\Delta m^2_{kj}L/(2E)} \tag{3}$$

with $c = 1$, $t \approx L$ (distance from source), $\Delta^2_{kj} = m^2_k - m^2_j$. In the two-flavor approximation we get

$$P_{\alpha\to\beta}(E, L) \approx \sin^2(2\theta)\sin^2\left(1.27\Delta m^2_{[eV^2]}L_{[m]}/E_{[MeV]}\right). \tag{4}$$

We note that the probability of measuring a particular flavor for a neutrino varies periodically as it propagates.

Various early measurements of neutrinos produced in the Sun, in the atmosphere, and by accelerators suggested that neutrinos might oscillate from one flavor (electron-, muon-, and tau-) to another expected as a consequence of non-zero neutrino mass. Starting 1998 we have a convincing evidence about the existence of neutrino masses due to SuperKamiokande (atmospheric neutrinos) [3], SNO (solar neutrinos) [4], KamLAND (reactor neutrinos) [5], MINOS (accelerator neutrinos) and other experiments. Neutrinos produced in the atmosphere arrived at the Super-Kamiokande detector from distances of about 40 km (if produced above it) to 12,000 km (if produced on the other side of the Earth). In accordance with theory of neutrino oscillations data were found to be dependent on the zenith angle. The neutrinos from accelerators or nuclear reactors on Earth were detected at significantly smaller distances. The behavior of neutrinos produced over a great distance at the Sun allows us observe effects, which would be invisible with a closer source. We note also that data agree well with the oscillation hypothesis regardless of energy.

With the discovery of neutrino oscillations quite a lot of information regarding the neutrino sector has become available. More specifically we know:

- The mixing angles $\theta_{12}$ and $\theta_{23}$, which are large, and we have both a lower and an upper bound on the small angle $\theta_{13}$.
- We know the mass squared differences:

$$\Delta^2_{\mathrm{SUN}} = \Delta^2_{12} = m_2^2 - m_1^2, \quad \text{and } \Delta^2_{\mathrm{ATM}} = |\Delta^2_{23}| = |m_3^2 - m_2^2|$$

  entering the solar and atmospheric neutrino oscillation experiments. Note that we do not know the absolute scale of the neutrino mass and the sign of $\Delta^2_{23}$.

Currently, the neutrino oscillation parameters are as follows: The MINOS value $\Delta m^2_{\mathrm{ATM}} = (2.43 \pm 0.13) \times 10^{-3}\mathrm{eV}^2$ [6], the global fit values $\Delta m^2_{\mathrm{SUN}} = (7.65^{+0.13}_{-0.20}) \times 10^{-5}\mathrm{eV}^2$ and $\sin^2\theta_{23} = 0.50^{+0.07}_{-0.06}$ [7], the solar-KamLAND value $\tan^2\theta_{12} = 0.452^{+0.035}_{-0.033}$ [5] and the recent T2K and DOUBLE CHOOZ observations $\theta_{13}$: $0.04 < \sin^2 2\theta_{13} < 0.34$ [8]. $\sin^2(2\theta_{13}) = 0.085 \pm 0.029(stat) \pm 0.042(syst)$ (68 % CL) [9].

## 4 The Elusive Absolute Scale of the Neutrino Mass

At present the structure of the neutrino mass spectrum is not known. There are two possible scenarios:

- Normal Spectrum (NS), $m_1 < m_2 < m_3$:

$$m_0 = m_1, \; m_2 = \sqrt{\Delta m^2_{\mathrm{SUN}} + m_0^2}\,, \; m_3 = \sqrt{\Delta m^2_{\mathrm{ATM}} + m_0^2},$$

with $\Delta m^2_{\rm SUN} = m_2^2 - m_1^2$ and $\Delta m^2_{\rm ATM} = m_3^2 - m_1^2$.

- Inverted Spectrum (IS), $m_3 < m_1 < m_2$:

$$m_0 = m_3, m_2 = \sqrt{\Delta m^2_{\rm ATM} + m_0^2}, m_1 = \sqrt{\Delta m^2_{\rm ATM} - \Delta m^2_{\rm SUN} + m_0^2}$$

with $\Delta m^2_{\rm SUN} = m_2^2 - m_1^2$ and $\Delta m^2_{\rm ATM} = m_2^2 - m_3^2$.

Here, $m_0 = m_1(m_3)$ is the lightest neutrino mass for NS (IS).

The absolute scale $m_0$ of neutrino mass can in principle be determined by the following observations:

- Neutrinoless double beta decay.

As we shall see later (Sect. 5) the effective Majorana neutrino mass $m_{\beta\beta}$ extracted in such experiments is given as follows:

$$m_{\beta\beta} = \sum_k^3 U_{ek}^2 m_k = c_{12}^2 c_{13}^2 e^{2i\alpha_1} m_1 + c_{13}^2 s_{12}^2 e^{2i\alpha_2} m_2 + s_{13}^2 m_3. \tag{5}$$

- The neutrino mass extracted from ordinary beta decay, e.g. from tritium $\beta$-decay [10].

$$m_\beta = \sqrt{\sum_k^3 |U_{ek}|^2 m_k^2} = \sqrt{c_{12}^2 c_{13}^2 m_1^2 + c_{13}^2 s_{12}^2 m_2^2 + s_{13}^2 m_3^2}. \tag{6}$$

assuming, of course, that the three neutrino states cannot be resolved.

- From astrophysical and cosmological observations (see, e.g., the recent summary [11]).

$$m_{\rm\ astro} = \sum_k^3 m_k. \tag{7}$$

The current limit on $m_{\rm\ astro}$ depends on the type of observation[11]. Thus CMB primordial gives 1.3 eV, CMB+distance 0.58 eV, galaxy distribution and lensing of galaxies 0.6 eV. On the other hand the largest photometric red shift survey yields 0.28 eV [12]. For purposes of illustration we will take a world average of $m_{\rm\ astro} =$ 0.71 eV.

The above results are exhibited in Fig. 1 for the tritium $\beta$-decay and cosmological limits as a function of the lowest neutrino mass and in Fig. 2 for the case of the $0\nu\beta\beta$-decay both for the NS and the IS scenarios. The allowed range values of $|m_{\beta\beta}|$ as a function of the lowest mass eigenstate $m_0$ is exhibited. For the values of the parameter $\sin^2 2\theta_{13}$ new Double Chooz data are used [9]. The IH allowed region for $|m_{\beta\beta}|$ is presented by the region between two parallel lines in the upper part of Fig. 2. The NH allowed region for $|m_{\beta\beta}| \approx$ few meV is compatible with $m_0$ smaller than 10 meV. The quasi-degenerate spectrum can be determined, if $m_0$ is known from future $\beta$-decay experiments KATRIN and MARE [10] or from cosmological observations.

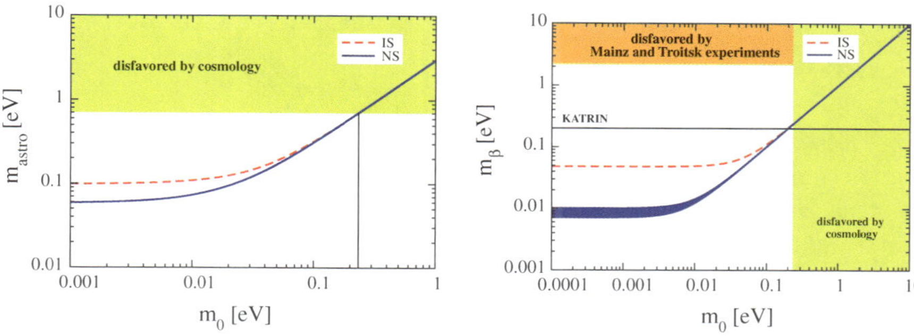


**Fig. 1** The neutrino mass limits in eV as a function of mass of the lowest eigenstate $m_0$ also in eV, extracted from cosmology (*left panel*) tritium $\beta$-decay (*right panel*). From the current upper limit of 2.2 eV of the Mainz and Troitsk experiments we deduce a lowest neutrino mass of 2.2 eV both for the NS and IS. From the astrophysical limit value of 0.71 eV the corresponding neutrino mass extracted is about 0.23 eV for the NS and IS

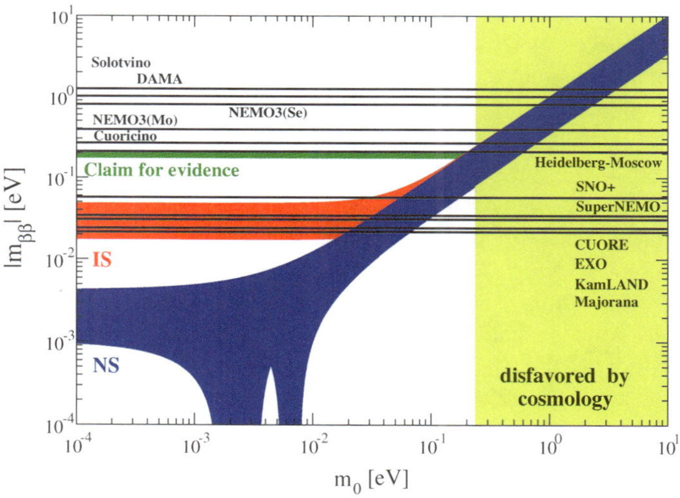


**Fig. 2** We show the allowed range of values for $|m_{\beta\beta}|$ as a function of the lowest mass eigenstate $m_0$ using the three standard neutrinos for the cases of normal (NS, $m_0 = m_1$) and inverted (IS, $m_0 = m_3$) spectrum of neutrino masses. Also shown are the current experimental limits and the expected future results [13] (QRPA NMEs with CD-Bonn short-range correlations and $g_A = 1.25$ are assumed [14]). Note that in the inverted hierarchy there is a lower bound, which means that in such a scenario the $0\nu\beta\beta$-decay should definitely be observed, if the experiments reach the required level. The same set of neutrino oscillation parameters as in Fig. 1 is considered

The lowest value for the sum of the neutrino masses, which can be reached in future cosmological measurements [12], is about (0.05–0.1) eV. The corresponding values of $m_0$ are in the region, where the IS and the NS predictions for $m_\beta$ differ significantly from each other.

## 5 Neutrinoless Double-Beta Decay and Double-Electron Capture

Investigations of neutrino oscillations in vacuum and in matter do not allow to distinguish massive Dirac from massive Majorana neutrinos. In order to reveal the Majorana nature of neutrinos is necessary to study processes in which the total lepton number is violated. The best sensitivity on small Majorana neutrino masses can be reached in the investigation of neutrinoless double-beta decay ($0\nu\beta\beta$-decay),

$$(A, Z) \rightarrow (A, Z+2) + 2e^{-}, \tag{8}$$

and the resonant neutrinoless double-electron capture ($0\nu$ECEC)

$$e^{-}_{b} + e^{-}_{b} + (A, Z) \rightarrow (A, Z-2)^{**}. \tag{9}$$

A double asterisk in Eq. (9) means that, in general, the final atom $(A, Z-2)$ is excited with respect to both the electron shell, due to formation of two vacancies for the electrons, and the nucleus.

There are few tenths of nuclear systems [15], which offer an opportunity to study the $0\nu\beta\beta$-decay and the most favorable are those with a large $Q_{\beta\beta}$-value. Neutrinoless double beta decay has not yet been confirmed. The strongest limits on the half-life $T^{0\nu}_{1/2}$ of the $0\nu\beta\beta$-decay were set in Heidelberg-Moscow ($^{76}$Ge, $1.9 \times 10^{25}$ y), [16], NEMO3 ($^{100}$Mo, $1.0 \times 10^{24}$ y) [17], CUORICINO ($^{130}$Te, $3.0 \times 10^{24}$ y) [18] and KamLAND-Zen ($^{136}$Xe, $5.7 \times 10^{24}$ y) [19] experiments. There exist, however, a claim of the observation of the $0\nu\beta\beta$-decay of $^{76}$Ge made by some participants of the Heidelberg-Moscow collaboration [20] with half-life $T^{0\nu}_{1/2} = 2.23^{+0.44}_{-0.31} \times 10^{25}$ years. This result will be checked by an independent experiment relatively soon. In the new germanium experiment GERDA [21], the Heidelberg-Moscow sensitivity will be reached in about one year of measuring time.

The inverse value of the $0\nu\beta\beta$-decay half-life for a given isotope (A,Z) is given by

$$(T^{0\nu}_{1/2})^{-1} = G_{0\nu}(Q_{\beta\beta}, Z)|M^{0\nu}|^2|m_{\beta\beta}|^2.$$

Here, $G_{0\nu}(Q_{\beta\beta}, Z)$ and $|M^{0\nu}|$ are, respectively, the known phase-space factor and the nuclear matrix element, which depends on the nuclear structure of the particular isotope under study.

The main aim of experiments on the search for $0\nu\beta\beta$-decay is the measurement of the effective Majorana neutrino mass $m_{\beta\beta}$. From the most precise experiments on the search for $0\nu\beta\beta$-decay [16, 18, 19] by using of nuclear matrix elements of Ref. [14] the following stringent bounds were inferred

$$|m_{\beta\beta}| < (0.20 - 0.32)\text{eV}\ (^{76}\text{Ge}),$$
$$< (0.33 - 0.46)\text{eV}\ (^{130}\text{Te}),$$
$$< (0.17 - 0.30)\text{eV}\ (^{136}\text{Xe}). \quad (10)$$

In future experiments, CUORE ($^{130}Te$), EXO, KamLAND-Zen ($^{136}Xe$), MAJORANA ($^{76}Ge$), SuperNEMO ($^{82}Se$), SNO+ ($^{150}Nd$), and others [22], a sensitivity

$$|m_{\beta\beta}| \simeq \text{a few}10^{-2}\text{eV} \quad (11)$$

is planned to be reached, what is the region of the IH of neutrino masses. In the case of the normal mass hierarchy $|m_{\beta\beta}|$ is too small in order to be probed in the $0\nu\beta\beta$-decay experiments of the next generation.

The mere observation of the resonant $0\nu$ECEC could also prove the Majorana nature of neutrinos as well as the violation of the total lepton number conservation. Recently, a new theoretical framework for the calculation of resonant $0\nu$ECEC transitions, namely the oscillation of stable and quasi-stationary atoms due to weak interaction with violation of the total lepton number and parity, was proposed in [23, 24].

The $0\nu$ECEC transition rate near the resonance is of Breit-Wigner form,

$$\Gamma_{ab}^{0\nu\text{ECEC}}(J^{\pi}) = \frac{|V_{ab}(J^{\pi})|^2}{\Delta^2 + \frac{1}{4}\Gamma_{ab}^2}\Gamma_{ab}, \quad (12)$$

where $J^{\pi}$ denotes angular momentum and parity of final nucleus. The degeneracy parameter can be expressed as $\Delta = Q - B_{ab} - E_{\gamma}$. $Q$ stands for a difference between the initial and final atomic masses in ground states and $E_{\gamma}$ is an excitation energy of the daughter nucleus. $B_{ab} = E_a + E_b + E_C$ is the energy of two electron holes, whose quantum numbers $(n, j, l)$ are denoted by indices $a$ and $b$ and $E_C$ is the interaction energy of the two holes. The binding energies of single electron holes $E_a$ are known with accuracy with few eV. The width of the excited final atom with the electron holes is given by

$$\Gamma_{ab} = \Gamma_a + \Gamma_b + \Gamma^*. \quad (13)$$

Here, $\Gamma_{a,b}$ is one-hole atomic width and $\Gamma^*$ is the de-excitation width of daughter nucleus, which can be neglected. Numerical values of $\Gamma_{ab}$ are about up to few tens eV.

For a capture of $s_{1/2}$ and $p_{1/2}$ electrons the explicit form of lepton number violating amplitude associated with nuclear transitions $0^+ \rightarrow J^{\pi} = 0^{\pm}, 1^{\pm}$ is given in [24]. By factorizing the electron shell structure and nuclear matrix element one gets

$$V_{ab}(J^{\pi}) = \frac{1}{4\pi} G_{\beta}^2 m_{\beta\beta} \frac{g_A^2}{R} \langle F_{ab} \rangle M^{0\nu ECEC}(J^{\pi}). \quad (14)$$

Here, $\langle F_{ab}\rangle$ is a combination of averaged upper and lower bispinor components of the atomic electron wave functions [24] and $M^{0\nu ECEC}(J^{\pi})$ is the nuclear matrix element. R is the nuclear radius and $g_A$ is the axial-vector coupling constant.

In the unitary limit some $0\nu$ECEC half-lives were predicted to be significantly below the $0\nu\beta\beta$-decay half-lives for the same value of $m_{\beta\beta}$.

## 6 Conclusion

As the most intriguing and fascinating fundamental particle, the neutrino is so important that neutrino physics has become one of the most significant branches of modern physics. If we are to understand 'why we are here' and the basic nature of the Universe in which we live, we must understand the basic properties of the neutrino. Once the fundamental properties of neutrinos became clear, neutrinos can be used to study various mechanisms in nature. We note that there is no place in the world that cannot be reached by neutrinos.

**Acknowledgments** The work of the author (FŠ) was supported by the VEGA Grant agency under the contract No. 1/0876/12.

## References

1. B. Pontecorvo, J. Exptl. Theoret. Phys. 33, 549 (1957) [Sov. Phys. JETP 6, 429 (1958)].
2. B. Pontecorvo, J. Exptl. Theoret. Phys. 34, 247 (1958) [Sov. Phys. JETP 7, 172 (1958)].
3. R. Wendell et al., The Super-Kamiokande collaboration. Phys. Rev. D **81**, 09200 (2010)
4. B. Aharmim et al., The SNO collaboration. Phys. Rev. C **81**, 055504 (2010)
5. A. Gando et al., The KamLAND collaboration. Phys. Rev. D **83**, 052002 (2011)
6. A. Habig et al., The MINOS collaboration. Mod. Phys. Lett. A **25**, 1219 (2010)
7. T. Schwetz, M. Tórtola, J.W.F. Valle, New J. Phys. **10**, 113011 (2008)
8. K. Abe et al., T2K collaboration. Phys. Rev. Lett. **107**, 041801 (2011)
9. H.D. Kerret, The Double Chooz collaboration. LowNu11, November 9–12, Seoul National University (Seoul, Korea, 2011).
10. E.W. Otten, C. Weinheimer, Rep. Prog. Phys. **71**, 086201 (2008)
11. K.N. Abazajian et al., Astropart. Phys. **35**, 177 (2011)
12. S.A. Thomas, F.B. Abdalla, O. Lahav, Phys. Rev. Let. **105**, 031301 (2010)
13. A.S. Barabash, AIP Conf. Proc. **1417**, 5 (2011)
14. F. Šimkovic, A. Faessler, H. Müther, V. Rodin, M. Stauf, Phys. Rev. C **79**, 055501 (2009)
15. V.I. Tretyak, Yu.G. Zdesenko, At. Dat. Nucl. Dat. Tabl. **80**, 83 (2002)
16. L. Baudis et al., The Heidelberg-Moscow collaboration. Phys. Rev. Lett. **83**, 41 (1999)
17. V.I. Tretyak, The NEMOIII collaboration. AIP Conf. Proc. **1417**, 125 (2011)
18. C. Arnaboldi et al., The CUORE collaboration. Phys. Lett. B **584**, 260 (2004)
19. A. Gando et al., The KamLAND-Zen collaboration. Phys. Rev. C **85**, 045504 (2012)
20. H.V. Klapdor-Kleingrothaus, I.V. Krivosheina, Mod. Phys. Lett. A **21**, 1547 (2006)
21. S. Schoenert, The GERDA collaboration. J. Phys. Conf. Ser. **203**, 012014 (2010)
22. F.T. Avignone, S.R. Elliott, J. Engel, Rev. Mod. Phys. **80**, 481 (2008)
23. F. Šimkovic, M.I. Krivoruchenko, Phys. Part. Nuc. **6**, 298 (2009)
24. M.I. Krivoruchenko, F. Šimkovic, D. Frekers, A. Faessler, Nucl. Phys. A **859**, 140 (2011)

# General $U(N)$ Gauge Transformations in the Realm of Covariant Hamiltonian Field Theory

**Jürgen Struckmeier and Hermine Reichau**

**Abstract** A consistent, local coordinate formulation of covariant Hamiltonian field theory is presented. While the covariant canonical field equations are equivalent to the Euler-Lagrange field equations, the covariant canonical transformation theory offers more general means for defining mappings that preserve the action functional—and hence the form of the field equations—than the usual Lagrangian description. Similar to the well-known canonical transformation theory of point dynamics, the canonical transformation rules for fields are derived from generating functions. As an interesting example, we work out the generating function of type $F_2$ of a general local $U(N)$ gauge transformation and thus derive the most general form of a Hamiltonian density $\mathscr{H}$ that is *form-invariant* under *local* $U(N)$ gauge transformations.

## 1 Covariant Hamiltonian Density

In field theory, the usual definition of a Hamiltonian density emerges from a Legendre transformation of a Lagrangian density $\mathscr{L}$ that only maps the time derivative $\partial_t \phi$ of a field $\phi(t, x, y, z)$ into a corresponding canonical momentum variable, $\pi_t$. Taking then the spatial integrals, we obtain a description of the field dynamics that corresponds to that of point dynamics. In contrast, a fully covariant Hamiltonian description treats space and time variables on equal footing [1, 2]. If $\mathscr{L}$ is a Lorentz scalar, this property is passed to the *covariant Hamiltonian*. Moreover, this description enables us to derive a consistent theory of canonical transformations in the realm of classical field theory.

J. Struckmeier(✉)
GSI Helmholtzzentrum für Schwerionenforschung GmbH, Planckstr. 1,
64291 Darmstadt, Germany
e-mail: j.struckmeier@gsi.de

H. Reichau
Frankfurt Institute of Advanced Studies (FIAS), Goethe-University Frankfurt am Main,
Ruth Moufang-Str. 1, 60438 Frankfurt am Main, Germany
e-mail: reichau@fias.uni-frankfurt.de

W. Greiner (ed.), *Exciting Interdisciplinary Physics*,
FIAS Interdisciplinary Science Series, DOI: 10.1007/978-3-319-00047-3_31,

## 1.1 Covariant Canonical Field Equations

The transition from particle dynamics to the dynamics of a *continuous* system is based on the assumption that a *continuum limit* exists for the given physical problem [3]. This limit is defined by letting the number of particles involved in the system increase over all bounds while letting their masses and distances go to zero. In this limit, the information on the location of individual particles is replaced by the *value* of a smooth function $\phi(\boldsymbol{x})$ that is given at a spatial location $x^1, x^2, x^3$ at time $t \equiv x^0$. The differentiable function $\phi(\boldsymbol{x})$ is called a *field*. In this notation, the index $\mu$ runs from 0 to 3, hence distinguishes the four independent variables of space-time $x^\mu \equiv (x^0, x^1, x^2, x^3) \equiv (ct, x, y, z)$, and $x_\mu \equiv (x_0, x_1, x_2, x_3) \equiv (ct, -x, -y, -z)$. We furthermore assume that the given physical problem can be described in terms of a set of $I = 1, \ldots, N$—possibly interacting—scalar fields $\phi^I(\boldsymbol{x})$ or vector fields $\boldsymbol{A}^I = (A^{I,0}, A^{I,1}, A^{I,2}, A^{I,3})$, with the index "$I$" enumerating the individual fields. In order to clearly distinguish scalar quantities from vector quantities, we denote the latter with boldface letters. Throughout the article, the summation convention is used. Whenever no confusion can arise, we omit the indexes in the argument list of functions in order to avoid the number of indexes to proliferate.

The Lagrangian description of the dynamics of a continuous system is based on the Lagrangian density function $\mathscr{L}$ that is supposed to carry the complete information on the given physical system. In a first-order field theory, the Lagrangian density $\mathscr{L}$ is defined to depend on the $\phi^I$, possibly on the vector of independent variables $\boldsymbol{x}$, and on the four first derivatives of the fields $\phi^I$ with respect to the independent variables, i.e., on the 1-forms (covectors)

$$\boldsymbol{\partial}\phi^I \equiv (\partial_{ct}\phi^I, \partial_x\phi^I, \partial_y\phi^I, \partial_z\phi^I).$$

The Euler-Lagrange field equations are then obtained as the zero of the variation $\delta S$ of the action integral

$$S = \int \mathscr{L}(\phi^I, \boldsymbol{\partial}\phi^I, \boldsymbol{x})\, d^4x \tag{1}$$

as (cf. for instance [3])

$$\frac{\partial}{\partial x^\alpha}\frac{\partial \mathscr{L}}{\partial(\partial_\alpha \phi^I)} - \frac{\partial \mathscr{L}}{\partial \phi^I} = 0. \tag{2}$$

To derive the equivalent *covariant* Hamiltonian description of continuum dynamics, we first define for each field $\phi^I(\boldsymbol{x})$ a 4-vector of conjugate momentum fields $\pi_I^\mu(\boldsymbol{x})$. Its components are given by

$$\pi_I^\mu = \frac{\partial \mathscr{L}}{\partial(\partial_\mu \phi^I)} \equiv \frac{\partial \mathscr{L}}{\partial\left(\frac{\partial \phi^I}{\partial x^\mu}\right)}. \tag{3}$$

The 4-vector $\boldsymbol{\pi}_I$ is thus induced by the Lagrangian $\mathscr{L}$ as the *dual counterpart* of the 1-form $\boldsymbol{\partial}\phi^I$. For the entire set of $N$ scalar fields $\phi^I(\boldsymbol{x})$, this establishes a set of $N$ conjugate 4-vector fields. With this definition of the 4-vectors of canonical momenta $\boldsymbol{\pi}_I(\boldsymbol{x})$, we can now define the Hamiltonian density $\mathscr{H}(\phi^I, \boldsymbol{\pi}_I, \boldsymbol{x})$ as the covariant Legendre transform of the Lagrangian density $\mathscr{L}(\phi^I, \boldsymbol{\partial}\phi^I, \boldsymbol{x})$

$$\mathscr{H}(\phi^I, \boldsymbol{\pi}_I, \boldsymbol{x}) = \pi_J^\alpha \frac{\partial \phi^J}{\partial x^\alpha} - \mathscr{L}(\phi^I, \boldsymbol{\partial}\phi^I, \boldsymbol{x}). \tag{4}$$

In oder for the Hamiltonian $\mathscr{H}$ to be valid, we must require the Legendre transformation to be *regular*, which means that for each index "$I$" the Hesse matrices $(\partial^2 \mathscr{L} / \partial(\partial^\mu \phi^I)\, \partial(\partial_\nu \phi^I))$ are non-singular. This ensures that by means of the Legendre transformation, the Hamiltonian $\mathscr{H}$ takes over the complete information on the given dynamical system from the Lagrangian $\mathscr{L}$. The definition of $\mathscr{H}$ by Eq. (4) is referred to in literature as the "De Donder-Weyl" Hamiltonian density.

Obviously, the dependencies of $\mathscr{H}$ and $\mathscr{L}$ on the $\phi^I$ and the $x^\mu$ only differ by a sign,

$$\left.\frac{\partial \mathscr{H}}{\partial x^\mu}\right|_{\text{expl}} = -\left.\frac{\partial \mathscr{L}}{\partial x^\mu}\right|_{\text{expl}}, \qquad \frac{\partial \mathscr{H}}{\partial \phi^I} = -\frac{\partial \mathscr{L}}{\partial \phi^I} = -\frac{\partial}{\partial x^\alpha} \frac{\partial \mathscr{L}}{\partial(\partial_\alpha \phi^I)} = -\frac{\partial \pi_I^\alpha}{\partial x^\alpha}.$$

These variables thus do not take part in the Legendre transformation of Eqs. (3), (4). Thus, with respect to this transformation, the Lagrangian density $\mathscr{L}$ represents a function of the $\partial_\mu \phi^I$ only and does *not depend* on the canonical momenta $\pi_I^\mu$, whereas the Hamiltonian density $\mathscr{H}$ is to be considered as a function of the $\pi_I^\mu$ only and does not depend on the derivatives $\partial_\mu \phi$ of the fields. In order to derive the second canonical field equation, we calculate from Eq. (4) the partial derivative of $\mathscr{H}$ with respect to $\pi_I^\mu$,

$$\frac{\partial \mathscr{H}}{\partial \pi_I^\mu} = \delta_J^I\, \delta_\mu^\alpha\, \frac{\partial \phi^J}{\partial x^\alpha} = \frac{\partial \phi^I}{\partial x^\mu} \qquad \Longleftrightarrow \qquad \frac{\partial \mathscr{L}}{\partial(\partial_\mu \phi^I)} = \pi_J^\alpha \delta_I^J\, \delta_\alpha^\mu = \pi_I^\mu.$$

The complete set of covariant canonical field equations is thus given by

$$\frac{\partial \mathscr{H}}{\partial \pi_I^\mu} = \frac{\partial \phi^I}{\partial x^\mu}, \qquad \frac{\partial \mathscr{H}}{\partial \phi^I} = -\frac{\partial \pi_I^\alpha}{\partial x^\alpha}. \tag{5}$$

This pair of first-order partial differential equations is equivalent to the set of second-order differential equations of Eq. (2). We observe that in this formulation of the canonical field equations, all coordinates of space-time appear symmetrically—similar to the Lagrangian formulation of Eq. (2). Provided that the Lagrangian density $\mathscr{L}$ is a Lorentz scalar, the dynamics of the fields is invariant with respect to Lorentz transformations. The covariant Legendre transformation (4) passes this property to the Hamiltonian density $\mathscr{H}$. It thus ensures *a priori* the relativistic invariance of the fields that emerge as integrals of the canonical field equations if $\mathscr{L}$—and hence $\mathscr{H}$—represents a Lorentz scalar.

## 2 Canonical Transformations in Covariant Hamiltonian Field Theory

The covariant Legendre transformation (4) allows us to derive a canonical transformation theory in a way similar to that of point dynamics. The main difference is that now the generating function of the canonical transformation is represented by a *vector* rather than by a scalar function. The main benefit of this formalism is that we are not dealing with plain transformations. Instead, we restrict ourselves *right from the beginning* to those transformations that preserve the form of the action functional. This ensures all eligible transformations to be *physical*. Furthermore, with a generating function, we not only define the transformations of the fields but also pinpoint simultaneously the corresponding transformation law of the canonical momentum fields.

### 2.1 Generating Functions of Type $F_1(\boldsymbol{\phi}, \boldsymbol{\Phi}, x)$

Similar to the canonical formalism of point mechanics, we call a transformation of the fields $(\boldsymbol{\phi}, \boldsymbol{\pi}) \mapsto (\boldsymbol{\Phi}, \boldsymbol{\Pi})$ *canonical* if the form of the variational principle that is based on the action functional (1) is maintained,

$$\delta \int_R \left( \pi_I^\alpha \frac{\partial \phi^I}{\partial x^\alpha} - \mathscr{H}(\boldsymbol{\phi}, \boldsymbol{\pi}, \boldsymbol{x}) \right) d^4x \stackrel{!}{=} \delta \int_R \left( \Pi_I^\alpha \frac{\partial \Phi^I}{\partial x^\alpha} - \mathscr{H}'(\boldsymbol{\Phi}, \boldsymbol{\Pi}, \boldsymbol{x}) \right) d^4x. \tag{6}$$

Equation (6) tells us that the *integrands* may differ by the divergence of a vector field $F_1^\mu$, whose variation vanishes on the boundary $\partial R$ of the integration region $R$ within space-time

$$\delta \int_R \frac{\partial F_1^\alpha}{\partial x^\alpha} d^4x = \delta \oint_{\partial R} F_1^\alpha dS_\alpha \stackrel{!}{=} 0.$$

The immediate consequence of the form invariance of the variational principle is the form invariance of the covariant canonical field equations (5)

$$\frac{\partial \mathscr{H}'}{\partial \Pi_I^\mu} = \frac{\partial \Phi^I}{\partial x^\mu}, \qquad \frac{\partial \mathscr{H}'}{\partial \Phi^I} = -\frac{\partial \Pi_I^\alpha}{\partial x^\alpha}.$$

For the integrands of Eq. (6)—hence for the Lagrangian densities $\mathscr{L}$ and $\mathscr{L}'$—we thus obtain the condition

$$\mathscr{L} = \mathscr{L}' + \frac{\partial F_1^\alpha}{\partial x^\alpha}$$

$$\pi_I^\alpha \frac{\partial \phi^I}{\partial x^\alpha} - \mathscr{H}(\boldsymbol{\phi}, \boldsymbol{\pi}, \boldsymbol{x}) = \Pi_I^\alpha \frac{\partial \Phi^I}{\partial x^\alpha} - \mathscr{H}'(\boldsymbol{\Phi}, \boldsymbol{\Pi}, \boldsymbol{x}) + \frac{\partial F_1^\alpha}{\partial x^\alpha}. \tag{7}$$

With the definition $F_1^\mu \equiv F_1^\mu(\boldsymbol{\phi}, \boldsymbol{\Phi}, \boldsymbol{x})$, we restrict ourselves to a function of exactly those arguments that now enter into transformation rules for the transition from the original to the new fields. The divergence of $F_1^\mu$ writes, explicitly,

$$\frac{\partial F_1^\alpha}{\partial x^\alpha} = \frac{\partial F_1^\alpha}{\partial \phi^I}\frac{\partial \phi^I}{\partial x^\alpha} + \frac{\partial F_1^\alpha}{\partial \Phi^I}\frac{\partial \Phi^I}{\partial x^\alpha} + \left.\frac{\partial F_1^\alpha}{\partial x^\alpha}\right|_{\text{expl}}. \tag{8}$$

The rightmost term denotes the sum over the *explicit* dependence of the generating function $F_1^\mu$ on the $x^\nu$. Comparing the coefficients of Eqs. (7) and (8), we find the local coordinate representation of the field transformation rules that are induced by the generating function $F_1^\mu$

$$\pi_I^\mu = \frac{\partial F_1^\mu}{\partial \phi^I}, \qquad \Pi_I^\mu = -\frac{\partial F_1^\mu}{\partial \Phi^I}, \qquad \mathscr{H}' = \mathscr{H} + \left.\frac{\partial F_1^\alpha}{\partial x^\alpha}\right|_{\text{expl}}. \tag{9}$$

The transformation rule for the Hamiltonian density implies that summation over $\alpha$ is to be performed. In contrast to the transformation rule for the Lagrangian density $\mathscr{L}$ of Eq. (7), the rule for the Hamiltonian density is determined by the *explicit* dependence of the generating function $F_1^\mu$ on the $x^\nu$. Hence, if a generating function does not explicitly depend on the independent variables, $x^\nu$, then the *value* of the Hamiltonian density is not changed under the particular canonical transformation emerging thereof.

Differentiating the transformation rule for $\pi_I^\mu$ with respect to $\Phi^J$, and the rule for $\Pi_J^\mu$ with respect to $\phi^I$, we obtain a symmetry relation between original and transformed fields

$$\frac{\partial \pi_I^\mu}{\partial \Phi^J} = \frac{\partial^2 F_1^\mu}{\partial \phi^I \partial \Phi^J} = -\frac{\partial \Pi_J^\mu}{\partial \phi^I}.$$

The emerging of symmetry relations is a characteristic feature of *canonical* transformations. As the symmetry relation directly follows from the second derivatives of the generating function, is does not apply for arbitrary transformations of the fields that do not follow from generating functions.

## 2.2 Generating Functions of Type $F_2(\boldsymbol{\phi}, \boldsymbol{\Pi}, \boldsymbol{x})$

The generating function of a canonical transformation can alternatively be expressed in terms of a function of the original fields $\phi^I$ and of the new *conjugate* fields $\Pi_I^\mu$. To derive the pertaining transformation rules, we perform the covariant Legendre transformation

$$F_2^{\mu}(\boldsymbol{\phi}, \boldsymbol{\Pi}, \boldsymbol{x}) = F_1^{\mu}(\boldsymbol{\phi}, \boldsymbol{\Phi}, \boldsymbol{x}) + \Phi^J \Pi_J^{\mu}, \qquad \Pi_I^{\mu} = -\frac{\partial F_1^{\mu}}{\partial \Phi^I}. \tag{10}$$

By definition, the functions $F_1^{\mu}$ and $F_2^{\mu}$ agree with respect to their $\phi^I$ and $x^{\mu}$ dependencies

$$\frac{\partial F_2^{\mu}}{\partial \phi^I} = \frac{\partial F_1^{\mu}}{\partial \phi^I} = \pi_I^{\mu}, \qquad \left.\frac{\partial F_2^{\alpha}}{\partial x^{\alpha}}\right|_{\text{expl}} = \left.\frac{\partial F_1^{\alpha}}{\partial x^{\alpha}}\right|_{\text{expl}} = \mathcal{H}' - \mathcal{H}.$$

The variables $\phi^I$ and $x^{\mu}$ thus do not take part in the Legendre transformation from Eq. (10). Therefore, the two $F_2^{\mu}$-related transformation rules coincide with the respective rules derived previously from $F_1^{\mu}$. As $F_1^{\mu}$ does not depend on the $\Pi_I^{\mu}$ whereas $F_2^{\mu}$ does not depend on the $\Phi^I$, the new transformation rule thus follows from the derivative of $F_2^{\mu}$ with respect to $\Pi_J^{\nu}$ as

$$\frac{\partial F_2^{\mu}}{\partial \Pi_I^{\nu}} = \Phi^J \frac{\partial \Pi_J^{\mu}}{\partial \Pi_I^{\nu}} = \Phi^J \delta_J^I \delta_{\nu}^{\mu}.$$

We thus end up with set of transformation rules

$$\pi_I^{\mu} = \frac{\partial F_2^{\mu}}{\partial \phi^I}, \qquad \Phi^I \delta_{\nu}^{\mu} = \frac{\partial F_2^{\mu}}{\partial \Pi_I^{\nu}}, \qquad \mathcal{H}' = \mathcal{H} + \left.\frac{\partial F_2^{\alpha}}{\partial x^{\alpha}}\right|_{\text{expl}}, \tag{11}$$

which is equivalent to the set (9) by virtue of the Legendre transformation (10) if the matrices $(\partial^2 F_1^{\mu}/\partial \phi^I \partial \Phi^J)$ are non-singular for all indexes "$\mu$". From the second partial derivations of $F_2^{\mu}$ one immediately derives the symmetry relation

$$\frac{\partial \pi_I^{\mu}}{\partial \Pi_J^{\nu}} = \frac{\partial^2 F_2^{\mu}}{\partial \phi^I \partial \Pi_J^{\nu}} = \frac{\partial \Phi^J}{\partial \phi^I} \delta_{\nu}^{\mu},$$

whose existence characterizes the transformation to be canonical.

# 3 Examples for Hamiltonian Densities in Covariant Field Theory

We present some simple examples Hamiltonian densities as they emerge from Lagrangian densities of classical Lagrangian field theory. It is shown that resulting canonical field equations are equivalent to the corresponding Euler-Lagrange equations.

## 3.1 Klein-Gordon Hamiltonian Density for Complex Fields

We first consider the Klein-Gordon *Lagrangian density* $\mathcal{L}_{\text{KG}}$ for a *complex* scalar field $\phi$ (see, for instance, Ref. [4]):

$$\mathscr{L}_{\mathrm{KG}}\left(\phi, \overline{\phi}, \partial^{\mu}\phi, \partial_{\mu}\overline{\phi}\right) = \left(\partial_{\alpha}\overline{\phi}\right)\left(\partial^{\alpha}\phi\right) - \Omega^{2}\,\overline{\phi}\phi.$$

Herein $\overline{\phi}$ denotes complex conjugate field of $\phi$. Both quantities are to be treated as independent. The Euler-Lagrange equations (2) for $\phi$ and $\overline{\phi}$ follow from this Lagrangian density as

$$\frac{\partial^{2}}{\partial x_{\alpha}\partial x^{\alpha}}\overline{\phi} = -\Omega^{2}\,\overline{\phi}, \qquad \frac{\partial^{2}}{\partial x_{\alpha}\partial x^{\alpha}}\phi = -\Omega^{2}\,\phi. \tag{12}$$

As a prerequisite for deriving the corresponding Hamiltonian density $\mathscr{H}_{\mathrm{KG}}$ we must first define from $\mathscr{L}_{\mathrm{KG}}$ the conjugate momentum fields,

$$\pi^{\mu} = \frac{\partial\mathscr{L}_{\mathrm{KG}}}{\partial\left(\partial_{\mu}\overline{\phi}\right)} = \frac{\partial\phi}{\partial x_{\mu}}, \qquad \overline{\pi}_{\mu} = \frac{\partial\mathscr{L}_{\mathrm{KG}}}{\partial(\partial^{\mu}\phi)} = \frac{\partial\overline{\phi}}{\partial x^{\mu}}.$$

The determinant of the Hesse matrix does not vanish for the actual Lagrangian $\mathscr{L}_{\mathrm{KG}}$ since

$$\det\left(\frac{\partial^{2}\mathscr{L}_{\mathrm{KG}}}{\partial(\partial^{\mu}\phi)\partial\left(\partial_{\nu}\overline{\phi}\right)}\right) = \det\left(\frac{\partial\overline{\pi}_{\mu}}{\partial\left(\partial_{\nu}\overline{\phi}\right)}\right) = \det\left(\delta_{\mu}^{\nu}\right) = 1.$$

This condition is always satisfied if the Lagrangian density $\mathscr{L}$ is *quadratic* in the derivatives of the fields. The Hamiltonian density $\mathscr{H}$ then follows as the Legendre transform of the Lagrangian density

$$\mathscr{H}(\pi^{\mu}, \overline{\pi}^{\mu}, \phi, \overline{\phi}) = \overline{\pi}^{\alpha}\frac{\partial\phi}{\partial x^{\alpha}} + \frac{\partial\overline{\phi}}{\partial x^{\alpha}}\pi^{\alpha} - \mathscr{L}.$$

The Klein-Gordon *Hamiltonian density* $\mathscr{H}_{\mathrm{KG}}$ is thus given by

$$\mathscr{H}_{\mathrm{KG}}(\pi_{\mu}, \overline{\pi}^{\mu}, \phi, \overline{\phi}) = \overline{\pi}_{\alpha}\pi^{\alpha} + \Omega^{2}\,\overline{\phi}\phi. \tag{13}$$

For the Hamiltonian density (13), the canonical field equations (5) provide the following set of coupled first order partial differential equations

$$\begin{aligned}
\frac{\partial\mathscr{H}_{\mathrm{KG}}}{\partial\pi^{\mu}} &= \frac{\partial\overline{\phi}}{\partial x^{\mu}} = \overline{\pi}_{\mu}, &\qquad \frac{\partial\mathscr{H}_{\mathrm{KG}}}{\partial\overline{\pi}_{\mu}} &= \frac{\partial\phi}{\partial x_{\mu}} = \pi^{\mu} \\
\frac{\partial\mathscr{H}_{\mathrm{KG}}}{\partial\phi} &= -\frac{\partial\overline{\pi}^{\alpha}}{\partial x^{\alpha}} = \Omega^{2}\overline{\phi}, &\qquad \frac{\partial\mathscr{H}_{\mathrm{KG}}}{\partial\overline{\phi}} &= -\frac{\partial\pi^{\alpha}}{\partial x^{\alpha}} = \Omega^{2}\phi.
\end{aligned}$$

In the first row, the canonical field equations for the scalar fields $\phi$ and $\overline{\phi}$ reproduce with the definitions of the momentum fields $\pi_{\mu}$ and $\overline{\pi}^{\mu}$ from the Lagrangian density $\mathscr{L}_{\mathrm{KG}}$. Eliminating the $\pi_{\mu}$, $\overline{\pi}^{\mu}$ from the canonical field equations then yields the Euler-Lagrange equations of Eq. (12).

## 3.2 Maxwell's Equations as Canonical Field Equations

The Lagrangian density $\mathscr{L}_{\mathrm{M}}$ of the electromagnetic field is given by

$$\mathscr{L}_{\mathrm{M}}(\boldsymbol{a}, \boldsymbol{\partial a}, \boldsymbol{x}) = -\tfrac{1}{4} f_{\alpha\beta} f^{\alpha\beta} - \frac{4\pi}{c} j^{\alpha}(\boldsymbol{x})\, a_{\alpha}, \quad f_{\mu\nu} = \frac{\partial a_{\nu}}{\partial x^{\mu}} - \frac{\partial a_{\mu}}{\partial x^{\nu}}. \tag{14}$$

Herein, the four components $a^{\mu}$ of the 4-vector potential $\boldsymbol{a}$ now take the place of the scalar fields $\phi^{I} \equiv a^{\mu}$ in the notation used so far. The Lagrangian density (14) thus entails a set of *four* Euler-Lagrange equations, i.e., an equation for each component $a_{\mu}$. The source vector $\boldsymbol{j} = (c\rho, j_x, j_y, j_z)$ denotes the 4-vector of electric currents combining the usual current density vector $(j_x, j_y, j_z)$ of configuration space with the charge density $\rho$. In a local Lorentz frame, i.e., in Minkowski space, the Euler-Lagrange equations (2) take on the form,

$$\frac{\partial}{\partial x^{\alpha}} \frac{\partial \mathscr{L}_{\mathrm{M}}}{\partial(\partial_{\alpha} a_{\mu})} - \frac{\partial \mathscr{L}_{\mathrm{M}}}{\partial a_{\mu}} = 0, \qquad \mu = 0, \ldots, 3. \tag{15}$$

With $\mathscr{L}_{\mathrm{M}}$ from Eq. (14), we obtain directly

$$\frac{\partial f^{\mu\alpha}}{\partial x^{\alpha}} + \frac{4\pi}{c} j^{\mu} = 0. \tag{16}$$

In Minkowski space, this is the tensor form of the inhomogeneous Maxwell equation. In order to formulate the equivalent Hamiltonian description, we first define, according to Eq. (3), the canonically field components $p^{\mu\nu}$ as the conjugate objects of the derivatives of the 4-vector potential $\boldsymbol{a}$

$$p^{\mu\nu} = \frac{\partial \mathscr{L}_{\mathrm{M}}}{\partial(\partial_{\nu} a_{\mu})} \equiv \frac{\partial \mathscr{L}_{\mathrm{M}}}{\partial a_{\mu,\nu}} \tag{17}$$

With the particular Lagrangian density (14), Eq. (17) means

$$\begin{aligned} f_{\alpha\beta} &= a_{\beta,\alpha} - a_{\alpha,\beta} \\ p^{\mu\nu} &= -\tfrac{1}{4}\left(\frac{\partial f_{\alpha\beta}}{\partial a_{\mu,\nu}} f^{\alpha\beta} + \frac{\partial f^{\alpha\beta}}{\partial a_{\mu,\nu}} f_{\alpha\beta}\right) = -\tfrac{1}{2} \frac{\partial f_{\alpha\beta}}{\partial a_{\mu,\nu}} f^{\alpha\beta} \\ &= -\tfrac{1}{2}\left(\delta^{\mu}_{\beta}\delta^{\nu}_{\alpha} - \delta^{\mu}_{\alpha}\delta^{\nu}_{\beta}\right) f^{\alpha\beta} = \tfrac{1}{2}(f^{\mu\nu} - f^{\nu\mu}) = f^{\mu\nu}. \end{aligned}$$

The tensor $p^{\mu\nu}$ thus matches exactly the electromagnetic field tensor $f^{\mu\nu}$ from Eq. (14) and hence inherits the skew-symmetry of $f^{\mu\nu}$ because of the particular dependence of $\mathscr{L}_{\mathrm{M}}$ on the $a_{\mu,\nu} \equiv \partial a_{\mu}/\partial x^{\nu}$.

As the Lagrangian density (14) now describes the dynamics of a *vector field*, $a_{\mu}$, rather than a set of scalar fields $\phi^{I}$, the canonical momenta $p^{\mu\nu}$ now constitute a second rank *tensor* rather than a vector. The Legendre transformation corresponding to Eq. (4) then comprises the product $p^{\alpha\beta}\partial_\beta a_\alpha$. The skew-symmetry of the momentum tensor $p^{\mu\nu}$ picks out the skew-symmetric part of $\partial_\nu a_\mu$ as the symmetric part of $\partial_\nu a_\mu$ vanishes identically calculating the product $p^{\alpha\beta}\partial_\beta a_\alpha$

$$p^{\alpha\beta}\frac{\partial a_\alpha}{\partial x^\beta} = \tfrac{1}{2}p^{\alpha\beta}\underbrace{\left(\frac{\partial a_\alpha}{\partial x^\beta} - \frac{\partial a_\beta}{\partial x^\alpha}\right)}_{=p_{\beta\alpha}} + \tfrac{1}{2}\,p^{\alpha\beta}\underbrace{\left(\frac{\partial a_\alpha}{\partial x^\beta} + \frac{\partial a_\beta}{\partial x^\alpha}\right)}_{\equiv 0}.$$

For a skew-symmetric momentum tensor $p^{\mu\nu}$, we thus obtain the Hamiltonian density $\mathscr{H}_\mathrm{M}$ as the Legendre-transformed Lagrangian density $\mathscr{L}_\mathrm{M}$

$$\mathscr{H}_\mathrm{M}(\boldsymbol{a},\boldsymbol{p},\boldsymbol{x}) = \tfrac{1}{2}p^{\alpha\beta}\left(\frac{\partial a_\alpha}{\partial x^\beta} - \frac{\partial a_\beta}{\partial x^\alpha}\right) - \mathscr{L}_\mathrm{M}(\boldsymbol{a},\boldsymbol{\partial a},\boldsymbol{x}).$$

From this Legendre transformation prescription and the corresponding Euler-Lagrange equations (15), the canonical field equations are immediately obtained as

$$\begin{aligned}
\frac{\partial\mathscr{H}_\mathrm{M}}{\partial p^{\mu\nu}} &= \frac{1}{2}\left(\frac{\partial a_\mu}{\partial x^\nu} - \frac{\partial a_\nu}{\partial x^\mu}\right)\\
\frac{\partial\mathscr{H}_\mathrm{M}}{\partial a_\mu} &= -\frac{\partial\mathscr{L}_\mathrm{M}}{\partial a_\mu} = -\frac{\partial}{\partial x^\alpha}\frac{\partial\mathscr{L}_\mathrm{M}}{\partial(\partial_\alpha a_\mu)} = -\frac{\partial p^{\mu\alpha}}{\partial x^\alpha}\\
\frac{\partial\mathscr{H}_\mathrm{M}}{\partial x^\nu} &= -\frac{\partial\mathscr{L}_\mathrm{M}}{\partial x^\nu}.
\end{aligned}$$

The Hamiltonian density for the Lagrangian density (14) follows as

$$\begin{aligned}
\mathscr{H}_\mathrm{M}(\boldsymbol{a},\boldsymbol{p},\boldsymbol{x}) &= -\tfrac{1}{2}p^{\alpha\beta}\,p_{\alpha\beta} + \tfrac{1}{4}p^{\alpha\beta}\,p_{\alpha\beta} + \frac{4\pi}{c}\,j^\alpha(\boldsymbol{x})\,a_\alpha\\
&= -\tfrac{1}{4}p^{\alpha\beta}\,p_{\alpha\beta} + \frac{4\pi}{c}\,j^\alpha(\boldsymbol{x})\,a_\alpha.
\end{aligned} \tag{18}$$

The first canonical field equation follows from the derivative of the Hamiltonian density (18) with respect to $p^{\mu\nu}$ and $p_{\mu\nu}$

$$\frac{1}{2}\left(\frac{\partial a_\mu}{\partial x^\nu} - \frac{\partial a_\nu}{\partial x^\mu}\right) = \frac{\partial\mathscr{H}_\mathrm{M}}{\partial p^{\mu\nu}} = -\tfrac{1}{2}p_{\mu\nu},\quad \frac{1}{2}\left(\frac{\partial a^\mu}{\partial x_\nu} - \frac{\partial a^\nu}{\partial x_\mu}\right) = \frac{\partial\mathscr{H}_\mathrm{M}}{\partial p_{\mu\nu}} = -\tfrac{1}{2}p^{\mu\nu}, \tag{19}$$

which reproduces the definition of $p_{\mu\nu}$ and $p^{\mu\nu}$ from Eq. (17).

The second canonical field equation is obtained calculating the derivative of the Hamiltonian density (18) with respect to $a_\mu$

$$-\frac{\partial p^{\mu\alpha}}{\partial x^\alpha} = \frac{\partial\mathscr{H}_\mathrm{M}}{\partial a_\mu} = \frac{4\pi}{c}j^\mu.$$

Inserting the first canonical equation, the second order field equation for the $a_\mu$ is thus obtained for the Maxwell Hamiltonian density (18) as

$$\frac{\partial f^{\mu\alpha}}{\partial x^\alpha} + \frac{4\pi}{c} j^\mu = 0,$$

which agrees, as expected, with the corresponding Euler-Lagrange equation (16).

### *3.3 The Proca Hamiltonian Density*

In relativistic quantum field theory, the dynamics of particles of spin 1 and mass $m$ is derived from the Proca Lagrangian density $\mathscr{L}_\mathrm{P}$,

$$\mathscr{L}_\mathrm{P} = -\tfrac{1}{4} f^{\alpha\beta} f_{\alpha\beta} + \tfrac{1}{2}\omega^2 a^\alpha a_\alpha, \qquad f_{\mu\nu} = \frac{\partial a_\nu}{\partial x^\mu} - \frac{\partial a_\mu}{\partial x^\nu}, \qquad \omega = \frac{mc}{\hbar}.$$

We observe that the kinetic term of $\mathscr{L}_\mathrm{P}$ agrees with that of the Lagrangian density $\mathscr{L}_\mathrm{M}$ of the electromagnetic field of Eq. (14). Therefore, the field equations emerging from the Euler-Lagrange equations (15) are similar to those of Eq. (16)

$$\frac{\partial f^{\mu\alpha}}{\partial x^\alpha} - \omega^2 a^\mu = 0. \tag{20}$$

The transition to the corresponding Hamilton description is performed by defining on the basis of the actual Lagrangian $\mathscr{L}_\mathrm{P}$ the canonical momentum field tensors $p^{\mu\nu}$ as the conjugate objects of the derivatives of the 4-vector potential $\boldsymbol{a}$

$$p^{\mu\nu} = \frac{\partial \mathscr{L}_\mathrm{P}}{\partial\left(\partial_\nu a_\mu\right)} \equiv \frac{\partial \mathscr{L}_\mathrm{P}}{\partial a_{\mu,\nu}}.$$

Similar to the preceding section, we find

$$p^{\mu\nu} = f^{\mu\nu}, \qquad p_{\mu\nu} = f_{\mu\nu}$$

because of the particular dependence of $\mathscr{L}_\mathrm{P}$ on the derivatives of the $a^\mu$. With $p^{\alpha\beta}$ being skew-symmetric in $\alpha$, $\beta$, the product $p^{\alpha\beta}\, a_{\alpha,\beta}$ picks out the skew-symmetric part of the partial derivative $\partial a_\alpha / \partial x^\beta$ as the product with the symmetric part vanishes identically. Denoting the skew-symmetric part by $a_{[\alpha,\beta]}$, the Legendre transformation prescription

$$\begin{aligned}\mathscr{H}_\mathrm{P} &= p^{\alpha\beta}\, a_{\alpha,\beta} - \mathscr{L}_\mathrm{P} = p^{\alpha\beta}\, a_{[\alpha,\beta]} - \mathscr{L}_\mathrm{P} \\ &= \tfrac{1}{2} p^{\alpha\beta} \left(\frac{\partial a_\alpha}{\partial x^\beta} - \frac{\partial a_\beta}{\partial x^\alpha}\right) - \mathscr{L}_\mathrm{P},\end{aligned}$$

leads to the Proca Hamiltonian density by following the path of Eq. (18)

$$\mathscr{H}_{\mathrm{P}} = -\tfrac{1}{4} p^{\alpha\beta} p_{\alpha\beta} - \tfrac{1}{2}\omega^2 a^\alpha a_\alpha. \tag{21}$$

The canonical field equations emerge as

$$\begin{aligned} a_{[\mu,\nu]} \equiv \frac{1}{2}\left(\frac{\partial a_\mu}{\partial x^\nu} - \frac{\partial a_\nu}{\partial x^\mu}\right) &= \frac{\partial \mathscr{H}_{\mathrm{P}}}{\partial p^{\mu\nu}} = -\tfrac{1}{2} p_{\mu\nu} \\ -\frac{\partial p^{\mu\alpha}}{\partial x^\alpha} &= \frac{\partial \mathscr{H}_{\mathrm{P}}}{\partial a_\mu} = -\omega^2 a^\mu. \end{aligned}$$

By means of eliminating $p^{\mu\nu}$, this coupled set of first order equations can be converted into second order equations for the vector field $\boldsymbol{a}(\boldsymbol{x})$,

$$\frac{\partial}{\partial x_\alpha}\left(\frac{\partial a_\mu}{\partial x^\alpha} - \frac{\partial a_\alpha}{\partial x^\mu}\right) - \omega^2 a_\mu = 0.$$

As expected, this equation coincides with the Euler-Lagrange equation (20).

## 3.4 The Dirac Hamiltonian Density

The dynamics of particles with spin $\frac{1}{2}$ and mass $m$ is described by the Dirac equation. With $\gamma^i$, $i = 1, \ldots, 4$ denoting the $4 \times 4$ Dirac matrices, and $\phi$ a four component Dirac spinor, the Dirac Lagrangian density $\mathscr{L}_{\mathrm{D}}$ is given by

$$\mathscr{L}_{\mathrm{D}} = i\overline{\phi}\gamma^\alpha \frac{\partial \phi}{\partial x^\alpha} - m\overline{\phi}\phi, \tag{22}$$

wherein $\overline{\phi} \equiv \phi^\dagger \gamma^0$ denotes the adjoint spinor of $\phi$. In the following we summarize some fundamental relations that apply for the Dirac matrices $\gamma^\mu$, and their duals, $\gamma_\mu$,

$$\begin{aligned} \{\gamma^\mu, \gamma^\nu\} &\equiv \gamma^\mu\gamma^\nu + \gamma^\nu\gamma^\mu = 2\eta^{\mu\nu}\mathbb{1} \\ \gamma^\mu\gamma_\mu &= \gamma_\mu\gamma^\mu = 4\,\mathbb{1} \\ \left[\gamma^\mu, \gamma^\nu\right] &\equiv \gamma^\mu\gamma^\nu - \gamma^\nu\gamma^\mu \equiv -2i\sigma^{\mu\nu} \\ \left[\gamma_\mu, \gamma_\nu\right] &\equiv \gamma_\mu\gamma_\nu - \gamma_\nu\gamma_\mu \equiv -2i\sigma_{\mu\nu} \\ \det\sigma^{\mu\nu} &= 1, \qquad \mu \neq \nu \end{aligned} \tag{23}$$

$$\begin{aligned} \sigma_{\mu\lambda}\sigma^{\lambda\nu} &= \sigma^{\nu\lambda}\sigma_{\lambda\mu} = \delta^\nu_\mu\,\mathbb{1} \\ \gamma^\mu\sigma_{\mu\nu} &= \sigma_{\nu\mu}\gamma^\mu = -\frac{i}{3}\gamma_\nu \\ \gamma_\mu\sigma^{\mu\nu} &= \sigma^{\nu\mu}\gamma_\mu = 3i\gamma^\nu \\ \gamma^\mu\sigma_{\mu\nu}\gamma^\nu &= -\frac{4i}{3}\mathbb{1} \end{aligned} \tag{24}$$

Herein, the symbol $\mathbb{1}$ stands for the $4 \times 4$ unit matrix, and the real numbers $\eta^{\mu\nu}$, $\eta_{\mu\nu} \in \mathbb{R}$ for an element of the Minkowski metric $(\eta^{\mu\nu}) = (\eta_{\mu\nu})$. The matrices $(\sigma^{\mu\nu})$ and $(\sigma_{\mu\nu})$ are to be understood as $4 \times 4$ block matrices, with each block $\sigma^{\mu\nu}$, $\sigma_{\mu\nu}$ representing a $4 \times 4$ matrix of complex numbers. Thus, $(\sigma^{\mu\nu})$ and $(\sigma_{\mu\nu})$ are actually $16 \times 16$ matrices of complex numbers.

The Dirac Lagrangian density $\mathscr{L}_{\mathrm{D}}$ can be rendered symmetric by combining the Lagrangian density Eq. (22) with its adjoint, which leads to

$$\mathscr{L}_{\mathrm{D}} = \frac{i}{2}\left(\overline{\phi}\gamma^{\alpha}\frac{\partial \phi}{\partial x^{\alpha}} - \frac{\partial \overline{\phi}}{\partial x^{\alpha}}\gamma^{\alpha}\phi\right) - m\overline{\phi}\phi. \tag{25}$$

The resulting Euler-Lagrange equations are identical to those derived from Eq. (22),

$$\begin{aligned} i\gamma^{\alpha}\frac{\partial \phi}{\partial x^{\alpha}} - m\phi &= 0 \\ i\frac{\partial \overline{\phi}}{\partial x^{\alpha}}\gamma^{\alpha} + m\overline{\phi} &= 0. \end{aligned} \tag{26}$$

As both Lagrangians (22) and (25) are *linear* in the derivatives of the fields, the determinant of the Hessian vanishes,

$$\det\left[\frac{\partial^2 \mathscr{L}_{\mathrm{D}}}{\partial\left(\partial_{\mu}\phi\right)\partial\left(\partial_{\nu}\overline{\phi}\right)}\right] = 0. \tag{27}$$

Therefore, Legendre transformations of the Lagrangian densities (22) and (25) are irregular. Nevertheless, as a Lagrangian density is determined only up to the divergence of an arbitrary vector function $F^{\mu}$ according to Eq. (7), one can construct an equivalent Lagrangian density $\mathscr{L}_{\mathrm{D}}'$ that yields identical Euler-Lagrange equations while yielding a regular Legendre transformation. The additional term [5] emerges as the divergence of a vector function $F^{\mu}$, which may be expressed in symmetric form as

$$F^{\mu} = \frac{i}{2m}\left(\overline{\phi}\,\sigma^{\mu\alpha}\frac{\partial \phi}{\partial x^{\alpha}} + \frac{\partial \overline{\phi}}{\partial x^{\alpha}}\sigma^{\alpha\mu}\,\phi\right).$$

The factor $m^{-1}$ was chosen to match the dimensions correctly. Explicitly, the additional term is given by

$$\begin{aligned} \frac{\partial F^{\beta}}{\partial x^{\beta}} &= \frac{i}{2m}\left(\partial_{\beta}\overline{\phi}\sigma^{\beta\alpha}\partial_{\alpha}\phi + \overline{\phi}\sigma^{\beta\alpha}\partial_{\beta}\partial_{\alpha}\phi + \partial_{\beta}\partial_{\alpha}\overline{\phi}\sigma^{\alpha\beta}\phi + \partial_{\alpha}\overline{\phi}\sigma^{\alpha\beta}\partial_{\beta}\phi\right) \\ &= \frac{i}{m}\frac{\partial \overline{\phi}}{\partial x^{\alpha}}\sigma^{\alpha\beta}\frac{\partial \phi}{\partial x^{\beta}}. \end{aligned}$$

Note that the double sums $\sigma^{\beta\alpha}\partial_\beta\partial_\alpha\phi$ and $\partial_\beta\partial_\alpha\overline{\phi}\sigma^{\alpha\beta}$ vanish identically, as we sum over a symmetric ($\partial_\mu\partial_\nu\phi = \partial_\nu\partial_\mu\phi$) and a skew-symmetric ($\sigma^{\mu\nu} = -\sigma^{\nu\mu}$) factor. Following Eq. (7), the equivalent Lagrangian density is given by $\mathscr{L}'_\mathrm{D} = \mathscr{L}_\mathrm{D} - \partial F^\beta/\partial x^\beta$, which means, explicitly,

$$\mathscr{L}'_\mathrm{D} = \frac{i}{2}\left(\overline{\phi}\gamma^\alpha\frac{\partial\phi}{\partial x^\alpha} - \frac{\partial\overline{\phi}}{\partial x^\alpha}\gamma^\alpha\phi\right) - \frac{i}{m}\frac{\partial\overline{\phi}}{\partial x^\alpha}\sigma^{\alpha\beta}\frac{\partial\phi}{\partial x^\beta} - m\overline{\phi}\phi. \tag{28}$$

Due to the skew-symmetry of the $\sigma^{\mu\nu}$, the Euler-Lagrange equations (2) for $\mathscr{L}'_\mathrm{D}$ yield again the Dirac equations (26). As desired, the Hessian of $\mathscr{L}'_\mathrm{D}$ is not singular,

$$\det\left[\frac{\partial^2\mathscr{L}'_\mathrm{D}}{\partial\left(\partial_\mu\overline{\phi}\right)\partial\left(\partial_\nu\phi\right)}\right] = -\frac{i}{m}\det\sigma^{\mu\nu} = -\frac{i}{m}, \qquad \nu\neq\mu, \tag{29}$$

hence, the Legendre transformation of the Lagrangian density $\mathscr{L}'_\mathrm{D}$ is now *regular*. It is remarkable that it is exactly a term which does *not* contribute to the Euler-Lagrange equations that makes the Legendre transformation of $\mathscr{L}'_\mathrm{D}$ *regular* and thus transfers the information on the dynamical system that is contained in the Lagrangian to the Hamiltonian description. The canonical momenta follow as

$$\begin{aligned}\overline{\pi}^\mu &= \frac{\partial\mathscr{L}'_\mathrm{D}}{\partial\left(\partial_\mu\phi\right)} = -\frac{i}{2}\overline{\phi}\gamma^\mu - \frac{i}{m}\frac{\partial\overline{\phi}}{\partial x^\alpha}\sigma^{\alpha\mu} \\ \pi^\mu &= \frac{\partial\mathscr{L}'_\mathrm{D}}{\partial\left(\partial_\mu\overline{\phi}\right)} = -\frac{i}{2}\gamma^\mu\phi - \frac{i}{m}\sigma^{\mu\alpha}\frac{\partial\phi}{\partial x^\alpha}.\end{aligned} \tag{30}$$

The Legendre transformation can now be worked out, yielding

$$\begin{aligned}\mathscr{H}_\mathrm{D} &= \overline{\pi}^\nu\partial_\nu\phi + \left(\partial_\nu\overline{\phi}\right)\pi^\nu - \mathscr{L}'_\mathrm{D} \\ &= -\frac{i}{m}\partial_\mu\overline{\phi}\,\sigma^{\mu\nu}\partial_\nu\phi + m\overline{\phi}\phi \\ &= \left(\overline{\pi}^\nu - \frac{i}{2}\overline{\phi}\gamma^\nu\right)\frac{\partial\phi}{\partial x^\nu} + m\overline{\phi}\phi.\end{aligned}$$

As the Hamiltonian density must always be expressed in terms of the canonical momenta rather then by the velocities, we must solve Eq. (30) for $\partial_\mu\phi$ and $\partial_\mu\overline{\phi}$. To this end, we multiply $\overline{\pi}^\mu$ by $\sigma_{\mu\nu}$ from the right, and $\pi^\mu$ by $\sigma_{\nu\mu}$ from the left,

$$\begin{aligned}\frac{\partial\overline{\phi}}{\partial x^\nu} &= im\left(\overline{\pi}^\mu - \frac{i}{2}\overline{\phi}\gamma^\mu\right)\sigma_{\mu\nu} \\ \frac{\partial\phi}{\partial x^\nu} &= im\,\sigma_{\nu\mu}\left(\pi^\mu + \frac{i}{2}\gamma^\mu\phi\right).\end{aligned} \tag{31}$$

The Dirac Hamiltonian density is then finally obtained as

$$\mathscr{H}_{\mathrm{D}} = im\left(\overline{\pi}^{\nu} - \frac{i}{2}\overline{\phi}\gamma^{\nu}\right)\sigma_{\nu\mu}\left(\pi^{\mu} + \frac{i}{2}\gamma^{\mu}\phi\right) + m\overline{\phi}\phi. \tag{32}$$

We may expand the products in Eq. (32) using Eqs. (24) to find

$$\mathscr{H}_{\mathrm{D}} = im\left(\overline{\pi}^{\mu}\sigma_{\mu\nu}\pi^{\nu} + \frac{1}{6}\overline{\pi}^{\nu}\gamma_{\nu}\phi - \frac{1}{6}\overline{\phi}\gamma_{\nu}\pi^{\nu}\right) + \frac{4}{3}m\,\overline{\phi}\phi. \tag{33}$$

In order to show that the Hamiltonian density $\mathscr{H}_{\mathrm{D}}$ describes the same dynamics as $\mathscr{L}_{\mathrm{D}}$ from Eq. (22), we set up the canonical equations

$$\begin{aligned}
\frac{\partial\overline{\phi}}{\partial x^{\nu}} &= \frac{\partial\mathscr{H}_{\mathrm{D}}}{\partial\pi^{\nu}} = im\left(\overline{\pi}^{\mu}\sigma_{\mu\nu} - \frac{1}{6}\overline{\phi}\,\gamma_{\nu}\right) \\
\frac{\partial\phi}{\partial x^{\mu}} &= \frac{\partial\mathscr{H}_{\mathrm{D}}}{\partial\overline{\pi}^{\mu}} = im\left(\sigma_{\mu\nu}\pi^{\nu} + \frac{1}{6}\gamma_{\mu}\phi\right).
\end{aligned}$$

Obviously, these equations reproduce the definition of the canonical momenta from Eqs. (30) in their inverted form given by Eqs. (31). The second set of canonical equations follows from the $\phi$ and $\overline{\phi}$ dependence of the Hamiltonian $\mathscr{H}_{\mathrm{D}}$,

$$\begin{aligned}
\frac{\partial\overline{\pi}^{\alpha}}{\partial x^{\alpha}} = -\frac{\partial\mathscr{H}_{\mathrm{D}}}{\partial\phi} &= -\frac{im}{6}\overline{\pi}^{\mu}\gamma_{\mu} - \frac{4m}{3}\overline{\phi} \\
&= -\frac{im}{6}\left(\frac{i}{2}\overline{\phi}\gamma^{\mu} - \frac{i}{m}\partial_{\nu}\overline{\phi}\,\sigma^{\nu\mu}\right)\gamma_{\mu} - \frac{4m}{3}\overline{\phi} \\
&= -m\overline{\phi} - \frac{i}{2}\partial_{\nu}\overline{\phi}\gamma^{\nu} \\
\frac{\partial\pi^{\alpha}}{\partial x^{\alpha}} = -\frac{\partial\mathscr{H}_{\mathrm{D}}}{\partial\overline{\phi}} &= \frac{im}{6}\gamma_{\mu}\pi^{\mu} - \frac{4m}{3}\phi \\
&= -\frac{im}{6}\gamma_{\mu}\left(\frac{i}{2}\gamma^{\mu}\phi + \frac{i}{m}\sigma^{\mu\nu}\partial_{\nu}\phi\right) - \frac{4m}{3}\phi \\
&= -m\phi + \frac{i}{2}\gamma^{\nu}\partial_{\nu}\phi.
\end{aligned}$$

The divergences of the canonical momenta follow equally from the derivatives of the first canonical equations, or, equivalently, from the derivatives of Eqs. (30),

$$\begin{aligned}
\partial_{\mu}\overline{\pi}^{\mu} &= \frac{i}{2}\partial_{\mu}\overline{\phi}\gamma^{\mu} - \frac{i}{m}\partial_{\mu}\partial_{\nu}\overline{\phi}\,\sigma^{\nu\mu} = \frac{i}{2}\partial_{\mu}\overline{\phi}\gamma^{\mu} \\
\partial_{\mu}\pi^{\mu} &= -\frac{i}{2}\gamma^{\mu}\partial_{\mu}\phi - \frac{i}{m}\sigma^{\nu\mu}\,\partial_{\mu}\partial_{\nu}\phi = -\frac{i}{2}\gamma^{\mu}\partial_{\mu}\phi.
\end{aligned}$$

The terms containing the second derivatives of $\phi$ and $\overline{\phi}$ vanish due to the skew-symmetry of $\sigma^{\mu\nu}$. Equating finally the expressions for the divergences of the canonical momenta, we encounter, as expected, the Dirac equations (26)

$$\begin{aligned}\frac{i}{2}\frac{\partial\overline{\phi}}{\partial x^{\alpha}}\gamma^{\alpha} &= -m\overline{\phi} - \frac{i}{2}\frac{\partial\overline{\phi}}{\partial x^{\nu}}\gamma^{\nu} \\ -\frac{i}{2}\gamma^{\alpha}\frac{\partial\phi}{\partial x^{\alpha}} &= -m\phi + \frac{i}{2}\gamma^{\nu}\frac{\partial\phi}{\partial x^{\nu}}.\end{aligned}$$

It should be mentioned that this section is similar to the derivation of the Dirac Hamiltonian density in Ref. [6]. We note that the additional term in the Dirac Lagrangian density $\mathcal{L}_{\mathrm{D}}'$ from Eq. (28)—as compared to the Lagrangian $\mathcal{L}_{\mathrm{D}}$ from Eq. (25)—entails additional terms in the energy-momentum tensor, namely,

$$T_{\mu}^{\nu\prime} - T_{\mu}^{\nu} \equiv j_{\mu}^{\nu}(\boldsymbol{x}) = -\frac{i}{m}\left(\partial_{\alpha}\overline{\phi}\sigma^{\alpha\nu}\partial_{\mu}\phi + \partial_{\mu}\overline{\phi}\sigma^{\nu\alpha}\partial_{\alpha}\phi - \delta_{\mu}^{\nu}\partial_{\alpha}\overline{\phi}\sigma^{\alpha\beta}\partial_{\beta}\phi\right).$$

We easily convince ourselves by direct calculation that the divergences of $T_{\mu}^{\nu\prime}$ and $T_{\mu}^{\nu}$ coincide,

$$\begin{aligned}\frac{\partial j_{\mu}^{\lambda}}{\partial x^{\lambda}} =& -\frac{i}{m}\Big(\partial_{\lambda}\partial_{\alpha}\overline{\phi}\sigma^{\alpha\lambda}\partial_{\mu}\phi + \partial_{\alpha}\overline{\phi}\sigma^{\alpha\lambda}\partial_{\lambda}\partial_{\mu}\phi + \partial_{\lambda}\partial_{\mu}\overline{\phi}\sigma^{\lambda\alpha}\partial_{\alpha}\phi \\ &+ \partial_{\mu}\overline{\phi}\sigma^{\lambda\alpha}\partial_{\lambda}\partial_{\alpha}\phi - \delta_{\mu}^{\lambda}\partial_{\lambda}\partial_{\alpha}\overline{\phi}\sigma^{\alpha\beta}\partial_{\beta}\phi - \delta_{\mu}^{\lambda}\partial_{\alpha}\overline{\phi}\sigma^{\alpha\beta}\partial_{\lambda}\partial_{\beta}\phi\Big) \\ =& -\frac{i}{m}\Big(\partial_{\alpha}\overline{\phi}\sigma^{\alpha\lambda}\partial_{\lambda}\partial_{\mu}\phi + \partial_{\lambda}\partial_{\mu}\overline{\phi}\sigma^{\lambda\alpha}\partial_{\alpha}\phi \\ &- \partial_{\mu}\partial_{\alpha}\overline{\phi}\sigma^{\alpha\beta}\partial_{\beta}\phi - \partial_{\alpha}\overline{\phi}\sigma^{\alpha\beta}\partial_{\mu}\partial_{\beta}\phi\Big) \\ =& \,0,\end{aligned}$$

which means that both energy-momentum tensors represent the same physical system. For each index $\mu$, $j_{\mu}^{\nu}(\boldsymbol{x})$ represents a conserved current vector which are all associated with the transformation from $\mathcal{L}_{\mathrm{D}}$ to $\mathcal{L}_{\mathrm{D}}'$.

# 4 Examples of Canonical Transformations in Covariant Hamiltonian Field Theory

The formalism of canonical transformations that was worked out in Sect. 2 is now shown to yield a generalized representation of Noether's theorem. Furthermore, a generalized theory of $U(N)$ gauge transformations is outlined.

## *4.1 Generalized Noether Theorem*

Canonical transformations are defined by Eq. (6) as the particular subset of general transformations of the fields $\phi^{I}$ and their conjugate momentum vector fields $\pi_{I}$

that preserve the action functional (6). Such a transformation depicts a symmetry transformation that is associated with a conserved four-current vector, hence with a vector whose space-time divergence vanishes [7]. In the following, we shall work out the correlation of this conserved current by means an *infinitesimal* canonical transformation of the field variables. The generating function $F_2^\mu$ of an *infinitesimal* transformation differs from that of an *identical* transformation by a infinitesimal parameter $\delta\varepsilon \neq 0$ times an as yet arbitrary function $g^\mu(\phi^I, \boldsymbol{\pi}_I, \boldsymbol{x})$,

$$F_2^\mu(\phi^I, \boldsymbol{\Pi}_I, \boldsymbol{x}) = \phi^J\, \Pi_J^\mu + \delta\varepsilon\, g^\mu(\phi^I, \boldsymbol{\pi}_I, \boldsymbol{x}). \tag{34}$$

To first order in $\delta\varepsilon$, the subsequent transformation rules follow from the general rules (11) as

$$\pi_I^\mu = \frac{\partial F_2^\mu}{\partial \phi^I} = \Pi_I^\mu + \delta\varepsilon\, \frac{\partial g^\mu}{\partial \phi^I}, \qquad \Phi^I\, \delta_\nu^\mu = \frac{\partial F_2^\mu}{\partial \Pi_I^\nu} = \phi^I\, \delta_\nu^\mu + \delta\varepsilon\, \frac{\partial g^\mu}{\partial \pi_I^\nu},$$
$$\mathscr{H}' = \mathscr{H} + \left.\frac{\partial F_2^\alpha}{\partial x^\alpha}\right|_{\mathrm{expl}} = \mathscr{H} + \delta\varepsilon\, \left.\frac{\partial g^\alpha}{\partial x^\alpha}\right|_{\mathrm{expl}},$$

hence

$$\delta\pi_I^\mu = -\delta\varepsilon\, \frac{\partial g^\mu}{\partial \phi^I}, \qquad \delta\phi^I\, \delta_\nu^\mu = \delta\varepsilon\, \frac{\partial g^\mu}{\partial \pi_I^\nu}, \qquad \delta\mathscr{H}|_{\mathrm{CT}} = \delta\varepsilon\, \left.\frac{\partial g^\alpha}{\partial x^\alpha}\right|_{\mathrm{expl}}. \tag{35}$$

As the transformation does not change the independent variables, $x^\mu$, both the original as well as the transformed fields refer to the same space-time event $\boldsymbol{x}$, hence $\delta x^\mu = 0$. Making use of the canonical field equations (5), the variation of $\mathscr{H}$ due to the variations (35) of the canonical field variables $\phi^I$ and $\pi_I^\mu$ emerges as

$$\begin{aligned}
\delta\mathscr{H} &= \frac{\partial \mathscr{H}}{\partial \phi^I}\, \delta\phi^I + \frac{\partial \mathscr{H}}{\partial \pi_I^\alpha}\, \delta\pi_I^\alpha = -\frac{\partial \pi_I^\beta}{\partial x^\alpha}\, \delta_\beta^\alpha\, \delta\phi^I + \frac{\partial \phi^I}{\partial x^\alpha}\, \delta\pi_I^\alpha \\
&= -\delta\varepsilon \left( \frac{\partial g^\alpha}{\partial \pi_I^\beta} \frac{\partial \pi_I^\beta}{\partial x^\alpha} + \frac{\partial g^\alpha}{\partial \phi^I} \frac{\partial \phi^I}{\partial x^\alpha} \right) = -\delta\varepsilon \left( \frac{\partial g^\alpha}{\partial x^\alpha} - \left.\frac{\partial g^\alpha}{\partial x^\alpha}\right|_{\mathrm{expl}} \right) \\
&= -\delta\varepsilon \frac{\partial g^\alpha}{\partial x^\alpha} + \delta\mathscr{H}|_{\mathrm{CT}}.
\end{aligned} \tag{36}$$

If and only if the infinitesimal transformation rule $\delta\mathscr{H}|_{\mathrm{CT}}$ for the Hamiltonian from Eqs. (35) coincides with the variation $\delta\mathscr{H}$ at $\delta x^\mu = 0$ from Eq. (36), then the set of infinitesimal transformation rules is consistent and actually defines a *canonical* transformation. We thus have

$$\delta\mathscr{H}|_{\mathrm{CT}} \stackrel{!}{=} \delta\mathscr{H} \quad \Longleftrightarrow \quad \frac{\partial g^\alpha}{\partial x^\alpha} \stackrel{!}{=} 0. \tag{37}$$

Thus, the divergence of the characteristic function $g^{\mu}(\boldsymbol{x})$ in the generating function (34) must vanish in order for the transformation (35) to be *canonical*, and hence to preserve the form of the action functional (6). The $g^{\mu}(\boldsymbol{x})$ then define a conserved four-current vector, commonly referred to as *Noether current*. The canonical transformation rules then furnish the corresponding infinitesimal one-parameter group of symmetry transformations

$$\frac{\partial g^{\alpha}(\boldsymbol{x})}{\partial x^{\alpha}} = 0 \tag{38}$$
$$\delta\pi_I^{\mu} = -\delta\varepsilon\,\frac{\partial g^{\mu}}{\partial\phi^I}, \qquad \delta\phi^I\,\delta_{\nu}^{\mu} = \delta\varepsilon\,\frac{\partial g^{\mu}}{\partial\pi_I^{\nu}}, \qquad \delta\mathscr{H} = \delta\varepsilon\,\left.\frac{\partial g^{\alpha}}{\partial x^{\alpha}}\right|_{\text{expl}}.$$

We can now formulate the generalized Noether theorem and its inverse in the realm of covariant Hamiltonian field theory as:

**Theorem 4.1 (generalized Noether)** *The characteristic vector function $g^{\mu}(\phi^I, \boldsymbol{\pi}_I, \boldsymbol{x})$ in the generating function $F_2^{\mu}$ from Eq. (34) must have* zero divergence *in order to define a canonical transformation. The subsequent transformation rules (38) then define an infinitesimal one-parameter group of symmetry transformations that preserve the form of the action functional (6).*

*Conversely, if a one-parameter symmetry transformation is known to preserve the form of the action functional (6), then the transformation is* canonical *and hence can be derived from a generating function. The characteristic* 4-*vector function $g^{\mu}(\phi^I, \boldsymbol{\pi}_I, \boldsymbol{x})$ in the corresponding* infinitesimal *generating function (34) then represents a conserved current, hence $\partial g^{\alpha}/\partial x^{\alpha} = 0$.*

In contrast to the usual derivation of this theorem in the Lagrangian formalism, we are not restricted to point transformations as the $g^{\mu}$ may be *any* divergence-free 4-vector function of the given dynamical system. In this sense, we have found a generalization of Noether's theorem.

### 4.1.1 Gauge Invariance of the Electromagnetic 4-Potential

For the Maxwell Hamiltonian $\mathscr{H}_{\mathrm{M}}$ from Eq. (18), the correlation of the 4-vector potential $a^{\mu}$ with the conjugate fields $p_{\mu\nu}$ is determined by the first field equation (19) as the generalized curl of $\boldsymbol{a}$. This means on the other hand that the correlation between $\boldsymbol{a}$ and the $p_{\mu\nu}$ is *not unique*. Defining a transformed 4-vector potential $\boldsymbol{A}$ according to

$$A_{\mu} = a_{\mu} + \frac{\partial\chi(\boldsymbol{x})}{\partial x^{\mu}}, \tag{39}$$

with $\chi = \chi(\boldsymbol{x})$ an arbitrary differentiable function of the independent variables. This means for the transformation of the $p_{\mu\nu}$

$$p_{\mu\nu} = \frac{\partial a_{\nu}}{\partial x^{\mu}} - \frac{\partial a_{\mu}}{x^{\nu}} = \frac{\partial A_{\nu}}{\partial x^{\mu}} - \cancel{\frac{\partial^2\chi(\boldsymbol{x})}{\partial x^{\nu}\partial x^{\mu}}} - \frac{\partial A_{\mu}}{\partial x^{\nu}} + \cancel{\frac{\partial^2\chi(\boldsymbol{x})}{\partial x^{\mu}\partial x^{\nu}}} = P_{\mu\nu}. \tag{40}$$

The transformations (39) and (40) can be regarded as a canonical transformation, whose generating function $F_2^{\mu}$ is given by

$$F_2^{\mu}(\boldsymbol{a}, \boldsymbol{P}, \boldsymbol{x}) = a_{\alpha} P^{\alpha\mu} + \frac{\partial}{\partial x^{\alpha}} \left( P^{\alpha\mu} \chi(\boldsymbol{x}) \right). \tag{41}$$

For a vector field $\boldsymbol{a}$ and its set of canonical conjugate fields $\boldsymbol{p}^{\mu}$, the general transformation rules (11) are rewritten as

$$p^{\nu\mu} = \frac{\partial F_2^{\mu}}{\partial a_{\nu}}, \qquad A_{\nu}\, \delta_{\beta}^{\mu} = \frac{\partial F_2^{\mu}}{\partial P^{\nu\beta}}, \qquad \mathscr{H}' = \mathscr{H} + \left.\frac{\partial F_2^{\alpha}}{\partial x^{\alpha}}\right|_{\text{expl}}, \tag{42}$$

which yield for the particular generating function of Eq. (41) the transformation prescriptions

$$\begin{aligned}
p^{\nu\mu} &= \frac{\partial a_{\alpha}}{\partial a_{\nu}} P^{\alpha\mu} = \delta_{\alpha}^{\nu} P^{\alpha\mu} = P^{\nu\mu} \\
A_{\nu}\, \delta_{\beta}^{\mu} &= a_{\alpha} \delta_{\nu}^{\alpha} \delta_{\beta}^{\mu} + \delta_{\nu}^{\alpha} \delta_{\beta}^{\mu} \frac{\partial \chi(\boldsymbol{x})}{\partial x^{\alpha}} \\
\Rightarrow \quad A_{\nu} &= a_{\nu} + \frac{\partial \chi(\boldsymbol{x})}{\partial x^{\nu}} \\
\mathscr{H}' - \mathscr{H} &= \frac{\partial^2 p^{\alpha\beta}}{\partial x^{\alpha} \partial x^{\beta}} \chi(\boldsymbol{x}) + \frac{\partial p^{\alpha\beta}}{\partial x^{\alpha}} \frac{\partial \chi(\boldsymbol{x})}{\partial x^{\beta}} + p^{\alpha\beta} \frac{\partial^2 \chi(\boldsymbol{x})}{\partial x^{\alpha} \partial x^{\beta}} \\
&= -\frac{\partial p^{\alpha\beta}}{\partial x^{\beta}} \frac{\partial \chi(\boldsymbol{x})}{\partial x^{\alpha}}.
\end{aligned}$$

The canonical transformation rules coincide with the correlations of Eqs. (39) and (40) defining the Lorentz gauge. The last equation holds because of the skew-symmetry of the canonical momentum tensor $p^{\nu\mu} = -p^{\mu\nu}$. In oder to determine the conserved Noether current that is associated with the canonical point transformation generated by $\boldsymbol{F}_2$ from Eq. (41), we need the generator of the corresponding *infinitesimal* canonical point transformation,

$$F_2^{\mu}(\boldsymbol{a}, \boldsymbol{P}, \boldsymbol{x}) = a_{\alpha} P^{\alpha\mu} + \varepsilon g^{\mu}(\boldsymbol{p}, \boldsymbol{x}), \qquad g^{\mu} = \frac{\partial}{\partial x^{\alpha}} \left[ p^{\alpha\mu} \chi(\boldsymbol{x}) \right].$$

Herein, $\varepsilon \neq 0$ denotes a small parameter. The pertaining infinitesimal canonical transformation rules are

$$\begin{aligned}
p^{\nu\mu} &= \frac{\partial F_2^{\mu}}{\partial a_{\nu}} = P^{\nu\mu}, \qquad A_{\nu} = a_{\nu} + \varepsilon \frac{\partial \chi(\boldsymbol{x})}{\partial x^{\nu}} \\
\delta \mathscr{H}|_{\text{CT}} &= \left.\frac{\partial F_2^{\alpha}}{\partial x^{\alpha}}\right|_{\text{expl}} = \mathscr{H}_{\text{M}}' - \mathscr{H}_{\text{M}} = -\varepsilon \frac{\partial p^{\alpha\beta}}{\partial x^{\beta}} \frac{\partial \chi(\boldsymbol{x})}{\partial x^{\alpha}}.
\end{aligned}$$

The coordinate transformation rules agree with Eqs. (39) and (40) in the finite limit. Because of $\delta p^{\nu\mu} \equiv P^{\nu\mu} - p^{\nu\mu} = 0$, the variation $\delta\mathcal{H}$ due to the variation of the canonical variables reduces to the term proportional to $\delta a_\nu \equiv A_\nu - a_\nu$,

$$\delta\mathcal{H} = \frac{\partial \mathcal{H}_{\mathrm{M}}}{\partial a_\alpha}\,\delta a_\alpha = -\varepsilon \frac{\partial p^{\alpha\beta}}{\partial x^\beta}\frac{\partial \chi(\boldsymbol{x})}{\partial x^\alpha}.$$

Hence, $\delta\mathcal{H}$ coincides with the corresponding canonical transformation rule $\delta\mathcal{H}|_{\mathrm{CT}}$, as required for the transformation to be canonical. With the requirement (37) fulfilled, the characteristic function $g^\mu(\boldsymbol{p}, \boldsymbol{x})$ in the infinitesimal generating function $F_2^\mu$ then directly yields the conserved 4-current $\boldsymbol{j}_{\mathrm{N}}(\boldsymbol{x})$, $j_{\mathrm{N}}^\mu = g^\mu$ according to Noether's theorem from Eq. (38)

$$\frac{\partial j_{\mathrm{N}}^\alpha(\boldsymbol{x})}{\partial x^\alpha} = 0, \qquad j_{\mathrm{N}}^\mu(\boldsymbol{x}) = \frac{\partial}{\partial x^\alpha}\left(p^{\alpha\mu}\chi(\boldsymbol{x})\right).$$

By calculating its divergence, we verify directly that $\boldsymbol{j}_{\mathrm{N}}(\boldsymbol{x})$ is indeed the conserved Noether current that corresponds to the symmetry transformation (39)

$$\begin{aligned}
\frac{\partial j_{\mathrm{N}}^\beta(\boldsymbol{x})}{\partial x^\beta} &= \frac{\partial}{\partial x^\beta}\left(\frac{\partial p^{\alpha\beta}}{\partial x^\alpha}\chi + p^{\alpha\beta}\frac{\partial \chi}{\partial x^\alpha}\right) \\
&= \frac{\partial^2 p^{\alpha\beta}}{\partial x^\alpha \partial x^\beta}\chi + \frac{\partial p^{\alpha\beta}}{\partial x^\alpha}\frac{\partial \chi}{\partial x^\beta} + \frac{\partial p^{\alpha\beta}}{\partial x^\beta}\frac{\partial \chi}{\partial x^\alpha} + p^{\alpha\beta}\frac{\partial^2 \chi}{\partial x^\alpha \partial x^\beta} \\
&= 0.
\end{aligned}$$

The first and the fourth term on the right hand side vanish individually due to $p^{\nu\mu} = -p^{\mu\nu}$. The second and the third terms cancel each other for the same reason.

## 4.2 General Local $U(N)$ Gauge Transformation

As an interesting example of a canonical transformation in the covariant Hamiltonian description of classical fields, the general local $U(N)$ gauge transformation is treated in this section. The main feature of the approach is that the terms to be added to a given Hamiltonian $\mathcal{H}$ in order to render it *locally* gauge invariant only depends on the *type of fields* contained in the Hamiltonian $\mathcal{H}$ and not on the particular form of the original Hamiltonian itself. The only precondition is that $\mathcal{H}$ must be invariant under the corresponding *global* gauge transformation, hence a transformation *not* depending explicitly on $\boldsymbol{x}$.

### 4.2.1 External Gauge Field

We consider a system consisting of a vector of $N$ complex fields $\phi_I$, $I = 1, \ldots, N$, and the adjoint field vector, $\boldsymbol{\phi}^\dagger$,

$$\boldsymbol{\phi} = \begin{pmatrix} \phi_1 \\ \vdots \\ \phi_N \end{pmatrix}, \qquad \boldsymbol{\phi}^\dagger = \left(\overline{\phi}_1 \cdots \overline{\phi}_N\right).$$

A general local linear transformation may be expressed in terms of a complex matrix $U(\boldsymbol{x}) = (u_{IJ}(\boldsymbol{x}))$ and its adjoint, $U^\dagger$ that may depend explicitly on the independent variables, $x^\mu$, as

$$\begin{aligned} \boldsymbol{\Phi} &= U\,\boldsymbol{\phi}, & \boldsymbol{\Phi}^\dagger &= \boldsymbol{\phi}^\dagger\,U^\dagger \\ \Phi_I &= u_{IJ}\,\phi_J, & \overline{\Phi}_I &= \overline{\phi}_J\,\overline{u}_{JI}. \end{aligned} \tag{43}$$

With this notation, $\phi_I$ may stand for a set of $I = 1, \ldots, N$ complex scalar fields $\phi_I$ or Dirac spinors. In other words, $U$ is supposed to define an isomorphism within the space of the $\phi_I$, hence to linearly map the $\phi_I$ into objects of the same type. The uppercase Latin letter indexes label the field or spinor number. Their transformation in iso-space are not associated with any metric. We, therefore, do not use superscripts for these indexes as there is not distinction between covariant and contravariant components. In contrast, Greek indexes are used for those components that *are* associated with a metric—such as the derivatives with respect to a space-time variable, $x^\mu$. As usual, summation is understood for indexes occurring in pairs.

We restrict ourselves to transformations that preserve the norm $\overline{\boldsymbol{\phi}}\boldsymbol{\phi}$

$$\begin{aligned} \overline{\boldsymbol{\Phi}}\boldsymbol{\Phi} &= \overline{\boldsymbol{\phi}}\,U^\dagger U\,\boldsymbol{\phi} = \overline{\boldsymbol{\phi}}\boldsymbol{\phi} & &\Longrightarrow & U^\dagger U &= \mathbb{1} = U U^\dagger \\ \overline{\Phi}_I \Phi_I &= \overline{\phi}_J \overline{u}_{JI}\,u_{IK}\phi_K = \overline{\phi}_K\phi_K & &\Longrightarrow & \overline{u}_{JI}\,u_{IK} &= \delta_{JK} = u_{JI}\,\overline{u}_{IK}. \end{aligned}$$

This means that $U^\dagger = U^{-1}$, hence that the matrix $U$ is supposed to be *unitary*. The transformation (43) follows from a generating function that—corresponding to $\mathcal{H}$—must be a real-valued function of the generally complex fields $\boldsymbol{\phi}$ and their canonical conjugates, $\boldsymbol{\pi}^\mu$,

$$\begin{aligned} F_2^\mu(\boldsymbol{\phi}, \overline{\boldsymbol{\phi}}, \boldsymbol{\Pi}^\mu, \overline{\boldsymbol{\Pi}}^\mu, \boldsymbol{x}) &= \overline{\boldsymbol{\Pi}}^\mu\,U\,\boldsymbol{\phi} + \overline{\boldsymbol{\phi}}\,U^\dagger\,\boldsymbol{\Pi}^\mu \\ &= \overline{\Pi}_K^\mu\,u_{KJ}\,\phi_J + \overline{\phi}_K\,\overline{u}_{KJ}\,\Pi_J^\mu. \end{aligned} \tag{44}$$

According to Eqs. (11) the set of transformation rules follows as

$$\begin{aligned} \overline{\pi}_I^\mu &= \frac{\partial F_2^\mu}{\partial \phi_I} = \overline{\Pi}_K^\mu u_{KJ}\delta_{IJ}, & \overline{\Phi}_I\delta_\nu^\mu &= \frac{\partial F_2^\mu}{\partial \Pi_I^\nu} = \overline{\phi}_K \overline{u}_{KJ}\delta_\nu^\mu\delta_{IJ} \\ \pi_I^\mu &= \frac{\partial F_2^\mu}{\partial \overline{\phi}_I} = \delta_{IK}\overline{u}_{KJ}\Pi_J^\mu, & \Phi_I\delta_\nu^\mu &= \frac{\partial F_2^\mu}{\partial \overline{\Pi}_I^\nu} = \delta_\nu^\mu\delta_{IK}u_{KJ}\phi_J. \end{aligned}$$

The complete set of transformation rules and their inverses then read in component notation

$$
\begin{aligned}
\Phi_I &= u_{IJ}\,\phi_J, & \overline{\Phi}_I &= \overline{\phi}_J\,\overline{u}_{JI}, & \Pi_I^\mu &= u_{IJ}\,\pi_J^\mu, & \overline{\Pi}_I^\mu &= \overline{\pi}_J^\mu\,\overline{u}_{JI} \\
\phi_I &= \overline{u}_{IJ}\,\Phi_J, & \overline{\phi}_I &= \overline{\Phi}_J u_{JI}, & \pi_I^\mu &= \overline{u}_{IJ}\,\Pi_J^\mu, & \overline{\pi}_I^\mu &= \overline{\Pi}_J^\mu u_{JI}.
\end{aligned}
\tag{45}
$$

We assume the Hamiltonian $\mathscr{H}$ to be *form-invariant* under the *global* gauge transformation (43), which is given for $U = \text{const.}$, hence for all $u_{IJ}$ *not* depending on the independent variables, $x^\mu$. In contrast, if $U = U(\boldsymbol{x})$, the transformation (45) is referred to as a *local* gauge transformation. The transformation rule for the Hamiltonian is then determined by the explicitly $x^\mu$-dependent terms of the generating function $F_2^\mu$ according to

$$
\begin{aligned}
\mathscr{H}' - \mathscr{H} = \left.\frac{\partial F_2^\alpha}{\partial x^\alpha}\right|_{\text{expl}} &= \overline{\Pi}_I^\alpha \frac{\partial u_{IJ}}{\partial x^\alpha}\,\phi_J + \overline{\phi}_I \frac{\partial \overline{u}_{IJ}}{\partial x^\alpha}\,\Pi_J^\alpha \\
&= \overline{\pi}_K^\alpha\,\overline{u}_{KI}\frac{\partial u_{IJ}}{\partial x^\alpha}\phi_J + \overline{\phi}_I \frac{\partial \overline{u}_{IJ}}{\partial x^\alpha}\,u_{JK}\pi_K^\alpha \\
&= \left(\overline{\pi}_K^\alpha\,\phi_J - \overline{\phi}_K\pi_J^\alpha\right)\overline{u}_{KI}\frac{\partial u_{IJ}}{\partial x^\alpha}.
\end{aligned}
\tag{46}
$$

In the last step, the identity

$$
\frac{\partial \overline{u}_{JI}}{\partial x^\mu}\,u_{IK} + \overline{u}_{JI}\,\frac{\partial u_{IK}}{\partial x^\mu} = 0
$$

was inserted. If we want to set up a Hamiltonian $\overline{\mathscr{H}}$ that is *form-invariant* under the *local*, hence $x^\mu$-dependent transformation generated by (44), then we must compensate the additional terms (46) that emerge from the explicit $x^\mu$-dependence of the generating function (44). The only way to achieve this is to *adjoin* the Hamiltonian $\mathscr{H}$ of our system with terms that correspond to (46) with regard to their dependence on the canonical variables, $\boldsymbol{\phi}$, $\overline{\boldsymbol{\phi}}$, $\boldsymbol{\pi}^\mu$, $\overline{\boldsymbol{\pi}}^\mu$. With a *unitary* matrix $U$, the $u_{IJ}$-dependent terms in Eq. (46) are *skew-hermitian*,

$$
\overline{\overline{u}_{KI}\,\frac{\partial u_{IJ}}{\partial x^\mu}} = \frac{\partial \overline{u}_{JI}}{\partial x^\mu}\,u_{IK} = -\overline{u}_{JI}\,\frac{\partial u_{IK}}{\partial x^\mu}, \qquad \overline{\frac{\partial u_{KI}}{\partial x^\mu}\,\overline{u}_{IJ}} = u_{JI}\frac{\partial \overline{u}_{IK}}{\partial x^\mu} = -\frac{\partial u_{JI}}{\partial x^\mu}\overline{u}_{IK},
$$

or in matrix notation

$$
\left(U^\dagger \frac{\partial U}{\partial x^\mu}\right)^\dagger = \frac{\partial U^\dagger}{\partial x^\mu}U = -U^\dagger\frac{\partial U}{\partial x^\mu}, \qquad \left(\frac{\partial U}{\partial x^\mu}U^\dagger\right)^\dagger = U\frac{\partial U^\dagger}{\partial x^\mu} = -\frac{\partial U}{\partial x^\mu}U^\dagger.
$$

The $u$-dependent terms in Eq. (46) can thus be compensated by a *Hermitian* matrix $(\boldsymbol{a}_{KJ})$ of "4-vector gauge fields", with each off-diagonal matrix element, $\boldsymbol{a}_{KJ}$, $K \neq J$, a complex 4-vector field with components $a_{KJ\mu}$, $\mu = 0, \ldots, 3$

$$
a_{KJ\mu} = \overline{a}_{JK\mu}.
$$

The number of independent gauge fields thus amount to $N^2$ real 4-vectors. The amended Hamiltonian $\bar{\mathscr{H}}$ thus reads

$$\bar{\mathscr{H}} = \mathscr{H} + \mathscr{H}_{\mathrm{a}}, \qquad \mathscr{H}_{\mathrm{a}} = iq\left(\overline{\pi}^{\alpha}_{K}\phi_J - \overline{\phi}_K\pi^{\alpha}_{J}\right)a_{KJ\alpha}. \tag{47}$$

With the real coupling constant $q$, the interaction Hamiltonian $\mathscr{H}_{\mathrm{a}}$ is thus real. Submitting the amended Hamiltonian $\bar{\mathscr{H}}$ to the canonical transformation generated by Eq. (44), the new Hamiltonian $\bar{\mathscr{H}}'$ emerges as

$$\begin{aligned}
\bar{\mathscr{H}}' &= \bar{\mathscr{H}} + \left.\frac{\partial F_2^{\alpha}}{\partial x^{\alpha}}\right|_{\mathrm{expl}} \\
&= \mathscr{H} + \left(\overline{\pi}^{\alpha}_{K}\phi_J - \overline{\phi}_K\pi^{\alpha}_{J}\right)\left(iq\,a_{KJ\alpha} + \overline{u}_{KI}\frac{\partial u_{IJ}}{\partial x^{\alpha}}\right) \\
&= \mathscr{H} + \left(\overline{\Pi}^{\alpha}_{K}\Phi_J - \overline{\Phi}_K\Pi^{\alpha}_{J}\right)\left(iq\,u_{KL}a_{LI\alpha}\overline{u}_{IJ} + \frac{\partial u_{KI}}{\partial x^{\alpha}}\overline{u}_{IJ}\right) \\
&\stackrel{!}{=} \mathscr{H} + \left(\overline{\Pi}^{\alpha}_{K}\Phi_J - \overline{\Phi}_K\Pi^{\alpha}_{J}\right)iq\,A_{KJ\alpha},
\end{aligned}$$

with the $A_{IJ\mu}$ defining the gauge fields of the transformed system. The *form* of the interaction Hamiltonian $\mathscr{H}_{\mathrm{a}}$ is thus maintained under the canonical transformation,

$$\bar{\mathscr{H}}' = \mathscr{H} + \mathscr{H}'_{\mathrm{a}}, \qquad \mathscr{H}'_{\mathrm{a}} = iq\left(\overline{\Pi}^{\alpha}_{K}\Phi_J - \overline{\Phi}_K\Pi^{\alpha}_{J}\right)A_{KJ\alpha}, \tag{48}$$

provided that the gauge fields obey the transformation rule

$$A_{KJ\mu} = u_{KL}\,a_{LI\mu}\,\overline{u}_{IJ} - \frac{i}{q}\frac{\partial u_{KI}}{\partial x^{\mu}}\,\overline{u}_{IJ}. \tag{49}$$

We observe that for any type of canonical field variables $\phi_I$ and for any Hamiltonian system $\mathscr{H}$, the transformation of the 4-vector gauge fields $\boldsymbol{a}_{IJ}(\boldsymbol{x})$ is uniquely determined according to Eq. (49) by the transformation matrix $U(\boldsymbol{x})$ for the $N$ fields $\phi_I$. In the notation of the 4-vector gauge fields $\boldsymbol{a}_{KJ}(\boldsymbol{x})$, $K, J = 1, \ldots, N$, the transformation rule is equivalently expressed as

$$\boldsymbol{A}_{KJ} = u_{KL}\,\boldsymbol{a}_{LI}\,\overline{u}_{IJ} - \frac{i}{q}\frac{\partial u_{KI}}{\partial \boldsymbol{x}}\,\overline{u}_{IJ},$$

or, in matrix notation

$$\hat{A}_{\mu} = U\hat{a}_{\mu}U^{\dagger} - \frac{i}{q}\frac{\partial U}{\partial x^{\mu}}U^{\dagger}, \qquad \hat{\boldsymbol{A}} = U\,\hat{\boldsymbol{a}}\,U^{\dagger} - \frac{i}{q}\frac{\partial U}{\partial \boldsymbol{x}}\,U^{\dagger}, \tag{50}$$

with $\hat{a}_{\mu}$ denoting the $N \times N$ matrices of the $\mu$-components of the 4-vectors $\boldsymbol{A}_{IK}(\boldsymbol{x})$, and, finally, $\hat{\boldsymbol{a}}$ the $N \times N$ matrix of gauge 4-vectors $\boldsymbol{a}_{IK}(\boldsymbol{x})$. The matrix $U(\boldsymbol{x})$ is *unitary*, hence belongs to the group $U(N)$

$$U^{\dagger}(\boldsymbol{x}) = U^{-1}(\boldsymbol{x}), \qquad |\det U(\boldsymbol{x})| = 1.$$

For $\det U(\boldsymbol{x}) = +1$, the matrix $U(\boldsymbol{x})$ is an element of the group $SU(N)$.

Equation (50) is the general transformation law for gauge bosons. $U$ and $\hat{a}_{\mu}$ do not commute if $N > 1$, hence if $U$ is a unitary matrix rather than a complex number of modulus 1. We are then dealing with a non-Abelian gauge theory. As the matrices $\hat{a}_{\mu}$ are Hermitian, the number of independent gauge 4-vectors $\boldsymbol{a}_{IK}$ amounts to $N$ real vectors on the main diagonal, and $(N^2 - N)/2$ independent complex off-diagonal vectors, which corresponds to a total number of $N^2$ independent real gauge 4-vectors for a $U(N)$ symmetry transformation, and hence $N^2 - 1$ real gauge 4-vectors for a $SU(N)$ symmetry transformation.

### 4.2.2 Including the Gauge Field Dynamics

With the knowledge of the required transformation rule for the gauge fields from Eq. (49), it is now possible to redefine the generating function (44) to also describe the gauge field transformation. This simultaneously defines the transformation of the canonical conjugates, $p_{JK}^{\mu\nu}$, of the gauge fields $a_{JK\mu}$. Furthermore, the redefined generating function yields additional terms in the transformation rule for the Hamiltonian. Of course, in order for the Hamiltonian to be invariant under local gauge transformations, the additional terms must be invariant as well. The transformation rules for the fields $\boldsymbol{\phi}$ and the gauge field matrices $\hat{\boldsymbol{a}}$ (Eq. (50)) can be regarded as a canonical transformation that emerges from an explicitly $x^{\mu}$-dependent and real-valued generating function vector of type $F_2^{\mu} = F_2^{\mu}(\boldsymbol{\phi}, \overline{\boldsymbol{\phi}}, \boldsymbol{\Pi}, \overline{\boldsymbol{\Pi}}, \hat{\boldsymbol{a}}, \hat{\boldsymbol{P}}, \boldsymbol{x})$,

$$F_2^{\mu} = \overline{\Pi}_K^{\mu}\, u_{KJ}\, \phi_J + \overline{\phi}_K\, \overline{u}_{KJ}\, \Pi_J^{\mu} + P_{JK}^{\alpha\mu}\left(u_{KL}\, a_{LI\alpha}\, \overline{u}_{IJ} - \frac{i}{q}\frac{\partial u_{KI}}{\partial x^{\alpha}}\, \overline{u}_{IJ}\right). \tag{51}$$

Accordingly, the subsequent transformation rules for canonical variables $\boldsymbol{\phi}, \overline{\boldsymbol{\phi}}$ and their conjugates, $\boldsymbol{\pi}^{\mu}, \overline{\boldsymbol{\pi}}^{\mu}$, agree with those from Eqs. (45). The rule for the gauge fields $a_{IK\alpha}$ emerges as

$$A_{KJ\alpha}\, \delta_{\nu}^{\mu} = \frac{\partial F_2^{\mu}}{\partial P_{JK}^{\alpha\nu}} = \delta_{\nu}^{\mu}\left(u_{KL}\, a_{LI\alpha}\, \overline{u}_{IJ} - \frac{i}{q}\frac{\partial u_{KI}}{\partial x^{\alpha}}\overline{u}_{IJ}\right),$$

which obviously coincides with Eq. (49), as demanded. The transformation of the momentum fields is obtained from the generating function (51) as

$$p_{IL}^{\alpha\mu} = \frac{\partial F_2^{\mu}}{\partial a_{LI\alpha}} = \overline{u}_{IJ}\, P_{JK}^{\alpha\mu}\, u_{KL}. \tag{52}$$

It remains to work out the difference of the Hamiltonians that are submitted to the canonical transformation generated by (51). Hence, according to the general rule from Eq. (11), we must calculate the divergence of the explicitly $x^{\mu}$-dependent terms of $F_2^{\mu}$

$$\left.\frac{\partial F_2^{\alpha}}{\partial x^{\alpha}}\right|_{\mathrm{expl}} = \overline{\Pi}_I^{\alpha}\,\frac{\partial u_{IJ}}{\partial x^{\alpha}}\,\phi_J + \overline{\phi}_I\,\frac{\partial \overline{u}_{IJ}}{\partial x^{\alpha}}\,\Pi_J^{\alpha} + P_{JK}^{\alpha\beta}\left(\frac{\partial u_{KL}}{\partial x^{\beta}} a_{LI\alpha}\overline{u}_{IJ} + u_{KL}a_{LI\alpha}\frac{\partial \overline{u}_{IJ}}{\partial x^{\beta}} - \frac{i}{q}\frac{\partial u_{KI}}{\partial x^{\alpha}}\frac{\partial \overline{u}_{IJ}}{\partial x^{\beta}} - \frac{i}{q}\frac{\partial^2 u_{KI}}{\partial x^{\alpha}\partial x^{\beta}}\overline{u}_{IJ}\right). \tag{53}$$

We are now going to replace all $u_{IJ}$-dependencies in (53) by canonical variables making use of the canonical transformation rules. The first two terms on the right-hand side of Eq. (53) can be expressed in terms of the canonical variables by means of the transformation rules (45), (49), and (52) that all emerge from the generating function (51)

$$\begin{aligned}\overline{\Pi}_I^{\alpha}\frac{\partial u_{IJ}}{\partial x^{\alpha}}\phi_J + \overline{\phi}_I\frac{\partial \overline{u}_{IJ}}{\partial x^{\alpha}}\Pi_J^{\alpha} &= \overline{\Pi}_I^{\alpha}\frac{\partial u_{IJ}}{\partial x^{\alpha}}\overline{u}_{JK}\Phi_K + \overline{\Phi}_K u_{KI}\frac{\partial \overline{u}_{IJ}}{\partial x^{\alpha}}\Pi_J^{\alpha}\\ &= \overline{\Pi}_I^{\alpha}\frac{\partial u_{IJ}}{\partial x^{\alpha}}\overline{u}_{JK}\Phi_K - \overline{\Phi}_K\frac{\partial u_{KI}}{\partial x^{\alpha}}\overline{u}_{IJ}\Pi_J^{\alpha}\\ &= iq\overline{\Pi}_I^{\alpha}\left(A_{IK\alpha} - u_{IL}a_{LJ\alpha}\overline{u}_{JK}\right)\Phi_K\\ &\quad - iq\overline{\Phi}_K\left(A_{KJ\alpha} - u_{KL}a_{LI\alpha}\overline{u}_{IJ}\right)\Pi_J^{\alpha}\\ &= iq\left(\overline{\Pi}_K^{\alpha}\Phi_J - \overline{\Phi}_K\Pi_J^{\alpha}\right)A_{KJ\alpha}\\ &\quad - iq\left(\overline{\pi}_K^{\alpha}\phi_J - \overline{\phi}_K\pi_J^{\alpha}\right)a_{KJ\alpha}.\end{aligned}$$

The second derivative term in Eq. (53) is *symmetric* in the indexes $\alpha$ and $\beta$. If we split $P_{JK}^{\alpha\beta}$ into a symmetric $P_{JK}^{(\alpha\beta)}$ and a skew-symmetric part $P_{JK}^{[\alpha\beta]}$ in $\alpha$ and $\beta$

$$P_{JK}^{\alpha\beta} = P_{JK}^{(\alpha\beta)} + P_{JK}^{[\alpha\beta]}, \qquad P_{JK}^{[\alpha\beta]} = \tfrac{1}{2}\left(P_{JK}^{\alpha\beta} - P_{JK}^{\beta\alpha}\right),$$
$$P_{JK}^{(\alpha\beta)} = \tfrac{1}{2}\left(P_{JK}^{\alpha\beta} + P_{JK}^{\beta\alpha}\right),$$

then the second derivative term vanishes for $P_{JK}^{[\alpha\beta]}$,

$$P_{JK}^{[\alpha\beta]}\frac{\partial^2 u_{KI}}{\partial x^{\alpha}\partial x^{\beta}} = 0.$$

By inserting the transformation rules for the gauge fields from Eqs. (49), the remaining terms of (53) for the skew-symmetric part of $P_{JK}^{\alpha\beta}$ are converted into

$$\begin{aligned}&P_{JK}^{[\alpha\beta]}\left(\frac{\partial u_{KL}}{\partial x^{\beta}}\,a_{LI\alpha}\,\overline{u}_{IJ} + u_{KL}\,a_{LI\alpha}\,\frac{\partial \overline{u}_{IJ}}{\partial x^{\beta}} - \frac{i}{q}\frac{\partial u_{KI}}{\partial x^{\alpha}}\frac{\partial \overline{u}_{IJ}}{\partial x^{\beta}}\right)\\ &= iq\left(p_{JK}^{[\alpha\beta]}\,a_{KI\alpha}\,a_{IJ\beta} - P_{JK}^{[\alpha\beta]}\,A_{KI\alpha}\,A_{IJ\beta}\right)\end{aligned}$$

$$= \tfrac{1}{2}iq\left[\left(p_{JK}^{\alpha\beta} - p_{JK}^{\beta\alpha}\right) a_{KI\alpha}\, a_{IJ\beta} - \left(P_{JK}^{\alpha\beta} - P_{JK}^{\beta\alpha}\right) A_{KI\alpha}\, A_{IJ\beta}\right]$$
$$= \tfrac{1}{2}iq\left[p_{JK}^{\alpha\beta}\left(a_{KI\alpha}\, a_{IJ\beta} - a_{KI\beta}\, a_{IJ\alpha}\right) - P_{JK}^{\alpha\beta}\left(A_{KI\alpha}\, A_{IJ\beta} - A_{KI\beta}\, A_{IJ\alpha}\right)\right].$$

For the symmetric part of $P_{JK}^{\alpha\beta}$, we obtain

$$P_{JK}^{(\alpha\beta)}\left(\frac{\partial u_{KL}}{\partial x^{\beta}} a_{LI\alpha}\,\overline{u}_{IJ} + u_{KL} a_{LI\alpha}\frac{\partial \overline{u}_{IJ}}{\partial x^{\beta}} - \frac{i}{q}\frac{\partial u_{KI}}{\partial x^{\alpha}}\frac{\partial \overline{u}_{IJ}}{\partial x^{\beta}} - \frac{i}{q}\frac{\partial^2 u_{KI}}{\partial x^{\alpha}\partial x^{\beta}}\overline{u}_{IJ}\right)$$
$$= P_{JK}^{(\alpha\beta)}\left(\frac{\partial A_{KJ\alpha}}{\partial x^{\beta}} - u_{KL}\frac{\partial a_{LI\alpha}}{\partial x^{\beta}}\overline{u}_{IJ}\right)$$
$$= \tfrac{1}{2}P_{JK}^{\alpha\beta}\left(\frac{\partial A_{KJ\alpha}}{\partial x^{\beta}} + \frac{\partial A_{KJ\beta}}{\partial x^{\alpha}}\right) - \tfrac{1}{2}p_{JK}^{\alpha\beta}\left(\frac{\partial a_{KJ\alpha}}{\partial x^{\beta}} + \frac{\partial a_{KJ\beta}}{\partial x^{\alpha}}\right).$$

In summary, by inserting the transformation rules into Eq. (53), the divergence of the explicitly $x^{\mu}$-dependent terms of $F_2^{\mu}$—and hence the difference of transformed and original Hamiltonians—can be expressed completely in terms of the canonical variables as

$$\left.\frac{\partial F_2^{\alpha}}{\partial x^{\alpha}}\right|_{\mathrm{expl}} = iq\Big[\left(\overline{\Pi}_K^{\alpha}\Phi_J - \overline{\Phi}_K\Pi_J^{\alpha}\right)A_{KJ\alpha} - \left(\overline{\Pi}_K^{\alpha}\overline{\pi}_K^{\alpha}\phi_J - \overline{\phi}_K\pi_J^{\alpha}\right)a_{KJ\alpha}$$
$$- \tfrac{1}{2}P_{JK}^{\alpha\beta}\left(A_{KI\alpha}\, A_{IJ\beta} - A_{KI\beta}\, A_{IJ\alpha}\right)$$
$$+ \tfrac{1}{2}p_{JK}^{\alpha\beta}\left(a_{KI\alpha}\, a_{IJ\beta} - a_{KI\beta}\, a_{IJ\alpha}\right)\Big]$$
$$+ \tfrac{1}{2}P_{JK}^{\alpha\beta}\left(\frac{\partial A_{KJ\alpha}}{\partial x^{\beta}} + \frac{\partial A_{KJ\beta}}{\partial x^{\alpha}}\right) - \tfrac{1}{2}p_{JK}^{\alpha\beta}\left(\frac{\partial a_{KJ\alpha}}{\partial x^{\beta}} + \frac{\partial a_{KJ\beta}}{\partial x^{\alpha}}\right).$$

We observe that *all* $u_{IJ}$-dependencies of Eq. (53) were expressed *symmetrically* in terms of the original and transformed complex scalar fields $\phi_J$, $\Phi_J$ and 4-vector gauge fields $\boldsymbol{a}_{JK}$, $\boldsymbol{A}_{JK}$, in conjunction with their respective canonical momenta. Consequently, a Hamiltonian of the form

$$\bar{\mathscr{H}} = \mathscr{H}(\boldsymbol{\pi}, \boldsymbol{\phi}, \boldsymbol{x}) + iq\left(\overline{\pi}_K^{\alpha}\phi_J - \overline{\phi}_K\pi_J^{\alpha}\right)a_{KJ\alpha}$$
$$- \tfrac{1}{2}iqp_{JK}^{\alpha\beta}\left(a_{KI\alpha}\, a_{IJ\beta} - a_{KI\beta}\, a_{IJ\alpha}\right) + \tfrac{1}{2}p_{JK}^{\alpha\beta}\left(\frac{\partial a_{KJ\alpha}}{\partial x^{\beta}} + \frac{\partial a_{KJ\beta}}{\partial x^{\alpha}}\right)$$

is then transformed according to the general rule (11)

$$\bar{\mathscr{H}}' = \bar{\mathscr{H}} + \left.\frac{\partial F_2^{\alpha}}{\partial x^{\alpha}}\right|_{\mathrm{expl}}$$

into the new Hamiltonian

$$\begin{aligned}\bar{\mathscr{H}}' =& \mathscr{H}(\boldsymbol{\Pi}, \boldsymbol{\Phi}, \boldsymbol{x}) + iq\left(\overline{\Pi}_K^{\alpha}\Phi_J - \overline{\Phi}_K \Pi_J^{\alpha}\right) A_{KJ\alpha} \\ & - \tfrac{1}{2} iq P_{JK}^{\alpha\beta}\left(A_{KI\alpha} A_{IJ\beta} - A_{KI\beta} A_{IJ\alpha}\right) + \tfrac{1}{2} P_{JK}^{\alpha\beta}\left(\frac{\partial A_{KJ\alpha}}{\partial x^{\beta}} + \frac{\partial A_{KJ\beta}}{\partial x^{\alpha}}\right).\end{aligned}$$

The entire transformation is thus *form-conserving* provided that the original Hamiltonian $\mathscr{H}(\boldsymbol{\pi}, \boldsymbol{\phi}, \boldsymbol{x})$ is also form-invariant if expressed in terms of the new fields, $\mathscr{H}(\boldsymbol{\Pi}, \boldsymbol{\Phi}, \boldsymbol{x})$, according to the transformation rules (45). In other words, $\mathscr{H}(\boldsymbol{\pi}, \boldsymbol{\phi}, \boldsymbol{x})$ must be form-invariant under the corresponding *global* gauge transformation.

In order to completely describe the dynamics of the gauge fields $\hat{\boldsymbol{a}}(\boldsymbol{x})$, we must further amend the Hamiltonian by a kinetic term that describes the dynamics of the free fields $\boldsymbol{a}_{IJ}$, namely

$$\tfrac{1}{4} p_{JK}^{\alpha\beta}\, p_{KJ\alpha\beta}.$$

We must check whether this additional term is also invariant under the canonical transformation generated by Eq. (51). From the transformation rule (52), we find

$$\begin{aligned} p_{JK}^{\alpha\beta} p_{KJ\alpha\beta} &= \left(\overline{u}_{JI}\, P_{IL}^{\alpha\beta}\, u_{LK}\right)\left(\overline{u}_{KM}\, P_{MN\alpha\beta}\, u_{NJ}\right) \\ &= \delta_{NI}\, P_{IL}^{\alpha\beta}\, \delta_{LM}\, P_{MN\alpha\beta} = P_{NL}^{\alpha\beta}\, P_{LN\alpha\beta}. \end{aligned} \tag{54}$$

Thus, the total amended Hamiltonian $\bar{\mathscr{H}}$ that is *form-invariant* under a local $U(N)$ symmetry transformation (43) of the fields $\boldsymbol{\phi}, \overline{\boldsymbol{\phi}}$ is given by

$$\begin{aligned} \bar{\mathscr{H}} =& \mathscr{H} + \mathscr{H}_{\mathrm{g}} \qquad (55)\\ \mathscr{H}_{\mathrm{g}} =& iq\left(\overline{\pi}_K^{\alpha}\phi_J - \overline{\phi}_K \pi_J^{\alpha}\right) a_{KJ\alpha} - \tfrac{1}{4} p_{JK}^{\alpha\beta}\, p_{KJ\alpha\beta} \\ & - \tfrac{1}{2} iq\, p_{JK}^{\alpha\beta}\left(a_{KI\alpha}\, a_{IJ\beta} - a_{KI\beta}\, a_{IJ\alpha}\right) + \tfrac{1}{2} p_{JK}^{\alpha\beta}\left(\frac{\partial a_{KJ\alpha}}{\partial x^{\beta}} + \frac{\partial a_{KJ\beta}}{\partial x^{\alpha}}\right). \end{aligned}$$

We reiterate that the original Hamiltonian $\mathscr{H}$ must be invariant under the corresponding *global* gauge transformation, hence a transformation of the form of Eq. (45) with the $u_{IK}$ *not* depending on $\boldsymbol{x}$ explicitly. In the Hamiltonian description, the partial derivatives of the fields in (55) do *not* constitute canonical variables and must hence be regarded as $x^{\mu}$-dependent coefficients when setting up the canonical field equations according to Eqs. (5).

### 4.2.3 Gauge-Invariant Lagrangian

In order to set up the corresponding gauge-invariant Lagrangian, the relation of the canonical momenta $p_{LM}^{\mu\nu}$ to the derivatives of the fields, $\partial a_{ML\mu}/\partial x^{\nu}$, must be known. This information is generally provided by the first canonical field equation (5). This means for the particular gauge-invariant Hamiltonian (55)

$$
\begin{aligned}
\frac{\partial a_{ML\mu}}{\partial x^{\nu}} &= \frac{\partial \mathscr{H}_{\mathrm{g}}}{\partial p_{LM}^{\mu\nu}} \\
&= -\tfrac{1}{2} p_{ML\mu\nu} - \tfrac{1}{2} iq \left( a_{MI\mu}\, a_{IL\nu} - a_{MI\nu}\, a_{IL\mu} \right) + \tfrac{1}{2} \left( \frac{\partial a_{ML\mu}}{\partial x^{\nu}} + \frac{\partial a_{ML\nu}}{\partial x^{\mu}} \right),
\end{aligned}
$$

hence

$$
p_{KJ\mu\nu} = \frac{\partial a_{KJ\nu}}{\partial x^{\mu}} - \frac{\partial a_{KJ\mu}}{\partial x^{\nu}} + iq \left( a_{KI\nu}\, a_{IJ\mu} - a_{KI\mu}\, a_{IJ\nu} \right). \tag{56}
$$

We observe that $p_{KJ\mu\nu}$ occurs to be skew-symmetric in the indexes $\mu$, $\nu$. Here, this feature emerges from the canonical formalism and does not have to be postulated, as required in the Lagrangian description.

The equivalent gauge-invariant Lagrangian $\bar{\mathscr{L}}$ is derived by means of the Legendre transformation

$$
\bar{\mathscr{L}} = \mathscr{L} + \mathscr{L}_{\mathrm{g}}, \qquad \mathscr{L}_{\mathrm{g}} = p_{JK}^{\alpha\beta} \frac{\partial a_{KJ\alpha}}{\partial x^{\beta}} - \mathscr{H}_{\mathrm{g}},
$$

with $p_{JK}^{\mu\nu}$ from Eq. (56). We thus have

$$
\begin{aligned}
p_{JK}^{\alpha\beta} \frac{\partial a_{KJ\alpha}}{\partial x^{\beta}} =& \tfrac{1}{2} p_{JK}^{\alpha\beta} \left( \frac{\partial a_{KJ\alpha}}{\partial x^{\beta}} - \frac{\partial a_{KJ\beta}}{\partial x^{\alpha}} \right) + \tfrac{1}{2} p_{JK}^{\alpha\beta} \left( \frac{\partial a_{KJ\alpha}}{\partial x^{\beta}} + \frac{\partial a_{KJ\beta}}{\partial x^{\alpha}} \right) \\
=& -\tfrac{1}{2} p_{JK}^{\alpha\beta}\, p_{KJ\alpha\beta} - \tfrac{1}{2} iq\, p_{JK}^{\alpha\beta} \left( a_{KI\alpha}\, a_{IJ\beta} - a_{KI\beta}\, a_{IJ\alpha} \right) \\
&+ \tfrac{1}{2} p_{JK}^{\alpha\beta} \left( \frac{\partial a_{KJ\alpha}}{\partial x^{\beta}} + \frac{\partial a_{KJ\beta}}{\partial x^{\alpha}} \right).
\end{aligned}
$$

The Lagrangian $\mathscr{L}_{\mathrm{g}}$ that renders $\bar{\mathscr{L}}$ locally gauge invariant is then

$$
\mathscr{L}_{\mathrm{g}} = -\tfrac{1}{4} p_{JK}^{\alpha\beta}\, p_{KJ\alpha\beta} - iq \left( \overline{\pi}_{K}^{\alpha} \phi_{J} - \overline{\phi}_{K} \pi_{J}^{\alpha} \right) a_{KJ\alpha}. \tag{57}
$$

As implied by the Lagrangian formalism, the dynamical variables are given by both the fields, $\overline{\phi}_K$, $\phi_J$, and $a_{KJ\alpha}$, in conjunction with their respective partial derivatives with respect to the independent variables, $x^{\mu}$. Therefore, the $\overline{\pi}_{K}^{\alpha}$, the $\pi_{J}^{\alpha}$, and the $p_{KJ\alpha\beta}$ in $\mathscr{L}_{\mathrm{g}}$ from Eq. (57) are now merely abbreviations for a combination of the Lagrangian dynamical variables, namely

$$
\begin{aligned}
\overline{\pi}_{K}^{\alpha} &= \frac{\partial \mathscr{L}}{\partial \left( \frac{\partial \phi_K}{\partial x^{\alpha}} \right)}, \qquad \pi_{J}^{\alpha} = \frac{\partial \mathscr{L}}{\partial \left( \frac{\partial \overline{\phi}_J}{\partial x^{\alpha}} \right)} \\
p_{KJ\alpha\beta} &= \frac{\partial a_{KJ\beta}}{\partial x^{\alpha}} - \frac{\partial a_{KJ\alpha}}{\partial x^{\beta}} + iq \left( a_{KI\beta}\, a_{IJ\alpha} - a_{KI\alpha}\, a_{IJ\beta} \right).
\end{aligned}
$$

Thence, for any *globally* gauge-invariant Lagrangian $\mathscr{L}(\phi^{I}, \boldsymbol{\partial}\phi^{I}, \boldsymbol{x})$, the amended Lagrangian $\bar{\mathscr{L}} = \mathscr{L} + \mathscr{L}_{\mathrm{g}}$ describes the associated physical system that is invariant under *local* gauge transformations. The equation of motion for the gauge fields $a_{KJ\mu}$ then follows from the Euler-Lagrange equation

$$\frac{\partial}{\partial x^\alpha}\frac{\partial \mathcal{L}_{\mathrm{g}}}{\partial\left(\frac{\partial a_{JK\mu}}{\partial x^\alpha}\right)}-\frac{\partial \mathcal{L}_{\mathrm{g}}}{\partial a_{JK\mu}}=0, \qquad \frac{\partial \mathcal{L}_{\mathrm{g}}}{\partial\left(\frac{\partial a_{JK\mu}}{\partial x^\alpha}\right)}=p_{KJ}^{\mu\alpha}.$$

To cite an example, a generalized Dirac Lagrangian (25) describing $N$ spin-$\frac{1}{2}$ fields, each of them being associated with the same mass $m$,

$$\mathcal{L}_{\mathrm{D}}=\frac{i}{2}\left(\overline{\phi}_I\gamma^\alpha\frac{\partial \phi_I}{\partial x^\alpha}-\frac{\partial \overline{\phi}_I}{\partial x^\alpha}\gamma^\alpha\phi_I\right)-m\,\overline{\phi}_I\phi_I,$$

is invariant under global gauge transformations. The corresponding locally gauge-invariant Lagrangian $\bar{\mathcal{L}}_{\mathrm{D}}$ is then

$$\bar{\mathcal{L}}_{\mathrm{D}}=\mathcal{L}_{\mathrm{D}}-\tfrac{1}{4}p_{JK}^{\alpha\beta}p_{KJ\alpha\beta}+q\left(\overline{\phi}_K a_{KJ\alpha}\gamma^\alpha\phi_J\right).$$

## 5 Conclusions

With the present chapter, we have worked out a consistent local inertial frame description of the canonical formalism in the realm of covariant Hamiltonian field theory. On that basis, the Noether theorem as well as the idea of gauge theory—to amend the Hamiltonian of a given system in order to render the resulting system locally gauge invariant—could elegantly and most generally be formulated as particular canonical transformations.

**Acknowledgments** To the memory of my (J.S.) colleague and friend Dr. Claus Riedel (GSI), who contributed vitally to this work. Furthermore, the authors are indebted to Prof. Dr. Dr. hc. mult. Walter Greiner from the Frankfurt Institute of Advanced Studies (FIAS) for his long-standing hospitality, his critical comments and encouragement.

## References

1. Th. De Donder, *Théorie Invariantive Du Calcul des Variations* (Gaulthier-Villars & Cie, Paris, 1930)
2. H. Weyl, Geodesic fields in the calculus of variation for multiple integrals. Ann. Math. **36**, 607 (1935)
3. J.V. José, E.J. Saletan, *Classical Dynamics* (Cambridge University Press, Cambridge, 1998)
4. W. Greiner, B. Müller, J. Rafelski, *Quantum Electrodynamics of Strong Fields* (Springer-Verlag, Berlin, 1985)
5. S. Gasiorowicz, *Elementary particle physics* (Wiley, New York, 1966)

6. J. von Rieth, The Hamilton-Jacobi theory of De Donder and Weyl applied to some relativistic field theories. J. Math. Phys. **25**, 1102 (1984)
7. E. Noether, Invariante Variationsprobleme. Nachr. Ges. Wiss. Göttingen, Math.-Phys. Kl. **57**, 235 (1918)

# Part V
# Atomic Physics

# Crystalline Undulator: Current Status and Perspectives

**A. Kostyuk, A. Korol, A. Solov'yov and W. Greiner**

**Abstract** Recent advances in the theory of novel sources of hard electromagnetic radiation,—a crystalline undulator (CU) and a Crystalline Undulator based Laser (CUL), are reviewed. The operating principle of CU is based on the channeling phenomenon. Channeling takes place if a particle enters a crystal at small angle to major crystallographic planes (or axes). The particle becomes confined by the planar or axial potential and move preferably along the plane or axis following its shape. If the planes or axes are periodically bent, the particles move along nearly sinusoidal trajectories. Similarly to what happens in an ordinary undulator, relativistic charged particles radiate electromagnetic waves in the forward direction. The advantage of CU is that due to extremely strong electrostatic fields inside the crystal the particles are steered much more effectively than by the field of the most advanced superconductive magnets. This allows one to make the period of CU two or even three orders of magnitude smaller then that of the conventional undulator. As a result the frequency of the radiation can reach the hard X-ray and gamma ray range. Similarly as it takes place in an ordinary free electron laser (FEL), the radiation becomes more powerful and coherent if the density of the particle beam is modulated along the beam direction with the period equal to the wavelength of the produced radiation.

A. Kostyuk (✉) · A. Korol · A. Solov'yov · W. Greiner
Frankfurt Institute for Advanced Studies, Goethe-Universität, Frankfurt am Main, Germany
e-mail: kostyuk@fias.uni-frankfurt.de

A. Korol
e-mail: korol@fias.uni-frankfurt.de

A. Solov'yov
e-mail: solovyov@fias.uni-frankfurt.de

W. Greiner
e-mail: greiner@fias.uni-frankfurt.de

W. Greiner (ed.), *Exciting Interdisciplinary Physics*,
FIAS Interdisciplinary Science Series, DOI: 10.1007/978-3-319-00047-3_32,

# 1 Introduction

The aim of this chapter is to review recent results in the theory of the crystalline undulator—a novel source of high energy photons. The *feasibility* of this idea was explicitly demonstrated in [1, 2]. The history of the idea as well as newer results was reviewed in [3].

The term 'crystalline undulator' stands for a system which consists of two essential parts: (a) a crystal whose crystallographic planes are bent periodically, and (b) a bunch of ultra-relativistic positively charged particles undergoing planar channeling in the crystal (see Fig. 1). In such a system there appears, in addition to a well-known channeling radiation, the radiation of an undulator type which is due to the periodic motion of channeling particles which follow the bending of the crystallographic planes.

The $(yz)$-plane in Fig. 1 is a cross section of a periodically bent crystal (PBCr), and the $z$-axis represents the beam direction. Two sets of filled circles denote the nuclei which belong to the periodically bent neighbouring planes spaced by the interplanar distance $d$. The planes form a periodically bent channel whose shape is defined by $y_{\rm B}(z) = a\cos(2\pi z/\lambda_{\rm u})$. The amplitude of the bending, $a$, is defined as a maximum displacement of the deformed midplane (the thin dashed line) from its position in a straight crystal. The quantity $\lambda_{\rm u}$ stands for a spatial period of the bending. The quantities $d$, $a$ and $\lambda_{\rm u}$ satisfy strong double inequality: $d \ll a \ll \lambda_{\rm u}$. Typically $d \sim 10^{-8}$ cm, $a \sim 10\dots10^2\,d$, and $a \sim 10^{-5}\dots10^{-4}\,\lambda_{\rm u}$.

The operational principle of a CU does not depend on the type of a projectile. Provided certain conditions are met (see Sect. 2), the particles, injected into the crystal, will undergo channeling in PBCr. Thus, the trajectory of a particle contains two elements, which are illustrated by Fig. 1 where the solid wavy line represents

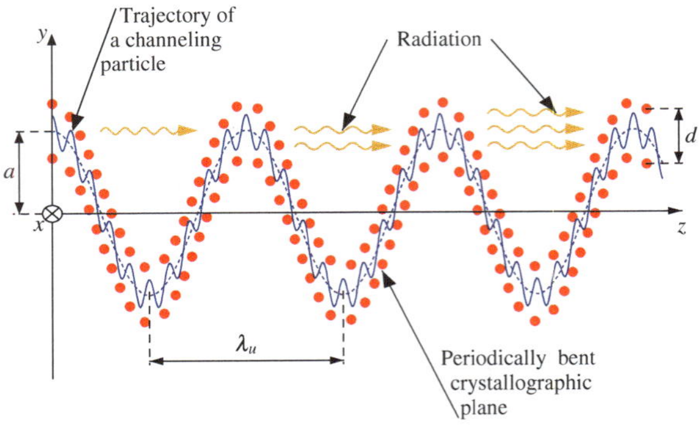


**Fig. 1** Schematic representation of the crystalline undulator. *Closed circles* mark the atoms belonging to two neighbouring crystallographic planes (separated by the interplanar distance $d$) which are periodically bent with period $\lambda_{\rm u}$ and amplitude $a$ (the $y$- and $z$-scales are incompatible!). *Thin solid line* illustrates the trajectory of the particle, which propagates along the centerline (the undulator motion) and, simultaneously, undergoes channeling oscillations

the trajectory of the particle. Firstly, there are oscillations inside the channel due to the action of the interplanar potential,—the channeling oscillations. This mode is characterized by a frequency $\Omega_{\rm ch}$ dependent on the projectile type, energy, and the parameters of the interplanar potential. Secondly, there are oscillations caused by the periodicity of the bent channel,—the undulator oscillations, whose frequency is $\omega_0 \approx 2\pi c/\lambda_{\rm u}$ ($c$ is the velocity of light which approximately is the velocity of an ultra-relativistic particle).

Spontaneous emission of photons which appears in this system is associated with both of these oscillations. Typical frequency of the emission due to the channeling oscillations is $\omega_{\rm ch} \approx 2\gamma^2\Omega_{\rm ch}$ where $\gamma$ is the relativistic Lorenz factor $\gamma = \varepsilon/mc^2$. The undulator oscillations give rise to the photons with frequency $\omega \approx 4\gamma^2\omega_0/(2+K^2)$ where the quantity $K$, a so-called undulator parameter, is related to the amplitude and the period of bending, $K = 2\pi\gamma(a/\lambda_{\rm u})$.

If strong inequality $\omega_0 \ll \Omega_{\rm ch}$ is met than the frequencies of the channeling radiation (ChR) and the undulator radiation (UR) are also well separated, $\omega \ll \omega_{\rm ch}$.

The scheme presented in Fig. 1 leads also to the possibility of generating a stimulated emission of the FEL type. Thus, it is meaningful to discuss a novel source of electromagnetic radiation in hard X and gamma range,—a *Crystalline Undulator Laser* (CUL) [1–9]. Specific features of CUL as well as quantitative estimates of the parameters of stimulated emission are presented below in Sect. 5.

There is essential feature which distinguish a seemingly simple scheme presented in Fig. 1 from a conventional undulator based on the action of periodic magnetic field on the projectile. In the latter the beam of particles and the photon flux move in vacuum whereas in the proposed scheme they propagate through a crystalline media. The interaction of both beams with the crystal constituents makes the problem much more complicated from theoretical, experimental and technological viewpoints.

## 2 Feasibility of CU

The conditions, which must be met to treat a *CU as a feasible scheme* for devising a new source of electromagnetic radiation, are as follows:

$$\begin{cases} C = 4\pi^2\varepsilon a/U'_{\rm max}\lambda_{\rm u}^2 < 1 & \text{–stable channeling,} \\ d < a \ll \lambda_{\rm u} & \text{–large-amplitude regime,} \\ N = L/\lambda_{\rm u} \gg 1 & \text{–large number of periods,} \\ L \sim \min\left[L_{\rm d}(C), L_{\rm a}(\omega)\right] & \text{–account for dechanneling and photon attenuation,} \\ \Delta\varepsilon/\varepsilon \ll 1 & \text{–low energy losses.} \end{cases} \tag{1}$$

The formulated conditions are of a general nature since they are applicable to any type of a projectile undergoing channeling in periodically bent channel (PBCh). Their application to the case of a specific projectile and/or a crystal channel allows one to analyze the feasibility of the CU by establishing the ranges of $\varepsilon$, $a$, $\lambda_{\rm u}$, $L$, $N$ and $\hbar\omega$ which can be achieved.

- A *stable channeling* of a projectile in a PBCh occurs if the maximum centrifugal force $F_{\rm cf}$ is less than the maximal interplanar force $U'_{\rm max}$, i.e. $C = F_{\rm cf}/U'_{\rm max} < 1$. Expressing $F_{\rm cf}$ through the energy $\varepsilon$ of the projectile, the period and amplitude of the bending one formulates this condition as it is written in (1).
- The operation of a CU should be considered in the *large-amplitude regime*. The limit $a/d > 1$ accompanied by the condition $C \ll 1$ is mostly advantageous, since in this case the characteristic frequencies of UR and ChR are well separated: $\omega_{\rm u}^2/\omega_{\rm ch}^2 \sim Cd/a \ll 1$. As a result, the channeling motion does not affect the parameters the UR, the intensity of which can be comparable or higher than that of ChR. A strong inequality $a \ll \lambda_{\rm u}$ ensured elastic deformation of the crystal.
- The term "undulator" implies that the *number of periods, $N$, is large*. Only then the emitted radiation bears the features of an UR (narrow, well-separated peaks in spectral-angular distribution). This is stressed by the third condition.
- A CU essentially differs from a conventional undulator, in which the beams of particles and photons move in vacuum, In CU the both beams propagate in crystalline medium and, thus, are affected by *the dechanneling and the photon attenuation*. The dechanneling effect stands for a gradual increase in the transverse energy of a channeled particle due to inelastic collisions with the crystal constituents [10]. At some point the particle gains a transverse energy higher than the planar potential barrier and leaves the channel. The average interval for a particle to penetrate into a crystal until it dechannels is called the dechanneling length, $L_{\rm d}$. In a straight channel this quantity depends on the crystal, on the energy and the type of a projectile. In a PBCh bent channel there appears an additional dependence on the parameter $C$. The intensity of the photon flux, which propagates through a crystal, decreases due to the processes of absorption and scattering. The interval within which the intensity decreases by a factor of $e$ is called the attenuation length, $L_a(\omega)$. This quantity is tabulated for a number of elements and for a wide range of photon frequencies (see, e.g., Ref. [11]).
  The fourth condition in (1) takes into account severe limitation of the allowed values of the length $L$ of a CU due to the dechanneling and the attenuation.
- Finally, let us comment on the last condition, which is of most importance for light projectiles, positrons and electrons. For sufficiently large photon energies ($\hbar\omega \gtrsim 10^1 \dots 10^2$ keV depending on the type of the crystal atom) the restriction due to the attenuation becomes less severe than due to the dechanneling effect. Then, the value of $L_{\rm d}(C)$ effectively introduces an upper limit on the length of a CU. Since for an ultra-relativistic particle $L_{\rm d} \propto \varepsilon$ (see, e.g., [12]), it seems natural that to increase the effective length one can consider higher energies. However, at this point another limitation manifests itself [13]. The coherence of UR is only possible when the energy loss $\Delta\varepsilon$ of the particle during its passage through the undulator is small, $\Delta\varepsilon \ll \varepsilon$. This statement, together with the fact, that for ultra-relativistic electrons and positrons $\Delta\varepsilon$ is mainly due to the photon emission, leads to the conclusion that $L$ must be much smaller than the radiation length $L_{\rm r}$, the distance over which a particle converts its energy into radiation.

For a positron-based CU a thorough analysis of the system (1) was carried out for the first time in Refs. [1–3, 13–15]. Later on, the feasibility of the CU utilizing the planar channeling of electrons was demonstrated [16]. Recently, similar analysis was carried out for heavy ultra-relativistic projectiles (muon, proton, ion) [17].

# 3 Positron-Based Crystalline Undulator

To illustrate the crystalline undulator radiation phenomenon, let us consider the spectra of spontaneous radiation emitted during the passage of positrons through PBCr.

The calculated spectra of the radiation emitted in the forward direction (with respect to the $z$-axis, see Fig. 1) in the case of $\varepsilon = 0.5\,\mathrm{GeV}$ planar channeling in Si along (110) crystallographic planes and for the photon energies from 45 keV to 1.5 MeV are presented in Fig. 2 [15]. The ratio $a/d$ was varied within the interval

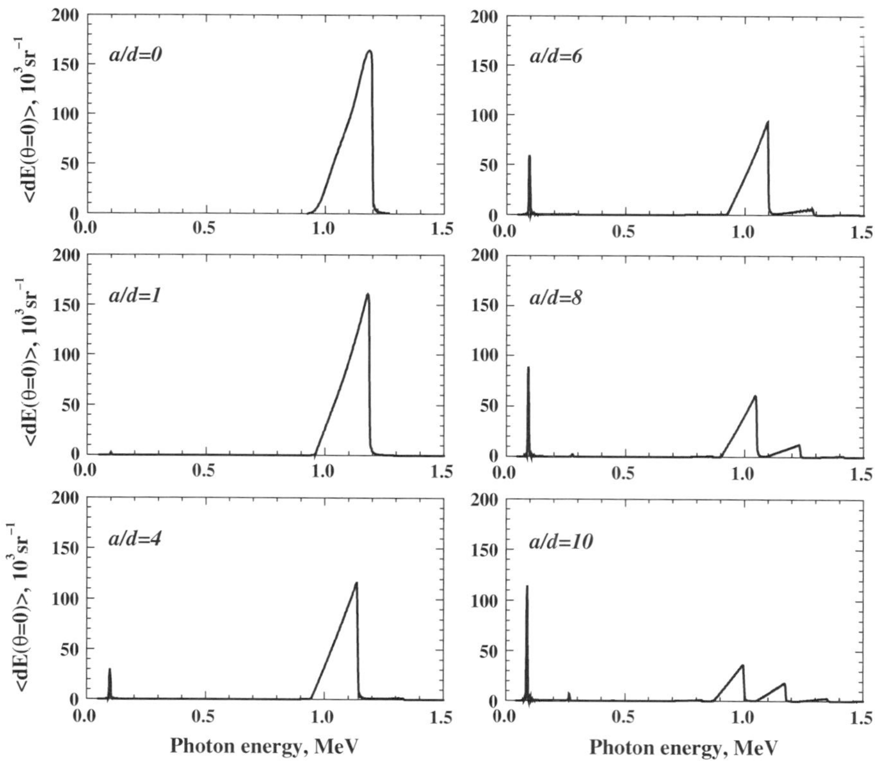


**Fig. 2** Spectral distribution of the total radiation emitted in the forward direction ($\vartheta = 0°$) for $\varepsilon = 0.5\,\mathrm{GeV}$ positron channeling in Si (110) crystallographic plane calculated at different $a/d$ ratios. Other parameters are given in the text

$a/d = 0 \dots 10$ (the interplanar spacing is 1.92 Å). The case $a/d = 0$ corresponds to the straight channel. The period $\lambda_u$ used for these calculations equals to $2.33 \times 10^{-3}$ cm. The number of undulator periods and crystal length were fixed at $N = 15$ and $L = N\,\lambda_u = 3.5 \times 10^{-2}$ cm.

The spectra correspond to the total radiation, which accounts for the two mechanisms, the undulator and the channeling. They were calculated using the quasi-classical method [18, 19]. Briefly, to evaluate the spectral distribution the following procedure was adopted (for more details see [15, 20, 21]). First, for each $a/d$ value the spectrum was calculated for individual trajectories of the particles. These were obtained by solving the relativistic equations of motion with both the interplanar and the centrifugal potentials taken into account. We considered two frequently used [22] analytic forms for the continuum interplanar potential, the harmonic and the Molière potentials calculated at the temperature $T = 150$ K to account for the thermal vibrations of the lattice atoms. The resulting radiation spectra were obtained by averaging over all trajectories. Figure. 2 correspond to the spectra obtained by using the Molière approximation for interplanar potential.

The first graph in Fig. 2 corresponds to the case of zero amplitude of the bending (the ratio $a/d = 0$) and, hence, presents the spectral dependence of the ordinary channeling radiation only. The asymmetric shape of the calculated channeling radiation peak, which is due to the strong anharmonic character of the Molière potential, bears close resemblance with the experimentally measured spectra [23]. The spectrum starts at $\hbar\omega \approx 960$ keV, reaches its maximum value at 1190 keV, and steeply cuts off at 1200 keV. This peak corresponds to the radiation into the first harmonic of the ordinary channeling radiation (see e.g. [24]), and there is almost no radiation into higher harmonics.

Increasing the $a/d$ ratio leads to the modifications in the radiation spectrum. The changes which occur are: (i) the lowering of the channeling radiation peak, (ii) the gradual increase of the intensity of UR due to the crystal bending.

The decrease in the intensity of the channeling radiation is related to the fact that the increase of the amplitude $a$ of the bending leads to lowering of the allowed maximum value of the channeling oscillations amplitude $a_{ch}$ (this is measured with respect to the centerline of the bent channel) [13, 25]. Hence, the more the channel is bent, the lower the allowed values of $a_{ch}$ are, and, consequently, the less intensive is the channeling radiation, which is proportional to $a_{ch}^2$ [18].

The UR related to the motion of the particle along the centerline of the PBCh bent channel is absent in the case of the straight channel (the graph $a/d = 0$), and is almost invisible for comparatively small amplitudes (see the graph for $a/d = 1$). Its intensity, which is proportional to $(a/d)^2$, gradually increases with the amplitude $a$. For large $a$ values ($a/d \sim 10$) the intensity of the first harmonic of the UR becomes larger than that of the ChR. The undulator peak is located at much lower energies, $\hbar\omega \approx 90$ keV, and has the width $\hbar\Delta\omega \approx 6$ keV which is almost 40 times less than the width of the peak of ChR.

# 4 Electron-Based Crystalline Undulator

Initially, it was proposed to use positron beams in the crystalline undulator. Such undulator has been considered in the previous section. Positively charged particles are repelled by the crystal nuclei and, therefore, they move between the crystal planes, where there are no atomic nuclei and the electron density is less than average. This reduces the probability of random collisions with the crystal constituents. Hence, the transverse momentum of the particle increases slowly and the particle travels a longer distance in the channelling regime and performs more undulator oscillations.

In contrast, negatively charged particles are attracted by the crystal nuclei and therefore they have to cross the crystallographic plane in the cause of the channeling oscillation. Therefore, the probability of random collisions of the projectile with the crystal constituents is strongly enhanced. The negative particles dechannel very quickly. Typically the dechanneling length for electrons is two orders of magnitude shorter then for positrons at the same conditions.

On the other hand, the electron beams are easier available and are usually of higher intensity and quality. Therefore, from the practical point of view, electron based crystalline undulator has its own advantages and deserves a thorough investigation.

It has been found [16] that an electron based crystalline undulator is also feasible. However, it requires the electron beam energy of a few tens of GeV.

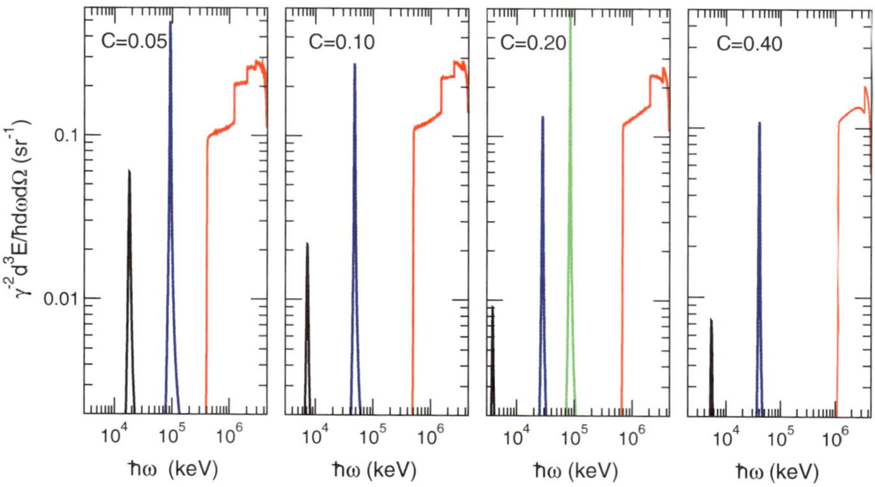


**Fig. 3** Spectral distributions (scaled by $\gamma^2$) of UR and ChR (wide peaks) emitted in the forward direction by a 50 GeV electron in Si (111). Each graph corresponds to the indicated value of parameter $C$. Narrow peaks stand for the spectral distribution of UR in the vicinity of the fundamental harmonics for nine different undulators. In each graph the leftest narrow peak corresponds to $N_d L_d/\lambda_u = 5$, the second peak—to $N_d = 10$, and the third peak (only for C $= 0.20$)—to $N_d = 15$

The graphs in Fig. 3 illustrate that by changing parameters of the undulator one can vary the frequency and the peak intensity of the UR over wide ranges. However, it is important to compare these quantities with the characteristics of the ChR.

The wide peak in each graph stands for the spectral distribution of the ChR in the forward direction. Figure 3 clearly demonstrates that by tuning the parameters of bending it is possible to separate the frequencies of the UR from those of the ChR, and to make the intensity of the former comparable or higher than of the latter.

The present technologies enable one to construct the periodically bent crystalline structures with the required parameters. These include making regularly spaced grooves on the crystal surface either by a diamond blade [26, 27] or by means of laser-ablation [28], deposition of periodic $Si_3N_4$ layers onto the surface of a Si crystal [27], growing of $Si_{1-x}Ge_x$ crystals[29] with a periodically varying Ge content $x$ [21, 30].

Similar to the case of a positron-based undulator [14], the parameters of high-energy electrons beams available at present [11] are sufficient to achieve the necessary conditions to construct the undulator and to create, on its basis, powerful radiation sources in the $\gamma$-region of the spectrum.

## 5 Crystalline Undulator Based Gamma Laser

The proposed gamma laser combines a crystalline undulator with a free electron laser (FEL) [8], see Fig. 4.

The first essential element of the apparatus is a CU in which the radiation with the wavelength $\lambda$ is formed. The CU has to be manufactured in such a way that charged particles, when enter the crystal from the appropriate direction, move inside the crystal along the periodically bent planes or axes in channeling regime as it described in the previous sections. The parameters of the crystal bending, the period and the amplitude, have to be chosen to satisfy the resonance condition for the desirable wavelength $\lambda$ of the produced radiation at given average particle energy $\varepsilon$ of the beam in the FEL undulator. The crystal length in the beam direction has to be comparable to or smaller than the attenuation length of the radiation with the wavelength $\lambda$ in the crystal material. Choosing the type of crystal and the plane or axis with largest *demodulation length* (see [4, 5] for the definition) is preferable: larger

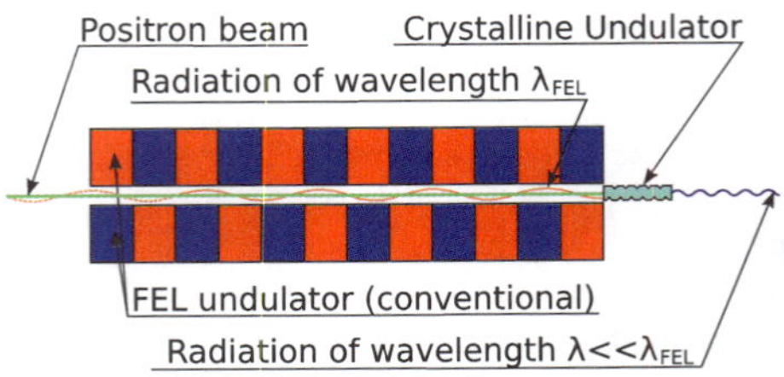


**Fig. 4** A scheme of the crystalline undulator-based gamma laser

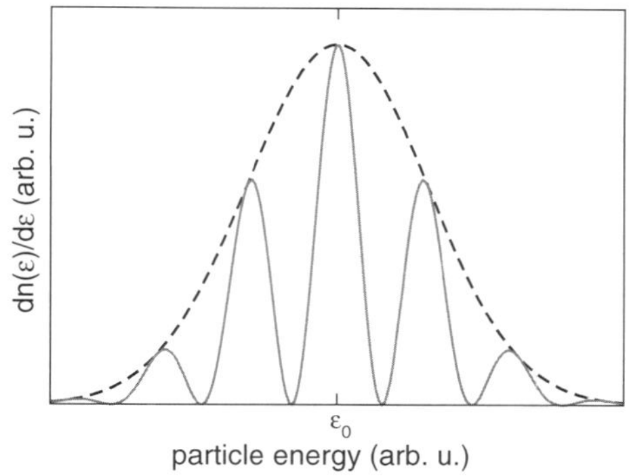


**Fig. 5** The Gaussian energy distribution (*dashed curve*) and the layered energy distribution (*solid curve*) of the beam particles

number of undulator periods per demodulation length improves the performance of the device. The use of a positron beam is preferable over an electron beam.

The second key element is conventional FEL which must be tuned to the wavelength $\lambda_{\rm FEL}$ larger than the wavelength of the produced radiation: $\lambda_{\rm FEL} > \lambda$. The ratio $\lambda_{\rm FEL}/\lambda$ has to be an integer.

The PBCr is placed at the exit from the FEL and exposed to the particle beam that has traveled some distance inside the FEL undulator.

The beam source of FEL has to be modified in such a way that the distribution of the particles with respect to their energy becomes *layered*, i.e. the distribution has a number of maxima separated by minima (an example is shown in Fig. 5). The distances between the maxima has to be optimized to enhance the radiation of CUL at the wavelength $\lambda$.

It is known from the theory of the conventional undulator that its radiation becomes much more intense and less divergent if the particle density in the beam is modulated with the period approximately equal to the wavelength of the produced radiation [31]. This phenomenon is known as coherent emission and is used in free electron lasers [32–35]. It has been demonstrated recently [4, 5] that the initially modulated beam can preserve its modulation at sufficient depth while channeling in the crystal. Therefore, coherent emission takes place also in the crystalline undulator, provided that it is fed by a modulated particle beam whose modulation period is close to the wavelength of the produced radiation. This effect is utilized in the proposed apparatus.

In the proposed device, the free electron laser is used as a source of the modulated beam. It is tuned to the wavelength $\lambda_{\rm FEL}$ which is larger than the wavelength of the produced radiation $\lambda$: $\lambda_{\rm FEL} > \lambda$. According to the theory of FEL (see, e.g., [34, 35]), the particle beam while traveling through the FEL undulator becomes modulated (micro-bunched) with the period $\lambda_{\rm FEL}$.

If the energy distribution of the particles in the beam that enters the undulator of FEL is *layered*, the shape of micro-bunches will have maxima and minima whose position depends on the position of the maxima and minima in the layered particle distribution with respect to the energy (see Ref. [36] for the details). In this case, the

Fourier expansion of the dependence of particle density in the beam on the spatial coordinate along the beam direction contains also a higher harmonic with the period approximately equal to the distances between maxima (or minima) of the bunch shape. If the crystalline undulator is tuned to this harmonic, an enhanced coherent emission will be observed.

The proposed gamma laser can be used in scientific laboratories, in particular for nuclear physics and plasma physics laboratories. It may be used in medicine, e.g. as a diagnostic tool or for cancer therapy. It can be used for nondestructive analysis of isotope composition of various objects. Other applications can be found in future.

**Acknowledgments** This work was supported by the Deutsche Forschungsgemeinschaft (DFG) and the European Commission.

## References

1. A.V. Korol, A.V. Solov'yov, W. Greiner, J. Phys. G Part. Nucl. **24**, L45–L53 (1998)
2. A.V. Korol, A.V. Solov'yov, W. Greiner, Int. J. Mod. Phys. E **8**, 49–100 (1999)
3. A.V. Korol, A.V. Solov'yov, W. Greiner, Int. J. Mod. Phys. E **13**, 867–916 (2004)
4. A. Kostyuk, A.V. Korol, A.V. Solov'yov, W. Greiner, J. Phys. B At. Mol. Opt. Phys. **43**, 151001 (2010)
5. A. Kostyuk, A.V. Korol, A.V. Solov'yov, W. Greiner, Nucl. Instrum. Meth. B **269**, 1482–1492 (2011)
6. A. Kostyuk, A.V. Korol, A.V. Solov'yov, W. Greiner, J. Phys. G Part. Nucl. **36**, 025107 (2009)
7. A.V. Korol, A.V. Solov'yov, W. Greiner, Lasing effect in crystalline undulators. Proc. SPIE **5974**, 59740O (2005)
8. W. Greiner, A.V. Korol, A. Kostyuk, A.V. Solov'yov, Vorrichtung und Verfahren zur Erzeugung electromagnetischer Strahlung. Application for German patent, Ref.: 10 2010 023 632.2, 14 June 2010
9. A. Kostyuk, A.V. Korol, A.V. Solov'yov, W. Greiner, Estimation of peak brilliance for a crystalline undulator laser. Unpublished (2012b)
10. J. Lindhard, Influence of crystal lattice on motion of energetic charged particles. K. Dan. Vidensk. Selsk. Mat. Fys. Medd. **34**, 1–64 (1965)
11. K. Nakamura K. et al. (Particle Data Group), Review of particle physics. J. Phys. G Part. Nucl. **37**, 075021 (2010)
12. U.I. Uggerhøj, Rev. Mod. Phys. **77**, 1131–1171 (2005)
13. A.V. Korol, A.V. Solov'yov, W. Greiner, Int. J. Mod. Phys. E **9**, 77–105 (2000)
14. A.V. Korol, A.V. Solov'yov, W. Greiner, Proc. SPIE **5974**, 597405 (2005) (see also physics/0412101)
15. W. Krause, A.V. Korol, A.V. Solov'yov, W. Greiner, J. Phys. G Part. Nucl. **26**, L87–L95 (2000)
16. M. Tabrizi, A.V. Korol, A.V. Solov'yov, W. Greiner, Phys. Rev. Lett. **98**, 164801 (2007)
17. A.V. Korol, A.V. Solov'yov, W. Greiner, *Channeling and Radiation in Periodically Bent Crystals* Springer Series on Atomic, Optical, and Plasma Physics, Vol. **69** (Springer, Berlin, 2013)
18. V.N. Baier, V.M. Katkov, V.M. Strakhovenko, *Electromagnetic Processes at High Energies in Oriented Single Crystals* (World Scientific, Singapore, 1998)
19. V.N. Baier, V.M. Katkov, Zh. Eksp. Teor. Fiz. **53**, 1478–1491 (1967) (English translation: Sov. Phys. JETP **26**, 854–860 (1968))
20. W. Krause, A.V. Korol, A.V. Solov'yov, W. Greiner, Nucl. Instrum. Meth. A **475**, 441–444 (2001)

21. A.V. Korol, W. Krause, A.V. Solov'yov, W. Greiner, Nucl. Instrum. Meth. A **483**, 455–460 (2002)
22. D.S. Gemmell, Rev. Mod. Phys. **46**, 129–227 (1974)
23. E. Uggerhøj, Rad. Eff. Defects Solids **25**, 3–21 (1993)
24. M.A. Kumakhov, F.F. Komarov, *Radiation from Charged Particles in Solids* (AIP, New York, 1989)
25. V.M. Biryukov, YuA Chesnokov, V.I. Kotov, *Crystal Channeling and its Application at High-Energy Accelerators* (Springer, Berlin, 1996)
26. S. Bellucci, S. Bini, V.M. Biryukov, YuA Chesnokov et al., Phys. Rev. Lett. **90**, 034801 (2003)
27. V. Guidi, A. Antonioni, S. Baricordi, F. Logallo, C. Malagù, E. Milan, A. Ronzoni, M. Stefancich, G. Martinelli, A. Vomiero, Nucl. Inst. Meth. B **234**, 40–46 (2005)
28. P. Balling, J. Esberg, K. Kirsebom, D.Q.S. Le, U.I. Uggerhøj, S.H. Connell, J. Härtwig, F. Masiello, A. Rommeveaux, Nucl. Instrum. Meth. B **267**, 2952–2957 (2009)
29. M.B.H. Breese, Nucl. Instrum. Meth. B **132**, 540–547 (1997)
30. U. Mikkelsen, E. Uggerhøj, Nucl. Instrum. Meth. B **160**, 435–439 (2000)
31. V.L. Ginzburg, Izv. Akad. Nauk SSSR **11**, 165 (1947) (in Russian)
32. J.M.J. Madey, Stimulated emission of radiation in periodically deflected electron beam. US Patent No. 3822410 (1974)
33. J.M.J. Madey, J. Appl. Phys. **42**, 1906–1913 (1971)
34. E.L. Saldin, E.A. Schneidmiller, M.V. Yurkov, *The Physics of Free-Electron Lasers* (Springer, Berlin, 1999)
35. P. Schmüser, M. Dohlus, J. Rossbach, *Ultraviolet and Soft X-Ray Free-Electron Lasers* (Springer, Berlin, 2008)
36. E.G. Bessonov, Nucl. Instrum. Meth. A **528**, 511–515 (2004)

# Crystals, Critical Fields, Collision Points, and a QED Analogue of Hawking Radiation

**Ulrik I. Uggerhøj**

## 1 Introduction

During the penetration of a crystal close to a crystallographic direction, the trajectory of the penetrating particle—due to the sequence of binary encounters—becomes indistinguishable from the trajectory obtained from 'smearing' (averaging) the charges along the string or plane, see Fig. 1.

From the resulting translational invariance in the longitudinal direction of the potential inside the crystal, follows a separation of the longitudinal and transverse motions, since the longitudinal momentum $p_{\parallel}$ is conserved. The result is a conserved 'transverse energy' and a transverse potential $U(r_{\perp})$ in which the particle moves:

$$U(r_{\perp}) = \frac{1}{d}\int_{-\infty}^{\infty} V(r_{\perp}, z)\mathrm{d}z \tag{1}$$

where $V(r_{\perp}, z)$ is the potential of the atom at the location of the projectile.

s

In the continuum model the transverse motion is given by:

$$\frac{d}{dt}\gamma m\dot{r}_{\perp} = -\frac{d}{dr_{\perp}}U(r_{\perp}(t)) \tag{2}$$

where the dot denotes differentiation with respect to time, $t$, and $r_{\perp}$ is the transverse coordinate. Using energy conservation and neglecting terms of order $1/\gamma^2$, the transverse energy reduces to

$$E_{\perp} = \frac{1}{2}pv\psi^2 + U(r_{\perp}) \tag{3}$$

U. I. Uggerhøj (✉)
Department of Physics and Astronomy, Aarhus University, Aarhus, Denmark
e-mail: ulrik@phys.au.dk

W. Greiner (ed.), *Exciting Interdisciplinary Physics*,
FIAS Interdisciplinary Science Series, DOI: 10.1007/978-3-319-00047-3_33,

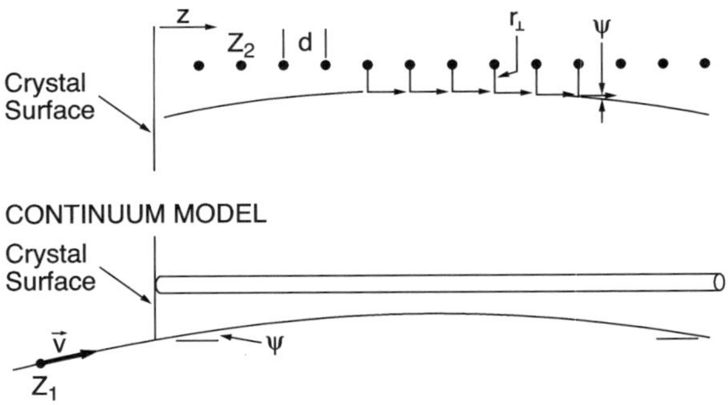


**Fig. 1** A schematical drawing of the discrete nature of the scattering centers in a crystal and the resulting continuum approximation. The target atoms with atomic number $Z_2$ and distance $d$ along the string, impose a curved trajectory on the penetrating particle with atomic number $Z_1$ through binary encounters over the transverse distance $r_\perp$. The resulting trajectory with entrance angle $\psi$ can be accurately described as if being the result of interaction with a string of continuous charge distribution, i.e. the charges $Z_2 e$ being 'smeared' along the direction of motion $z$

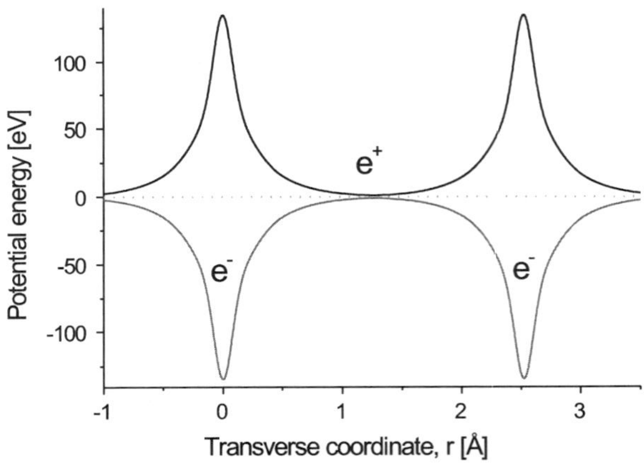


**Fig. 2** The transverse potential energy for positrons and electrons, in the continuum approximation, for diamond along the ⟨110⟩ axis at room temperature. The main regions for channeled $e^-$ and $e^+$ are indicated. The Doyle-Turner approximation for the atomic potential has been used [1]

where $\psi$ is the particle angle to the axis [2, 3].

As can be seen from Fig. 2, the axial potential in diamond varies roughly 50 V over a transverse distance of 0.1 Å, corresponding to an electric field

$$\mathscr{E}_d \simeq 5 \cdot 10^{10}\ \mathrm{V/cm} \tag{4}$$

This extremely strong, and macroscopically continuous, electric field arises from the coherent interaction of the screened nuclear fields along the direction of motion.

It should be emphasized that although this field also appears in combination with channeled particles, it is not a necessary condition that the particle is channeled—the continuum approximation generally applies at angles one to two orders of magnitude higher than the critical angle for channeling.

For a thorough introduction to strong fields in crystals at high energies, see e.g. [4, 5].

## 1.1 The Critical Field

In the present connection, 'strong' means comparable to the quantum mechanical critical field in a Lorentz-invariant expression. In atomic physics, the scale for 'strong' fields is set by the electric field a $1s$ electron is exposed to in atomic hydrogen, the atomic unit $\mathcal{E}_a = e/a_0^2 = 5.14 \cdot 10^9$ V/cm, and the magnetic field that gives rise to the same force on the $1s$ electron $B_a = \mathcal{E}_a/c\alpha = 2.35 \cdot 10^5$ T. In quantum electrodynamics, on the other hand, the strength of the electric field is measured in units of the critical field $\mathcal{E}_0$ (and the corresponding magnetic field $B_0$), obtained from a combination of the electron charge and mass, the velocity of light and Planck's (reduced) constant (with values from [6])

$$\mathcal{E}_0 = \frac{m^2c^3}{e\hbar} = 1.323285 \cdot 10^{16}\,\mathrm{V/cm} \quad , \quad B_0 = 4.414005 \cdot 10^9\,\mathrm{T} \tag{5}$$

These scales of strength are thus related as $\mathcal{E}_a = \alpha^3 \mathcal{E}_0$ and $B_a = \alpha^2 B_0$. As we shall see shortly, classical strong fields are even stronger $\mathcal{E}_c = \mathcal{E}_0/\alpha$ and $B_c = B_0/\alpha$.

The critical field $\mathcal{E}_0$ is frequently referred to as the Schwinger field [7], although it was treated as early as 1931 by Sauter [8, 9], following a supposition by Bohr on the magnitudes of fields relevant in the Klein paradox [10].

Rewriting the expression for the critical field to $\mathcal{E}_0 = mc^2/e\lambda\!\!\!^-$, where $\lambda\!\!\!^-_c = \hbar/mc$ is the reduced Compton wavelength, it appears that in a critical field a (virtual) electron may obtain an energy corresponding to the electron rest energy $mc^2$ while moving over a distance corresponding to the uncertainty of its location $\lambda\!\!\!^-_c$. Thus, the strong gradient of the potential combined with quantum uncertainty, as e.g. also seen in Zitterbewegung, may produce new particles—a QED phenomenon analogous to the Hawking radiation, discussed below.

In a classical analogue, a similar field strength is obtained from the field at a distance of a classical electron radius from the 'center' of the electron, $\mathcal{E}_c = e/r_e^2$. This field 'on the surface' of a classical electron is likewise where $e$ transported over $r_e$ yields $mc^2$, $\mathcal{E}_c = mc^2/er_e = \mathcal{E}_0/\alpha$, and, as must be required of a classical field, it does not contain $\hbar$. It is approximately equal to the Born-Infeld limiting field strength $b_l$. The latter was introduced 'dogmatically' by a Lagrangian $L = -b^2(1 - \sqrt{1 - (\mathcal{E}^2 - B^2)/b_l^2})$ (inspired by the relativistic $mc^2(1 - \sqrt{1 - v^2/c^2})$ where $c$ is the limiting speed) and described a transition to non-linear, classical electrodynamics [11]. The classical strong field is thus 137 times larger than the quantum one, i.e. a tunneling process reduces the necessary field strength required to produce a pair in quantum theory [12].

## 1.2 Electrodynamical Invariants

We now consider the general case of a charged particle interacting with an electromagnetic field, following e.g. [13, 14]. Three dimensionless invariants can be constructed from the electromagnetic field strength tensor, $F_{\mu\nu}$, and the momentum four-vector $p^\nu$ (or, in the case of a photon, $\hbar k^\nu$):

$$\chi^2 = \frac{(F_{\mu\nu} p^\nu)^2}{m^2 c^4 \mathcal{E}_0^2} \tag{6}$$

$$\Xi = \frac{F_{\mu\nu}^2}{\mathcal{E}_0^2} = \frac{2(\mathbf{B}^2 - \boldsymbol{\mathcal{E}}^2)}{\mathcal{E}_0^2} \tag{7}$$

$$\Gamma = \frac{e_{\lambda\mu\nu\rho} F^{\lambda\mu} F^{\nu\rho}}{\mathcal{E}_0^2} = \frac{8\boldsymbol{\mathcal{E}} \cdot \mathbf{B}}{\mathcal{E}_0^2} \tag{8}$$

where $e_{\lambda\mu\nu\rho}$ is the antisymmetric unit tensor and contraction is indicated by repeated indices. For an ultra-relativistic particle moving across fields $\mathcal{E} \ll \mathcal{E}_0$, $B \ll B_0$ with an angle $\theta \gg 1/\gamma$ the invariants fulfill $\chi \gg \Xi, \Gamma$ and $\Xi, \Gamma \ll 1$. The relation of $\chi$ to the fields $\boldsymbol{\mathcal{E}}$ and $\mathbf{B}$ is given by [14]

$$\chi^2 = \frac{1}{\mathcal{E}_0^2 m^2 c^4}((\mathbf{p}c \times \mathbf{B} + E \cdot \boldsymbol{\mathcal{E}})^2 - (\mathbf{p}c \cdot \boldsymbol{\mathcal{E}})^2) \tag{9}$$

For an ultrarelativistic particle moving perpendicularly to a pure electric or pure magnetic field this reduces to

$$\chi = \frac{\gamma \mathcal{E}}{\mathcal{E}_0} \text{ or } \chi = \frac{\gamma B}{B_0} \tag{10}$$

Due to $\mathcal{E}_0$ being proportional to $m^2$, we note that $\chi$ scales with $1/m^2$ such that e.g. the coherent production of muon pairs from electrons becomes appreciable only at energies $207^2 \approx 4 \cdot 10^4$ times larger than electron-positron pairs.

For the emission of radiation it is the trajectory that is decisive. Therefore, it is insignificant if the field responsible for the path is electric or magnetic and as a consequence they are frequently used indiscriminately in radiation emission. Since $\chi$ is invariant, $\gamma B$ (or $\gamma \mathcal{E}$) is the same in any reference system and thus it is reasonable to transform to the electron frame. In this reference system, by definition the Lorentz factor of the electron is 1 and the field present in the frame of the laboratory is boosted by $\gamma = E/mc^2$, where $E$ is the energy of the electron in the laboratory. This means that the field *in the rest-frame of the electron* can become critical for achievable $\gamma$-values.

## 2 Quantum Synchrotron Radiation

Concerning the recoil in the emission process, a classical calculation of the synchrotron radiation emission in a magnetic field leads to a spectrum which extends to $\omega_c \simeq 3\gamma^3 eB/2p = 3\gamma^3\omega_B/2$ [15, 16], i.e.

$$\frac{\hbar\omega_c}{E} \simeq \frac{3\gamma B}{2B_0} = \frac{3\gamma\kappa_f}{2} = \frac{3\chi}{2} \tag{11}$$

which for sufficiently large $\gamma$ exceeds 1. Here $\omega_B = eBv/pc$ is the cyclotron angular frequency and $\chi$ is the strong field parameter. Thus, for $\gamma$ values beyond a certain point, the classically calculated radiation spectrum extends beyond the available energy [4, 17–19]. In this case a quantum treatment taking recoil into account becomes necessary:

> "...the condition for quantum effects to be unimportant is that the momenta of the radiated quanta be small compared with the electron momentum" [17].

As a result of the quantum correction, the total radiated intensity for the classical emission is according to Schwinger reduced by a factor

$$I/I_{\text{cl}} = 1 - 55\sqrt{3}\lambda\!\!\!^{-}_c\omega_B\gamma^2/16c \tag{12}$$

due to first order quantum corrections when $\chi \ll 1$ [17]. Including the second order term the reductions for small values of $\chi$ are [14]

$$I/I_{\text{cl}} = 1 - 55\sqrt{3}\chi/16 + 48\chi^2 \qquad \chi \ll 1 \tag{13}$$

and asymptotically for large values of $\chi$

$$I/I_{\text{cl}} \simeq 1.2\chi^{-4/3} \qquad \chi \gg 1 \tag{14}$$

Furthermore, an approximate expression ("accuracy better than 2 % for arbitrary $\chi$" [20, eq. (4.57)])

$$I/I_{\text{cl}} \simeq (1 + 4.8(1+\chi)\ln(1+1.7\chi) + 2.44\chi^2)^{-2/3} \tag{15}$$

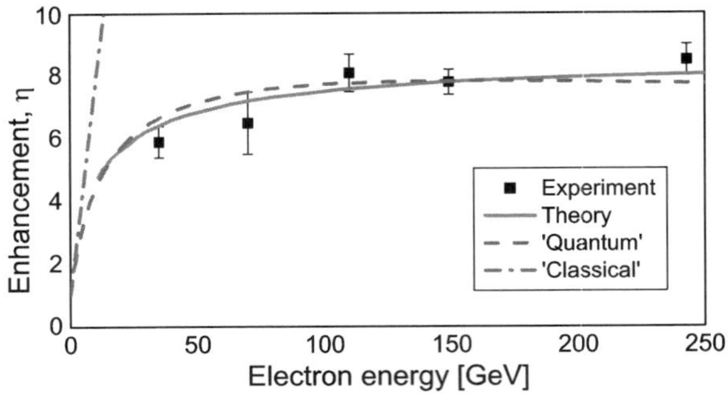


**Fig. 3** Experiment [23] and theory [24] for radiation emission from electrons penetrating a tungsten crystal near the ⟨111⟩ axis. For comparison, a *curve* based on Eq. (15) with a slightly arbitrary, but realistic $\overline{\chi} = 0.02 \cdot E[\text{GeV}]$ (and vertical scale obtained as the best fit) is shown as the *dashed line* ('quantum'), and the corresponding classical expression as the *dash-dotted line* ('classical'). The enormous difference between the 'classical' and 'quantum' curves directly show the strong quantum suppression in the experimentally accessible regime

gives a compact analytical expression applicable e.g. in computer codes. From this, it is clear that the emission of synchrotron radiation is affected already at fairly small values of $\chi$. A graphical representation compared to measured values is given in [21, 22].

In Fig. 3 is shown results for radiation emission from electrons impinging on a tungsten crystal close to the ⟨111⟩ axis. As a consequence of the strong deflection upon the passage of the string of nuclei composing the axis, the electron is forced to emit radiation as in a constant field, as described above. This happens much like in normal synchrotron radiation emission, only in a much more intense field, $\lesssim 10^{11}$ V/cm, corresponding to 30.000 T. As a result of the high peak value of the $\chi$ parameter ($\chi_{\text{W},\langle 111\rangle} \simeq 0.03 \cdot E[\text{GeV}]$), the radiation emission is subjected to strong quantum suppression. In the limit $\chi \ll 1$ the enhancement would be linear with increasing energy, as shown by the dash-dotted line. This is the case because synchrotron radiation emission is quadratic in energy and radiation from an amorphous foil is linear in energy, but due to the strong quantum suppression, the enhancement is reduced to the level shown by the dashed line, as also expected from equation (15). The good agreement between experimental values and theory shown in Fig. 3, combined with the equality of beamstrahlung and strong field theory shown in Fig. 4 provide a strong experimental indication that QED theory as applied to beamstrahlung—discussed in the following section—in the regime $1 \lesssim \Upsilon \lesssim 10$ is correct.

The accuracy of the experimental values is 5–10 %, enough to ascertain the validity of the theoretical approach. The quantum synchrotron behaviour of radiation emission in a strong field, is thus experimentally well confirmed at relatively high values of $1 \lesssim \chi \lesssim 10$.

# 3 Beam-Beam Interactions—Beamstrahlung

In the construction of linear colliders an important phenomenon is the emission of intense radiation due to the interaction of particles in one bunch with the electromagnetic field from the opposing bunch. This leads to the synchrotron radiation equivalent of particle deflection in the field of the bunch, instead of in a magnetic dipole: Beamstrahlung. As the emission of beamstrahlung has a direct and significant impact on the energy of the colliding particles, it is a decisive factor for e.g. the energy-weighted luminosity. Conversely, beamstrahlung emission may provide a method for luminosity measurement. It is therefore important to know if beamstrahlung theory is correct for the conceptual and technical design of the collision region—the center about which the rest of the machine is based.

The Lorentz factor $\gamma$ in this case is understood as the Lorentz factor of each of the oppositely directed beams, measured in the laboratory system. Then relativistic velocity addition $v' = (v - V)/(1 - vV/c^2) = 2v/(1 + v^2/c^2)$ with $V = -v$ and $\gamma' = 1/\sqrt{1 - v'^2/c^2}$ yields the Lorentz factor $\gamma'$ of one beam seen from a particle in the other beam of $\gamma' = 2\gamma^2 - 1$, usually shortened to $2\gamma^2$ in the ultrarelativistic limit. Thus, in the restframe of a particle in one bunch the field of the other bunch is boosted by a factor $\simeq 2\gamma^2$ and may approach or even exceed critical field values. The emission of beamstrahlung can be expressed as a function of $\chi$ (often called $\Upsilon$ in the accelerator physics community) which for the Stanford Linear Collider (SLC) is small $\simeq 10^{-3}$ but of the order unity for the next generation linear colliders [25]. For the planned Compact LInear Collider (CLIC) at CERN, the collision point is designed such that $\overline{\Upsilon} \simeq 4$. Quantum corrections to the emission of beamstrahlung therefore become crucial.

## 3.1 Quantum Treatment of Beamstrahlung

The discussion of quantum effects in radiation emission from energetic particles in collision with a counterpropagating bunch was started in the mid-80's [26–28]. In particular the suppression of the intensity stemming from the strong field deflection was of interest. It later continued with treatments of pair creation [29, 30]. Of particular relevance to the connection of beamstrahlung with emission from electrons penetrating crystals is the contribution by Baier, Katkov and Strakhovenko to the early development of the theory of beamstrahlung [31].

Shortly after the first publications on the relevance of quantum theory to beamstrahlung, Blankenbecler and Drell contributed a full quantum treatment of the problem, based on the eikonal approximation [32]. The scaling parameter in their approach is given by

$$C = \frac{m^2 c^3 RL}{4Ne^2\gamma^2\hbar} \tag{16}$$

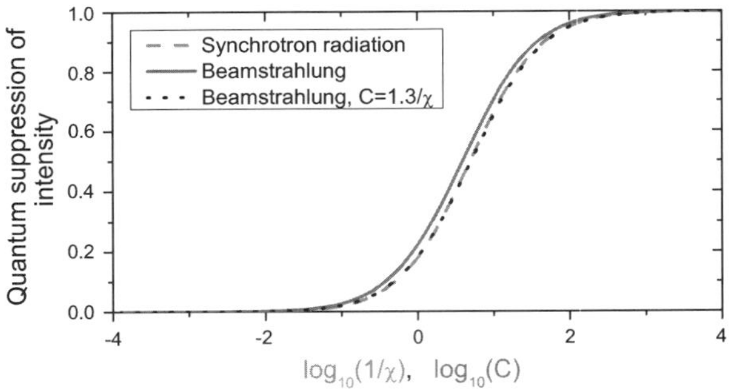


**Fig. 4** The quantum suppression of radiation emission intensity, according to Eqs. (15) and (17)

representing the electric field from a homogeneously charged cylinder of length $L$ and radius $R$ holding $N$ charges, in units of $\mathcal{E}_0 = m^2c^3/\hbar e$. The applicability of this scaling parameter was later elaborated upon by Solov'yov and Schäfer [33, 34]. From this model of a beam, the form factor $F = \delta/\delta_{\mathrm{classical}}$, describing the quantal energy loss in units of the classical, was derived and approximated by:

$$F(C) = \left[1 + \frac{1}{b_1}[C^{-4/3} + 2C^{-2/3}(1 + 0.20C)^{-1/3}]\right]^{-1} \tag{17}$$

with $b_1 = 0.83$, see also [33]. Clearly, as stated by Blankenbecler and Drell, in the classical regime $\hbar \to 0$ in Eq. (16) such that $C$ tends to infinity, and therefore the form factor tends to 1 according to Eq (17), as must be required.

As a result of the quantum correction, the total radiated intensity for the classical emission is reduced as given by equation (15).

In Fig. 4 is shown graphs based on Eqs. (17) and (15), where it has been assumed that $C = 1/\chi$. The curves are very similar, over the entire range of more than five orders of magnitude in $\chi$. The expressions originating from the same phenomenon becomes even more evident by adjusting to $C = 1.3/\chi$ which results in the curves being indistinguishable on the plot. This is not a fortuitous coincidence: In Blankenbecler and Drell's theory, the bunch is treated as a homogeneously charged cylinder of length $L$ and radius $R$ holding $N$ charges. At the distance $r$ from the center of this cylinder the electric field is $2\gamma Ne/Lr$, which leads to an average (over $r$) field

$$\mathcal{E} = \frac{\gamma Ne}{LR} \tag{18}$$

which can be combined with Eq. (16) to give

$$C = \frac{m^2c^3}{\gamma\mathcal{E}e\hbar} = \frac{\mathcal{E}_0}{\gamma\mathcal{E}} = \frac{1}{\chi} \tag{19}$$

so it is legitimate to interchange $C$ and $1/\chi$. The additional factor 1.3 that brings the curves into almost exact agreement, is due to the radiation intensity being non-linear in $C$, i.e. averaging over the field encountered and then calculating the intensity from this field may be different from calculating the intensity from the fields encountered and then averaging.

Early studies by Chen and Yokoya [25] showed that field gradient effects are small for a collider operating near $\Upsilon = \chi = 1/C = 1$, i.e. also for the planned CLIC at CERN where the expected value is as mentioned $\Upsilon \simeq 4$. It is therefore to a high degree of accuracy sufficient to use Eq. (17) derived for the homogeneously charged cylinder in calculations for beamstrahlung.

Nevertheless, due to the emission of quantum beamstrahlung, the 'useful' luminosity $\mathscr{L}_1$ (where $\mathscr{L}_1$ is defined as the luminosity for that part of the beam where the energy is still at least 99 % of the initial) becomes about 40 % of the nominal, due to the loss of energy in the beamstrahlung process. In a classical calculation the useful luminosity would have been at least an order of magnitude smaller. For such future colliders, $\gamma\gamma$-collisions, resulting e.g. also in hadronic interactions, may be generated from the beams themselves and the advantage of using leptonic beams ('clean' collisions) is to some extent lost. The beamstrahlung problem is unavoidable since single passage (as opposed to circular machines) forces small beam cross sections to give high luminosity. And since $\Upsilon \propto N\gamma/(\sigma_x + \sigma_y)\sigma_z$, with $\sigma$ denoting the beam size, high energies and high luminosity means a high value for $\Upsilon$. However, the problem may be partly alleviated by applying special bunch structures ('flat' beams, maximizing $\sigma_x + \sigma_y$) to avoid rapid beam deterioration from strong field effects [35].

Finally, it should be mentioned that effects of the spin of the particle become very important in the beamstrahlung emission. As polarimeters cannot be positioned at the intersection point of the crossing beams, reliable models for the degree of polarization after emission (immediately before the collision) must be developed for measurements with polarized beams to make sense.

## 4 Hawking Radiation and Unruh Effect

A fascinating analogy exists between the critical field and the Hawking radiation from a black hole: The gravitational acceleration at the Schwarzschild radius $R_S = 2GM/c^2$ equals $g(R_S) = c^4/4GM$ where $G$ is Newton's constant and $M$ the mass of the black hole. From the equivalence principle, locally the gravitational field is analogous to an accelerating frame of reference. The word *locally* is crucial in this context: A gravitational field and an accelerating frame are closely related, but they are not equal. It is impossible to 'transform away' a gravitational field by shifting to an accelerating frame due to the existence of tidal forces as expressed by the Riemann curvature tensor. But locally, e.g. for one test particle only, there is equivalence. Setting this gravitational acceleration equal to the acceleration of an electron in a critical field $g_0 = e\mathscr{E}_0/m = c^2/\lambda\!\!\!^{-}_c$ the condition $\lambda\!\!\!^{-}_c = 2R_S$ is obtained. In words: the black hole emits particles with (reduced) Compton wavelengths that are as large as or larger than the hole itself. This is approximately (within a factor 2 as for the calculation of the deflection of light by use of the equivalence principle) equal to the answer obtained in a full analysis of the Hawking radiation [36].

Likewise, the equivalence between the temperature of the Hawking radiation from a black hole and the temperature of the vacuum in a constantly accelerated frame [37, 38] has been widely discussed—the so-called Unruh effect. As channeled particles are subject to enormous fields and accelerations, outlines for possible detection schemes using strong crystalline fields have been put forward [39, 40]. In [39] it is estimated that a planar channeled positron with $\gamma \gtrsim 10^8$ will emit Unruh radiation as intense as the incoherent bremsstrahlung. These estimates, however, do not discuss the subtleties connected to the inherently non-constant acceleration for a channeled particle.

The Unruh effect gives rise to a Planckian photon spectral distribution at a temperature

$$T = \frac{\hbar a}{2\pi k_B c} \tag{20}$$

where $a$ is the acceleration and $k_B$ the Boltzmann constant.

Several other methods have been proposed to pursue the problem of measuring the Unruh temperature experimentally. According to Baier—who was an expert also in the field of radiative polarization [41]—the Unruh mechanism is a possible *interpretation* ("theoretical game") of the radiative depolarization in a storage ring [42], as originally suggested by Bell. For an overview of the suggested experimental methods and a review of the literature on the subject see e.g. [43, 44] (these chapters are mainly sophisticated theory chapters, but do contain references to experimental methods). In this connection, it may be mentioned that even the lightest charged composite, the positronium negative ion $Ps^-$, will have an essentially unaffected hyperfine structure when exposed to Unruh radiation during state-of-the-art acceleration in a high-gradient radio-frequency cavity [45].

The above mentioned analogy between the critical field and the Hawking radiation from a black hole becomes even more compelling by interpreting the field as a temperature as is done for the Unruh effect, Eq. (20): $T_0 = e\mathcal{E}_0\hbar/2\pi m k_B c$ [46] and inserting $g(R_S) = c^4/4GM$ instead of $g_0 = e\mathcal{E}_0/m$ in $T_0$ from which the correct Hawking temperature appears [47]:

$$T_0 = \frac{\hbar c^3}{8\pi G M k_B} \tag{21}$$

The Hawking radiation can thus be viewed as a critical field phenomenon, where the electromagnetic critical field is replaced by a gravitational field. Generally speaking, the uncertainty of the location of the particles is given by their (reduced) Compton wavelength, as evidenced e.g. by the Zitterbewegung. Thus, the interpretation that a quantum fluctuation—a virtual pair—can become real due to the presence of the critical field, where the rest mass energy is created over exactly this length, is valid in both cases. As the gravitational field at the Schwarzschild radius $g(R_S)$ is larger for small holes, short Compton wavelength—'hot'—radiation may be emitted, which is why light black holes possess a higher temperature than heavy ones—they possess a higher gradient.

It must be emphasized, though, that it is an analogy, not a one-to-one correspondence between electrodynamics and geometrodynamics. In the former case, for instance, the invariant $\Xi$, Eq. (7), is much smaller than one.

From the above qualitative considerations, it is clear that the QED analogy of Hawking radiation, critical field radiation, is of high importance to be investigated experimentally. This is perhaps even more the case as long as the gravitational version is not within observational reach in the foreseeable (perhaps even imaginable) future.

**Acknowledgments** I gratefully acknowledge the initiatives of Profs. Greiner, Newman and Vilakazi to arrange not only an exceptionally interesting symposium with inspiring delegates from many subjects and countries, but also for arranging it in a very exciting environment. Finally, I wish to thank Dr. Weber and his family and staff at Makutsi Safari Farm for their dedication to making the symposium the success it was.

## References

1. P.A. Doyle, P.S. Turner, Relativistic Hartree-Fock X-ray and electron scattering factor. Acta Crystallogr. Sect. A **24**, 390–397 (1968)
2. E. Bonderup, J.U. Andersen, R.H. Pantell, Channeling radiation. Ann. Rev. Nucl. Part. Sci. **33**, 453–504 (1983)
3. J. Bak, J.A. Ellison, B. Marsh, F.E. Meyer, O. Pedersen, J.B.B. Petersen, E. Uggerhøj, K. Østergaard, S.P. Møller, A.H. Sørensen, M. Suffert, Channeling radiation from 2–55 GeV/c electrons and positrons: (i). planar case. Nucl. Phys. B **254**, 491–527 (1985)
4. A.H. Sørensen, Channeling, bremsstrahlung and pair creation in single crystals. NATO ASI Ser. **255**, 91–118 (1991)
5. E. Uggerhøj, U.I. Uggerhøj, Strong crystalline fields–a possibility for extraction from the LHC. Nucl. Instrum. Methods Phys. Res., Sect. B **234**(1–2), 31–39 (2005), Relativistic Channeling and Related Coherent Phenomena in Strong Fields
6. Particle Data Group, C. Amsler, M. Doser, M. Antonelli et al., Review of particle physics, Phys. Lett. B **667**, 1–5 (2008)
7. J. Schwinger, On gauge invariance and vacuum polarization. Phys. Rev. **82**, 664–679 (1951)
8. F. Sauter, Über das Verhalten eines Elektrons im homogenen elektrischen Feld nach der relativistischen Theorie Diracs. Z. Phys. **69**, 742–764 (1931)
9. F. Sauter, Zum "Kleinschen Paradoxon". Z. Phys. **73**, 547–552 (1931)
10. O. Klein, Die Reflexion von Elektronen an einem Potentialsprung nach der relativistischen Dynamik von Dirac. Z. Phys. **53**, 157–165 (1929)
11. M. Born, L. Infeld, Foundations of the new field theory. Proc. R. Soc. Lond. A **144**, 425–451 (1934)
12. R.P. Feynman, A Relativistic cut-off for classical electrodynamics. Phys. Rev. **74**, 939–946 (1948)
13. A.I. Nikishov, V.I. Ritus, Quantum processes in the field of a plane electromagnetic wave and in a constant field. Zh. Eksp. Teor. Fiz. **46**, 776–796 (1964)
14. V.B. Berestetskii, E.M. Lifshitz, L.P. Pitaevskii, *Relativistic Quantum Theory* (Elsevier, UK, 1971)
15. L.I. Schiff, Quantum effects in the radiation from accelerated relativistic electrons. Am. J. Phys. **20**, 474–478 (1952)

16. J.D. Jackson, *Classical Electrodynamics* (Wiley, New York, 1975)
17. J. Schwinger, The quantum correction in the radiation by energetic accelerated electrons. Proc. Natl. Acad. Sci. USA **40**, 132–136 (1954)
18. W. Tsai, A. Yildiz, Motion of an electron in a homogeneous magnetic field–modified propagation function and synchrotron radiation. Phys. Rev. D **8**, 3446–3460 (1973)
19. W. Tsai, Magnetic bremsstrahlung and modified propagation function. Spin-0 charged particles in a homogeneous magnetic field. Phys. Rev. D **8**, 3460–3469 (1973)
20. V.N. Baier, V.M. Katkov, V.M. Strakhovenko, *Electromagnetic Processes at High Energies in Oriented Single Crystals* (World Scientific, Singapore, 1998)
21. V.N. Baier, V.M. Katkov, V.M. Strakhovenko, Quantum effects in radiation emitted by ultrahigh energy electrons in aligned crystals. Phys. Lett. A **114**, 511–515 (1986)
22. A. Belkacem, G. Bologna, M. Chevallier, N. Cue, M.J. Gaillard, R. Genre, J. Kimball, R. Kirsch, B. Marsh, J.P. Peigneux, J.C. Poizat, J. Remillieux, D. Sillou, M. Spighel, C.R. Sun, Strong field interactions of high energy electrons and photons in ge crystals. Nucl. Instrum. Methods Phys. Res., Sect. B **33**, 1–10 (1988)
23. K. Kirsebom, Y.V Kononets, U. Mikkelsen, S.P. Møller, E. Uggerhøj, T. Worm, K. Elsener, C. Biino, N. Doble, S. Ballestrero, P. Sona, R.O. Avakian, K.A. Ispirian, S.P. Taroian, S.H. Connell, J.P.F. Sellschop, Z.Z. Vilakazi, R. Moore, M.A. Parker, A. Baurichter, V.M. Strakhovenko, Radiation emission and its influence on the motion of multi-GeV electrons and positrons incident on a single diamond crystal. Nucl. Instr. Meth. B **174**, 274–296 (2001)
24. V.M. Katkov, V.N. Baier, Coherent and incoherent radiation from high-energy electron and the lpm effect in oriented single crystal. Phys. Lett. A **353**, 91–97 (2006)
25. P. Chen, K. Yokoya, Field-gradient effect in quantum beamstrahlung. Phys. Rev. Lett. **61**, 1101–1104 (1988)
26. T. Himel, J. Siegrist, Quantum effects in linear collider scaling laws, SLAC-PUB-3572, 1–7 (1985)
27. K. Yokoya, Quantum correction to beamstrahlung due to the finite number of photons. Nucl. Instr. Meth. A **1–16**, 251 (1986)
28. R.J. Noble, Beamstrahlung from colliding electron-positron beams with negligible disruption. Nucl. Instr. Meth. A **256**, 427–433 (1987)
29. P. Chen, V.I. Telnov, Coherent pair creation in linear colliders. Phys. Rev. Lett. **63**, 1796–1799 (1989)
30. R. Blankenbecler, S.D. Drell, N. Kroll, Pair production from photon-pulse collisions. Phys. Rev. D **40**, 2462–2476 (1989)
31. V.N. Baier, V.M. Katkov, V.M. Strakhovenko, Radiation from relativistic particles colliding in a medium in the presence of an external field. Zh. Eksp. Teor. Fiz. **94**, 125–139 (1988)
32. R. Blankenbecler, S.D. Drell, Quantum treatment of beamstrahlung. Phys. Rev. D **36**, 277–288 (1987)
33. A.V. Solov'yov, A. Schäfer, C. Hofmann, Quasiclassical approximation for the beamstrahlung process. Phys. Rev. E **47**, 2860–2867 (1993)
34. A.V. Solov'yov, A. Schäfer, Pair creation by photons in the field of an electron or positron pulse of high density. Phys. Rev. E **48**, 1404–1409 (1993)
35. R. Blankenbecler, S.D. Drell, Quantum beamstrahlung: Prospects for a photon-photon collider. Phys. Rev. Lett. **61**, 2324–2327 (1988)
36. P.C.W. Davies, Thermodynamics of black holes. Rep. Prog. Phys. **41**, 1313–1355 (1978)
37. W.G. Unruh, Notes on black-hole evaporation. Phys. Rev. D **14**, 870–892 (1976)
38. P.C.W. Davies, Scalar particle production in Schwarzschild and Rindler metrics. J. Phys. A **8**, 609–616 (1975)
39. K.A. Ispirian, S.M. Darbinian, A.T. Margarian, New mechanism for Unruh radiation of channeled particles (1989)
40. K.T. McDonald, *Hawking-Unruh Radiation and Radiation of a Uniformly Accelerated Charge* (California, Quantum Aspects of Beam Physics, 1998), p. 643
41. V.N. Baier, Radiative polarization of electrons in storage rings. Usp. Fiziol. Nauk **105**, 441–478 (1972)

42. V.N. Baier, V.M. Katkov, Electroproduction of electron-positron pair in oriented crystal at high energy. Phys. Lett. A **373**, 1874–1879 (2009)
43. H.C. Rosu, *Unruh Effect as Particular Frenet-Serret Vacuum Radiation and Detection Proposals* (Quantum Aspects of Beam Physics, California, 2004), pp. 164–175
44. L.C.B. Crispino, A. Higuchi, G.E.A. Matsas, The Unruh effect and its applications. Rev. Mod. Phys. **80**, 787 (2008)
45. U.I. Uggerhøj, Relativistic $ps^-$ and ps. Phys. Rev. A **73**, 052705 (2006)
46. B. Müller, W. Greiner, J. Rafelski, Interpretation of external fields as temperature. Phys. Lett. A **63**, 181–183 (1977)
47. B. Müller, W. Greiner, J. Rafelski, *Quantum Electrodynamics of Strong Fields* (Springer, Berlin, 1985)

# QED Calculations on Highly Charged Ions, Using a Unified MBPT-QED Approach

**Ingvar Lindgren, Sten Salomonson, Daniel Hedendahl and Johan Holmberg**

**Abstract** There is presently a great interest in studying QED effects in highly charged ions by means of large accelerators, and the best information is usually gained from the study of multi-electron ions. It might then be possible to detect the combined effect of QED and electron correlation, which has so far never been observed. That could be possible also from accurate laser or X-ray data. For the corresponding theoretical analysis it will then be necessary to treat the effects of QED and electron correlation simultaneously in a coherent manner. This is not possible with presently available techniques but will require the new generation of atomic calculations that is now being developed at our laboratory. The calculations have to be performed in the Coulomb gauge, and a procedure for renormalization in that gauge has very recently been tested for hydrogen-like ions. Work is now in progress to perform unified MBPT-QED calculations on multi-electron systems.

## 1 Introduction

QED effects in highly charged ions have for a long time been studied in large accelerators, like that at GSI in Darmstadt. The efforts have mainly been concentrated on hydrogen-like ions, as illustrated by the experimental and theoretical results for $U^{91+}$ in Table 1. As can be seen from the table, the nuclear-size effect is here considerable,

---

I. Lindgren (✉) · S. Salomonson · D. Hedendahl · J. Holmberg
Physics Department, University of Gothenburg, Gothenburg, Sweden
e-mail: ingvar.lindgren@physics.gu.se

S. Salomonson
e-mail: sten.salomonson@physics.gu.se

D. Hedendahl
e-mail: daniel.hedendahl@combine.se

J. Holmberg
e-mail: johan.holmberg@physics.gu.se

W. Greiner (ed.), *Exciting Interdisciplinary Physics*,
FIAS Interdisciplinary Science Series, DOI: 10.1007/978-3-319-00047-3_34,

**Table 1** Ground-state Lamb shift of hydrogenlike uranium (in eV, mainly from Ref. [23].)

| Correction | Value | Reference |
|---|---|---|
| Nuclear size | 198.82 | |
| First-order self energy | 355.05 | [22, 24] |
| Vacuum polarization | −88.59 | [28] |
| Second-order effects | −1.57 | |
| Nuclear recoil | 0.46 | |
| Nuclear polarization | −0.20 | |
| Total theory | 463.95 | |
| Experimental | 460.2 (4.6) | |

and the finer details of the QED part are hidden in the uncertainty of the nuclear-size effect.

In order to gain more information of QED effects in highly charged systems, we believe that it would be more appropriate to study ions with more than one electron, like He- and Li-like ions, where the nuclear effect is considerably reduced. In Table 2 we show some old results for Li-like uranium. The experiment as well as the old calculations indicated were the first of its kind, and considering the fact that these data are 20 years old, the results are impressive.

The price one has to pay in studying multi-electron ions is that the *electron correlation* has to be taken into account. But this will also make it possible, in principle, to detect the *combined* effect of QED and electron correlation, which has never been observed.

The effect of electron correlation itself can be effectively treated by highly developed methods of Many-Body Perturbation Theory (MBPT) and Coupled-Cluster Approach (CCA) (see the book of Lindgren and Morrison for a review [17]). For QED alone several methods are available, all of which, however, for practical reasons are limited to second order, and hence the electron correlation is in several cases treated insufficiently.

**Table 2** $2s - 2p_{1/2}$ Lamb shift of lithiumlike uranium (in eV)

| Correction | Ref. [4] | Ref. [27] | Ref. [32] |
|---|---|---|---|
| Relativistic MBPT | 322.41 | 322.32 | 322.10 |
| 1. Order self energy | −53.94 | −54.32 | |
| 1. Order vacuum polarization | (12.56) | 12.56 | |
| 1. Order self energy + vac. pol. | −41.38 | −41.76 | −41.77 |
| 2. Order self energy + vac. pol. | | 0.03 | 0.17 |
| Nuclear recoil | (0.10) | (−0.08) | −0.07 |
| Nuclear polarization | (0.10) | (0.03) | −0.07 |
| Total theory | 280.83(10) | 280.54(15) | 280.48(20) |
| Experimental | 280.59(9) | | |

**Table 3** The transition $1s2s\,^1S_0 - 1s2p\,^3P_1$ for He-like Si (in $\mathrm{cm}^{-1}$)

| | | Reference |
|---|---|---|
| Expt'l | 7230.585(6) | Myers et al. [8] |
| Theory | 7231.1 | Plante et al. [29] |
| | 7229(2) | Artemyev et al. [3] |

The standard many-body procedures can be complemented by first-order QED energies, but this is as far as one can go for the time being. In order to get further and observe higher-order effects, it will be necessary to include the QED effects in the electronic wave function, which is not possible with the present techniques. Such a procedure, however, is now being developed by our group in Gothenburg, and the ideas behind the procedure are described in some publications [19, 20] as well as in a recent monography [15]. The procedure is based upon the Covariant-Evolution-Operator (CEO) method for QED calculations, earlier developed at our laboratory [16, 18]. This has a structure that is similar to that of *many-body perturbation theory* (MBPT), and can therefore serve as a basis for a unified theory. This represents a *new generation* of atomic calculations that goes beyond existing many-body and QED methods.

An additional problem in developing a unified MBPT-QED procedure is that the calculations should be performed in the Coulomb gauge in order to take full advantage of the developments in MBPT. Numerical calculations with renormalization in that gauge has to our knowledge never been performed. During the last year, however, we have successfully implemented a procedure for dimensional regularization in Coulomb gauge, and this represents a major breakthrough in this kind of work [10, 11, 13].

Very accurate measurements can also be performed on lighter ions by means of laser techniques, as demonstrated particularly by Myers and his group in Florida [8, 25, 26]. In Table 3 we show the result for a transition in helium-like silicon, compared with available theoretical results. The calculation of Plante et al. is of MBPT type with first-order QED effects added and that of Artemyev et al. is a second-order QED calculation. These calculations represent the present status of the field, and the comparison shows that this can by no means match the experimental accuracy. To go beyond that, the new approach is needed, where the QED effects are included directly into the atomic wave function rather than added just as an energy correction. This combined MBPT-QED effect could in this case be estimated to of the order of a tenth of a $\mathrm{cm}^{-1}$, which obviously has to be considered in order to explain the experimental data.

As a second example we consider the accurate K$\alpha$ data for the copper atom (see Table 4). The calculations are performed with multi-configuration Dirac-Fock (MCDF) with first-order QED energy correction, which represents the best that can be performed today. The experimental data are one order of magnitude more accurate, and it is clear that the combined QED-MBPT effect, not included in the calculation and expected to be to the order 0.01–0.1 eV, could here be detected. The situation is similar for several other K$\alpha$ lines.

**Table 4** K$\alpha$ X-ray data for copper (in eV)

| | K$\alpha_1$ | K$\alpha_2$ | Reference |
|---|---|---|---|
| Expt'l | 8047.8237(26) | 8027.8416(26) | Deslattes et al. [7] |
| Theory | 8047.86(4) | 8027.92(4) | Chantler et al. [6] |

In the following we shall first briefly describe the standard procedures for relativistic MBPT and for QED calculations and finally consider the unification of the two.

## 2 Relativistic MBPT

Relativistic MBPT is normally based upon the *projected Dirac-Coulomb-Breit Hamiltonian* [31][1]

$$H = \Lambda_+ \Big[ \sum_{i=1}^{N} h_D(i) + V_C + V_B \Big] \Lambda_+, \tag{1}$$

where the first term represents the sum of single-particle Dirac Hamiltonians, the second term the Coulomb interaction and the third term the *instantaneous Breit interaction* between the electrons,

$$V_B = -\frac{e^2}{8\pi} \sum_{i<1} \Big[ \frac{\boldsymbol{\alpha}_i \cdot \boldsymbol{\alpha}_j}{r_{ij}} + \frac{(\boldsymbol{\alpha}_i \cdot \boldsymbol{r}_{ij})(\boldsymbol{\alpha}_j \cdot \boldsymbol{r}_{ij})}{r_{ij}^3} \Big], \tag{2}$$

where $\boldsymbol{\alpha}_i$ is the Dirac alpha matrix vector for particle $i$. The projection operators, $\Lambda_\pm$, are inserted in order to avoid negative-energy states that can lead to singularities (Breit-Ravenhall disease [5]). The Breit interaction represents the magnetic interaction as well as the leading effect of the *retardation*, caused by the fact that the electromagnetic radiation propagates with the finite speed of light.

The Hamiltonian is expressed using the *Coulumb gauge*, which is standard in all MBPT calculations, relativistic as well as non-relativistic. In the relativistic case this gauge has the great advantage before all other gauges that the Hamiltonian is correct to order $\alpha^2$ Rydbergs (or $\alpha^4 mc^2$), where $\alpha$ is the fine-structure constant. This is referred to as the *No-Virtual-Pair Approximation* (NVPA). Effects beyond this approximation—of order $\alpha^3$ Rydbergs and higher—are conventionally referred to as *QED effects*.

Relativistic as well as non-relativistic MBPT calculations in physics and quantum chemistry are normally based upon the *Bloch equation*. In the general case we consider a set of *target states* satisfying the eigenvalue equation

[1] Relativistic units are used: $c = m = \hbar = \varepsilon_0 = 1$.

$$H|\Psi^\alpha\rangle = E^\alpha|\Psi^\alpha\rangle \quad (\alpha = 1\ldots d), \tag{3}$$

and for each target state there exists a *model state*, $|\Psi_0^\alpha\rangle$ $(\alpha = 1\ldots d)$, which form a *model space*. The projection operator for the model space is denoted by $P$ and for the complementary space by $Q = 1 - P$. We apply *intermediate normalization* (IN), where the model states are the projection of the target states on the model space,

$$\langle\Psi_0^\alpha|\Psi^\alpha\rangle = 1; \quad |\Psi_0^\alpha\rangle = P|\Psi^\alpha\rangle \quad (\alpha = 1\ldots d). \tag{4}$$

A single *wave operator*, $\Omega$, transforms all model states to the corresponding target states,

$$\Omega|\Psi_0^\alpha\rangle = |\Psi^\alpha\rangle \quad (\alpha = 1\ldots d). \tag{5}$$

The Hamiltonian is normally partitioned into

$$H = H_0 + V, \tag{6}$$

where $H_0$ is a *model Hamiltonian*, containing the sum of single-electron Hamiltonians, and $V$ is a *perturbation*. Then the wave operator satisfies the commonly used form of the *generalized Bloch equation* [14, 17]

$$\boxed{\big[\Omega, H_0\big]P = Q\big(V\Omega - \Omega PV\Omega\big)P.} \tag{7}$$

By solving this equation iteratively, the most important correlation effects can be treated essentially to *all orders of perturbation theory*.

By expressing the wave operator in exponential form, a very effective calculation scheme, known as the exponential Ansatz or Coupled-Cluster Approach, can be constructed. We shall not consider this any further here but refer interested readers to the standard texts [17].

## 3 Combining MBPT with QED

The QED effects defined above are of three kinds, (i) retardation, (ii) virtual electron-positron pairs, and (iii) radiative effects (electron self energy, vacuum polarization, and vertex correction). The radiative effects are singular and have to be renormalized.

Two major problems have to be solved in order to combine the QED and MBPT procedures into a unified theory,

- firstly, the QED effects have to be included into the relativistic wave function in a systematic way, and
- secondly, the radiative QED effects have to be renormalized, using the Coulomb gauge.

We shall start with the first problem, which we have for some time demonstrated can be handled by means of the covariant-evolution operator. Then we shall return to the second problem that we have only recently found a solution to.

## 3.1 Covariant Evolution Operator

Mainly three methods for QED calculations on bound states have been developed. The standard procedure is the well-known *S-matrix formulation*, and more recently two other methods have been developed, the *Two-times Green's-function technique*, developed by the St Petersburg group [30], and the *Covariant-Evolution-Operator (CEO) method*, developed by our group [18]. The latter two methods have the advantage over the S-matrix formulation that they can be applied to *quasi-degenerate* systems, such as closely spaced fine-structure levels. The CEO has in addition the important advantage that it has a structure that is akin to that of MBPT and therefore is suitable as a basis for a combined MBPT-QED procedure. In the following we shall first consider the original CEO procedure.

The standard *time-evolution operator*, $U(t, t_0)$, describes in the interaction picture the evolution of the non-relativistic time-dependent state vector

$$|\chi(t)\rangle = NU(t, t_0)\,|\chi(t_0)\rangle \qquad (t > t_0), \tag{8}$$

where $N$ is a normalization constant (the evolution operator generally does not preserve the normalization). In addition, the evolution operator can contain singularities.

The perturbation we shall use here is the interaction between the electrons and the radiation field that can be expressed by means of the energy density

$$\mathscr{H}(x) = -\hat{\psi}^{\dagger}(x)e\alpha^{\mu}A_{\mu}(x)\hat{\psi}(x), \tag{9}$$

where $\hat{\psi}(x)$, $\hat{\psi}^{\dagger}(x)$ represent electron-field operators and $A_{\mu}(x)$ the electro-magnetic field. This corresponds to the time-dependent perturbation

$$v(t) = \int \mathrm{d}^3x\, \mathscr{H}(t, \boldsymbol{x}), \tag{10}$$

which represents the emission/absorption of a *single* photon. This operates in the *extended Fock space* with a variable number of photons. *Two* such interactions are needed to form the exchange of a virtual photon between the electrons (see Fig. 1).

The evolution operator for single-photon exchange between the electrons is illustrated in the first diagram in Fig. 2. Since this operator evolves only in the positive direction, it is *not covariant* and consequently cannot be used in a relativistic treatment. The Green's function, on the other hand, illustrated by the second diagram, has electron propagators on the free ends and is therefore covariant. By attaching free-electron lines to the graphical representation of the Green's function, we will represent an evolution operator that is covariant, the *Covariant Evolution Operator* (CEO) (third diagram). The covariant evolution operator is the evolution operator for the relativistic state vector in analogy with Eq. (8).

In a ladder of single-photon exchange, involving only positive-energy states and interacting to the far right on an unperturbed state of energy $E_0$, the general potential becomes (see Fig. 3)

$$\langle rs|V_{\mathrm{sp}}(E_0)|ab\rangle = \Big\langle rs\Big|\int_0^\infty \mathrm{d}k\, f(k)\Big[\frac{1}{E_0-\varepsilon_r-\varepsilon_u-(k-\mathrm{i}\gamma)_r}+\frac{1}{E_0-\varepsilon_s-\varepsilon_t-(k-\mathrm{i}\gamma)_s}\Big]\Big|ab\Big\rangle. \tag{11}$$

Here, $\varepsilon_x$ represent the orbital energies, $(x)_r$ represents an expression with the sign of the orbital $r$, $f(k)$ is a gauge-dependent function of the photon momentum, and $\gamma$ is an adiabatic-damping factor. Note, that this potential depends on the initial energy, $E_0$.

## 3.2 Green's Operator

The CEO becomes singular (or quasi-singular) when a state degenerate (or quasi-degenerate) with the initial state is involved. In order to eliminate the singularities, we define a *Green's operator* by the relation

$$\boxed{U(t,t_0)P = \mathcal{G}(t,t_0)\cdot PU(0,t_0)P,} \tag{12}$$

where the operator to the left of the heavy dot does not operate beyond the dot. The Green's operator is regular.

It can be shown that the state vector at arbitrary time is given by

$$\big|\chi^\alpha(t)\big\rangle = N^\alpha U(t,-\infty)\big|\Phi^\alpha\big\rangle = N^\alpha \mathcal{G}(t,-\infty)\cdot PU(0,-\infty)\big|\Phi^\alpha\big\rangle = \mathcal{G}(t,-\infty)|\Psi_0^\alpha\rangle, \tag{13}$$

where

$$N^\alpha PU(0,-\infty)\big|\Phi^\alpha\big\rangle = P|\Psi^\alpha\rangle = |\Psi_0^\alpha\rangle \tag{14}$$

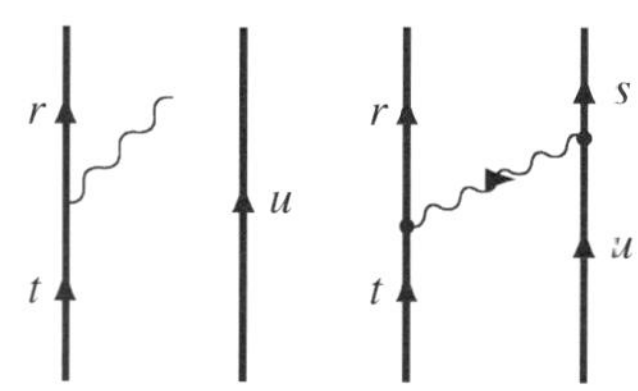


**Fig. 1** The single-photon exchange between the electrons is in the CEO represented by two perturbations (10)

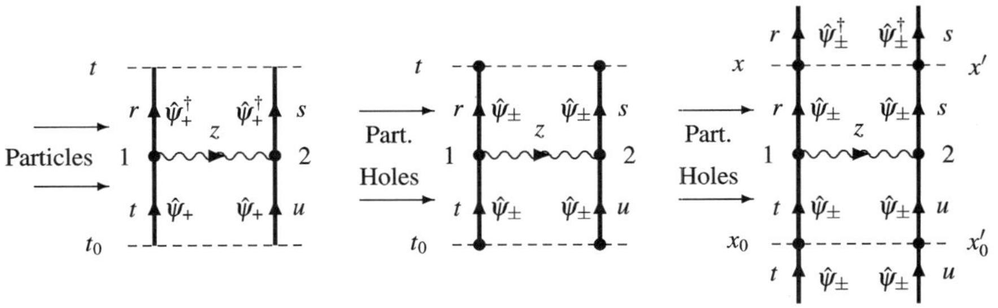


**Fig. 2** Comparison between the standard evolution operator, the Green's function and the covariant evolution operator for single-photon exchange in the equal-time approximation (Fig. 8.1, Ref. [15])

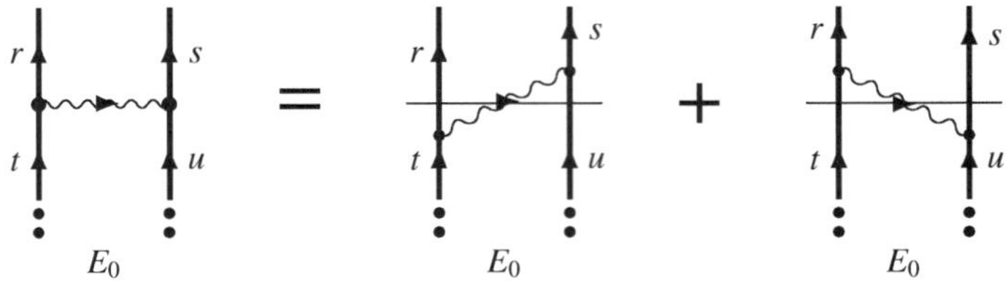


**Fig. 3** The evolution-operator diagram for single-photon exchange

is the model state (Eq. 4). Therefore *the Green's operator acts as a wave operator for the relativistic state vector at all times*. In particular, the covariant analogue of the MBPT wave operator (5) becomes

$$\boxed{\Omega = \mathcal{G}(0, -\infty),} \tag{15}$$

which is also a Fock-space operator. This gives the connection between the CEO formalism and standard MBPT.

## 3.3 The Many-Body Hamiltonian

According to the Gell-Mann–Low theorem [9, 18] the state vector $|\Psi^\alpha\rangle$ satisfies a relativistic "Schrödinger-like" eigenvalue equation in the extended Fock space

$$\boxed{H|\Psi^\alpha\rangle = \big(H_0 + V_{\rm F}\big)|\Psi^\alpha\rangle = E^\alpha|\Psi^\alpha\rangle,} \tag{16}$$

The perturbation, $V_{\rm F}$, is then given by the Coulomb interaction, $V_{\rm C}$, and the transverse part of the perturbation (10). The total many-body Hamiltonian then becomes in second quantization

$$H = \int d^3x \, \hat{\psi}^\dagger(x)\Big(\boldsymbol{\alpha}\cdot\hat{\boldsymbol{p}} + \beta m + v_{\rm ext}(x) - e\alpha^\mu A_\mu(x)\Big)\hat{\psi}(x) + H_{\rm Rad} + \frac{1}{2}\iint d^3x_1\, d^3x_2\, \hat{\psi}^\dagger(x_1)\,\hat{\psi}^\dagger(x_2) V_{\rm C}\, \hat{\psi}(x_2)\,\hat{\psi}(x_1), \tag{17}$$

where $v_{\rm ext}(x)$ is the external (usually nuclear) potential. This is *Fock-space operator*, operating in the extended space with unpaired photons. Since the number of photons is not conserved, also the radiation-field Hamiltonian, $H_{\rm Rad}$, is included.

## 3.4 QED Potential

Of the various QED effects defined above, the retardation effect is taken care of in the photon exchange (Eq. 11). Virtual pairs can be treated together with single-photon exchange by generalizing the potential to [15, Eq. 8.11]

$$\big\langle rs\big|V_{\rm sp}{}^{\rm VP}(\mathcal{E})\big|tu\big\rangle = \Big\langle rs\Big|\int d\kappa\, f(\kappa)\Big[\pm\frac{t_\pm r_\mp}{\varepsilon_t-\varepsilon_r\pm\kappa} \pm\frac{t_\pm s_\pm}{\mathcal{E}-\varepsilon_t-\varepsilon_s\mp\kappa} \pm\frac{u_\pm r_\pm}{\mathcal{E}-\varepsilon_r-\varepsilon_u\mp\kappa} \pm\frac{u_\pm s_\mp}{\varepsilon_u-\varepsilon_s\pm\kappa}\Big]\Big|tu\Big\rangle, \tag{18}$$

where $t_\pm$ etc. represent *projection operators* for particle/hole states, respectively. The upper or lower sign should be used consistently in each term, inclusive the sign in the front. Vacuum-polarization effects on electron propagators can be represented by a potential, as discussed in several publications [28].

The single-photon exchange is in the CEO procedure represented by two single-particle interactions (10). The first-order electron self energy can then be evaluated by closing the second interaction on the same electron (cf. Fig. 1). In both cases—before closing the photon—one or several instantaneous interactions (Coulomb or Breit) could be inserted, leading to the "QED potential", illustrated in Fig. 4. (We return later to the question of renormalization). For the time being only a single retarded photon can be handled numerically. It should be noted that by including one or several instantaneous Breit interactions, most of the higher-order effects are actually included. This approximation is therefore quite accurate.

The QED potential can be iterated, and this leads to a procedure that is equivalent to the famous *Bethe-Salpeter equation*. In fact, the procedure is even more general than the standard Bethe-Salpeter equation, since it is applicable also to a multi-dimensional model space, in analogy with the MBPT procedure (Sect. 2). Therefore, it can also handle the quasi-degenerate case, like fine-structure separations. In addition, the procedure leads automatically to a perturbative expansion.

The procedure discussed above has been implemented numerically and applied to helium-like ions in the doctoral thesis of Daniel Hedendahl [10] (see also [15, Ch. 10]). This is illustrated in Fig. 5. It is found that for light and medium-heavy

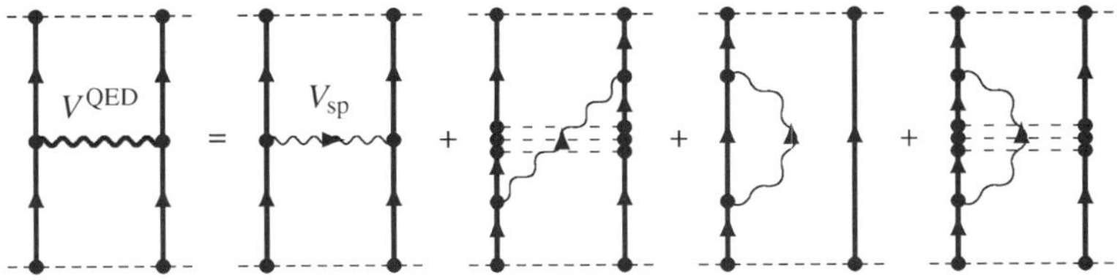


**Fig. 4** Feynman diagram representing the "QED potential", $V^{\rm QED}$. The *dotted lines* represent instantaneous Coulomb and (if needed) Breit interactions. (Fig. 8.10, Ref. [15])

elements first-order QED together with electron correlation beyond second order dominate over the second-order QED. Therefore, it is important for such systems to include the combined QED-MBPT effect. By including also the instantaneous Breit interaction (2), the accuracy can be further improved. These calculations are the first of the new type with QED effects included in the atomic wave function, rather than just as energy correction.

### 3.5 Renormalization

We now turn to the second major problem, mentioned above, namely that of renormalization of the radiative parts in the Coulomb gauge.

Formulas have been derived by Adkins some time ago for the dimensional regularization in Coulomb gauge of the lowest-order free-electron self energy and vertex correction [1, 2]. To or knowledge, however, this procedure has never been tested in numerical applications.

The formulas of Adkins are not apt for numerical work. The self-energy part has been reformulated by Malenfant [21] by performing some of the integrations

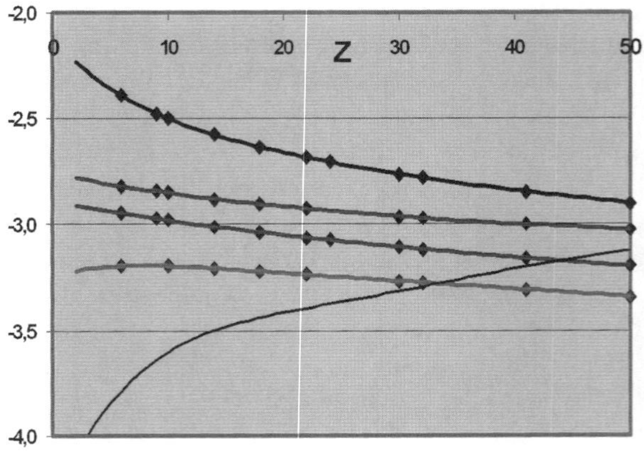


**Fig. 5** The effect of first-order non-radiative QED with electron correlation. The *lines* represent in order from the *top*: singly unretarded Breit, single retarded Breit without and with Coulomb crossings, and single Breit with virtual pairs, all with electron correlation beyond two-photon exchange. The *bottom line* shows as comparison second-order retarded Breit

**Table 5** Comparison between Coulomb-gauge and Feynman-gauge results for the self energy of hydrogen like ions (in eV, from Ref. [11])

| Z | Coulomb gauge | Feynman gauge |
|---|---|---|
| 18 | 1.216901(3) | 1.21690(1) |
| 54 | 50.99727(2) | 50.99731(8) |
| 66 | 102.47119(3) | 102.4713(1) |
| 92 | 355.0430(1) | 355.0432(2) |

analytically. In an analogous way one of us (JH) has recently modified the more complicated vertex-correction part [12]. The new renormalization procedure has been tested on the self-energy of hydrogen like ions, where very accurate calculations have been performed also in the Feynman gauge [11] . The results of the calculations in the two gauges are compared in Table 5.

The agreement between the calculations is excellent, and it is seen that the accuracy is actually higher in the Coulomb gauge for light and medium-heavy elements. This is due to the fact that the main uncertainty originates from many-potential term, which is considerably smaller in that case.

The successful renormalization of the self-energy in Coulomb gauge represents a major breakthrough in this work and opens up the way for complete MBPT-QED calculations.

## 4 Summary and Outlook

We have for the first time developed a unified covariant theory of MBPT and QED, which represents a new generation of atomic calculation that goes beyond presently available many-body and QED procedures. This has been applied to highly charged helium-like ions, and so far only non-radiative effects have been included. It has been demonstrated that these combined effects are particularly important for light and medium-heavy ions. We have also very recently developed a procedure for dimensional regularization of the self energy and vertex correction in the Coulomb gauge, and successfully demonstrated this on hydrogen-like ions. This will make it possible to include also the radiative QED parts into the many-body wave function, which are generally even more important than the non-radiative ones. It is expected that the combined QED-correlation effects, which have so far never been detected, can in the future be found in various kinds of atomic data.

## References

1. G. Adkins, One-loop renormalization of Coulomb-gauge QED. Phys. Rev. D **27**, 1814–20 (1983)
2. G. Adkins, One-loop vertex function in Coulomb-gauge QED. Phys. Rev. D **34**, 2489–92 (1986)

3. A.N. Artemyev, V.M. Shabaev, V.A. Yerokhin, G. Plunien, G. Soff, QED calculations of the n = 1 and n = 2 energy levels in He-like ions. Phys. Rev. A **71**, 062104 (2005)
4. S. Blundell, Calculations of the screened self-energy and vacuum polarization in Li-like, Na-like, and Cu-like ions. Phys. Rev. A **47**, 1790–1803 (1993)
5. G.E. Brown, D.G. Ravenhall, On the interaction of two electrons. Proc. R. Soc. Lond. Ser. A **208**, 552–59 (1951)
6. C.T. Chantler, J.A. Lowe, I.P. Grant, Multiconfiguration Dirac-Fock calculations in open-shell atoms: convergence methods and satellite spectrs of the copper K $\alpha$ photoemission spectra). Phys. Rev. A **82**, 052505 (2010)
7. R.D. Deslattes, E.G. Kessler, P. Indelicato, L. deBilly, E. Lindroth, J. Aron, X-ray transition energies: new approach to a comprehensive evaluation. Rev. Mod. Phys. **75**, 15–19 (2003)
8. T.R. DeVore, D.N. Crosby, E.G. Myers, Improved measurement of the $1s2s\,^1S_0 - 1s2p\,^3P_1$ interval in heliumlike silicon. Phys. Rev. Lett. **100**, 243001 (2008)
9. M. Gell-Mann, F. Low, Bound states in quantum field theory. Phys. Rev. **84**, 350–354 (1951)
10. D. Hedendahl, Towards a Relativistic Covariant Many-Body Perturbation Theory, Ph.D. thesis, University of Gothenburg, Gothenburg, Sweden, 2010
11. D. Hedendahl, J. Holmberg, Coulomb-gauge self-energy calculation for high-Z hydrogenic ions. Phys. Rev. A **85**, 012514 (2012)
12. J. Holmberg, *Master's thesis*, University of Gothenburg, Gothenburg, Sweden, 2011
13. J. Holmberg, Scalar vertex operator for bound-state QED in Coulomb gauge. Phys. Rev. A **84**, 062504 (2012)
14. I. Lindgren, The Rayleigh-Schrödinger perturbation and the linked-diagram theorem for a multi-configurational model space. J. Phys. B **7**, 2441–70 (1974)
15. I. Lindgren, *Relativistic many-body theory: a new field-theoretical approach* (Springer, New York, 2011)
16. I. Lindgren, B. Åsén, S. Salomonson, A.M. Mårtensson-Pendrill, QED procedure applied to the quasidegenerate fine-structure levels of He-like ions. Phys. Rev. A **64**, 062505 (2001)
17. I. Lindgren, J. Morrison, *Atomic Many-Body Theory*, 2nd edn. (Springer, Berlin, 1986) (Reprinted 2009)
18. I. Lindgren, S. Salomonson, B. Åsén, The covariant-evolution-operator method in bound-state QED. Phys. Rep. **389**, 161–261 (2004)
19. I. Lindgren, S. Salomonson, D. Hedendahl, Many-body-QED perturbation theory: connection to the two-electron Bethe-Salpeter equation. Einstein centennial review paper. Can. J. Phys. **83**, 183–218 (2005)
20. I. Lindgren, S. Salomonson, D. Hedendahl, Many-body procedure for energy-dependent perturbation: merging many-body perturbation theory with QED. Phys. Rev. A **73**, 062502 (2006)
21. J. Malenfant, Renormalized Coulomb-gauge self-energy function. Phys. Rev. D **35**, 1525–1527 (1987)
22. P. Mohr, Self-energy correction to one-electron energy levels in a strong Coulomb field. Phys. Rev. A **46**, 4421–24 (1992)
23. P.J. Mohr, G. Plunien, G. Soff, QED corrections in heavy atoms. Phys. Rep. **293**, 227–372 (1998)
24. P.J. Mohr, G. Soff, Nuclear size correction to the electron self energy. Phys. Rev. Lett. **70**, 158–161 (1993)
25. E.G. Myers, H.S. Margolis, J.K. Thompson, M.A. Farmer, J.D. Silver, M.R. Tarbutt, Precision measurement of the $1s2p\,^3P_2 -^3 P_1$ fine structure interval in heliumlike fluorine. Phys. Rev. Lett. **82**, 4200–4203 (1999)
26. E.G. Myers, M.R. Tarbutt, Measurement of the $1s2p^3P_0 -^3 P_1$ fine structure interval in heliumlike magnesium. Phys. Rev. A **61**, 010501R (1999)
27. H. Persson, I. Lindgren, L.N. Labzowsky, G. Plunien, T. Beier, G. Soff, Second-order self-energy-vacuum-polarization contributions to the Lamb shift in highly charged few-electron ions. Phys. Rev. A **54**, 2805–13 (1996)
28. H. Persson, I. Lindgren, S. Salomonson, P. Sunnergren, Accurate vacuum-polarization calculations. Phys. Rev. A **48**, 2772–78 (1993)

29. D.R. Plante, W.R. Johnson, J. Sapirstein, Relativistic all-order many-body calculations of the n = 1 and n = 2 states of heliumlike ions. Phys. Rev. A **49**, 3519–3530 (1994)
30. V.M. Shabaev, Two-times Green's function method in quantum electrodynamics of high-Z few-electron atoms. Phys. Rep. **356**, 119–228 (2002)
31. J. Sucher, Foundations of the relativistic theory of many electron atoms. Phys. Rev. A **22**, 348–362 (1980)
32. V.A. Yerokhin, A.N. Artemyev, V.M. Shabaev, M.M. Sysak, O.M. Zherebtsov, G. Soff, Evaluation of the two-photon exchange graphs for the $2p_{1/2} - 2s$ transition in Li-like Ions. Phys. Rev. A **64**, 032109 (2001)

# Supercritical QED and Time-Delayed Heavy Ion Collisions

**Joachim Reinhardt and Walter Greiner**

**Abstract** The theory of Quantum Electrodynamics predicts the "spontaneous" production of electron-positron pairs in the presence of strong electric fields. Collisions of heavy ions with a combined nuclear charge exceeding the value of 172 are expected to provide the required supercritical field at least transiently. Extensive experimental searches performed about two decades ago, mainly at GSI, have confirmed the expected strong enhancement of pair production in high-Z collisions. The short time scales involved, however, have prevented an unequivocal confirmation of the mechanism of supercritical pair production. We revisit this problem in the view of recent results from nuclear reaction theory. If reactions with a prolonged lifetime approaching $10^{-20}$ s can be selected using suitable coincidence conditions, it should be possible to experimentally verify the vacuum decay of QED.

## 1 Supercritical Fields: The charged Vacuum in QED

The study of Quantum Electrodynamics of strong fields dates back to the early days of quantum mechanics. (For extensive references we refer the reader to the book [1]). Soon after Dirac's formulation of the relativistic quantum theory of electrons, O. Klein discovered that scattering off an electrostatic potential barrier $V_0$ which exceeds the height of $2mc^2$ leads to anomalous behaviour of the transmission and reflection coefficients. In 1931 F. Sauter derived the transmission coefficient for

J. Reinhardt (✉)
Institute for Theoretical Physics, Johann Wolfgang Goethe-University, Frankfurt am Main, Germany
e-mail: jr@th.physik.uni-frankfurt.de

W. Greiner
Frankfurt Institute for Advanced Studies, Johann Wolfgang Goethe-University, Frankfurt am Main, Germany
e-mail: greiner@fias.uni-frankfurt.de

W. Greiner (ed.), *Exciting Interdisciplinary Physics*,
FIAS Interdisciplinary Science Series, DOI: 10.1007/978-3-319-00047-3_35,

"tunneling through the gap of the Dirac equation" for a potential barrier of finite width $a$ as $T \simeq \exp\left[-\pi mc^2/(V_0/a)(\hbar/mc)\right] = \exp(-E_{\rm cr}/E)$. This expression displays a nonanalytic dependence on the electrical field strength $E$ and is exponentially suppressed unless $E$ reaches the value of the *critical field strength*

$$E_{\rm cr} = \frac{\pi m^2 c^3}{e\hbar} \simeq 1.3\ 10^{18}\,\frac{\rm V}{\rm m}\ . \tag{1}$$

It soon became clear, in particular through the work Heisenberg and his student Euler, that the vacuum of QED is a dynamical polarizable medium in which an external field can induce the production of virtual and (provided the field is strong enough) also real electron-positron pairs. An elegant formulation of this vacuum instability was given in 1951 by Schwinger using his eigentime formalism.

Although the existence of the "Schwinger" pair production mechanism is generally accepted, it has eluded experimental verification. Extended ultrastrong electrostatic fields $E \simeq E_{\rm cr}$ are unattainable in the laboratory. On an atomic scale, however, this is not true: The electrical field strength at the surface of a nucleus exceeds $E_{\rm cr}$ by about three orders of magnitude. Nevertheless in ordinary atoms pair creation does not occur because the created electron would not fit into the narrow well of the Coulomb potential. This changes when atomic structure is extrapolated from the known region of chemical elements by about a factor of two. When $Z$ approaches the value of the inverse fine structure constant $1/\alpha \simeq 137$ the inner-shell electrons gain tremendously in binding energy. The lowest state $1s_{1/2}$—and also the next higher $2p_{1/2}$ state—traverses the gap between the positive and negative energy continuum solutions of the Dirac equation. The total energy $E_{1s}$ becomes negative at $Z = 150$ and is predicted to reach the value $-m$ (i.e., a binding energy of $2m$=1.022 MeV) at the *critical nuclear charge* $Z_{\rm cr} \simeq 172$.

What happens at and beyond this critical charge was clarified in the early 1970s by our group at Frankfurt [2] and by another group in Moscow [3]. For a detailed overview of vacuum properties in the presence of supercritical fields see [1]. If the strength of the Coulomb potential exceeds the critical value, i.e., $Z > Z_{\rm cr}$, the spectrum of the stationary Dirac equation undergoes a qualitative change. The $1s$ state leaves the discrete spectrum and merges with the lower continuum of the Dirac equation which it enters as a narrow *resonance*. In Dirac's hole picture (which can be corroborated by arguments based on second-quantized field theory) the lower continuum (the Dirac sea) is occupied with electrons. If an empty bound state enters the continuum it will get filled by a sea electron which can *tunnel* through the classically forbidden gap of the Dirac equation, leaving behind a hole, i.e., a positively charged positron, which escapes to infinity. The process has been termed "spontaneous pair creation" of "decay of the vacuum" of QED. Obviously this process is closely related to the well-known Klein paradox and to the "Schwinger formula" for pair creation in a constant electric field. The difference is that in supercritical atoms the strong field is confined to a small region in space which can harbour only a small number of created electrons.

Spontaneous pair creation occurs already at the tree level of QED and it survives when higher-order processes of quantum field theory are taken into account. It was shown that the level shifts caused by vacuum polarization and electron self energy amount to less than $10^{-3}$ of the total K-shell binding energy (see [1]).

## 2 Dynamics of the Electron-Positron Field

In close collisions of two very heavy nuclei, the supercritical electric field of a combined nucleus with charge $Z = Z_1 + Z_2$ is generated transiently. To study this problem, the Dirac equation with two Coulomb centers was solved ([4, 5], see also [6] and references therein). It was found that the lowest molecular electron level ($1s\sigma$) can be traced down to the lower continuum of the Dirac equation and reaches a binding energy of $2m$ at a critical two-center distance $R_{\rm cr}$. Example in U+U collisions the critical distance is approximately $R_{\rm cr} \approx 30$ fm. However, in a heavy-ion collision the nuclei move on their Rutherford trajectories $\mathbf{R}(t)$ which causes the wave functions and binding energies to vary rapidly with time and also leads to strong dynamically induced transitions. This makes it necessary to solve the *time dependent two-center Dirac equation*.

A considerable number of approaches has been developed to attack this problem. The brute-force way is to solve the time-dependent two-center Dirac equation numerically as a system of coupled partial differential equations, either in coordinate space or in momentum space. Calculations of this kind are very demanding, in particular in view of the long range of the Coulomb potential and of the small size of pair production amplitudes. A physically inspired way to proceed is by expanding the time-dependent wave function into a complete set of basis states. The time dependent Dirac equation is thus converted to a set (in principle infinite, but truncated in practical calculations) of coupled ordinary differential equations.

At high energies (see, e.g., [7]) an expansion in terms of atomic basis states is adequate. At bombarding energies not much above the nuclear Coulomb barrier (i.e. $E/A \simeq 6$ MeV) the ion velocity is comparatively slow, $v \simeq 0.1c$, so that the relativistic inner-shell electrons have time to adjust to the nuclear Coulomb field. This gives justification for an adiabatic description of the collision in terms of *superheavy quasimolecules*. Then it is useful to expand the time-dependent wave function

$$\Psi_i(\mathbf{r},t) = \sum_j a_{ij}(t)\varphi_j(\mathbf{R}(t),\mathbf{r})\,\mathrm{e}^{-\mathrm{i}\chi_j(t)} \quad \text{where} \quad \chi_j(t) = \int_{t_0}^{t} dt \langle\varphi_j|H|\varphi_j\rangle\,. \tag{2}$$

in the *adiabatic basis* of molecular states $\varphi_j$, i.e., eigenstates of the stationary two-center Dirac Hamiltonian $H$ at a given internuclear distance $\mathbf{R}$

$$\Big(H(\mathbf{R},\mathbf{r}) - E_j(\mathbf{R})\Big)\varphi_j(\mathbf{R},\mathbf{r}) = 0\,, \tag{3}$$

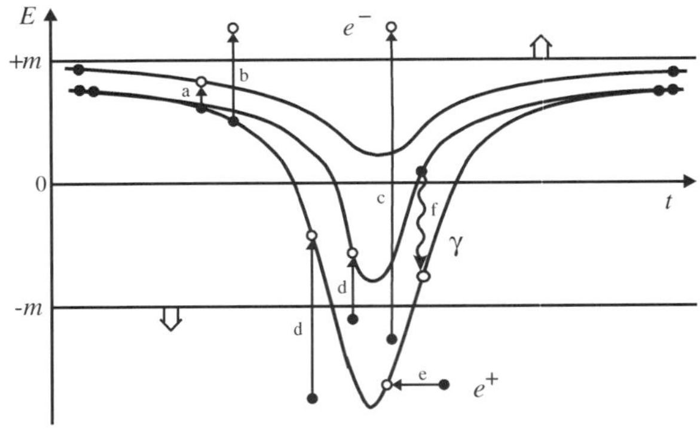


**Fig. 1** The time-dependent energy levels in a supercritical heavy ion collision. The *arrows* symbolize various electron excitation mechanisms. **a** bound-bound electron excitation; **b** ionization; **c** direct (free-free) electron-positron pair production; **d** induced (bound-free) pair production; **e** spontaneous pair production; **f** quasimolecular X-ray emission

The summation extends over bound states and the two sets of continuum states. The time-dependent expansion coefficients $a_{ij}(t)$, which satisfy the boundary condition $a_{ij}(-\infty) = \delta_{ij}$, are determined by solving a truncated set of ordinary differential equations, the coupled channel equations

$$\dot{a}_{ik} = -\sum_{j\neq k} a_{ij}\,\langle\varphi_k|\frac{\partial}{\partial t}|\varphi_j\rangle\; e^{i(\chi_k-\chi_j)} \tag{4}$$

where $\partial/\partial t$ acts on the parametric time dependence of the basis wave functions.

To account for the nature of the Dirac vacuum, a Fermi level $F$ has to be specified up to which all states are occupied initially. Neglecting electron correlations all the necessary information to describe the physical observables is contained in the set of single-particle transition amplitudes $a_{ij}(t \to \infty)$. The number of produced electrons $N_i$ or holes (positrons) $\overline{N}_j$, resp., is given by the summation

$$N_i = \sum_{k<F} |a_{ki}(+\infty)|^2 \quad (i > F) \qquad \text{and} \qquad \overline{N}_j = \sum_{k>F} |a_{kj}(+\infty)|^2 \quad (j < F) \tag{5}$$

while correlated electron-hole pairs are described by [8]

$$N_{ij} = N_i\,\overline{N}_j + \left|\sum_{k>F} a^*_{ki}(+\infty)a_{kj}(+\infty)\right|^2 \quad (i > F\;,\; j < F) \tag{6}$$

For the description of inner-shell excitation and pair production in very heavy systems it was found sufficient to include only states with angular momentum $j = 1/2$ ($s_{1/2}$ and $p_{1/2}$) which are most strongly affected by the relativistic "collapse of the wavefunction". An inspection of the exact solutions of the two-center Dirac equation

[4, 5] has shown that the problem can be greatly simplified if the full two-center potential is approximated its lowest-order term in the multipole expansion (the monopole approximation). This framework has been employed with considerable success to calculate *K-hole production*, *δ-electron* and *positron emission*, and *quasimolecular X-ray radiation*, for a review see [9].

## 3 Supercritical Heavy-Ion Collisions

In supercritical collisions, the combined nuclear charge is sufficiently large to let the quasimolecular 1s-state enter the Dirac sea at a critical distance $R_{\rm cr}$. Then in the adiabatic picture the $1s$ state vanishes from the bound spectrum, becoming admixed to the lower continuum. To treat the dynamics in this case, method was developed [10] which introduces a normalizable wavefunction $\tilde{\varphi}_r$ for the supercritical 1s-state by artificially cutting off the "oscillating tail" of the resonance wave function.[1] With the help of a projection technique then a matching set of modified continuum states $\tilde{\varphi}_E$ can be constructed by solving the equation

$$(H - E)\tilde{\varphi}_E = \langle\tilde{\varphi}_r|H|\tilde{\varphi}_E\rangle\,\tilde{\varphi}_r \qquad \text{for} \quad (E < -m)\ . \tag{7}$$

Since $\tilde{\varphi}_r$ and $\tilde{\varphi}_E$ do not diagonalize the two-center Hamiltonian, there exists a non-vanishing static coupling between the truncated 1s-resonance state and the modified negative energy continuum which describes the spontaneous decay of a vacancy in the supercritical 1s-state. The decay width is $\Gamma = 2\pi|\langle\tilde{\varphi}_r|H|\tilde{\varphi}_E\rangle|^2$. The coupled channel Eq. (4) then contain the coherent superposition of an "induced" and a "spontaneous" coupling matrix element

$$\langle\varphi_{1s}|\partial/\partial t|\varphi_E\rangle \longrightarrow \langle\tilde{\varphi}_r|\partial/\partial t|\tilde{\varphi}_E\rangle + i\langle\tilde{\varphi}_r|H|\tilde{\varphi}_E\rangle\ . \tag{8}$$

For elastic collisions without nuclear contact the contribution from the "spontaneous" coupling constitutes only a small fraction of the produced positrons. Furthermore, the time-energy uncertainty relation leads to a large collisional broadening. Accordingly, the calculations do not yield any perceptible change in the shape of the predicted positron spectra for such collisions when going from subcritical to supercritical systems. However, the positron cross section for $Z > 137$ will to grow at a very rapid rate that can be roughly parametrized by an effective power law

$$\sigma_{e^+}(Z) \propto Z^n, \qquad n \approx 20\ . \tag{9}$$

---

[1] Other definitions of the resonance wave function are possible. Example it can be defined as a state with complex energy $E_{\rm res} = E_r + i\Gamma/2$ using a complex rescaling of the coordinate $\mathbf{r} \to \mathbf{r}\,e^{i\theta}$ [11, 12].

The large value of the exponent demonstrates the nonperturbative nature of pair production in collisions of high-$Z$ ions. The fact that this prediction has been verified in the experiments at GSI (see below) is a major confirmation of our understanding of quantum electrodynamics in strong Coulomb fields.

## 4 Positron Production Experiments

For nearly two decades the study of atomic excitation processes and in particular of positron creation in heavy-ion collisions had been a major research topic at GSI (Darmstadt) [13–18]. To summarize, the measured positron production rates and their dependence on nuclear charge, collision energy, and impact parameter are in quantitative agreement with parameter-free theoretical predictions based on the formalism sketched above. Figure 2 showing measurements of the EPOS group demonstrates that excellent agreement was achieved when the calculated spectra of QED positrons and the background positrons from the pair conversion caused by nuclear excitation were added up. In particular the strong rise of the QED positron yield (dotted lines in Fig. 2) as a function of the combined nuclear charge $Z = Z_1 + Z_2$ is in accordance with the expected behaviour (9) and clearly demonstrates the nonperturbative action of the strong Coulomb field.

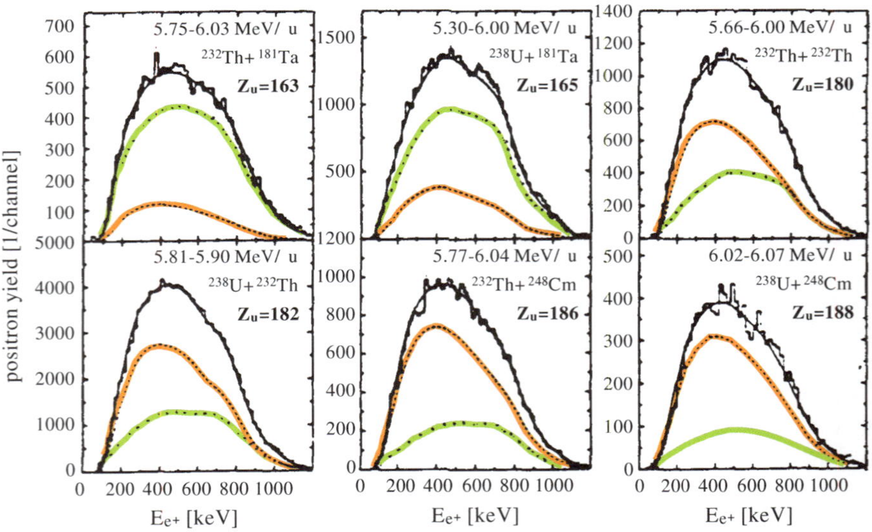


**Fig. 2** Total positron spectra for various collision systems measured by the EPOS collaboration; *dotted lines*: predicted QED pair production; *dashed lines*: nuclear pair conversion; *solid lines*: sum of both. With increasing nuclear charge $Z$ a strong rise of the QED process (dynamical plus induced positron production) is observed

An unexpected feature, however, had perplexed experimentalists and theorists alike for more than a decade: Narrow line structures of unexplained origin were discovered on top of the well understood continuous spectra by the ORANGE and EPOS groups at GSI. These lines first were associated with the spontaneous positron emission line that had been predicted by theory for supercritical collision systems with long nuclear time delay [19, 20]. However, this explanation had to be ruled out because the structures did not show the expected strong $Z$-dependence. More exotic speculations were put forward, e.g., the creation and subsequent two-body decay of some unidentified neutral object $X^0 \to e^+ + e^-$, supported by the reported observation of line structures in the coincident electron-positron pair spectra, but no convincing model did emerge. Interest in the "positron puzzle" largely evaporated when attempts at an independent verification by the APEX collaboration at Argonne National Laboratory [21–23] as well as further experiments at GSI with improved detectors [24, 25] failed to reproduce the former results. The prevalent opinion now is that the earlier observations were caused in part by statistical fluctuations and in part also by lines from nuclear pair conversion.

It should be added that all positron experiments so far were performed by colliding partly stripped projectile ions with solid state targets. Under these conditions bound-free pair productions ("electron capture from the vacuum") is suppressed by *Pauli blocking*: The K shell initially is occupied and can contribute to pair creation only after being emptied through ionization, which at best takes place in a few percent of the collisions. Bound-free pair creation thus could be studied much more cleanly if the required holes were brought in right from the start, i.e., by using *fully stripped ions*. Heavy ions can be stripped bare quite easily at high beam energies. In fact, bound-free pair creation has been first observed at the Bevalac using 1 GeV/nucleon $U^{92+}$ beams [26].

The consequences of the transition to collisions of fully stripped ions have been studied in [27]. The positron production cross section is predicted to grow by up to two orders of magnitude. Nearly all created electrons end up in bound states (mostly $1s$ and $2p_{1/2}$). Much of the advantage can be achieved already with fully ionized projectiles impinging on stationary neutral target atoms since here half of the projectile K-holes are transferred to the $1s\sigma$ state. It is expected that collisions of fully stripped ions at energies in the region of the Coulomb barrier will be possible in the future using new experimental possibilities at the GSI-FAIR facility.

## 5 Collisions with Nuclear Time Delay

When in the course of a heavy-ion collision the two nuclei come into contact, a nuclear reaction can occur that entails a certain *delay time*. For light and medium-heavy nuclei the nuclear attraction can overcome the repulsive Coulomb force, thus allowing for rather long reaction times (up to infinity, if fusion occurs). For very heavy nuclei, however, the Coulomb interaction is the dominant force between the nuclei, so that delay times are typically much shorter, of the order $1-2$ zs. (The zeptosecond, 1 zs $= 10^{-21}$ s, is the appropriate unit for the following discussion.)

A delay in the collision due to a nuclear reaction can lead to interesting modifications in atomic excitation processes. Two main observable effects in such collisions are: (a) interference patterns in the spectrum of $\delta$-electrons, and (b) a change in the number of K-vacancies formed. These phenomena have become known as the *atomic clock effect*, for an overview see [9]

In a simple schematic model for the atomic clock effect, a classical trajectory $R(t)$ for the relative motion of two nuclei is used and the only effect of the nuclear reaction is to introduce a time delay $T$ between the incoming and outgoing branches of the trajectory, most simply described by setting $\dot{R}(t) = 0$ for $0 \le t \le T$. It is easy to show that in first-order perturbation theory the resulting excitation amplitude with time delay then reads (assuming elastic scattering)

$$a_{ik}^T(\infty) = a_{ik}(0) - a_{ik}^*(0)\, \mathrm{e}^{\mathrm{i}(E_k - E_i)T} \,, \tag{10}$$

where the energies have to be taken at the distance of nuclear contact. The excitation probability, obtained by squaring this amplitude, is obviously an oscillating function. There are effects smearing out these oscillations and what remains is a partially destructive interference between the incoming and outgoing branches of the trajectory, observable as a decrease in the K-vacancy yield or a steepening of the slope of the low-energy part of the $\delta$-electron spectrum. There have been several experimental studies [28–33] of the effect of nuclear reactions on delta electron emission; evidence for rather short nuclear delay times of the order 1 zs was found.

Nuclear time-delays are of particular interest for supercritical collisions since they have the potential to shift the balance between induced and spontaneous positron production. In a supercritical system an additional coupling between the resonant bound state and the positron continuum states arises, cf. Eq. (8). In the schematic model this leads to (again invoking perturbation theory for simplicity)

$$a_E^T(\infty) = a_E(0) - a_E^*(0)\, \mathrm{e}^{\mathrm{i}(E - E_r)T} - \langle \tilde{\varphi}_r | H | \tilde{\varphi}_E \rangle \, \frac{\mathrm{e}^{\mathrm{i}(E - E_r)T} - 1}{E - E_r} \,, \tag{11}$$

where $E_r$ is the energy of the supercritical state when the nuclei are in contact. The extra term vanishes for $T = 0$, but grows rapidly with increasing $T$. For delay times considerably greater than $10^{-21}$ s the additional term in Eq. (11) begins to dominate over the first two terms, causing the emergence of a *peak in the positron spectrum* at the energy of the supercritical bound state. This is illustrated in Fig. 3 which compares the spectra of positrons produced in subcritical ($Z = 164$) and supercritical ($Z = 184$) collisions assuming delay times from 0 to 10 zs. Our coupled channel results [19, 20] recently were confirmed by an independent calculation solving the time dependent Dirac equation using a mapped Fourier grid matrix method [34].

In supercritical systems the positron yield is predicted to increase strongly when the delay time exceeds about 3 zs. Even a small admixture of collisions with very long reaction times might become visible in the positron spectrum and serve as a

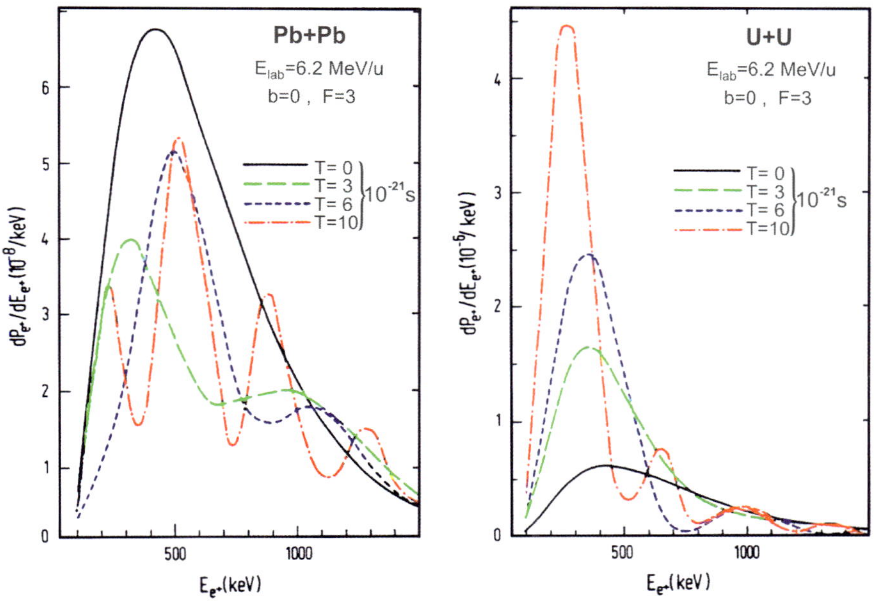


**Fig. 3** Positron spectra in central Pb+Pb and U+U collisions at $E_{\mathrm{lab}}/A = 6.2$ MeV assuming various nuclear delay times. The subcritical system displays destructive interference while in the supercritical system spontaneous positron production leads to the build-up of a *peak* in the spectrum [19, 20]

proof for spontaneous positron creation, since this mechanism acts as a "magnifying glass" for long delay times [19, 20, 35].

The most obvious mechanism for obtaining long reaction times would be the presence of an attractive pocket in the internuclear potential in which the dinuclear system could be trapped. In lighter nuclear systems this phenomenon is known to lead to the formation of nuclear molecules which can undergo several revolutions. For heavy systems like U+U it appears that the conditions for a potential pocket are not met because of the strong Coulomb repulsion.

Up to now no conclusive theoretical or experimental evidence exists for long nuclear delay times in very heavy collision systems. However, based on widely different methods, several recent works from nuclear reaction theory, which we will briefly review in the following, hint at the possibility of prolonged reaction times.

Maruyama et al. [36] have performed dynamical microscopic simulations for Au + Au collisions using the method of constrained molecular dynamics (CoMD). This model is based on solving the classical equations of motions for the nucleons, described by Gaussian wave packets, complemented with a constraint which to some extent accounts for quantum effects, i.e., Pauli blocking and Fermi motion. In their CoMD simulations at low energies the authors observe a reaction type somewhat between deep-inelastic and molecular resonance scattering, where an elongated di-nuclear system persists for a considerable life-time. The available experimental information does not constrain the parameters of the model sufficiently so that no unique predictions can be made. For certain parameter sets the CoMD model predicts reaction times of more than 10 zs, being largest at $E_{\mathrm{lab}} = 10$ MeV/u.

Tian et al. [37] have performed another microscopic study based on quantum molecular dynamics. Their ImQMD model treats the mean field and collision term properly and approximates antisymmetrization using a phase space occupation constraint. Shell effects, the spin-orbit force and groundstate deformation are not included. Caused by strong dissipation and the shape of the single-particle potential, the ImQMD model predicts a prolonged lifetime of the giant dinuclear system. Example for $^{238}$U+$^{238}$U collisions at $E_{\mathrm{lab}}/A = 9$ MeV the average lifetime of the composite nuclear system is predicted to be 4 zs.

A different kind of microscopic simulation of $^{238}$U+$^{238}$U collisions was presented by Golabek and Simenel [38, 39]. These authors performed a fully 3D Time Dependent Hartee-Fock (TDHF) calculation using a Skyrme energy density functional. On this basis, collision times of up to 4 zs were predicted, with a maximum at $E_{\mathrm{lab}}/A = 10$ MeV. The temporal evolution was found to be quite sensitive to the relative orientation of the strongly deformed uranium nuclei, with belly-to-belly configurations showing the longest reaction time.

Zagrebaev et al. [40–42] studied the problem using a macroscopic dynamical model based on the solution of Langevin-type equations for the motion of the two interacting overlapping nuclei, including dynamical deformation and orientation. To describe the production of superheavy elements, a description of the de-excitation of the two excited primary fragments via fission or light-particle emissions was added. The method has been applied to collisions of $^{238}$U+$^{238}$U, $^{238}$U+$^{248}$Cm, and $^{232}$Th+$^{250}$Cf, leading to a good description of available experimental data on the production of superheavy elements.

Analyzing the time-dependence of the reaction $^{238}$U+$^{248}$Cm at 800 MeV c.m. energy ($E_{\mathrm{lab}} = 6.6$ MeV/u) Zagrebaev et al. [40–42] found a sizable fraction of events with nuclear contact lasting for 10 zs or longer, see Fig. 4. Although the employed internuclear potential does not exhibit a pocket, it is found that owing to nuclear viscosity the system moves through the multidimensional potential surface with almost zero kinetic energy. Bombarding energies directly at the Coulomb barrier are found to be best suited to achieve long reaction times. The predicted absolute cross section for long-lasting reactions ($>$ 10 zs) is be about 0.5 millibarn.

Since the majority of nuclear reactions proceed on a short time-scale, a trigger for delayed collisions will be needed when searching for signs of spontaneous positron production. Zagebaev et al. [40–42] suggest looking for the most strongly damped collisions (highest loss of total kinetic energy), (Fig. 4c) in conjunction with small deflection angles (Fig. 4d). A further promising trigger is the selection of large mass transfer (Fig. 4b). The emerging fragment will be highly excited and in most cases will undergo fission. However, there is a chance that fragments in the region of the doubly magic lead nucleus $^{208}$Pb (30 nucleons transferred from uranium) will cool down by nucleon evaporation and survive. Zagrebaev et al. suggest searching for outcoming Pb-like nuclei at c.m. angles less than 60° as the most promising trigger to select for reactions with long delay times.

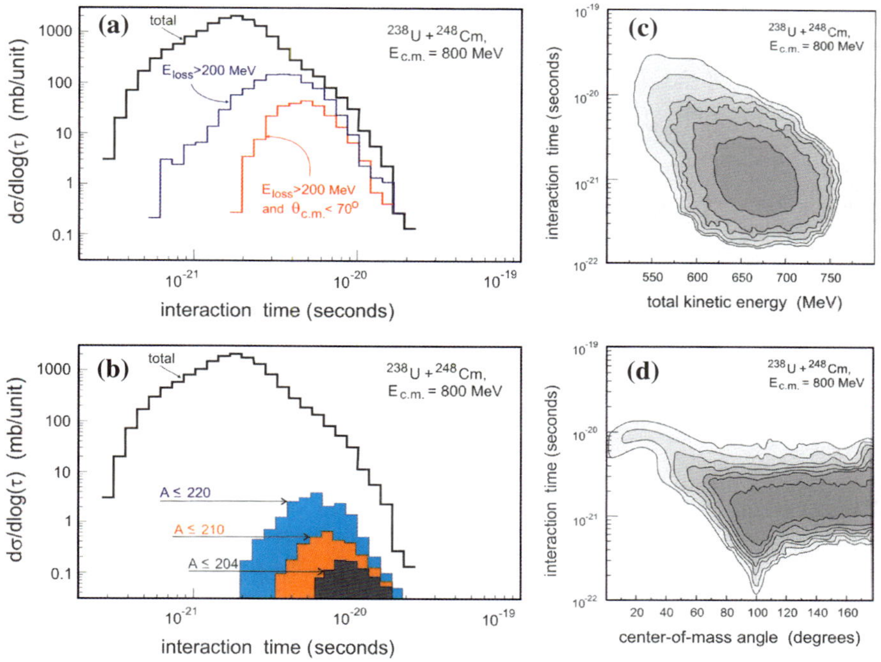


**Fig. 4** *Left:* **a** Calculated distribution of reaction times in $^{238}$U+$^{248}$Cm collisions at 800 MeV c.m. energy. The two *lower curves* refer to a selection collisions with large total kinetic energy loss $E_{\text{loss}} > 200$ MeV and in addition a selection in *scattering angle* $\theta_{\text{cm}} < 70^0$. **b** As *above*, but with cuts in the mass number $A$ of the emitted fragment. *Right*: Contour plots of reaction times versus **c** total kinetic energy or **d** c.m. angle. The lines are drawn on a logarithmic scale over one order of magnitude, the quasi-elastic peak is removed. Results taken from Zagrebaev et al. [40–42]

Because of the small cross sections and large backgrounds involved, performing positron spectroscopy under these conditions requires a dedicated experimental effort. From the theoretical side, one should perform calculations of positron production using realistic time-dependent nuclear charge distributions predicted by nuclear reaction models, including trigger conditions for long reaction times and averaging over many collisions. Whether this will produce a robust signal which stands a chance to overcome the inevitable experimental backgrounds is an open question. However, such studies may hold the key for finally identifying the decay of the vacuum of QED.

## 6 Summary

Quantum Electrodynamics of strong fields offers a "clean" laboratory where a fundamental quantum field theory can be studied theoretically and tested through experiment. A particularly interesting prediction is the decay of the neutral vacuum in the presence of strong external Coulomb fields. Collisions of very heavy ions provide an opportunity to realize this situation experimentally, although impeded by the short duration of these collisions. Inner-shell hole production, $\delta$-electron emission, and positron creation all are sensitive to the strong electric fields. These processes have been studied experimentally in great detail and are well described by theory. There is clear evidence for a rapid growth in binding energy and strong localization of inner shell orbitals in high-$Z$ systems.

The goal to detect the process of spontaneous positron creation and thus the instability of the QED vacuum in the presence of a supercritial field, however, remains elusive. To overcome the problem posed by the short time scale of supercriticality ($\tau \simeq 10^{-21}$ s) one would need to select collisions in which a nuclear reaction with sufficient time delay occurs. Recent results from nuclear reaction theory give hope that such reactions indeed do occur.

## References

1. W. Greiner, B. Müller, J. Rafelski, *Quantum Electrodynamics of Strong Fields* (Springer, Berlin, 1985)
2. B. Müller, J. Rafelski, W. Greiner, Z. Phys. **257**(62), 183 (1972)
3. Ya. Zel'dovich, V.S. Popov, Sov. Phys. Usp. **14**, 673 (1972)
4. B. Müller, J. Rafelski, W. Greiner, Phys. Lett. **47B**, 5 (1973)
5. B. Müller, W. Greiner, Z. Naturforsch. **31a**, 1 (1976)
6. V.S. Popov, Phys. At. Nucl. **64**, 367 (2001)
7. J. Eichler, W.E. Meyerhof, *Relativistic Atomic Collisions* (Academic Press, San Diego, 1995)
8. J. Reinhardt, B. Müller, W. Greiner, G. Soff, Phys. Rev. Lett. **43**, 1307 (1979)
9. U. Müller, G. Soff, Phys. Rep. **246**, 101 (1994)
10. J. Reinhardt, B. Müller, W. Greiner, Phys. Rev. A **24**, 103 (1981)
11. J. Hoppe, J. Reinhardt, UFTP preprint 151/1985
12. E. Ackad, M. Horbatsch, Phys. Rev. A **75**, 022508 (2007)
13. H. Backe et al., Phys. Rev. Lett. **40**, 1443 (1978)
14. C. Kozhuharov et al., Phys. Rev. Lett. **42**, 376 (1979)
15. J. Schweppe et al., Phys. Rev. Lett. **51**, 2261 (1983)
16. T. Cowan et al., Phys. Rev. Lett. **54**, 1761 (1985)
17. M. Clemente et al., Phys. Lett. B **137**, 41 (1984)
18. H. Tsertos et al., Phys. Lett. B **162**, 273 (1985)
19. J. Reinhardt, U. Müller, B. Müller, W. Greiner, Z. Phys. A **303**, 173 (1981)
20. U. Müller, G. Soff, T. de Reus, J. Reinhardt, B. Müller, W. Greiner, Z. Phys. A **313**, 263 (1983)
21. I. Ahmad et al. (APEX collaboration), Phys. Rev. Lett. **75**, 2658 (1995)
22. I. Ahmad et al., Phys. Rev. Lett. **78**, 618 (1997)
23. I. Ahmad et al., Phys. Rev. C **60**, 064601 (1999)
24. R. Ganz et al. (EPOS collaboration), Phys. Lett. B **389**, 4 (1996)
25. U. Leinberger et al. (ORANGE collaboration), Phys. Lett. B **394**, 26 (1997)
26. A. Balkacem, H. Gould, B. Feinberg, R. Bossingham, W.E. Meyerhof, Phys. Rev. Lett. **71**, 1514 (1993)
27. U. Müller, T. de Reus, J. Reinhardt, B. Müller, W. Greiner, G. Soff, Phys. Rev. A **37**, 1449 (1988)
28. H. Backe et al., Phys. Rev. Lett. **50**, 1838 (1983)
29. M. Krämer et al., Phys. Lett. B **201**, 215 (1988)

30. M. Krämer et al., Phys. Rev. C **40**, 1662 (1989)
31. M. Rhein et al., Phys. Rev. Lett. **69**, 1340 (1992)
32. M. Rhein et al., Phys. Rev. C **49**, 250 (1994)
33. J. Stroth et al., Z. Phys. A **357**, 441 (1997)
34. E. Ackad, M. Horbatsch, Phys. Rev. A **A8**, 062711 (2008)
35. J. Rafelski, B. Müller, W. Greiner, Z. Phys. A **285**, 49 (1978)
36. T. Maruyama, A. Bonasera, M. Papa, S. Chiba, Eur. Phys. J. A **14**, 191 (2002)
37. J. Tian, X. Wu, K. Zhao, Y. Zhang, Phys. Rev. C **77**, 064603 (2008)
38. C. Golabek, C. Simenel, Phys. Rev. Lett. **103**, 042701 (2009)
39. C. Simenel, C. Golabek, D.J. Kedziora, EPJ Web of Conferences **17**, 09002 (2011)
40. V.I. Zagrebaev, Yu.Ts. Oganessian, M.G. Itkis, W. Greiner, Phys. Rev. C **73**, 031602(R) (2006)
41. V. Zagebaev, W. Greiner, J. Phys. G: Nucl. Part. Phys **34**, 1 (2007)
42. V. Zagebaev, W. Greiner, Int. J. Mod. Phys. E **16**, 969 (2007)

# Nuclear Muon Capture in Hydrogen Isotopes

**Claude Petitjean**

**Abstract** We present two precision experiments in nuclear muon capture which are performed at PSI's 600 MeV proton accelerator. The muon capture rates in hydrogen and deuterium are measured using the lifetime method and time projection chambers as active muon stopping targets. The MuCap experiment—muon capture on the proton—is in its final analysis. We present a result of the singlet $\mu p$ capture rate $\Lambda_S$ which can be directly related to $g_P$, the pseudoscalar form factor in weak interactions, and which is predicted by low energy heavy baryon chiral perturbation theory. The MuSun experiment measures the doublet capture rate in the $\mu d$ atom using a new cryogenic time projection chamber at 34 K. By effective field theory a unique low energy constant LEC can be determined which calibrates the rate of the main $pp$ fusion reaction of the sun.

## 1 Introduction

The Paul Scherrer Institute operates an isochronous sector focusing ring cyclotron which accelerates a 2.4 mA proton beam to 590 MeV with a 100 % macro duty cycle. With the beam power of 1.3 MW, it represents one of the most powerful proton accelerators of the world. On carbon production targets, low energy pion and muon beams with highest intensities and luminosities are produced. The proton beam dump acts as an intense neutron spallation source (SINQ), and on a split beamline a new *UCN* facility provides ultra cold neutrons at highest luminosities.

The pion and muon beams are used for a rich research program for experiments in fundamental particle physics, in solid state physics and materials research. The main goal of the particle experiments is to test the standard model and to search for physics beyond the standard model.

C. Petitjean (✉)
MuCap and MuSun Collaboration, Paul Scherrer Institute, CH-5232 Villigen, Switzerland
e-mail: claude.petitjean@psi.ch

W. Greiner (ed.), *Exciting Interdisciplinary Physics*,
FIAS Interdisciplinary Science Series, DOI: 10.1007/978-3-319-00047-3_36,

A major program in particle physics are investigations of nuclear muon capture reactions in hydrogen nuclei:

$$\mu p_{(F=0)} \rightarrow \nu_\mu + n \qquad (\text{rate } \Lambda_S), \tag{1}$$
$$\mu d_{(F=1/2)} \rightarrow \nu_\mu + n + n \qquad (\text{rate } \Lambda_D). \tag{2}$$

These experiments—if performed with high precision—allow in the case of $\mu p$ capture the determination of the fundamental electro-weak coupling constant $g_P$ predicted by chiral perturbation theory, and in $\mu d$ capture the determination of the low energy constant *LEC* in effective field theory. At low energies, LEC fixes the reaction rates in the two-nucleon system and thus calibrates the *pp* fusion reaction of the sun.

## 2 The MuCap Experiment

The MuCap experiment [1, 2] aims at a high precision measurement of the singlet muon capture rate $\Lambda_S$ on the proton from the lower hyperfine state F = 0, Eq. 1, to a precision of 1 %. The rate has been calculated precisely by *Low Energy Heavy Baryon Chiral Perturbation Theory* (HBChPT) [3–6]. $\Lambda_S$ can be derived from the two fundamental coupling constants (form factors) $g_A$ and $g_P$, but because $g_A$ is well determined from neutron lifetime experiments, the measurement of $\Lambda_S$ represents a direct determination of $g_P$, the *induced pseudoscalar coupling constant*. Since $g_P$ is only sensitive to large momentum transfer (of order pion mass squared), muon capture on the proton is the only direct way to determine this fundamental constant. The MuCap experiment presents therefore a unique and rigorous test of HBChPT [7].

Experimentally, $\mu p$ capture proves to be an extremely difficult experiment for several reasons:

- The output channel of reaction 1 consists of neutral particles, $\nu_\mu$ and n. While the neutrino disappears, the neutron rate cannot be determined with the required absolute precision of 1 %.
- Muons in a hydrogen target form $\mu p$ atoms which are tiny neutral objects and behave like neutrons. In collisions with other hydrogen nuclei they form mesic molecules $p\mu p$ which exist in two configurations, ortho or para states, having quite different capture probabilities than the $\mu p$ atom. The $p\mu p$ ortho state decays into the para state, unfortunately with a badly known rate. This makes it impossible to evaluate the $\mu p$ capture rate, if too many $p\mu p$ molecules are formed. It necessitates to use hydrogen targets at very low density, where $p\mu p$ formation is small.
- Muons in hydrogen may bounce against nuclei from impurities with large Z and can get quickly transferred. In nuclei with $Z \gg 1$ they get captured with a high rate and so distort the lifetime measurement. To avoid distortion effects from impurities, the hydrogen gas must be kept ultra-clean to a level of $\leq$10 ppb.

Based on these considerations, we have designed the MuCap experiment with the setup shown in Fig. 1. The central part is a time projection chamber (TPC) embedded in a cylindrical Alu pressure vessel closed by two steel flanges. The 40 l volume is filled with 10 bar of ulta-pure protium gas (pure ${}^{1}H_{1}$ isotope). Low energy muons enter through a thin scintillator $\mu$SC, two x-y multi-wire proportional chambers (MWPC) $\mu$PC and a 0.5 mm Beryllium window into the chamber volume. Each muon is registered by $\mu$SC and $\mu$PC and tracked in 3-d by the TPC to its stopping point.

The TPC vessel is surrounded by two cylindrical MWPC's ePC1, ePC2 (each with three wire planes) and by a scintillator hodoscope consisting of 16 double layer segments to detect the electrons from muon decay. The electrons can be tracked back with the MWPC's to the muon stop, allowing clean identification of $\mu$-e pairs and of their time differences between $\mu$ stop and decay electron.

The strategy of this setup is to perform a lifetime measurement of the $\mu$p system. The exponential time distribution yields the $\mu p$ decay rate $\lambda_{\mu p}$ $(= 1/\tau_{\mu p})$, and the difference of this rate with the free $\mu^{+}$ decay rate $\lambda_{\mu}^{+}$ $(= 1/\tau_{\mu}^{+})$ yields just the $\mu p$ capture rate $\Lambda_S$:

$$\Lambda_S = \lambda_{\mu p} - \lambda_{\mu^+}. \tag{3}$$

Lifetimes can be measured with high accuracy and allow a precise determination of the $\mu p$ capture rate. $\Lambda_S$ is small of order $10^{-3}$ with respect to $\lambda_{\mu p}$, therefore a

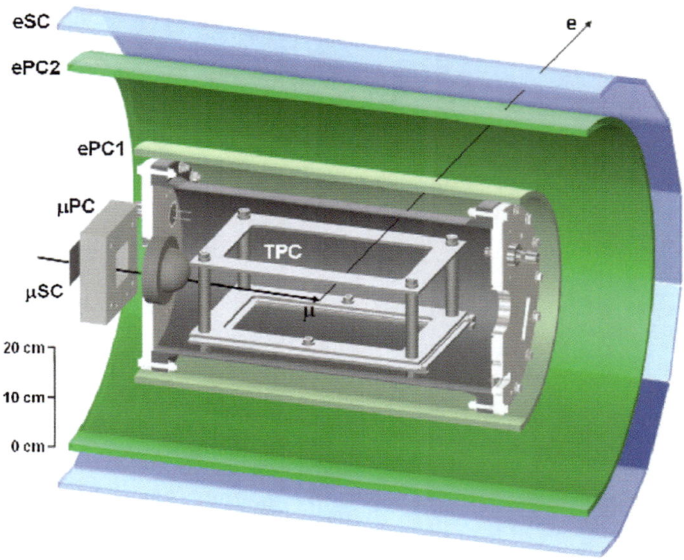


**Fig. 1** Cut out drawing of the MuCap detector

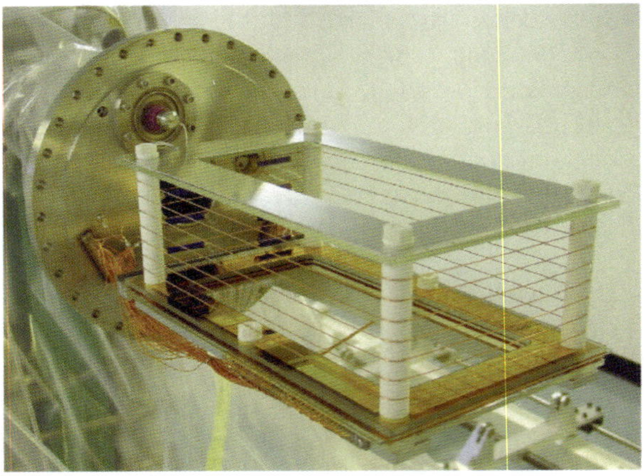

**Fig. 2** The MuCap TPC before mounting the Alu cylinder forming a pressure vessel. The 40 l volume gets filled with 10 bar of ultra-pure protium gas. The sensitive volume is 15 x 12 x 30 $cm^3$. A homogenous electrical drift-field of 2 kV/cm spans from the top cathode at −30 kV to a set of MWPC planes at the bottom. The negative charges from ionizing muon tracks are collected and amplified by the MWPC, and read out in both horizontal dimensions to preamplifiers located directly behind the TPC vessel. The drift time (22 $\mu s$ for 12 cm drift length) specifies the third (vertical) coordinate

statistics of $\geq 10^{10}$ single decay events must be accumulated to reach the precision level of 1 % for $\Lambda_S$.

## 2.1 The Time Projection Chamber

Figure 2 shows the time projection chamber *TPC* which sits in the center of the MuCap apparatus. The TPC was specially developed for the MuCap experiment [8]. It is filled with ultra-pure, deuterium-depleted hydrogen (protium) at 10 bar at ambient room temperature and acts as active muon stop detector. All materials in the TPC are metallic or consist of metal-oxydes and keramics. The whole chamber can be baked out, which strongly reduces any outgasing. It is absolutely essential to achieve the highest possible gas purity at ppb levels. In order to maintain the high gas purity, a gas circulation system was built pushing the gas continuously through Zeolite filters at liquid nitrogen temperature [9]. The purity was directly monitored via capture events observed in the TPC, but also by taking and analyzing hydrogen samples with gas chromatography. We also monitored water contents using a humidity sensor placed in the gas stream. Impurity levels of $\leq$7 ppb ($N_2$, $O_2$) and $\sim$10 ppb ($H_2O$) were reached in the final runs.

## 2.2 Performance and Results

The MuCap experiment was successfully operated during several data taking runs between 2004 and 2007. A muon beam was used in kicker mode with a flux of $\sim$20 kHz. After reductions due to pileup protection, deadtimes, efficiencies and various other cuts, a rate of $\sim$3.5 kHz good muon stops in a conservative fiducial volume were obtained. In total 1.2 $10^{10}$ fully tracked $\mu p$ decay events were collected of which a statistics of 1.6 $10^9$ is already published in [2]. Figure 3 shows the exponential time spectrum with fits using various electron definitions. These data produced the following results:

$\mu p$ decay rate: $\lambda_{\mu p} = 455'851.4 \pm 12.5_{\text{stat}} \pm 8.5_{\text{syst}} \quad \text{s}^{-1}$ .

After deduction of the $\mu^+$ decay rate $\lambda_{\mu^+} = 455'162.2 \pm 4.4$ [10] and after correcting for the bound state effect [11, 12] (12.3 $\text{s}^{-1}$) and $p\mu p$ formation (23.5 $\text{s}^{-1}$) we get according to Eq. 3

$$\Lambda_S^{MuCap} = 725 \pm 17\, s^{-1} \; . \tag{4}$$

in good agreement with the theory value [13–15]: $\Lambda_S^{Th} = 710.6\ \text{s}^{-1}$, and

$$g_P^{MuCap} = 7.3 \pm 1.1 \; , \tag{5}$$

also in agreement with $g_P^{Th} = 8.26$ [3, 4]. In the very near future we will publish the full statistics. We expect, that our final experimental accuracy shall be improved by a factor of $\sim$2.5.

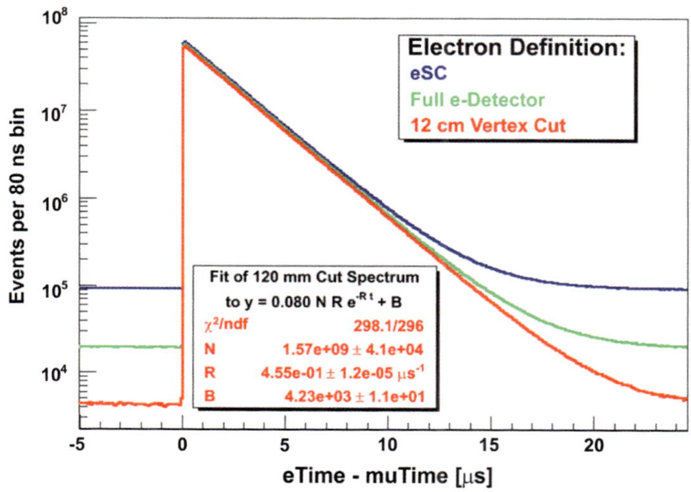


**Fig. 3** Lifetime *curve* of the first analyzed $\mu p$ capture data set

## 3 The MuSun Experiment

MuSun, the sister experiment to MuCap, was proposed in 2008 [16]. It aims at a ~1 % precision measurement of the nuclear muon capture rate $\Lambda_D$ on the deuteron, reaction Eq. 2. Since the weak interaction form factors $g_A$ and $g_P$ are precisely determined by the MuCap experiment and from the neutron lifetime, respectively, MuSun determines the strong interaction part of the two-nucleon system. It is of particular interest, that the two-nucleon problem can be treated at low energies with modern *effective field theory* to high precision [17, 18], and there is just one low energy constant *LEC* which fixes the interaction strength. Therefore, a measurement of $\Lambda_D$ can determine LEC and as a consequence fix other two-nucleon reactions like $pp$ fusion, the sun's main energy production reaction, which is of very high astrophysical interest. That's why the $\mu d$ capture experiment was baptized MuSun. In addition, LEC also determines the strength of $\nu - d$ interactions, important for the neutrino experiments at SNO, the Sudbury Neutrino Observatory [19].

### 3.1 Experimental Challenges

The MuSun experiment applies the same technique as MuCap to measure the capture rate $\Lambda_D$. A TPC filled with deuterium is used as active muon stop detector in combination with a lifetime measurement of the $\mu d$ system. But there are additional challenges which make the MuSun experiment more difficult. They are mainly due to the $d\mu d$ kinetic cycles shown in Fig. 4, leading to muon catalyzed $dd$ fusion [20]. The $d\mu d$ molecules formed in collisions $\mu d + d$ undergo one of the following $dd$ fusion reactions after $\sim 10^{-9}$ sec:

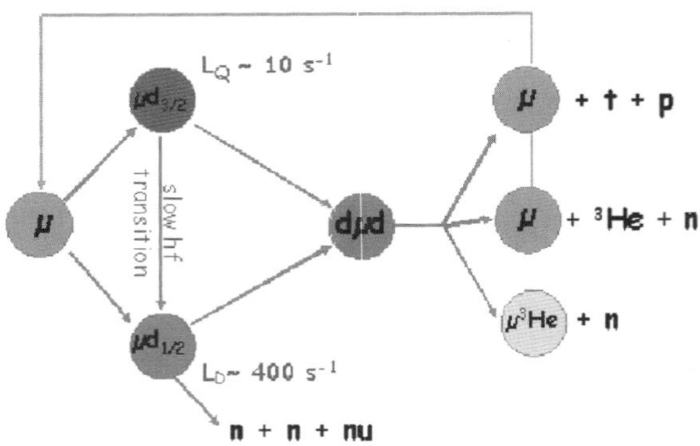


**Fig. 4** Kinetics scheme of muons in pure deuterium. More than 90 % of the muons get recycled after fusion

$$d\mu d \rightarrow^3 He(0.8\,\text{MeV}) + n(2.5\,\text{MeV}) + \mu \quad (6)$$
$$d\mu d \rightarrow \mu^3 He(0.8\,\text{MeV}) + n(2.5\,\text{MeV}) \quad (7)$$
$$d\mu d \rightarrow p(3.0\,\text{MeV}) + t(1.0\,\text{MeV}) + \mu. \quad (8)$$

The implications caused by these reactions are as follows:

- The $\mu d$ atom exists in two hyperfine (hfs) states F = 1/2 (doublet) and F = 3/2 (quartet), of which only the doublet state is sensitive to muon capture ($\Lambda_D \sim 400s^{-1}$ vs. $\Lambda_Q \sim 10s^{-1}$). Thus, the hfs population needs to be precisely known in the capture measurement. Otherwise no interpretation of the result is possible.
- Initially the $\mu d$ hfs states are populated by 2/3 (quartet) and 1/3 (doublet). In inelastic collisions ($\mu d + d$) the (F = 3/2) state gets depopulated with rate $\phi\lambda_{3/2,1/2}$, where $\phi$ is the deuterium gas density relative to liquid $H_2$ and $\lambda_{3/2,1/2} = 3.6 \cdot 10^7\ s^{-1}$ is the normalized spinflip rate. This transition can be observed at cryogenic temperatures T < 50 K via the resonant *dd* fusion reactions which at low T are resonant only for the (F = 3/2) level. e.g., by measuring the time distribution of the fusion neutrons, the development of the $\mu d$ hfs states can be exactly determined, allowing the correct evaluation of $\Lambda_D$.
- The $\mu d$ spinflip rate $\phi\lambda_{3/2,\,1/2}$ is slow at densities $\phi \sim 1\,\%$ (used in MuCap). We need $\phi\lambda_{3/2,\,1/2} \gg \lambda_\mu$ to get the $\mu d$ hfs population in the lower state (F = 1/2). Thus, a deuterium gas density $\phi \gg 1\%$ needs to be chosen. Figgure 5 shows the development of the muonic states with time at conditions T = 30 K, $\phi = 5\,\%$.
- MuSun has elevated purity requirements due to very large transfer rates to $N_2$ and $O_2$ molecules. Typically $c_N$, $c_O$ ~1 ppb are needed which necessitates careful cleaning and monitoring. On the other hand $c_{H_2O}$ is negligible at cryo temperatures.

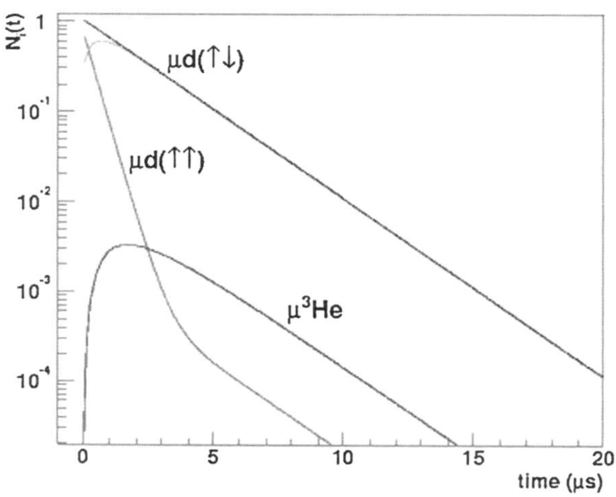


**Fig. 5** Development of $\mu d$ and $\mu^3 He$ populations in deuterium at 30 K, gas density $\phi = 5\,\%$ of liquid $H_2$. The $\mu d_{(F=3/2)}$ state gets depopulated by spin-flip processes

- The purity of the deuterium gas can be monitored by observing $\mu Z$ capture events of muons transferred to higher-Z nuclei (as in the case of MuCap). However this observation is strongly impeded in deuterium gas due the charged *dd* fusion products. It may require other ways of impurity controls.

As a consequence, the MuSun experiment must be performed with a completely new, redesigned TPC working at cryogenic temperatures and at larger gas densities than MuCap.

## 3.2 The Cryo TPC

Figure 6 shows a drawing of the new cryogenic TPC that was constructed in Gatchina/ Russia. It is mounted in a crogenic pressure vessel which sits in the center of the MuCap Alu cylinder that is now used as insulation vacuum chamber. The muon beam enters through a 0.5 mm Be window and stops mostly in the central volume. The cryo vessel is cooled to temperatures 30–50 K by liquid neon circulating from a refrigeration system through pipes in the vacuum. The drift space is a 72 mm vertical drift down to an anode area of 90 x 120 $mm^2$ divided up in 48 pads. The cathode plate at the top is on −80 kV, the grid on −3.3 kV. The chamber is operated at pressures of 5–10 bar corresponding at 34 K to a gas density $\phi$ ~5–9 %. In the electrical field of 11 kV/cm electron charges drift with 5 mm/μs to the anode pads on ground. The signals are amplified outside the TPC and fed into waveform digitizers. The deuterium gas is continuously circulated in pipes through the vacuum vessel to the MuCap purification system [9], cf. Chap. 2.1.

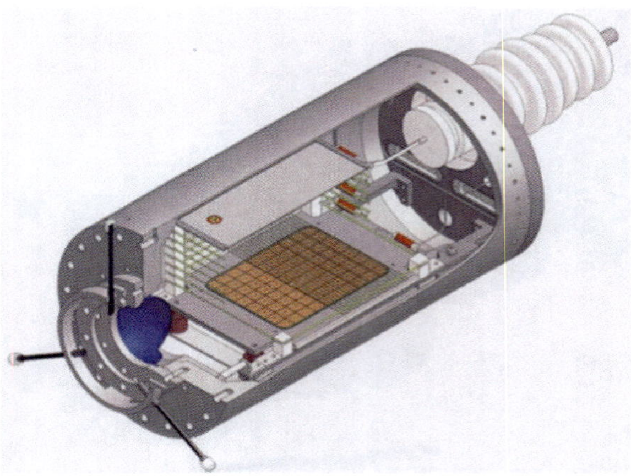

**Fig. 6** 3-d design drawing of the cryogenic TPC for MuSun

## *3.3 Performance of the Cryo TPC*

The Cryo TPC was successfully operated with muons during 2010/11 in several runs with improving performance. A statistics of $\sim 5 \cdot 10^9$ good $\mu e$ decay events was collected which is about 1/3 of the data required for reaching the anticipated precision. Figure 7 shows a typical muon stop which was followed by a delayed $dd \rightarrow p + t$ fusion event. Fusion processes occur in about in 6 % of the $\mu$-stops. Figure 8 shows the analyzed energy distribution of the time separated fusion events. The big peak on the left is from $^3$He recoils at ~0.4 MeV (half of the original energy since the other half gets lost by a recombination effect). The small peak indicated as $^3$He$\mu$ is from fusion events where the muon sticks to the $^3$He nucleus. It has only one charge and therefore less recombination [20]. The peak to the right is at ~3.7 MeV representing p+t recoils from reaction 8.

The overall performance of the TPC was quite satisfactory, but there is still a nymber of systematic problems to be solved before a precise $\mu d$ capture rate can be evaluated:

- The time distribution of fusion events is due to the $\mu d$ spinflips different from the muon lifetime curve. Therefore $\mu$ stop events with fusions may exhibit correlations leading time distortions. With careful Monte Carlo studies any such correlations must be recognized and corrected.
- Impurities above the 1 ppb level also lead to distortions of the $\mu d$ lifetime curve. We have installed Ge detectors to observe muonic X-rays from muons transferred to impurities. While in MuCap $H_2O$ was the most critical impurity to be kept under control, it is at 30–40 K nitrogen and oxygen that still exhibit a dangerous vapour

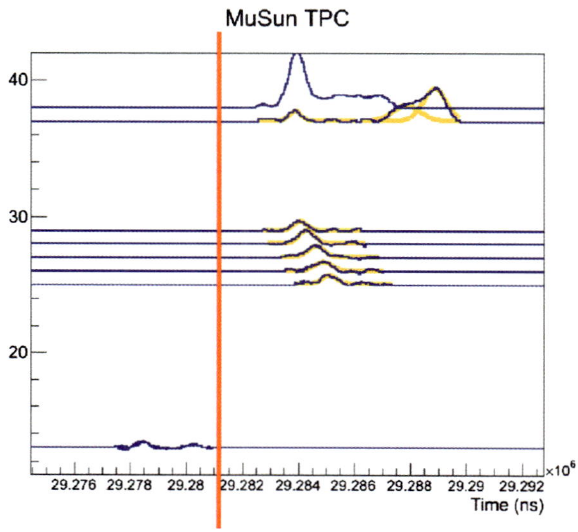


**Fig. 7** Event display showing pad signals of a muon stop followed by a delayed p+t fusion event. The peak on the top anode is the 3 MeV proton recoil which travels ~16 mm

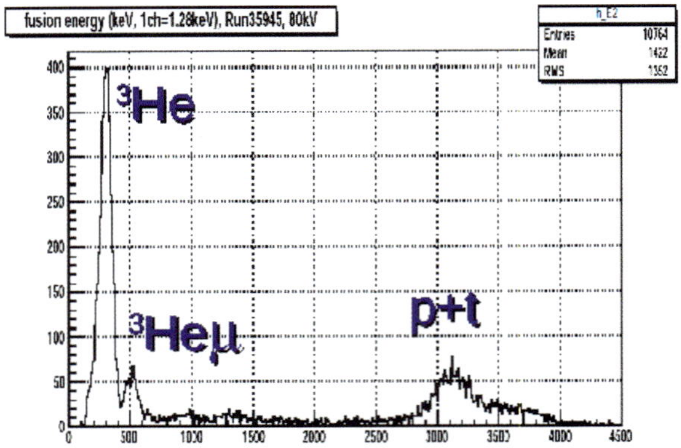


**Fig. 8** Energy distribution of observed separated fusion events

pressure, and which must be monitored. We have used gas chromatography which reaches the sensitivity of ppb levels.

In 2013–2014 we expect to perform the final data taking and to collect $2 - 3 \cdot 10^{10}$ good $\mu d$ capture events.

## References


1. D.V. Balin et al., High precision measurement of the singlet $\mu$p capture rate in $H_2$ gas, PSI proposal R-97-05.2 (2001), http://muon.npl.washington.edu/exp/MuCap/documents.html
2. V.A. Andreev et al., Phys. Rev. Lett. **99**, 032002 (2007)
3. V. Bernard, N. Kaiser, U.-G. Meissner, Phys. Rev. D **50**, 6899 (1994)
4. V. Bernard, L. Elouadrhiri, U.-G. Meissner, J. Phys. G **28**, R1 (2002)
5. J. Govaerts, J.L. Lucio-Martinez, Nucl. Phys. A **678**, 110 (2000)
6. T. Gorringe, H.W. Fearing, Rev. Mod. Phys. **76**, 31 (2003)
7. P. Kammel, K. Kubodera, Annu. Rev. Nucl. Part. Sci. **60**, 32753 (2010)
8. J. Egger, M. Hildebrandt, C. Petitjean, Nucl. Instr. Meth. A **628**, 199 (2011)
9. V.A. Ganzha et al., Nucl. Instr. Meth. A **578**, 485 (2007)
10. D.B. Chitwood et al., Phys. Rev. Lett. **99**, 032001 (2007)
11. H. Überall, Phys. Rev. **119**, 365 (1960)
12. H.C. von Baeyer, D. Leitner, Phys. Rev. A **19**, 1371 (1979)
13. V. Bernard, T.R. Hemmert, U.-G. Meissner, Nucl. Phys. A **686**, 290 (2001)
14. S. Ando, F. Myhrer, K. Kubodera, Phys. Rev. C **63**, 015203 (2000)
15. A. Czarnecki, W.J. Marciano, A. Shirlin, Phys. Rev. Lett. **99**, 032003 (2007)
16. MuSun Collaboration (http://muon.npl.washington.edu/exp/MuSun): V.A. Andreev, E.J. Barnes, R.M. Carey, V.A. Ganzha, A. Gardestig, T. Gorringe, F.E. Gray, D.W. Hertzog, M. Hildebrandt, L. Ibanez, P. Kammel, B. Kiburg, S.A. Kizilgul, S. Knaack, P.A. Kravtsov, A.G. Krivshich, K. Kubodera, B. Lauss, M. Levchenko, X. Luo, K.R. Lynch, E.M. Maev, O.E. Maev, F. Mulhauser, M.H. Murray, F. Myhrer, A. Nadtochy, C. Petitjean, G.E. Petrov, J. Phillips, R. Prieels, N. Raha, G.N. Schapkin, N. Schroeder, G.G. Semenchuk, M.A. Soroka, V. Tishchenko, A.A. Vasilyev, A.A. Vorobyov, N. Voropaev, M.E. Vznuzdaev, F. Wauters, P. Winter

17. L.E. Marinucci et al., Phys. Rev. C **83**, 014002
18. L.E. Marcucci, A. Kievsky, S. Rosati, R. Schiavilla, M. Viviani, Phys. Rev. Lett. **108**, 052502 (2012)
19. B. Aharmim et al., Phys. Rev. C **75**, 045502 (2007)
20. D.V. Balin et al., PNPI preprint 2729 (2007). Phys. Elem. Part. At. Nucl. **42**, 185 (2011)

# A Safari Through Density Functional Theory

**Reiner M. Dreizler and Cora S. Lüdde**

**Abstract** Density functional theory is widely used to treat quantum many body problems in many areas of physics and related fields. A brief survey of this method covering foundations, functionals and applications is presented here.

## 1 The Astonishing Rise of Density Functional Theory

The first text books on density functional theory contain at most 750 references ([1, 2]). In 2011 a search via Google yields the astonishing number of about 5 200 000 entries if one searches for 'density functional theory'. A selection of the topics, that are included in the Google listing, is

- Ground (and some excited) state properties of nuclei, atoms, molecules, clusters and solids
- Surface physics (adsorption, absorption)
- Properties of plasmas
- Superconducting systems
- Multicomponent systems
- Laser excitation
- DFT in molecular dynamics approaches
- Collision problems, e.g., ion-atom collisions
- Nano tubes, quantum dots

One may add that the Nobel Prize in Chemistry of the year 1998 has been awarded to W. Kohn and J. A. Pople for the development of density functional theory.

R. M. Dreizler (✉) · C. S. Lüdde
Institute of Theoretical Physics, Goethe University, Frankfurt/M, Germany
e-mail: dreizler@th.physik.uni-frankfurt.de

C. S. Lüdde
e-mail: cluedde@th.physik.uni-frankfurt.de

W. Greiner (ed.), *Exciting Interdisciplinary Physics*,
FIAS Interdisciplinary Science Series, DOI: 10.1007/978-3-319-00047-3_37,

All this raises the question: 'Why is density functional theory so popular?' The short answer is: It allows the reformulation of a quantum many body problem in terms of a set of equivalent one particle problems, which can be handled more easily.

In view of the complexity of the many body problem one might well ask: Is this really possible? The answer to this question is: An *exact* mapping is in principle possible, but—and this is the reason for the popularity of the method—in reality in terms of approximations that yield excellent results.

## 2 Basic Version: Ground State Properties

The starting point of the discussion is a standard many body Hamiltonian with kinetic energy, motion of the particles in an external potential and an interaction between all pairs of particles[1]

$$\hat{H} = \hat{T} + \hat{V} + \hat{W}.$$

The foundation of density functional theory is provided by the proof of the Hohenberg-Kohn Theorem [3], which states that the groundstate expectation value of any observable is a unique functional of the groundstate density

$$A_0[n_0] = \langle \psi_0[n_0] \,|\hat{A}|\, \psi_0[n_0] \rangle.$$

Furthermore: the functionals for the kinetic $T[n_0]$ and the interaction energies $W[n_0]$ are universal. This means: the kinetic energy functional has the same form for any (nonrelativistic fermion) system and the interaction energy functional is the same for any Coulomb system, however complicated the functionals are. The groundstate energy of a specific system (e.g. atom, molecule, solid) is therefore characterised by the one particle potential

$$V[n_0] = \langle \psi_0[n_0] \,|\hat{V}|\, \psi_0[n_0] \rangle = \int \mathrm{d}^3r \; v(\mathbf{r}) n_0(\mathbf{r}),$$

in the sense that there exists a unique, bijective mapping $V[n] \Longleftrightarrow n$.

On the basis of the theorem one may, given an energy functional, attempt to determine the groundstate energy variationally. The variational problem

$$\frac{\delta}{\delta n(\mathbf{r})} \left\{ E_0[n] - \mu \int \mathrm{d}^3r \; n(\mathbf{r}) \right\} = 0,$$

with the subsidiary condition of a fixed particle number $N$, allows the determination of the ground state density $n_0(\mathbf{r})$ and the ground state energy $E_0$. This direct path to the ground state energy was attempted by the precursors of modern density functional

[1] Questions of additional degrees of freedom as, e.g., spin will be suppressed here.

theory, the Thomas-Fermi model (and its extensions, see e.g. [1]). So far only results of modest accuracy have been obtained. The reason is: the kinetic energy functional $T[n]$ is not known well enough.

A way around the difficulties with the kinetic energy was suggested by Kohn and Sham [4]. The Kohn-Sham method can be summarized in the following fashion[2]:

- Step 1: Represent the ground state density (again suppressing questions of spin or other degrees of freedom)

$$n(\mathbf{r}) \equiv n_0(\mathbf{r}) = N \int d^3r_2 \ldots \int d^3r_N \, \Psi_0^\dagger(\mathbf{r}, \mathbf{r}_2, \ldots)\Psi_0(\mathbf{r}, \mathbf{r}_2, \ldots)$$

  in terms of a set of orbital functions

$$n(\mathbf{r}) = \sum_{i=1}^{N} \varphi_i^*(\mathbf{r})\varphi_i(\mathbf{r}).$$

- Step 2: Rearrange the expression for the groundstate energy

$$E_0[n] = T[n] + W[n] + \int d^3r \, v(\mathbf{r})n(\mathbf{r})$$

  after addition and subtraction of the *noninteracting* kinetic energy

$$T_s[n] = -\frac{\hbar^2}{2m} \sum_{i=1}^{N} \int d^3r \, \varphi_i^*(\mathbf{r})\nabla\varphi_i(\mathbf{r})$$

  and the Hartree energy

$$W_H(n) = \frac{1}{2} \int d^3r_1 \int d^3r_2 \, n(\mathbf{r}_1)w(\mathbf{r}_1, \mathbf{r}_2)n(\mathbf{r}_2)$$

  in the form

$$E_0[n] = T_s[n] + W_H[n] + V[n] + E_{xc}[n]$$

  with the *exchange-correlation energy*

$$E_{xc}[n] = T[n] - T_s[n] + W[n] - W_H[n].$$

  This central energy term is given by the difference between the full kinetic energy and its noninteracting counterpart as well as the difference between the full interaction energy and the Hartree energy. In other words, it contains all serious many

---

[2] The actual argumentation is more subtle. The subtlety is discussed under the heading of v-representability, which concerns the question of the existence of functional derivatives of kinetic energy functionals [5].

body aspects of the problem. On the other hand, through the rearrangement, the large and easy to obtain parts of the ground state energy have been separated neatly from the smaller, but for all finer points important, exchange-correlation contribution.

- Step 3: By variation with respect to the orbitals (and the subsidiary condition of orthonormality—which turns out to be sufficient) one arrives at the Kohn-Sham orbital equations

$$\left\{-\frac{\hbar^2}{2m}\nabla + v(\mathbf{r}) + v_H([n];\mathbf{r}) + v_{xc}([n];\mathbf{r})\right\}\varphi_i(\mathbf{r}) = \varepsilon_i\varphi_i(\mathbf{r}).$$

  This single particle equation contains three potential terms. Next to the given external potential and the Hartree potential one has to deal with the *exchange-correlation potential* $v_{xc}([n];\mathbf{r})$, which is defined by the functional derivative of the exchange-correlation energy

$$v_{xc}([n];\mathbf{r}) = \frac{\delta E_{xc}[n]}{\delta n(\mathbf{r})}.$$

The Kohn-Sham equations constitute the map of the many body problem on effective single particle problems. They invite the following comments:

- The full, effective Kohn-Sham potential

$$v_{KS}([n];\mathbf{r}) = v(\mathbf{r}) + v_H([n];\mathbf{r}) + v_{xc}([n];\mathbf{r})$$

  is determined by the density. The Kohn-Sham problem is, as the Hartree-Fock problem, a selfconsistency problem: an initial guess of the density (given the density dependence of the potentials) has to be iterated until selfconsistency.
- The Hartree approximation is recovered for $v_{xc} = 0$, the Hartree-Fock approximation—cum grano salis—for $v_{xc} = v_x$. The Kohn-Sham scheme deals, however, with correlation effects beyond these approximations.
- One welcome feature of the Kohn-Sham potential $v_{KS}$ is the fact, that it is multiplicative. It is much simpler to handle such a potential for geometrically complicated systems as compared to the Hartree-Fock potential, which includes a 'nonlocal' exchange term.
- Although it might be very tempting to endow some physical significance to the Kohn-Sham orbitals, the genesis of the approach does not support this point of view. As a matter of principle, the orbitals are just mathematical constructs aimed at generating the ground state density. In the same vein: the determinant constructed from the lowest energy Kohn-Sham orbitals does not represent the groundstate of the problem at hand.

# 3 Survey of xc-Functionals

The xc-energy functionals or potentials are at the heart of the ensuing discussion. Most functionals found in the literature address nonrelativistic, many particle Coulomb systems. This field of activity will serve for the introduction of the four types of functionals which have been used to date. Some sample results are found in Sect. 5.

The functionals, that are most often used and easiest to apply, are based on the local density approximation.

## *3.1 LDA: Local Density Approximation*

The LDA assumes that xc-energy densities ($e_{xc} = E_{xc}$/volume) obtained for homogeneous systems (the electron gas, nuclear matter) can be transfered locally to situations which are not so homogeneous

$$e_{xc}^{hom}(n) \longrightarrow e_{xc}^{LDA}(n(\mathbf{r})) \quad \text{so that} \quad E_{xc}^{LDA}[n] = \int \mathrm{d}^3r \; e_{xc}^{LDA}(n(\mathbf{r})).$$

The global functional dependence of the energy density on the constant density is supposed to be (approximately) correct locally. One may then use results from many body perturbation theory or Monte Carlo simulations of the homogeneous systems for the formulation of the functionals. This approach leads to functionals, which lead to very acceptable results, so that one is bound to ask for the reason. The reason for the success of the LDA is a cancelation of errors between the exchange and the correlation contributions. This, in turn, is due to the fact that the corresponding exchange and correlation holes (correlation functions between *pairs* of particles) of the LDA satisfy exact sum rules

$$\rho_x(\mathbf{r}_1, \mathbf{r}_2) \leq 0 \quad \int \mathrm{d}^3r' \rho_x(\mathbf{r}, \mathbf{r}+\mathbf{r}') = -1 \quad \int \mathrm{d}^3r' \rho_c(\mathbf{r}, \mathbf{r}+\mathbf{r}') = 0$$

on the average.

There are, nonetheless, deficiencies of the LDA. The major one is a problem with selfinteraction effects, due to a lack of cancelation of the selfinteraction between the direct and the exchange energies. This leads to an incorrect asymptotic limit of the xc-potential, for instance for neutral Coulomb systems

$$\lim_{r\to\infty} v_{xc}^{LDA} \longrightarrow -\varepsilon^{-\gamma r} \quad \text{instead of} \quad \lim_{r\to\infty} v_{xc}^{exact} \longrightarrow -\frac{1}{r}.$$

If one particle is removed far from the system, one should observe a Coulomb potential as an effect of the hole left behind. This feature is also the reason for a poor representation of negative ions.

## 3.2 GE: Gradient Expanded Functionals

If the homogeneous system does not provide an acceptable input, it is natural to look at corrections due to inhomogeneities. An expansion of the xc-energy to low order in derivatives of the density can be written as

$$\begin{aligned} E_{xc}[n] &= E_{xc}^{(0)}[n] + E_{xc}^{(2)}[n] + E_{xc}^{(4)}[n] + \dots \\ &= \int \mathrm{d}^3 r \left\{ e_{xc}^{LDA}(n(\mathbf{r})) + B_{xc}^{(2)}(n(\mathbf{r}))(\nabla n(\mathbf{r}))^2 + \dots \right\} . \end{aligned}$$

The technique used to evaluate the details is many body perturbation theory. For instance, the coefficient of the second order contribution in the exchange-only limit $B_x^{(2)}$ is determined by evaluation of the three diagrams of the irreducible polarisation insertion

Even the evaluation of these contributions did not turn out to be straightforward due to the singular structure of the Coulomb interaction at small momenta. The same must be said for low order correlation contribution. After much work, the gradient expansion was found to converge slowly and lead to unsatisfactory results. One of the reasons for this disappointing feature is the fact that the sum rules for the exchange and correlation hole (see above) are not satisfied if evaluated with input obtained by gradient expansion [6].

## 3.3 GGA: Generalised Gradient Functionals

The study of the gradient expansion of the xc-energy demonstrates that low order correction beyond the LDA contribution can be expressed in terms of two quantities, the density and a second order density gradient, usually called $s$

$$s(\mathbf{r}) = \left( \frac{\nabla n(\mathbf{r}) \cdot \nabla n(\mathbf{r})}{n(\mathbf{r})^{8/3}} \right)^{1/2} .$$

A standard ansatz takes the form

$$\Delta E_{xc}^{GGA}[n] = \int \mathrm{d}^3 r\; e_{xc}^{LDA}(n(\mathbf{r}))\, f_{xc}(n(\mathbf{r}), s(\mathbf{r})).$$

For Coulomb problems there exists a large number of well and not so well founded suggestions for the function $f_{xc}$. Parameters are, for instance, fitted by optimal reproduction of atomic data. The resulting functions can then, relying on the universality of the functionals in question, be used for other Coulomb systems, from molecules to solids. Again it is found that the sum rules for the pair correlation holes play an important part. They can be enforced by a 'real space cut-off', which is achieved technically by limiting the integrations to a suitable section of space [7].

## *3.4 OF: Orbital Functionals*

Orbital or implicit functionals are introduced with the argument that the ground state density is also a functional of the Kohn-Sham orbitals. It might, for this reason, be worthwhile to use the orbitals as the basic variables. It is then possible to define the Kohn-Sham exchange (with Kohn-Sham rather than Hartree-Fock orbitals)

$$E_x^{KS} = -\frac{1}{2}\sum_{i,j}\int \mathrm{d}^3 r_1 \int \mathrm{d}^3 r_2 \varphi_i^*(\mathbf{r}_1)\varphi_j^*(\mathbf{r}_2) W(\mathbf{r}_1, \mathbf{r}_2)\varphi_i(\mathbf{r}_2)\varphi_i(\mathbf{r}_1),$$

so that the exchange-only limit is selfinteraction-free. The calculation of the xc-potential is now more involved, as the application of the chain rule for the functional derivatives

$$\frac{\delta}{\delta n(\mathbf{r})} = \int \mathrm{d}^3 r' \sum_j \frac{\delta}{\delta \varphi_j(\mathbf{r}')}\frac{\delta \varphi_j(\mathbf{r}')}{\delta n(\mathbf{r})} + \text{herm. conj.}$$

leads to integral equations for the xc-potential. The appearance of the integral equations is the reason for the alternative name of the orbital approach: Optimised Potential Method. For instance in the x-only limit one obtains the equation

$$\int \mathrm{d}^3 r_1 v_x^{OF}(\mathbf{r}_1) K(\mathbf{r}_1, \mathbf{r}_2) = Q(\mathbf{r}_2)\ ,$$

where the kernel $K$ and the inhomogeneous term $Q$ depend explicitly on occupied as well as unoccupied Kohn-Sham orbitals. The solution of the integral equations for the effective potentials has to be repeated within each selfconsistency cycle. As this is quite time-consuming, efficient shortcuts [8] on the basis of closure have been invented and tested. The correlation contribution is usually evaluated in terms of a perturbative approach a posteriori.

# 4 Extensions

Extensions of the basic theory have been formulated and explored in many directions. As examples one might name

- Time-dependent density functional theory [9] for the exploration of excitation and collision processes: The basis is in this case the Runge-Gross Theorem [10], which can be viewed as a nontrivial extension of the Hohenberg-Kohn Theorem. The action functional

$$A[n(t)] = \langle\psi[n(t)]|\hat{A}(t)|\psi[n(t)]\rangle$$

  replaces the energy functional. The extension is, among others reasons, nontrivial, as retarded and advanced time structures have to be separated properly.
- Relativistic density functional theory [11, 12], e.g. for the investigation of systems involving heavy elements: the formulation (basic existence theorem, structure of the functionals, orbital equations) has to be based on a proper quantum field theoretical background (as quantum electrodynamics) in order to deal with questions of renormalisation, vacuum polarisation etc. A relativistic four-current replaces the density. In addition viable schemes have to be explored with the aim of handling the solution of the underlying single particle Dirac-Kohn-Sham equations with particle as well as hole states.
- Current/Spin-polarised density functional theory by inclusion of spin degrees of freedom.
- Thermal density functional theory based on use of the free energy and the grand potential.
- The list could be extended (see e.g. [5]).

# 5 Some Applications and Results

In this chapter a small selection of results, partly in the form of tables and partly by illustrations are presented, with minimal comments. As usual, the devil hides in the detail. Details are not be outlined here. Therefore interested readers are encouraged to consult the references for explicit methods used and for further examples.

## *5.1 Nuclei*

The question of the energy situation in heavy or superheavy nuclei has been investigated with an approach termed Quantum-Hadro-Dynamics [13], that is a non-renormalisable, field-theoretical meson-exchange model. The nucleons interact via the exchange of massive scalar and vector mesons. The coupling constants are

**Table 1 Binding energy/nucleon (MeV) and charge radius (fm) of element 114**, [14]

| | Binding energy (MeV) | | | Charge radius (fm) | | |
|---|---|---|---|---|---|---|
| | HF | LDA | Exp | HF | LDA | Exp |
| $^{16}O$ | 5.11 | 7.63 | 7.98 | 2.74 | 2.74 | 2.73 |
| $^{40}Ca$ | 6.46 | 8.26 | 8.55 | 3.46 | 3.52 | 3.46 |
| $^{48}Ca$ | 6.72 | 8.53 | 8.67 | 3.45 | 3.53 | 3.45 |
| $^{90}Zr$ | 7.11 | 8.73 | 8.71 | 4.23 | 4.33 | 4.23 |
| $^{208}Pb$ | 6.49 | 7.87 | 7.87 | 5.47 | 5.60 | 5.47 |
| 114 | | 7.10 | | | 6.32 | |

**Table 2** Mean absolute deviation of atomisation energies (kcal/mol) for the 32 'standard' molecules, [15]

| HF | MP2 | MP2+ | LSD | GGA/LSD | GGA |
|---|---|---|---|---|---|
| 85.9 | 22.4 | 28.8 | 35.7 | 4.4 | 5.6 |

determined so that they reproduce the properties of lighter nuclei. The model is then used to predict properties of heavier nuclei and superheavies. The results for the binding energy per nucleon (in [MeV]) and the charge radius (in [fm]) of a density functional calculation with the QHD model at the LDA level [14] are shown in Table 1.

## 5.2 Molecules

An investigation of various properties of 32 'standard molecules' by Johnson et al. [15] convinced the community of Chemists of the usefulness and predictive power of density functional theory. Table 2 shows results for the mean absolute deviation of atomisation energies (in units of kcal/mol), comparing three standard many body methods (Hartree-Fock and many body perturbation theory to second order, direct and augmented by selected fourth order contributions) with three different density functional approaches (local-spin-density, generalised gradient approximation for exchange plus local-spin-density for correlation and a generalised gradient functional for both exchange and correlation).

The same picture emerges on the level of individual molecules, for instance at the atomisation energies (in units of eV) of carbohydrates [16] as a function of the number of bonds ($N$) shown in Table 3.

Direct insight into the distribution of the electrons in the ground state of a molecule can be gleened by looking at the gradients of the density, as illustrated for instance in the two parts of Fig. 1. The figure shows the variation of the two gradient functions $s$ and $d$

**Table 3** Atomisation energies (eV) of selected carbohydrates, [16]

| | N | Exp | HF | LDA | GGA |
|---|---|---|---|---|---|
| $H_2$ | 1 | 4.75 | 3.63 | 4.89 | 4.55 |
| $C_2$ | 1 | 6.36 | 0.73 | 7.51 | 6.55 |
| $C_2H_2$ | 3 | 17.69 | 13.00 | 20.02 | 18.09 |
| $CH_4$ | 4 | 18.04 | 14.39 | 20.09 | 18.33 |
| $C_2H_4$ | 5 | 24.65 | 18.71 | 27.51 | 24.92 |
| $C_2H_6$ | 7 | 31.22 | 24.16 | 34.48 | 31.24 |
| $C_6H_6$ | 12 | 59.67 | 45.19 | 68.42 | 61.34 |

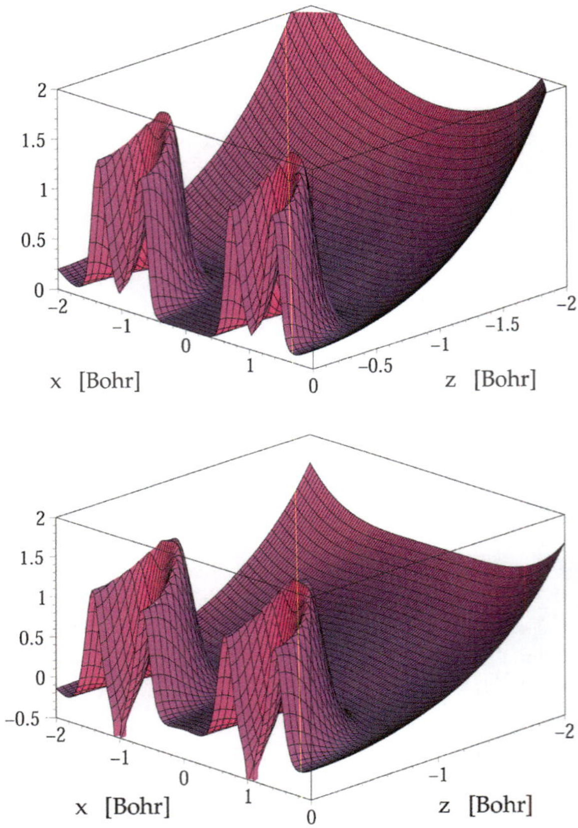


**Fig. 1** Characteristic gradients $s$ and $d$ of the density in the $N_2$-molecule, [5]

$$s = \left(\frac{\nabla n \cdot \nabla n}{n^{8/3}}\right)^{1/2} \quad \text{and} \quad d = \frac{\Delta n}{n^{5/3}}$$

for the nitrogen molecule $N_2$ (over the x-y plane). The density has been obtained by a Kohn-Sham calculation with exact exchange [5]. The two nuclei are located at $x = \pm 1.035$ Bohr, $y = z = 0$.

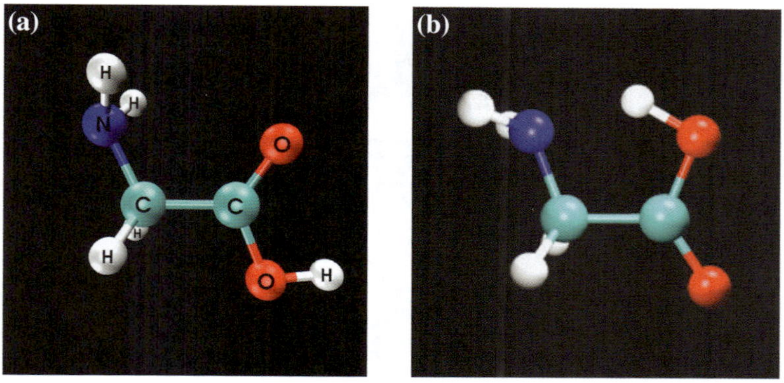


Glycine Ip

Glycine IIp

**Fig. 2** The conformers glycine Ip and glycine IIp

An example for the treatment of more complex molecules is the calculation of the structure of small amino acids like glycine. The chemical formula

$$\mathbf{NH_2\,CH_2\,COOH}$$

does not tell the full story, as there exist a number of conformers as for instance glycine Ip and glycine IIp shown[3] in Fig. 2.

These molecules were treated like solids with a supercell technique (with 23 Å) using a plane wave expansion of the orbitals and a pseudopotential approach of the Troullier-Martins type. Results for the various bond lengths and bond angles in glycine Ip are collected in Table 4. LDA results [17] are compared with experimentally determined values [18]. A corresponding calculation for glycine IIp, presented in Table 5, emphasises the variation of results with the size of the basis (cut-off of the plane wave basis at 60 versus 100 Rydbergs). This illustrates on one side the accuracy that can be obtained with reasonably modest means (within the LDA) and (in comparison with the Table 4) the difference in the structure of the two conformers. The results of [17] can be summarised in the form: the structure of all the conformers of all small amino acids is well reproduced by density functional theory. The effect of contributions of gradient corrections is not very dramatic.

The total groundstate energies of the two conformers in LDA are found to be (in Hartree)

$$E_0(Ip) = -55.8006\text{H} \quad E_0(IIp) = -55.9047\text{H}.$$

The energy difference $E_0(IIp) - E_0(Ip) = -0.05\text{H} = -0.14$ eV $= -3.12$ kcal/mol does however not agree with the experimental value $E_0(IIp) - E_0(Ip) = +0.06$ eV $= +1.4$ kcal/mol. The conformer $Ip$ has the lower energy.

[3] p stands for the fact that the heavy constituents are planar.

**Table 4** Structure of glycine Ip, [17]

| Ip | Bond lengths (Ångstrøm) | | | Bond angles (degrees) | |
|---|---|---|---|---|---|
| Bond | Expt | LDA | Angle | Expt | LDA |
| N–H | (1.001) | 1.020 | H–N–H | (110.3) | 106.5 |
| N–C | 1.467 | 1.431 | H–N–C | (113.3) | 111.0 |
| C–H | (1.081) | 1.101 | N–C–C | 112.1 | 115.0 |
| C–C | 1.526 | 1.506 | H–C–H | (107.0) | 104.4 |
| C–O | 1.355 | 1.332 | C–C–O | 111.6 | 114.9 |
| C = O | 1.205 | 1.200 | C–C = O | 125.1 | 125.7 |
| O–H | (0.966) | 0.976 | C–O–H | (112.3) | 104.7 |

**Table 5** Structure of glycine IIp, [17]

| IIp | Bond lengths (Ångstrøm) | | | Bond angles (degrees) | |
|---|---|---|---|---|---|
| Bond | LDA60 | LDA100 | Angle | LDA60 | LDA100 |
| N–H | 1.026 | 1.019 | H–N–H | 107.9 | 108.7 |
| N–C | 1.453 | 1.449 | H–N–C | 113.1 | 113.7 |
| C–H | 1.103 | 1.099 | N–C–C | 109.9 | 110.4 |
| C–C | 1.518 | 1.518 | H–C–H | 106.2 | 106.1 |
| C–O | 1.338 | 1.324 | C–C–O | 112.2 | 112.0 |
| C = O | 1.228 | 1.203 | C–C = O | 122.5 | 123.1 |
| O–H | 1.046 | 1.023 | C–O–H | 100.0 | 101.8 |

## *5.3 Solids*

Different options for density functional potentials in solids reproduce in general the gross features but differ sufficiently in detail, so that they give rise to different band structures. This point is illustrated in Fig. 3, which shows the exchange potential obtained by a plane wave pseudo potential calculations for Si [5]. The plot illustrates the difference between two GGA potentials, one LDA potential and the more involved (and more accurate) OPM potential. The potentials are plotted along the [111] direction of a diamond structure.

The band structure resulting from the OPM potential is displayed in Fig. 4, together with rather close results obtained with the simpler, very efficient KLI approximation.

## *5.4 Final Remark*

Obviously it is only possible to scratch the surface of a vast field in such a short communication. For further reading and study we naturally recommend the

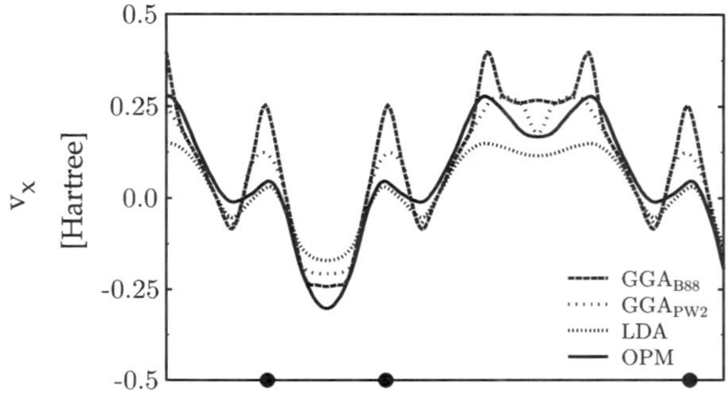


**Fig. 3** Exchange potential of Si, plotted along the [111] direction for a diamond structure, [5]

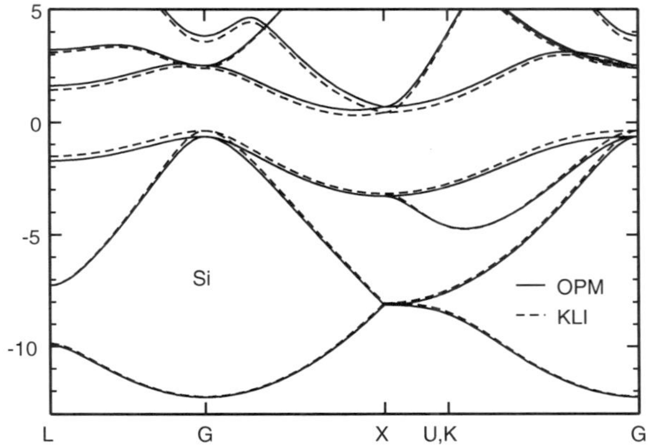


**Fig. 4** Band structure of Si obtained for the OPM and for the KLI approximation, [5]

book [5], which presents a fuller overview, addresses many finer points and provides the explanation of the figures, which are included here. The book also contains references to additional texts and reviews.

## References


1. R.G. Parr, W. Yang, *Density Functional Theory of Atoms and Molecules* (Oxford University Press, New York, 1989)
2. R.M. Dreizler, E.K.U. Gross, *Density Functional Theory* (Springer, Berlin, 1990)
3. P. Hohenberg, W. Kohn, Phys. Rev. B **136**, 864 (1964)
4. W. Kohn, L.J. Sham, Phys. Rev. A **140**, 1133 (1965)
5. E. Engel, R.M. Dreizler, *Density Functional Theory, an Advanced Course* (Springer, Heidelberg, 2011)
6. J.P. Perdew, in *Density Functional Theory, NATO ASI Series B*, vol. 337, ed. by E.K.U. Gross, R.M. Dreizler (Plenum, New York, 1995), p. 51
7. J.P. Perdew, K. Burke, M. Ernzerhof, Phys. Rev. Lett. **77**, 3865 (1996)

8. J.B. Krieger, Y. Li, G.J. Iafrate, Phys. Lett. A **146**, 256 (1990)
9. M.A.L. Marques, C.A. Ulrich, R. Nogeira, A. Rubio, K. Burke, E.K.U. Gross (eds.), *Time Dependent Density Functional Theory, Lecture Notes in Physics*, vol. 706 (Springer, Berlin, 2006)
10. E. Runge, E.K.U. Gross, Phys. Rev. Lett. **52**, 997 (1984)
11. E. Engel, R.M. Dreizler, in *Density Functional Theory II, Topics in Current Chemistry*, vol. 181, ed. R.F. Nalewajski (Springer, Berlin, 1996), p. 1
12. E. Engel, in *Relativistic Electron Structure Theory, Part I: Fundamentals*, ed. P. Schwerdtfeger (Elsevier, Amsterdam, 2002), p. 524
13. C. Speicher, R.M. Dreizler, E. Engel, Ann. Phys. (NY) **213**, 312 (1992)
14. R.N. Schmid, E. Engel, R.M. Dreizler, Phys. Rev. C **52**, 164 and 2804, (1995)
15. B.G. Johnson, P.M.W. Gill, J.A. Pople, J. Chem. Phys. **97**, 7846 (1992)
16. J.P. Perdew, J.A. Chevary, S.H. Vosko, K.A. Jackson, M.R. Pederson, D.J. Singh, C. Fiolhais, Phys. Rev. B **46**, 6671 (1992)
17. G. Iseri, *Diploma Thesis* (Goethe University, Frankfurt/M, 2003)
18. K. Iijima, K. Tanaka, S. Onuma, J. Mol. Struct. **246**, 257 (1991)

# Part VI
# Theoretical Biology

# Light-Activated Magnetic Compass in Birds

**Ilia A. Solov'yov and Walter Greiner**

**Abstract** Migrating birds fly thousand miles without having a map, or a GPS unit. But they may carry their own sensitive navigational tool, which allows them "see" the Earth's magnetic field. Here we review the important physical and chemical constraints on a possible compass sensor and discuss the suggestion that radical pairs in a photoreceptor cryptochrome might provide a biological realization for a magnetic compass. Finally, we review the current evidence supporting a role for radical pair reactions in the magnetic compass of birds.

## 1 Introduction

Migratory birds travel spectacular distances each year, navigating and orienting by a variety of means, most of which are poorly understood. Among them is a remarkable ability to perceive the intensity and direction of the Earth's magnetic field [1–3]. Biologically credible mechanisms for the detection of such a weak field (25–65 $\mu$T) are scarce and in recent years two proposals have emerged as front-runners. One, essentially classical, centers on clusters of magnetic iron-containing particles in the upper beak which appear to act as a magnetometer for determining geographical position [4–11]. However, the idea that bird orientation is guided by magnetic-sensing structures in the animals' beaks has been challenged by the suggestion that the iron-containing cells are macrophages, which have no link to the brain [12], i.e., a kind of immune cell that are also involved in iron homeostasis.

I. A. Solov'yov (✉)
Beckman Institute for Advanced Science and Technology, Urbana-Champaign, Illinois, USA
e-mail: ilia@illinois.edu

W. Greiner
Frankfurt Institute for Advanced Studies, Goethe-Universität, Frankfurt, Germany
e-mail: greiner@fias.uni-frankfurt.de

W. Greiner (ed.), *Exciting Interdisciplinary Physics*,
FIAS Interdisciplinary Science Series, DOI: 10.1007/978-3-319-00047-3_38,

The other mechanism relies on the quantum spin dynamics of transient photoinduced radical pairs [13–29]. Originally suggested by Schulten in 1978 [17] as the basis of the avian magnetic compass sensor, this mechanism gained support from the subsequent observation that the compass is light-dependent [30]. The radical pair hypothesis began to attract increased interest following the proposal in 2000 that free radical chemistry could occur in the bird's retina initiated by photoexcitation of cryptochrome, a specialized photoreceptor protein [20].

The quantum evolution of highly non-equilibrium electron spin states of pairs of transient spin-correlated radicals is conjectured to change the yields of their reaction products in 'wet, warm and noisy' biological surroundings even though the Zeeman interaction with the geomagnetic field is more than six orders of magnitude smaller than the thermal energy per molecule ($k_B T$). The classical thermodynamic effect of such minuscule interactions on the positions of chemical equilibria and the rates of activated reactions would be entirely negligible. The radical pair mechanism is the only well-established way in which an external magnetic field can influence a chemical reaction [31–34].

The origin of the magnetic field effect (MFE) can be understood by reference to the simple reaction scheme shown in Fig. 1: (1) A pair of radicals $A^{\bullet}B^{\bullet}$ is formed (e.g. by an electron transfer reaction) in an entangled state which may be either singlet (spin quantum number, $S = 0$) or triplet ($S = 1$) depending on the spin of the precursor molecule(s), which is conserved in the reaction. (2) The radical pair is able to recombine from both the S and T states to form chemically distinct products ($^{S}P$ and $^{T}P$ in Fig. 1) with rate constants $k_S$ and $k_T$ respectively. (3) S and T radical pairs coherently interconvert under the influence of local magnetic fields arising from hyperfine interactions of the electron spins with magnetic nuclei in the two radicals. As a consequence, the fractional yields of the two products are determined not only by $k_S$ and $k_T$ but also by the extent and timing of the magnetically controlled S$\leftrightarrow$T interconversion step. (4) This step is also, crucially, enhanced or hindered by electron Zeeman interactions with an external magnetic field. Thus, the fractional yields of the two products and the lifetime of the radical pair become magnetic field-dependent. If the radical pair is immobilized, the tensorial nature of the hyperfine interactions implies a directionality in the response to an external magnetic field which could form the basis of a compass sensor [26, 35–38]. The theory of the radical pair mechanism is well developed and has been successfully used over the last 40 years for the quantitative interpretation of a variety of *in vitro* experimental data—not just

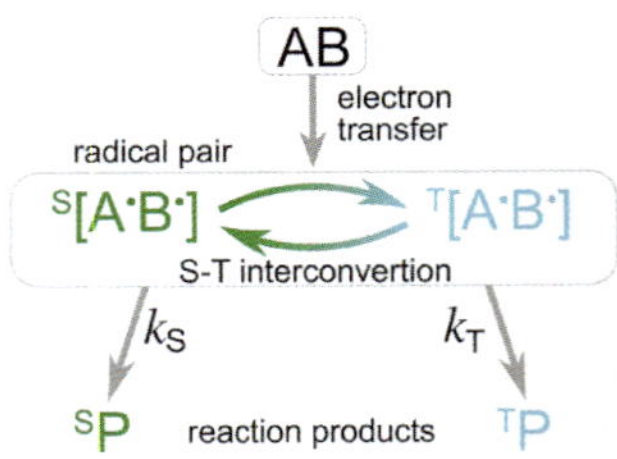


**Fig. 1** A simple radical pair reaction scheme

MFEs, but also electron and nuclear spin polarizations [39] and magnetic isotope effects [40].

There is no doubt whatsoever that radical pair MFEs rely on coherent quantum dynamics. When a radical pair is formed in a spin-conserving reaction from a singlet or triplet precursor, it is created in a non-stationary coherent superposition of the eigenstates of its spin Hamiltonian. As a consequence, the spin state of the radical pair oscillates coherently at frequencies and with amplitudes determined by the internal and external magnetic interactions. The frequencies typically fall in the $10^7$–$10^9$ Hz range and can be significantly faster than the spin relaxation processes (often $<10^7$ $s^{-1}$) that cause decoherence and loss of spin-correlation. In many cases, there is ample time for weak magnetic interactions to influence the spin dynamics before the radicals react, and therefore to affect the product yields. The clearest experimental demonstrations of this fundamentally quantum mechanical behavior, without which there would be no significant response to an external magnetic field, are the observations of quantum beats in the recombination luminescence of radical ion pairs in non-polar solvents [41–43] and the detection by EPR (electron paramagnetic resonance) of zero-quantum coherences in radical pairs in photosynthetic reaction centers [44–47].

In the following, we review the important physical and chemical constraints on a possible radical-pair-based compass sensor and discuss the suggestion that radical pairs in cryptochromes might provide a biological realization for a magnetic compass. We then summarize pertinent *in vitro* experimental data, and discuss their relevance to detecting the direction of the Earth's magnetic field. Finally, we review the current evidence supporting a role for radical pair reactions in the magnetic compass of birds.

## 2 Requirements for a Magnetic Compass

To form the basis of an effective compass magnetoreceptor, a radical pair reaction must satisfy a number of conditions [13], which fall into five broad overlapping areas: chemical, magnetic, kinetic, structural and dynamic. (1) The radical pair must be formed in a coherent superposition of its electron-nuclear spin states and at least one of the S and T states should undergo a spin-selective reaction that the other cannot. (2) There should be suitable anisotropic hyperfine interactions. (3) The lifetime of the radical pair must be long enough to allow the weak magnetic field to affect the spin-dynamics, and the rate constants $k_S$ and $k_T$ should not be too dissimilar. (4) The Zeeman interaction can only modulate the S$\leftrightarrow$T interconversion if inter-radical spin-spin (exchange and dipolar) interactions are sufficiently weak. (5) To deliver directional information, the radical pairs must be aligned and immobilized and the spin system should relax sufficiently slowly. These criteria are interlinked and not automatically mutually compatible. For example, (3) places an upper limit on the separation of the radicals, while (4) may require larger inter-radical distances. Both (3) and (4) may constrain the chemistry, e.g. by requiring the magnetically sensitive radical pair to be formed by sequential electron transfers rather than in a single step [13, 21]. Motional modulation of anisotropic hyperfine interactions is a major source of spin-relaxation in radicals so that (2) may be incompatible with (5) unless the radicals are strongly immobilized.

## 3 Cryptochrome Magnetoreception

Ritz et al. [20] proposed in 2000 that radical pairs formed photochemically in the protein cryptochrome could form the basis of the compass magnetoreceptor. No other candidate molecule has been put forward in the intervening years. Cryptochromes occur in several of the organisms for which magnetic field effects have been reported, including fruit flies, plants and migratory birds and have been shown to act as photoreceptors in a variety of species [48]. In plants, they serve as photosensors for a number of developmental responses such as hypocotyl growth, leaf expansion, induction of flowering time, and entrainment of the circadian clock. In insects, cryptochromes act as circadian photoreceptors.

Light-induced cryptochrome signalling appears to proceed via electron transfer involving a chain of three tryptophan amino acids (the Trp-triad) and the cofactor, flavin adenine dinucleotide (FAD) [49, 52–54], shown in Fig. 2a. Photo-excitation of

**(a)** **(b)**

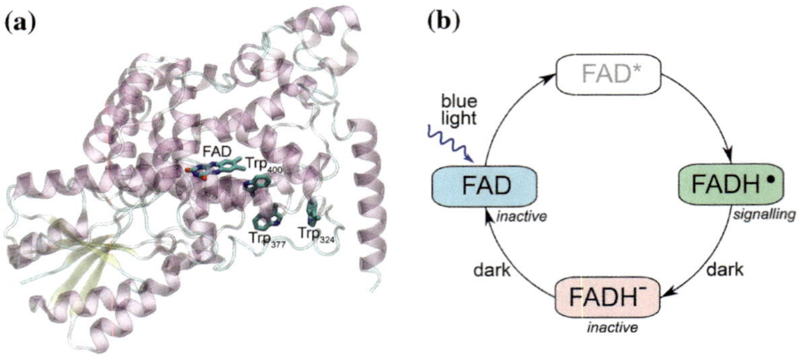


**Fig. 2** **a** Structure of cryptochrome, the protein implicated in avian magnetoreception. Cryptochrome internally binds the FAD (flavin adenine dinucleotide) cofactor which governs the functioning of the protein. The signalling state is achieved via a light-induced photoreduction pathway involving a chain of three tryptophan amino acids, indicated as $Trp_{400}$, $Trp_{377}$ and $Trp_{324}$ using the amino acid sequence numbers for *Arabidopsis thaliana* cryptochrome-1. **b** The cryptochrome photocycle. The signaling function of cryptochrome is controlled by the oxidation state of its flavin cofactor, which can exist in three interconvertible redox forms, FAD, $FADH^\bullet$ (or $FAD^{\bullet -}$), and $FADH^-$ (or $FADH_2$) [49–51]. The FAD form is inactive and is thought to be the resting state of the protein in the dark. Blue light triggers photoreduction of FAD to establish a photo-equilibrium that favors $FADH^\bullet$ over FAD and $FADH^-$. The semiquinone radical $FADH^\bullet$ state is the signalling state of the protein. $FADH^\bullet$ can be further reduced to the inactive $FADH^-$ form. The FAD$\rightarrow FADH^\bullet$ and $FADH^\bullet \rightarrow FADH^-$ reactions may be affected by an external magnetic field. The excited state of the flavin cofactor, FAD* is a short-lived intermediate in the photocycle

the FAD in its fully oxidized state leads to the formation of three consecutive radical pairs by donation of an electron along the Trp-triad to the FAD to form the $FADH^{\bullet}$ radical as illustrated in Fig. 2b. It is this state that is thought to be responsible for biological signalling. Any factor that increases (decreases) the yield of this state of the protein should result in an increased (decreased) cryptochrome signal for a given light intensity. In principle, an external magnetic field could alter the yield of the signalling state via its effect on the flavin-tryptophan radical pair [26, 21]. *In vitro*, the $FADH^{\bullet}$ state of cryptochrome has a lifetime of about 1–10 ms with respect to reversion to the FAD state [25, 52, 54].

Experimental evidence provides some support for a magnetosensing role for cryptochrome. Growth of *Arabidopsis thaliana* seedlings in a 500 μT magnetic field has been reported to enhance cryptochrome activity, such that the plants responded as if they had been exposed to higher intensities of blue light than was in fact the case [55]. Magnetically enhanced cryptochrome activity was manifested in shorter hypocotyls and higher anthocyanin levels compared to control plants grown under identical blue-light intensities in weaker magnetic fields. However, none of these effects could be replicated in a subsequent study which also failed to detect responses using substantially stronger magnetic fields where radical pair effects might be expected to be more pronounced [56]. Related effects have been found for the circadian clocks of fruit flies in which cryptochrome acts as a photoreceptor [57]. In response to blue light, cryptochrome activity increases the circadian period in *Drosophila*, an effect that was found to be more pronounced in the presence of a weak magnetic field, indicating enhanced cryptochrome signalling. Cryptochrome knock-out mutants showed no magnetic field sensitivity, while flies overexpressing cryptochrome in the clock neurons showed enhanced magnetic responses compared to wild type. A recent investigation of behavioral responses of *Drosophila* in applied magnetic fields has also implicated cryptochrome [58, 59]. In these experiments, flies were trained to associate the magnetic field with a food source, and learned to use it as an orientational cue. These responses were absent in cryptochrome-deficient flies.

In the context of avian magnetoreception, it is noteworthy that cryptochromes have been found in birds' retinas [15, 16, 62]. There are some genetic indications of an involvement of cryptochromes in magnetoreception in birds [63], but the lack of transgenic birds has hitherto precluded more clear-cut evidence. Theoretical considerations also provide support for the cryptochrome hypothesis. For example, the theory of electron transfer reactions [64] indicates that a radical pair in a protein environment could have a lifetime as long as 1 μs if the edge-to-edge inter-radical separation, $r_e$ were less than about 1.5 nm [13]. This appears to be consistent with the crystal structure of *Arabidopsis thaliana* cryptochrome in which $r_e = 1.47$ nm for the FAD cofactor and the terminal residue of the tryptophan triad [65]. A further, related kinetic constraint can be derived from the reasonable assumption that the magnetically responsive radical pair should be formed in less than 1 ns (so as to have a high quantum yield and a pure initial spin state). Estimates, also based on Marcus theory, suggest that this could be achieved if every electron transfer step involved in the formation of the pair had a donor-acceptor separation $r_e < 1.0$ nm, a condition

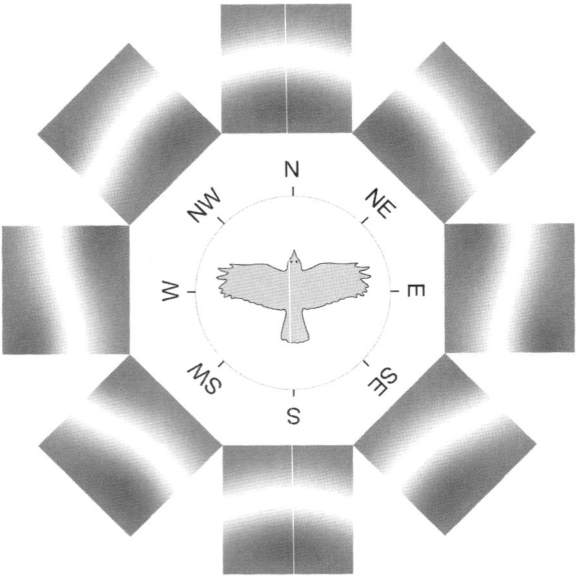


**Fig. 3** Schematic illustration of the visual modulation pattern that might be induced by the geomagnetic field for a bird flying in the eight cardinal directions (N, NE, E, SE, S, SW, W, and NW). The geomagnetic field inclination angle is 66° (appropriate for Frankfurt am Main, Germany) [60, 61]. For details, see [37]

which again is consistent with the FAD/Trp triad structure in *Arabidopsis thaliana* cryptochrome [13].

Finally, we turn to the degree of molecular ordering that would allow an array of cryptochromes to show a significant directional response to a $50\,\mu T$ magnetic field and therefore to act as a compass sensor. Assuming that the magnetic signal-transduction mechanism is linked into the rhodopsin-mediated visual detection system, so that the bird literally sees a representation of the Earth's magnetic field, one can derive a filter function to model the transformation of the visual field produced by a cryptochrome-based magnetoreceptor [37]. Figure 3 shows example of visual modulation patterns simulated in this way for a bird flying horizontally in eight cardinal directions. Such calculations indicate that even modest uniaxial molecular alignment could be sufficient to yield a directional response suitable for compass detection [35–37].

# 4 Evidence for a Radical Pair Mechanism in Birds

## *4.1 Lack of Evidence for Alternative Mechanisms*

Magnetoreception has long been postulated to be based on magnetite or other biogenic magnetic iron-oxide particles. Simple detection of iron-oxide contents in an animal is by itself not sufficient to indicate a role for iron oxides in magnetoreception, unless this is supported by corroborating behavioral observations. In birds, an iron-oxide system has been found in the beaks in the vicinity of the ophthalmic nerve [6, 11, 66]. However, under conditions in which birds show normal magnetic compass orientation in the seasonally appropriate migratory direction, their magnetic orientation responses are unaffected by anesthetization [67] of the beak or lesioning of the trigeminal branch of the ophthalmic nerve [29]. These results show clearly that birds can detect the direction of the magnetic field without using the iron-oxide system in the beak, thus indicating the existence of another, as of yet undiscovered, magnetoreception system. Moreover, the idea that bird orientation is guided by magnetic-sensing structures in the animals' beaks has been recently challenged by the suggestion that the iron-containing cells are macrophages, which have no link to the brain [12].

If this undiscovered system were based on iron-oxide particles, one would expect that a strong magnetic pulse would re-magnetize or re-organize the magnetic material and therefore affect magnetoreception of an iron-oxide based system. In a behavioral test, the bird beak system was anesthetized and a strong magnetic pulse applied prior to testing magnetic compass responses. In these experiments, the birds showed unimpaired magnetic compass orientation, strongly suggesting that the undiscovered magnetoreception system is not based on a mechanism involving iron oxides [68]. It is very likely that the beak iron-oxide system plays some role in magnetoreception, but there is scant evidence suggesting that magnetoreception in birds can occur *only* with the help of iron-oxide based mechanisms: birds can orient magnetically without using the only known iron-oxide system in their beaks and application of a strong magnetic pulse, the standard indirect behavioral test for identifying an iron-oxide based system, fails to indicate another iron-oxide based system.

## 4.2 Neurobiology

The radical pair mechanism postulates that magnetic field effects are perceived as an indirect effect on light sensing. The most likely place for the receptors to be located would be in the eye(s), so as to harness the power and speed of the visual processing system. As mentioned above, the candidate photo-magnetoreceptor molecule cryptochrome has indeed been found in avian retinas. Thus, the question arises whether brain centers have been identified that receive visual inputs and are involved in processing magnetic information. Using genetic markers, a brain area termed Cluster N has been identified in European robins that is most active during magnetic compass orientation experiments at night [69], when European robins migrate, and much less active when the eyes are closed [28, 69]. Cluster N is part of the tecto-fugal visual processing pathway and neuronal tracing has shown that it receives input from the eyes through only one synaptic transition [14, 27]. European robins with bilateral Cluster N lesions cannot perform magnetic compass orientation [29] but are capable of sun and star compass orientation, demonstrating that Cluster N is involved in processing magnetic information. It is unclear whether this area is involved in processing compass information in birds whose compass operates during daytime and seems to show the same functional properties as the compass of night migrants [70].

## *4.3 Radiofrequency Effects on Magnetic Orientation*

An oscillating magnetic field with a frequency that matches an energy-level splitting between radical pair spin states is expected to affect S↔T interconversion, as in the *in vitro* experiments. Such fields could therefore change the sensitivity of a radical pair to the geomagnetic field. Analogous to the application of a strong magnetic pulse to modify the response of an iron-oxide based compass system, one thus expects that the presence of a resonant oscillating field will modify the response of a radical-pair based compass system, leading to re-orientation or disorientation in behavioral experiments when such a field is applied. Frequencies of resonances with typical hyperfine couplings and the free electron Larmor frequency fall into the range 1–100 MHz and one expects such fields to affect magnetic compass orientation. The lack of knowledge of the chemical nature of the hypothetical radical pairs in animal compass systems precludes more accurate predictions.

Figure 4 shows the experimental arrangement used to investigate the effects of oscillating magnetic fields on the orientation of European robins [24, 71, 72]. In all conditions, the oscillating magnetic field was superimposed on a static magnetic field

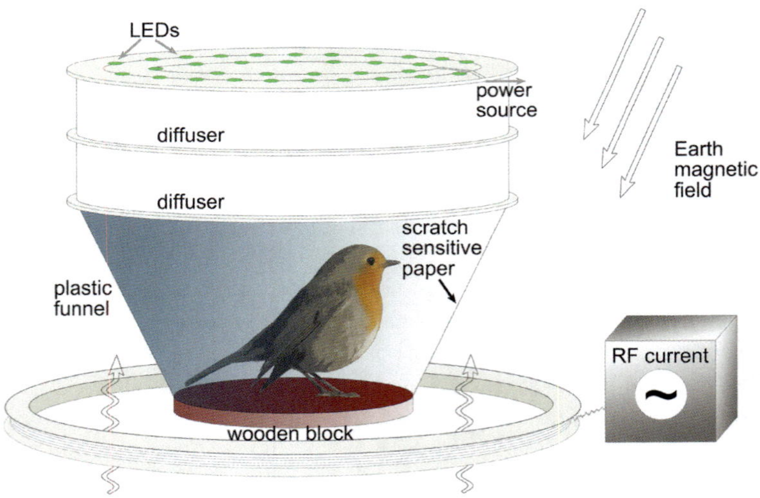


**Fig. 4** Schematic representation of the experimental arrangement used in Frankfurt to investigate the effects of radiofrequency magnetic fields on the orientation of European robins in the Earth's magnetic field. The birds' responses were recorded in funnel-shaped cages illuminated by diffuse light from above. In addition to the local geomagnetic field, an oscillating magnetic field was applied in each experimental condition. The funnels were lined with coated paper on which the birds left scratches as they moved. Analysis of the distribution of scratch marks allowed the birds' degree of orientation to be determined. For details see [24, 71]

of either 46 μT (geomagnetic field) or an amplified static field of doubled intensity. The linearly polarized oscillating field was vertical, thus forming a 24° angle with the static magnetic field. At an intensity of about 1 % of the geomagnetic field, oscillating fields disrupt orientation of European robins at frequencies between from 0.65 up to 7 MHz, the highest frequency realized in the experimental setup. At frequencies below 30 kHz, the oscillating fields did not affect the robins' orientation. Bimodal orientation results at 0.1 and 0.5 MHz, suggested a transition region between oriented and disoriented behavior. These results suggest that the radical pair lifetime or the spin relaxation time, whichever is shorter, is in the range 2–10 μs [24]. Clearly, an oscillating field with a period longer than the spin-relaxation time would be effectively static, and addition of a weak static magnetic field at 1 % of the geomagnetic intensity is not expected to have a significant effect.

Perhaps the most noteworthy feature of the oscillating field effects is that there is a dramatically stronger disruptive effect at 1.315 MHz, corresponding to the spin-only (i.e. $g = 2$) electron Larmor frequency in the geomagnetic field of 46 μT. At this frequency, a 15 nT RF field led to disoriented behavior, whereas about 30 times stronger fields were necessary to disorient birds at other frequencies. These observations suggest that one of the electron spins is magnetically isolated, i.e. that it is located on a radical with no hyperfine interactions [24]. This suggestion is bolstered by the observation that doubling the static field intensity also doubles the frequency at which a 15 nT field leads to disorientation, as expected for the Zeeman resonance of a $g = 2$ radical. A particularly strong disruptive effect of oscillating magnetic fields at the spin-only Larmor frequency has been observed in all species for which effects of oscillating fields on magnetic compass orientation have been found, namely in migratory European robins, non-migratory chickens [70] and Zebra finches [73], as well as in cockroaches [74]. This suggests that the magnetically sensitive radical pair reaction has a similar chemical nature in different species. Radicals with an isolated electron spin are unusual in organic environments, as they need to be devoid of hydrogen or nitrogen atoms. The chemical nature of this postulated radical remains unknown. Superoxide and dioxygen have been suggested as possible candidates [22, 24], but cannot be reconciled with known physical properties [75].

The existence of disruptive effects is a first indication supporting the radical-pair mechanism, but it is crucial that additional control conditions be tested to rule out that the change in orientation is due to an unrelated non-specific cause, e.g. a change in motivation due to the presence of the oscillating fields. Oscillating fields had no effect on the magnetic compass of mole rats, a blind, subterranean animal whose compass is probably based on iron-oxide materials [76], indicating that effects of oscillating fields appear not to affect iron-based systems. A key control observation is that the angle of the oscillating fields with respect to the geomagnetic field determines whether birds are oriented or disoriented [71, 72]. Birds were disoriented when the oscillating fields formed a 48° (or 24°) angle with the geomagnetic field, but not when they were collinear with the geomagnetic field. The choice of 48° is particularly meaningful as a control condition, because at this angle, the oscillating field is applied at the same angle relative to the horizontal plane (in which the birds move during the experiments) as in the 0° condition. There is no reason why the birds' motivation should be affected differently by non-specific effects of oscillating fields of equal intensity, frequency, and direction with respect to the horizontal. It appears much more likely that oscillating fields produce a resonance effect, in which case it is indeed expected that a collinear oscillating field will leave radical pair reactions unaffected [24].

## 5 Conclusion

The last decade has seen a number of studies from different fields that support the photo-magnetoreceptor and cryptochrome hypotheses. Man-made radical pair reactions have been designed that proved to be sensitive to Earth-strength magnetic fields [23]. Behavioral experiments using radiofrequency fields support the existence of a radical pair mechanism in birds. Studies at the protein level suggest that cryptochromes have properties conducive to magnetic sensing, such as formation of long-lived radical pairs. Magnetic field effects have been observed in several genetic organisms and were absent when cryptochromes were deleted. A visual brain area has been identified that is active during magnetic orientation behavior and without which birds become disoriented in magnetic orientation experiments. At this point, the radical pair hypothesis is not proven. However, support for this hypothesis has strengthened significantly, in particular for migratory birds. If it can be shown conclusively that birds use a radical-pair based compass, this would be a dramatic example of the use of a coherent quantum-mechanical process in a biological system and with clear biological relevance.

**Acknowledgments** We thank Klaus Schulten (University of Illinois at Urbana Champaign) and Peter Hore (University of Oxford) for inspiring remarks. I.S. acknowledges support as a Beckman Fellow.

## References

1. H. Mouritsen, T. Ritz, Curr. Opin. Neurobiol. **15**, 406 (2005)
2. R. Wiltschko, W. Wiltschko, BioEssays **28**, 157 (2006)
3. S. Johnsen, K. Lohmann, Phys. Today **61**, 29 (2008)
4. J. Kirschvink, J. Gould, Biosystems **13**, 181 (1981)
5. J. Kirschvink, M. Walker, C. Diebel, Curr. Opin. Neurobiol. **11**, 462 (2001)
6. G. Fleissner, B. Stahl, P. Thalau, G. Falkenberg, G. Fleissner, Naturwissenschaften **94**, 631 (2007)
7. I. Solov'yov, W. Greiner, Biophys. J. **93**, 1493 (2007)
8. I. Solov'yov, W. Greiner, Eur. Phys. J. D **51**, 161 (2009)
9. M. Walker, J. Theor. Biol. **250**, 85 (2008)
10. I. Solov'yov, W. Greiner, Phys. Rev. E **80**, 041919 (2009)
11. G. Falkenberg, G. Fleissner, K. Schuchardt, M. Kuehbacher, P. Thalau, H. Mouritsen, D. Heyers, G. Wellenreuther, G. Fleissner, PLoS ONE **5**, e9231 (2010)
12. C. Treiber et al., Nature **484**, 367 (2012)

13. C. Rodgers, P. Hore, Proc. Natl. Acad. Sci. U S A **106**, 353 (2009)
14. D. Heyers, M. Manns, H. Luksch, O. Güntürkün, H. Mouritsen, PLoS ONE **2**, e937 (2007)
15. H. Mouritsen, U. Janssen-Bienhold, M. Liedvogel, G. Feenders, J. Stalleicken, P. Dirks, R. Weiler, Proc. Natl. Acad. Sci. U S A **101**, 14294 (2004)
16. A. Möller, S. Sagasser, W. Wiltschko, B. Schierwater, Naturwissenschaften **91**, 585 (2004)
17. K. Schulten, C. Swenberg, A. Weller, Zeitschrift für Physikalische Chemie **NF111**, 1 (1978)
18. K. Schulten, in *Festkörperprobleme*, vol. 22, ed. by J. Treusch (Vieweg, Braunschweig, 1982), pp. 61–83
19. K. Schulten, A. Windemuth, in *Biophysical Effects of Steady Magnetic Fields, Proceedings in Physics*, vol. 11, ed. by G. Maret, N. Boccara, J. Kiepenheuer (Springer, Berlin, 1986), pp. 99–106
20. T. Ritz, S. Adem, K. Schulten, Biophys. J. **78**, 707 (2000)
21. I. Solov'yov, D. Chandler, K. Schulten, Biophys. J. **92**, 2711 (2007)
22. I. Solov'yov, K. Schulten, Biophys. J. **96**, 4804 (2009)
23. K. Maeda, K. Henbest, F. Cintolesi, I. Kuprov, C. Rodgers, P. Liddell, D. Gust, C. Timmel, P. Hore, Nature **453**, 387 (2008)
24. T. Ritz, R. Wiltschko, P. Hore, C. Rodgers, K. Stapput, P. Thalau, C. Timmel, W. Wiltschko, Biophys. J. **96**, 3451 (2009)
25. M. Liedvogel, K. Maeda, K. Henbest, E. Schleicher, T. Simon, C. Timmel, P. Hore, H. Mouritsen, PLoS ONE **2**, e1106 (2007)
26. F. Cintolesi, T. Ritz, C.W.M. Kay, C.R. Timmel, P.J. Hore, Chem. Phys. **294**, 707 (2003)
27. M. Liedvogel, G. Feenders, K. Wada, N. Troje, E. Jarvis, H. Mouritsen, Eur. J. Neurosci. **25**, 1166 (2007)
28. G. Feenders, M. Liedvogel, M. Rivas, M. Zapka, H. Horita, E. Hara, K. Wada, H. Mouritsen, E. Jarvis, PLoS ONE **3**, e1768 (2008)
29. M. Zapka, D. Heyers, C. Hein, S. Engels, N.L. Schneider, J. Hans, S. Weiler, D. Dreyer, D. Kishkinev, J. Wild, H. Mouritsen, Nature **461**, 1274 (2009)
30. W. Wiltschko, U. Munro, H. Ford, R. Wiltschko, Cell. Mol. Life Sci. **49**, 167 (1993)
31. B. Brocklehurst, Chem. Soc. Rev. **31**, 301 (2002)
32. C. Timmel, K. Henbest, Philos. Trans. Royal Soc. A **362**, 2573 (2004)
33. C. Rodgers, Pure Appl. Chem. **81**, 19 (2009)
34. U. Steiner, T. Ulrich, Chem. Rev. **89**, 51 (1989)
35. J. Lau, N. Wagner-Rundell, C. Rodgers, N. Green, P. Hore, J. Royal Soc. Interf. **7**, S257 (2010)
36. E. Hill, T. Ritz, J. Royal Soc. Interf. **7**, S265 (2010)
37. I. Solov'yov, H. Mouritsen, K. Schulten, Biophys. J. **99**, 40 (2010)
38. C.R. Timmel, F. Cintolesi, B. Brocklehurst, P.J. Hore, Chem. Phys. Lett. **334**, 387 (2001)
39. L. Muus, P. Atkins, K. McLauchlan, J. Pedersen, *Chemically Induced Magnetic Polarization* (D. Reidel, Dordrecht, 1977)
40. K. Salikhov, *Magnetic Isotope Effect in Radical Reactions* (Springer-Verlag, Vienna, 1996)
41. V. Grigoryants, B. Tadjikov, O. Usov, Y. Molin, Chem. Phys. Lett. **246**, 392 (1995)
42. V.A. Bagryansky, O.M. Usov, V.I. Borovkov, T.V. Kobzeva, Y.N. Molin, Chem. Phys. Lett. **255**, 237 (2000)
43. V. Bagryansky, V. Borovkov, Y. Molin, Russ. Chem. Rev. **76**, 493 (2007)
44. R. Bittl, G. Kothe, Chem. Phys. Lett. **177**, 547 (1991)
45. G. Kothe, S. Weber, R. Bittl, E. Ohmes, M. Thurnauer, J. Norris, Chem. Phys. Lett. **186**, 474 (1991)
46. Z. Wang, J. Tang, J. Norris, J. Magn. Res. **97**, 322 (1992)
47. S. Dzuba, M. Bosch, A. Hoff, Chem. Phys. Lett. **248**, 427 (1996)
48. C. Lin, T. Todo, Genome Biol. **6**, 220 (2005)
49. J.P. Bouly, E. Schleicher, M. Dionisio-Sese, F. Vandenbussche, D. Van Der Straeten, N. Bakrim, S. Meier, A. Batschauer, P. Galland, R. Bittl, M. Ahmad, J. Biol. Chem. **282**, 9383 (2007)
50. K. O'Day, PLoS Biol. **6**, 1359 (2008)

51. Y.T. Kao, C. Tan, S.H. Song, N. Öztürk, J. Li, L. Wang, A. Sancar, D. Zhong, J. Am. Chem. Soc. **130**, 7695 (2008)
52. B. Giovani, M. Byrdin, M. Ahmad, K. Brettel, Nat. Struct. Biol. **10**, 489 (2003)
53. N. Hoang, E. Schleicher, S. Kacprzak, J.P. Bouly, M. Picot, W. Wu, A. Berndt, E. Wolf, R. Bittl, M. Ahmad, PLoS Biol. **6**, 1559 (2008)
54. T. Biskup, E. Schleicher, A. Okafuji, G. Link, K. Hitomi, E. Getzoff, S. Weber, Angewandte Chemie Int. Ed. **48**, 404 (2009)
55. M. Ahmad, P. Galland, T. Ritz, R. Wiltschko, W. Wiltschko, Planta **225**, 615 (2007)
56. S.R. Harris, K.B. Henbest, K. Maeda, J.R. Pannell, C.R. Timmel, P.J. Hore, H. Okamoto, J. Royal Soc. Interf. **6**, 1193 (2009)
57. T. Yoshii, M. Ahmad, C. Helfrich-Foerster, PLoS Biol. **7**, 813 (2009)
58. R. Gegear, A. Casselman, S. Waddell, S. Reppert, Nature **454**, 1014 (2008)
59. R. Gegear, L. Foley, A. Casselman, S. Reppert, Nature **463**, 804 (2010)
60. W. Wiltschko, R. Wiltschko, J. Comp. Physiol. A **177**, 363 (1995)
61. R. Wiltschko, U. Munro, H. Ford, K. Stapput, W. Wiltschko, J. Exp. Biol. **211**, 3344 (2008)
62. C. Nießner, S. Denzau, J.C. Gross, L. Peichl, H.J. Bischof, G. Fleissner, W. Wiltschko, R. Wiltschko, PLoS ONE **6**, e20091 (2011)
63. R. Freire, U. Munro, L. Rogers, S. Sagasser, R. Wiltschko, W. Wiltschko, Animal Cogn. **11**, 547 (2008)
64. C. Moser, P. Dutton, Biochimica et Biophysica Acta **1101**, 171 (1992)
65. C. Brautigam, B. Smith, Z. Ma, M. Palnitkar, D. Tomchick, M. Machius, J. Deisenhofer, Proc. Natl. Acad. Sci. U S A **101**, 12142 (2004)
66. G. Fleissner, E. Holtkamp-Rötzler, M. Hanzlik, M. Winklhofer, G. Fleissner, N. Petersen, W. Wiltschko, J. Comp. Neurol. **458**, 350 (2003)
67. R. Beason, P. Semm, J. Exp. Biol. **199**, 1241 (1996)
68. W. Wiltschko, U. Munro, H. Ford, R. Wiltschko, Proc. Royal Soc. Lond. B. (Biological Sciences) **273**, 2815 (2006)
69. H. Mouritsen, G. Feenders, M. Liedvogel, K. Wada, E. Jarvis, Proc. Natl. Acad. Sci. U S A **102**, 8339 (2005)
70. R. Wiltschko, K. Stapput, T. Ritz, P. Thalau, W. Wiltschko, HFSP J. **1**, 41 (2007)
71. T. Ritz, P. Thalau, J. Phillips, R. Wiltschko, W. Wiltschko, Nature **429**, 177 (2004)
72. P. Thalau, T. Ritz, K. Stapput, R. Wiltschko, W. Wiltschko, Naturwissenschaften **92**, 86 (2005)
73. N. Keary, T. Ruploh, J. Voss, P. Thalau, R. Wiltschko, W. Wiltschko, H.J. Bischof, Frontiers Zool. **6**, 25 (2009)
74. M. Vacha, T. Puzová, M. Kvícalová, J. Exp. Biol. **212**, 3473 (2009)
75. H. Hogben, O. Efimova, N. Wagner-Rundell, C. Timmel, P. Hore, Chem. Phys. Lett. **490**, 118 (2009)
76. P. Thalau, T. Ritz, H. Burda, R. Wegner, R. Wiltschko, J. Royal Soc. Interf. **3**, 583 (2006)

# Statistical Mechanical Theory of Protein Folding in Water Environment

**Alexander V. Yakubovich, Andrey V. Solov'yov and Walter Greiner**

**Abstract** We present a statistical mechanics formalism for the theoretical description of the process of protein folding↔unfolding transition in water environment. The formalism is based on the construction of the partition function of a protein obeying two-stage-like folding kinetics. Using the statistical mechanics model of solvation of hydrophobic hydrocarbons we obtain the partition function of infinitely diluted solution of proteins in water environment. The calculated dependencies of the protein heat capacities upon temperature are compared with the corresponding results of experimental measurements for staphylococcal nuclease.

## 1 Introduction

Proteins are biological polymers consisting of elementary structural units, amino acids. Being synthesized at ribosome proteins are exposed to the cell interior where they fold into their unique three dimensional structure. The process of forming the protein's three dimensional structure is called the process of protein folding. The correct folding of protein is of crucial importance for the protein's proper functioning. Despite numerous works devoted to investigation of protein folding this process is still not entirely understood. The current state-of-the-art in experimental and theoretical studies of the protein folding process are described in recent reviews and references therein [1–5].

A. V. Yakubovich (✉) · A. V. Solov'yov · W. Greiner
Frankfurt Institute for Advanced Studies, Goethe University, 60438 Ruth-Moufang Str. 1, Frankfurt am Main, Germany
e-mail: yakubovich@fias.uni-frankfurt.de

A. V. Yakubovich
Author A.Y. on leave from A.F. Ioffe Physical Technical Institute, Polytechnicheskaya 26, 194021 Saint-Petersburg, Russia

W. Greiner (ed.), *Exciting Interdisciplinary Physics*,
FIAS Interdisciplinary Science Series, DOI: 10.1007/978-3-319-00047-3_39,

In this chapter we develop a novel theoretical method for the description of the protein folding process which is based on the statistical mechanics principles. Considering the process of protein folding as a first order type phase transition in a finite system, we present a statistical mechanics model for treating the folding↔unfolding phase transition in single-domain proteins. The suggested method is based on the theory developed for the helix↔coil transition in polypeptides discussed in [6–14]. A way to construct a parameter-free partition function for a system experiencing $\alpha$-helix↔random coil phase transition *in vacuo* was studied in [6]. In [8] we have calculated potential energy surfaces (PES) of polyalanines of different lengths with respect to their twisting degrees of freedom. This was done within the framework of classical molecular mechanics. The calculated PES were then used to construct a parameter–free partition function of a polypeptide and to derive various thermodynamical characteristics of alanine polypeptides as a function of temperature and polypeptide length.

In this chapterr we construct the partition function of a protein *in vacuo*, which is the further generalization of the formalism developed in [9], accounting for folded, unfolded and prefolded states of the protein. This way of the construction of the partition function is consistent with nucleation-condensation scenario of protein folding, which is a very common scenario for globular proteins [15] and implies that at the early stage of protein folding the native-like hydrophobic nucleus of protein is formed, while at the later stages of the protein folding process all the rest of amino acids also attain the native-like conformation.

For the correct description of the protein folding in water environment it is of primary importance to consider the interactions between the protein and the solvent molecules. The hydrophobic interactions are known to be the most important driving forces of protein folding [16]. In the present work we present a way how one can construct the partition function of the protein accounting for the interactions with solvent, i.e. accounting for the hydrophobic effect. The most prominent feature of our approach is that it is developed for concrete systems in contrast to various generalized and toy-models of protein folding process.

We treat the hydrophobic interactions in the system using the statistical mechanics formalism developed in [17] for the description of the thermodynamical properties of the solvation process of aliphatic and aromatic hydrocarbons in water.

However, accounting solely for hydrophobic interactions is not sufficient for the proper description of the energetics of all conformational states of the protein and one has to take electrostatic interactions into account. In the present work the electrostatic interactions are treated within a similar framework as described in [18].

We have applied the developed statistical mechanics model of protein folding for a globular protein, namely staphylococcal nuclease. This protein has simple two-stage-like folding kinetics and demonstrate two folding↔unfolding transitions, refereed as heat and cold denaturation [19, 20]. The comparison of the results of the theoretical model with that of the experimental measurements shows the applicability of the suggested formalism for an accurate description of various thermodynamical characteristics in the system, e.g. heat denaturation, cold denaturation, increase of the reminiscent heat capacity of the unfolded protein, etc.

Our chapter is organized as follows. In Sect. 2.1 we present the formalism for the construction of the partition function of the protein in water environment and justify the assumptions made on the system's properties. In Sect. 3 we discuss the results obtained with our model for the description of folding↔unfolding transition in staphylococcal nuclease. In Sect. 4 we summarize the chapter and suggest several ways for a further development of the theoretical formalism.

# 2 Theoretical Methods

## 2.1 Partition Function of a Protein

To study thermodynamic properties of the system one needs to investigate its potential energy surface with respect to all the degrees of freedom. For the description of macromolecular systems, such as proteins, efficient model approaches are necessary.

The most relevant degrees of freedom in the protein folding process are the twisting degrees of freedom along its backbone chain [6, 7, 9–11, 13, 14, 21, 22]. These degrees of freedom are defined for each amino acid of the protein except for the boundary ones and are described by two dihedral angles $\varphi_i$ and $\psi_i$ (for definition of $\varphi_i$ and $\psi_i$ see e.g. [6, 7, 9–11, 13, 14]).

The degrees of freedom of a protein can be classified as stiff and soft ones. We call the degrees of freedom corresponding to the variation of bond lengths, angles and improper dihedral angles as stiff, while degrees of freedom corresponding to the angles $\varphi_i$ and $\psi_i$ are soft degrees of freedom [6]. The stiff degrees of freedom can be treated within the harmonic approximation, because the energies needed for a noticeable structural rearrangement with respect to these degrees of freedom are about several eV, which is significantly larger than the characteristic thermal energy of the system (kT), being at room temperature equal to 0.026 eV [12–14, 23–25].

A Hamiltonian of a protein is constructed as a sum of the potential, kinetic and vibrational energy terms. Assuming the harmonic approximation for the stiff degrees of freedom it is possible to derive the following expression for the partition function of a protein *in vacuo* being in a particular conformational state $j$ [6]:

$$Z_j = A_j (kT)^{3N-3-\frac{l_s}{2}} \int_{\varphi \in \Gamma_j} \dots \int_{\psi \in \Gamma_j} e^{-\varepsilon_j(\{\varphi,\psi\})/kT} \mathrm{d}\varphi_1 \dots \mathrm{d}\varphi_n \mathrm{d}\psi_1 \dots \mathrm{d}\psi_n, \quad (1)$$

where $T$ is the temperature and $k$ is the Boltzmann constant. $N$ in Eq. (1) is the total number of atoms in the protein, $l_s$ is the number of soft degrees of freedom. $A_j$ in Eq. (1) is defined as follows:

$$A_j = \frac{V_j \cdot M^{3/2} \cdot \sqrt{I_j^{(1)} I_j^{(2)} I_j^{(3)}} \prod_{i=1}^{l_s} \sqrt{\mu_i^s}}{(2\pi)^{\frac{l_s}{2}} \pi \hbar^{3N} \prod_{i=1}^{3N-6-l_s} \omega_i}. \quad (2)$$

$A_j$ is a factor which depends on the mass of the protein $M$, its three main momenta of inertia $I$, specific volume $V$, the frequencies of the stiff normal vibrational modes $\omega_i$ and on the generalized masses $\mu^s$ corresponding to the soft degrees of freedom [6]. $\varepsilon$ in Eq. (1) describes the potential energy of the system corresponding to the variation of soft degrees of freedom.The contribution of the kinetic energy associated with soft degrees of freedom is also accounted for in Eq. 1. Integration in Eq. (1) is performed over a certain part of a phase space of the system (a subspace $\Gamma_j$) corresponding to the soft degrees of freedom $\varphi$ and $\psi$. The form of the partition function in Eq. (1) allows one to avoid the multidimensional integration over the whole coordinate space and to reduce the integration only to the relevant parts of the phase space. $\varepsilon_j$ in Eq. (1) denotes the potential energy surface of the protein as a function of twisting degrees of freedom in the vicinity of protein's conformational state $j$. Note, that in general the proper choice of all the relevant conformations of a protein and the corresponding set of $\Gamma_j$ is not a trivial task.

One can expect that the factors $A_j$ in Eq. (1) depend on the chosen conformation of the protein. However, due to the fact that the values of specific volumes, momenta of inertia and frequencies of normal vibration modes of the system in different conformations are expected to be close [9, 26], the values of $A_j$ in all conformations become nearly equal, at least in the zero order harmonic approximation, i.e. $A_j \equiv A$. Another simplification of the integration in Eq. (1) comes from the statistical independence of amino acids. We assume that within each conformational state $j$ all amino acids can be treated statistically independently, i.e. the particular conformational state of $i$-th amino acid characterized by angles $\varphi_i \in \Gamma_j$ and $\psi_i \in \Gamma_j$ does not influence the potential energy surface of all other amino acids, and vice versa. This assumption is well applicable for rigid conformational states of the protein such as native state. Indeed, for the ensemble of bound harmonic oscillators the partition function of the system depends only on the oscillators masses and spring strength, but not on the particular way of bindings in the system. Therefore the partition function of the protein in the rigid native state (i.e. all atoms vibrate harmonically) will depend only on the shape of the potential energy surfaces of amino acids in the vicinity of their minima, and not on the inter-amino acid interactions. For the native state of a protein all atoms of the molecule move in harmonic potential in the vicinity of their equilibrium positions. However, in unfolded states of the protein the flexibility of the backbone chain leads to significant variations of the distances between atoms, and consequently to a significant variation of interactions between atoms. Accurate accounting (both analytical and computational) for the interactions between distant atoms in the unfolded state of a protein is extremely difficult (see Ref. [27] for analytical treatment of interactions in unfolded states of a protein). In this work we assume that all amino acids in unfolded state of a protein move in the identical mean field created by all the amino acids and leave the corrections to this approximation for further considerations.

With the above mentioned assumptions the partition function of a protein $Z_p$ (without any solvent) reads as:

$$Z_p = A \cdot (kT)^{3N-3-\frac{l_s}{2}} \sum_{j=1}^{\xi} \prod_{i=1}^{a} \int_{-\pi}^{\pi} \int_{-\pi}^{\pi} \exp\left( -\frac{\varepsilon_i^{(j)}(\varphi_i, \psi_i)}{kT} \right) d\varphi d\psi, \quad (3)$$

where the summation over $j$ includes all $\xi$ statistically relevant conformations of the protein, $a$ is the number of amino acids in the protein and $\varepsilon_i^{(j)}$ is the potential energy surface as a function of twisting degrees of freedom $\varphi_i$ and $\psi_i$ of the $i$-th amino acid in the $j$-th conformational state of the protein. The exact construction of $\varepsilon_i^{(j)}(\varphi_i, \psi_i)$ for various conformational states of a particular protein will be discussed below. We consider the angles $\varphi$ and $\psi$ as the only two soft degrees of freedom in each amino acid of the protein, and therefore the total number of soft degrees of freedom of the protein $l_s = 2a$.

Partition function in Eq. (3) can be further simplified if one assumes (i) that each amino acid in the protein can exist only in two conformations: the native state conformation and the random coil conformation; (ii) the potential energy surfaces for all the amino acids are identical. This assumption is applicable for both the native and the random coil state. It is not very accurate for the description of thermodynamical properties of single amino acids, but is reasonable for the treatment of thermodynamical properties of the entire protein. The judgement of the quality of this assumption could be made on the basis of comparison of the results obtained with its use with experimental data. Such comparison is performed in Sect. 3 of this work.

Amino acids in a protein being in its native state vibrate in a steep harmonic potential. Here we assume that the potential energy profile of an amino acid in the native conformation should not be very sensitive to the type of amino acid and thus can be taken as e.g. the potential energy surface for an alanine amino acid in the $\alpha$-helix conformation [8]. This assumption is well justified for proteins with the rigid helix-rich native structure. The staphylococcal nuclease, which we study here has definitely high $\alpha$-helix content. Using the same arguments the potential energy profile for an amino acid in unfolded protein state can be approximated by e.g. the potential of alanine in the unfolded state of alanine polypeptide (see Ref. [8] for discussion and analysis of alanine's potential energy surfaces).

Indeed, for an unfolded state of a protein it is reasonable to expect that once neglecting the long-range interactions all the differences in the potential energy surfaces of various amino acids arise from the steric overlap of the amino acids's side chains. This is clearly seen on alanine's potential energy surface at values of $\varphi > 0°$ presented in Ref. [8]). But the part of the potential energy surface at $\varphi > 0°$ gives a minor contribution to the entropy of amino acid at room temperature. This fact allows one to neglect all the differences in potential energy surfaces for different amino acids in an unfolded protein, at least in the zero order approximation.

For the description of the folding $\leftrightarrow$ unfolding transition in small globular proteins obeying simple two-state-like folding kinetics we assume that the protein can exist in one of three states: completely folded state, completely unfolded state and partially folded state where some amino acids from the flexible regions with no prominent secondary structure are in the unfolded state, while other amino acids are in the folded conformation. With this assumption the partition function of the protein reads as:

$$Z_p = Z_0 + \sum_{i=a-\kappa}^{a} \frac{\kappa!}{(i-(a-\kappa))!(a-i)!} Z_i, \tag{4}$$

where $Z_i$ is defined in Eq. (1), $Z_0$ is the partition function of the protein in completely unfolded state, $a$ is the total number of amino acids in a protein and $\kappa$ is the number of amino acids in flexible regions. The factorial term in Eq. (4) accounts for the states in which various amino acids from flexible regions independently attain the native conformation. The summation in Eq. (4) is performed over all partially folded states of the protein, where $a-\kappa$ is the minimal possible number of folded amino acids. The factorial term describes the number of ways to select $i-(a-\kappa)$ amino acids from the flexible region of the protein consisting of $\kappa$ amino acids attaining native-like conformation.

Omitting the contribution of protein's center of mass motion, the partition function of the system can be written as:

$$Z_p = \tilde{Z}_p \cdot A(kT)^{3N-3-a}, \tag{5}$$

where

$$\tilde{Z}_p = Z_u^a + \sum_{i=a-\kappa}^{a} \frac{\kappa! Z_b^i Z_u^{a-i} \exp(i \cdot E_0/kT)}{(i-(a-\kappa))!(a-i)!} \tag{6}$$

$$Z_b = \int_{-\pi}^{\pi} \int_{-\pi}^{\pi} \exp\left(-\frac{\varepsilon_b(\varphi, \psi)}{kT}\right) d\varphi d\psi \tag{7}$$

$$Z_u = \int_{-\pi}^{\pi} \int_{-\pi}^{\pi} \exp\left(-\frac{\varepsilon_u(\varphi, \psi)}{kT}\right) d\varphi d\psi. \tag{8}$$

Here $\varepsilon_b(\varphi, \psi)$ (b stands for *bound*) is the potential energy surface of the amino acid in the native conformation and $\varepsilon_u(\varphi, \psi)$ (u - *unbound*) is the potential energy surface of amino acid in the random coil conformation. $E_0$ is the energy difference between the two conformational states of the amino acid. The potential energy profile of an amino acid is calculated as a function of its twisting degrees of freedom $\varphi$ and $\psi$. Let us denote by $\varepsilon_b^0$ and $\varepsilon_u^0$ the global minima on the potential energy surfaces of the amino acid in folded and in unfolded conformations respectively. The absolute value of the energy of the amino acid can be written as $\varepsilon^0 + \varepsilon(\varphi, \psi)$. By $E_0$ in Eq. (8) is denoted the energy difference between global minima of the potential energy surfaces of the amino acid in folded and in unfolded conformations, i.e. $E_0 = \varepsilon_u^0 - \varepsilon_b^0$. We count the absolute value of the potential energy from the energy of the global minima on the potential energy surface of the amino acid in unfolded conformation, $\varepsilon_u^0 = 0$ and each potential energy function $\varepsilon_b(\varphi, \psi)$ and $\varepsilon_u(\varphi, \psi)$ is counted form a reference energy value equal to $\varepsilon_b^0$ and $\varepsilon_u^0$ respectively.

The potential energy surfaces for amino acids as functions of angles $\varphi$ and $\psi$ were calculated and thoroughly analyzed in [8].

In nature proteins perform their function in the aqueous environment. The correct theoretical description of the folding↔unfolding transition in water environment should account for solvent effects.

## 2.2 Partition function of a protein in water environment

In this section we evaluate $E_0$ and construct the partition function for the protein in water environment.

The partition function of the infinitely diluted solution of proteins $Z$ can be constructed as follows:

$$Z = \sum_{j=1}^{\xi} Z_p^{(j)} \cdot Z_{\text{water}}^{(j)}, \tag{9}$$

where $Z_{\text{water}}^{(j)}$ is the partition function of all water molecules in the $j$-th conformational state of a protein and $\tilde{Z}_p^{(j)}$ is the partition function of the protein in its $j$-th conformational state, in which we further omit the factor describing the contribution of stiff degrees of freedom in the system. This is done in order to simplify the expressions, because stiff degrees of freedom provide a constant contribution to the heat capacity of the system since the heat capacity of the ensemble of harmonic oscillators is constant, i.e. $\tilde{Z}_p \equiv Z_p$.

There are two types of water molecules in the system: (i) molecules in pure water and (ii) molecules interacting with the protein. We assume that only the water molecules being in the vicinity of the protein's surface are involved in the folding↔unfolding transition, because they are affected by the variation of the hydrophobic surface of a protein. This surface is equal to the protein's solvent accessible surface area (SASA) of the hydrophobic amino acids. The number of interacting molecules is proportional to SASA and include only the molecules from the first protein's solvation shell. This area depends on the conformation of the protein. The main contribution to the energy of the system caused by the variation of the protein's SASA associated with the side-chains of amino acids because the contribution to the free energy assosiated with solvation of protein's backbone is small [28]. Thus, in this work we pay the main attention to the accounting for the SASA change arising due to the solvation of side chains.

We treat all water molecules as statistically independent, i.e. the energy spectra of the states of a given molecule and its vibrational frequencies do not depend on a particular state of all other water molecules. Thus, the partition function of the whole system $Z$ can be factorized and reads as:

$$Z = \sum_{j=1}^{\xi} Z_p \cdot Z_s^{Y_c(j)} Z_w^{N_0 - Y_c(j)}, \tag{10}$$

where $\xi$ is the total number of states of a protein, $Z_s$ is the partition function of a water molecule affected by the interaction with the protein and $Z_w$ is the partition function of a water molecule in pure water. $Y_c$ is the number of affected water molecules in the $j$-th conformational state of a protein. $N_0$ is the total number of water molecules in the system. Since we are focused on the *change* of the thermodynamic properties of the system in the course of protein unfolding, in the partition function we do not account for water molecules that do not interact with the protein in any of its conformational states, i.e. $N_0 = max\{Y_c(j)\}$.

To construct the partition function of water we follow the formalism developed in [17] and refer only to the most essential details of that work. The partition function of a water molecule in pure water reads as:

$$Z_s = \sum_{l=0}^{4} \xi_l f_l \exp(-E_l/kT), \tag{11}$$

where the summation is performed over 5 possible states of a water molecule (the states in which water molecule has 4, 3, 2, 1 and 0 hydrogen bonds with the neighboring molecules). $E_l$ are the energies of these states and $\xi_l$ are the combinatorial factors being equal to $1, 4, 6, 4, 1$ for $l = 0, 1, 2, 3, 4$, respectively. They describe the number of choices to form a given number of hydrogen bonds. $f_l$ in Eq. (11) describe the contribution to the partition function arising to to the translation and libration oscillations of the molecule. In the harmonic approximation $f_l$ are equal to:

$$f_l = \left[1 - \exp(-h\nu_l^{(T)}/kT)\right]^{-3} \left[1 - \exp(-h\nu_l^{(L)}/kT)\right]^{-3}, \tag{12}$$

where $\nu_l^{(T)}$ and $\nu_l^{(L)}$ are translation and libration motions frequencies of a water molecule in its $l$-th state, respectively. These frequencies are calculated in Ref. [17] and are presented in Table (1). The contribution of the internal vibrations of the water molecule is not accounted for in Eq. (11) since frequencies of these vibrations can be considered as equal in all energetic states of the molecule.

The partition function of a water molecule from the protein's first solvation shell reads as:

**Table 1** Parameters of the partition function of water

| Number of hydrogen bonds | u | 1 | 2 | 3 | 4 |
|---|---|---|---|---|---|
| Energy level, $E_i$ (cal/mol) | 6670 | 4970 | 3870 | 2030 | 0 |
| Translational frequencies, $\nu_i^{(T)}$, cm$^{-1}$ | 26 | 86 | 61 | 57 | 210 |
| Librational frequencies, $\nu_i^{(L)}$, cm$^{-1}$ | 197 | 374 | 500 | 750 | 750 |

$$Z_s = \sum_{l=0}^{4} \xi_l f_l \exp(-E_l^{\text{shell}}/kT), \tag{13}$$

where $f_l$ are defined as in Eq. (12) and $E_l^{\text{shell}}$ denotes the energy levels of a water molecule interacting with aliphatic hydrocarbons of protein's amino acids. For simplicity we treat all side-chains of the hydrophobic core of a protein as being consisted of aliphatic hydrocarbons since most of the protein's hydrophobic amino acids consists of aliphatic-like hydrocarbons.

In our theoretical model we also account for the electrostatic interaction of protein's charged groups with the water.

The presence of electrostatic field around the protein leads to the reorientation of $H_2O$ molecules in the vicinity of the charged groups due to the interaction of the $H_2O$ molecules dipole moments with the electrostatic field. The corresponding factor in the partition function of $H_2O$ molecules reads as:

$$Z_{\text{elec}} = \left( \frac{1}{4\pi} \int \exp\left( -\frac{E \cdot d \cos\theta}{kT} \right) \sin\theta \mathrm{d}\theta \mathrm{d}\varphi \right)^{\alpha}, \tag{14}$$

where $E$ is the strength of the electrostatic field, $d$ is the absolute value of the $H_2O$ molecule dipole moment and $\alpha$ is the effective number of such molecules that are affected by electrostatic interaction. Note that the effects of electrostatic interaction turn out to be more pronounced in the folded state of the protein. This happens because in the unfolded state of a protein opposite charges of amino acid's side chains are in average closer in space due to the flexibility of the backbone chain, while in the folded state the positions of the charges are fixed by the rigid structure of a protein.

Taking the integral in Eq. (14), the correction to the partition function of a water molecule in pure water reads as:

$$Z_w' = \left( \sum_{l=0}^{4} \left[ \xi_l f_l \exp(-E_l/kT) \right] \right) \left( \frac{kT \sinh\left[\frac{Ed}{kT}\right]}{Ed} \right)^{\alpha}. \tag{15}$$

This equation shows how the electrostatic field enters the partition function. In general, $E$ depends on the position in space with respect to the protein. However, here we neglect this dependence and instead we treat the parameter $E$ as an average, characteristic electrostatic field created by the protein.

Having constructed the partition function of the system, we can evaluate with its use the thermodynamic characteristics of the system, such as entropy, free energy, heat capacity, etc. In this work we focus on the analysis of the dependence of protein's heat capacity on temperature and compare the predictions of our model with available experimental data.

## 3 Results and Discussion

In this section we calculate the dependencies of the heat capacity on temperature for a globular protein staphylococcal nuclease and compare the results obtained with experimental data from [19, 20].

Staphylococcal nuclease is relatively small globular protein consisting of 149 amino acids. It is a relatively nonspecific enzyme that digests single-stranded and double-stranded nucleic acids, but is more active on single-stranded substrates [31]. The structure native structure of the protein in shown in the centre in Fig. 1. Under certain experimental conditions (salt concentration and pH) the staphylococcal nuclease experience two folding↔unfolding transitions, which induce two peaks in the dependency of heat capacity on temperature (see Fig. 1). The peaks at lower temperature are due to the cold denaturation of the proteins. The peaks at higher temperatures arise due to the ordinary folding↔unfolding transition. The availability of experimental data for the heat capacity profiles of the mentioned protein, the presence of the cold denaturation and simple two-stage-like folding kinetics are the reasons for selecting this particular protein as case study for the verification of the developed theoretical model.

To calculate the SASA of staphylococcal nuclease in the folded state the 3D structure of the protein was obtained from the Protein Data Bank [32] (PDB ID 1EYD). Using CHARMM27 [25] forcefield and NAMD program [33] we performed the structural optimization of the protein and calculated SASA with the solvent probe radius 1.63 Å.

The value of SASA of the side-chains in the folded protein conformation is equal to 6858 Å$^2$. In order to calculate SASA for an unfolded protein state all the angles $\varphi$ and $\psi$ have been taken to be equal to 180°, i.e. as in a fully stretched conformation. Then, the optimization of the structure with the fixed angles $\varphi$ and $\psi$ was performed. The optimized geometry of the stretched molecule has a minor dependence on the value of dielectric susceptibility of the solvent, therefore the value of dielectric susceptibility was chosen to be equal to 20, in order to mimic the screening of charges by solvent. SASA of the side-chains in the stretched conformation of the protein tuned out to be equal to 15813 Å$^2$. The volume of one mole of water is 18 $cm^3$ therefore the volume of one molecule is ~30 Å$^3$. To estimate the width of the solvation shell we consider the water as being in dense honeycomb hexagonal packing on the surface of the solute with a probe radius of the water molecule being equal to 1.4 Å [34]. The width of the first solvation shell of the solute equals to ∼3.26Å. Knowing the volume of a single water molecule, and the width of the solvation shell, one can obtain the total number of water molecules those interaction with the protein changes during the protein unfolding as follows:

$$N_{H_2O} = \frac{S_{unf} - S_{fol}}{V_0/h_w}, \tag{16}$$

where $\frac{15813-6858}{30/(2\cdot1.63)} \approx 973$. Therefore, there are 973 "pure" water molecules in the system with the folded protein that interact with the hydrophobic surface of the inner amino acids when the protein gets unfolded.

To account for the effects caused by the electrostatic interaction of water molecules with the charged groups of the protein it is necessary to evaluate the strength of the electrostatic field $E$ in Eq. (15). The strength of the field can be estimated as $E \cdot d = kT/2$, where $d$ is the dipole moment of a water molecule, $k$ is Bolzmann constant and T = 300 K is the room temperature. According to this estimate the energy of electrostatic interaction of water molecules is equal to half of the thermal energy per degree of freedom of a molecule. This value of the field corresponds to the distance ∼10 Åwith respect to a singly-charged atom. This distance was calculated using the distance-dependent dielectric susceptibility derived in Ref. [35] as follows:

$$\varepsilon(r) = (\varepsilon_\infty - \varepsilon_0)\left(\coth(\alpha(r - r_0)) - \frac{1}{\alpha(r - r_0)}\right) + \varepsilon_0, \tag{17}$$

where $\varepsilon(r)$ is the distance-dependent dielectric susceptibility, $r$ is the distance from the charge, $\varepsilon_\infty$ and $\varepsilon_0$ are dielectric susceptibilities of water (78) and protein (2) respectively, $\alpha = 0.45$ Å$^{-1}$ and $r_0 = 2.9$ Å. These values of $\alpha$ and $r_0$ were proposed in [35] for the treatment of biomolecules.

The number of water molecules that interact with the electrostatic field can be obtained as a number of the molecules being in the sphere around an amino acid's charge site. For simplicity we do no account for the spatial distribution of the charge in the charged amino acid and treat it point-like. The dipole moments of $H_2O$ molecules within the first and the second solvation shells of a charged atom are strongly polarized by the atom's electrostatic field [36]. Therefore, the orientation of the dipole moments of these $H_2O$ molecules does not depend on the conformational state of the protein. The number of water molecules, $N_w$, which *change* their orientation during the of protein folding process can be calculated as follows:

$$N_w = \frac{1}{\rho}\left(\frac{4}{3}\pi R^3 - \frac{4}{3}\pi(r_{\text{ion}} + 2\cdot 2r_w)^3\right), \tag{18}$$

where $r_{\text{ion}}$=1 Åis the radius of the charged atom, $r_w$=1.63 Åis the radius of a water molecule, $\rho$=30 nm$^{-3}$ is the density of water molecules and $R$ is the radius of the sphere around the charged atom. This radius can be calculated as follows. The change of the free energy of a water molecule $\Delta F_E$ associated with the electrostatic interaction of $H_2O$ molecules can be calculated from Eq. (15):

$$\Delta F_E = -kT \ln\left(\frac{kT \sinh\left[\frac{E(r)d}{kT}\right]}{E(r)d}\right), \tag{19}$$

where the strength of electrostatic field $E(r)$ depends on the distance $r$ from the charged atom. The averaged over the sphere $\Delta F_E$ should be equal to the $\Delta F_E$ calculated for water molecules at the distance of ~10 Åform a singly-charged atom. Thus,

$$\frac{3\int_{r_{\min}}^{R} \Delta F_E(r) r^2 dr}{R^3 - r_{\min}^3} = -kT \ln\left(\frac{kT \sinh\left(\frac{E(r_0)d}{kT}\right)}{E(r_0)d}\right), \tag{20}$$

where $r_{\min} = r_{\text{ion}} + 2 \cdot 2r_w$ and $r_0$ =10 Å. $R$ can be calculated from Eq. (20) using series expansion $\ln\left(\frac{\sinh(x)}{x}\right) \approx x^2/6$. The largest real root of Eq. (20) equals to 20.5 Å. Substituting the radius of sphere $R = 20.5$ Åto Eq. (18) one obtains the number of water molecules per charged residue of the protein $N_w \approx 1200$. Staphylococcal nuclease has 8 charged residues at physiological conditions [37], thus there are 9600 $H_2O$ molecules in the system that interact with the electrostatic field of the folded protein. The value of $\alpha$ in Eq. (15) is calculated as a ratio of the number of electrostatically interacting $H_2O$ molecules to the number of molecules that interact with the hydrophobic surface of the protein: $\alpha = \frac{9600}{973} \approx 10$.

Note that number of molecules interacting with the electrostatic field $N_w$ and the strength of the electrostatic field $E$ should be considered as the *effective* parameters of our model. In this work we do not perform accurate accounting for the spatial dependence of the electrostatic field. Instead, we introduce the parameters $\alpha$ and $E$ that can be interpreted as effective values of the number of $H_2O$ molecules and the strength of the electrostatic field correspondingly. Let us stress that the number of water molecules $\alpha$ and the strength of the field $E$ are not independent parameters of our model because by choosing the higher value of $E$ and smaller $\alpha$ or vice versa one can derive the same heat capacity profile. Therefore, below we focus on the investigation of the dependence of the protein heat capacity on E at the fixed value of $\alpha$ equal to 10.

Another important parameter of the model is the energy difference between the two states of the protein normalized per one amino acid, $E_0$ introduced in Eq. (8). This parameter describes both the energy loss due to the separation of the hydrophobic groups of the protein which attract in the native state of the protein via Van-der-Waals interaction and the energy gain due to the formation of Van-der-Waals interactions of hydrophobic groups of the protein with $H_2O$ molecules in the protein's unfolded state. Also, the difference of the electrostatic energy of the system in the folded and unfolded states is accounted for in $E_0$. The difference of the electrostatic energy

**Table 2** Values of $E_0$ for staphylococcal nuclease at different values of pH of the solvent

| pH value | 7.0 | 5.0 | 4.5 | 3.88 | 3.23 |
|---|---|---|---|---|---|
| $E_0$ (kcal/mol) | −1.0635 | −1.07 | −1.077 | −1.093 | −1.2 |

may depend on various characteristics of the system, such as concentration of ions in the solvent and its pH, on the exact location of the charged sites in the native conformation of the protein and on the probability distribution of distances between charged amino acids in the unfolded state. Thus, exact calculation of $E_0$ is a rather difficult and separate task which we do not intend to solve in this work. Instead, in the current study the energy difference between the two phases of the protein is considered as a parameter of the model. We treat $E_0$ as being dependent on external properties of the system, in particular on the pH value of the solution.

Another characteristic of the protein folding↔unfolding transition is its cooperativity. In the model it is described by the parameter $\kappa$ in Eq. (4). $\kappa$ describes the number of amino acids in the flexible regions of the protein. The staphylococcal nuclease possesses a prominent two-stage folding kinetics, therefore only 5–10% of amino acids is in the protein's flexible regions. Thus, the value of $\kappa$ for this protein is small can be estimated as being equal to $149 \cdot 7\,\% \approx 10$ amino acids.

The values of $E_0$ for staphylococcal nuclease at different values of pH are presented in Table 2. Note $E_0$ is negative similarly to hydrophobic molecule butane (see Ref. [17] for details).

The dependence of heat capacity on temperature calculated for staphylococcal nuclease at different pH are presented in Fig. 1 by solid lines.

The results of experimental measurements form Ref. [19] are presented by symbols. From Fig. 1 it is seen that staphylococcal nuclease experience two folding ↔ transitions in the range of pH between 3.78 and 7.0. At the pH value 3.23 no peaks in the heat capacity is present. It means that the protein exists in the unfolded state over the whole range of experimentally accessible temperatures.

In our calculation we have adjusted the absolute value of the heat capacity. However, the absolute value of the heat capacity is not a parameter of the model because the experimentally measured absolute value of the heat capacity depends not entirely on the properties of the protein but also on the properties of the solution, ion concentration, etc.

Comparison of the theoretical results with experimental data shows that our theoretical model reproduces experimental behavior better for the solvents with higher pH. The heat capacity peak arising at higher temperatures due to the standard folding↔unfolding transition is reproduced very well for pH values being in the region 4.5–7.0. The deviations at low temperatures can be attributed to the inaccuracy of the statistical mechanics model of water in the vicinity of the freezing point.

The accuracy of the statistical mechanics model for low pH values around 3.88 is also quite reasonable. The deviation of theoretical curves from experimental ones

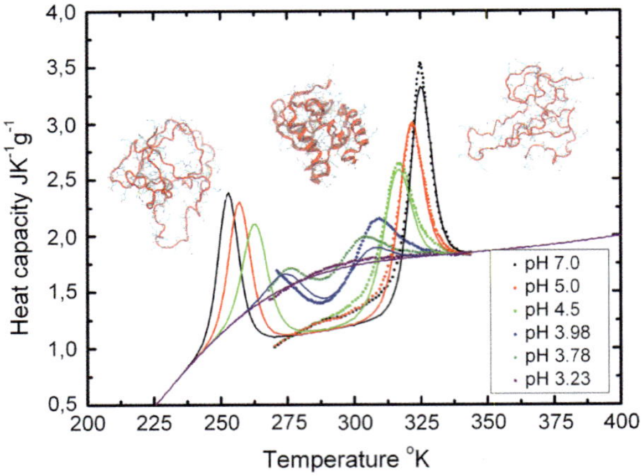


**Fig. 1** Dependencies of the heat capacity on temperature for staphylococcal nuclease (PDB ID 1EYD) at different values of pH. *Solid lines* show results of the developed theoretical model. Symbols present experimental data from Ref. [19]. Structure of the protein in native and unfolded conformations are shown in temperature regions where the corresponding conformation exists.

likely arise due to the alteration of the solvent properties at high concentration of protons or due to the change of partial charge of amino acids at pH values being far from the physiological conditions.

Despite some difference between the predictions of the developed model and the experimental results arising at certain temperatures and values of pH the overall performance of the model can be considered as extremely good for such a complex process as structural folding transition of a large biological molecule.

## 4 Conclusions

We have developed a novel statistical mechanics model for the description of folding↔unfolding processes in globular proteins obeying simple two-stage-like folding kinetics. The model is based on the construction of the partition function of the system as a sum over all statistically significant conformational states of a protein. The partition function of each state is a product of partition function of a protein in a given conformational state, partition function of water molecules in pure water and a partition function of $H_2O$ molecules interacting with the protein.

The introduced model includes a number of parameters responsible for certain physical properties of the system. The parameters were obtained from available experimental data and three of them (energy difference between two phases, cooperativity of the transition and the average strength of the protein's electrostatic field) were considered as being variable depending on a particular protein and pH of the solvent.

We have compared the predictions of the developed model with the results of experimental measurements of the dependence of the heat capacity on temperature for staphylococcal nuclease. The experimental results were obtained at various pH of solvent. The suggested model is capable to reproduce well within a single framework a large number of peculiarities of the heat capacity profile, such as the temperatures of cold and heat denaturations, the corresponding maximum values of the heat capacities, the temperature range of the cold and heat denaturation transitions, the difference between heat capacities of the folded and unfolded states of the protein.

The good agreement of the results of calculations obtained using the developed formalism with the results of experimental measurements demonstrates that it can be used for the analysis of thermodynamical properties of many biomolecular systems. Further development of the model can be focused on its advance and application for the description of the influence of mutations on protein stability, analysis of assembly and stability of protein complexes, protein crystallization process, etc.

**Acknowledgments** A.Y. thanks Stiftung Polytechnische Gesellschaft Frankfurt am Main for financial support.

## References

1. V. Muñoz, Conformational dynamics and ensembles in protein folding. Annu. Rev. Biophys. Biomol. Struct. **36**, 395–412 (2007)
2. K.A. Dill, S.B. Ozkan, M.S. Shell, T.R. Weikl, The protein folding problem. Annu. Rev. Biophys. **37**, 289–316 (2008)
3. J.N. Onuchic, P.G. Wolynes, Theory of protein folding. Curr. Op. Struct. Biol. **14**, 70–75 (2004)
4. E. Shakhnovich, Protein folding thermodynamics and dynamics: Where physics, chemistry, and biology meet. Chem. Rev. **106**, 1559–1588 (2006)
5. N.V. Prabhu, K.A. Sharp, Protein-solvent interactions. Chem. Rev. **106**, 1616–1623 (2006)
6. A. Yakubovich, I. Solov'yov, A. Solov'yov, W. Greiner, Ab initio theory of helix↔coil phase transition. Eur. Phys. J. D. **46**, 215–225, (2007)(arXiv:0704.3079v1 [physics.bio-ph])
7. A. Yakubovich, I. Solov'yov, A. Solov'yov, W. Greiner, Ab initio description of phase transitions in finite bio-nano-systems. Europhys. News **38**, 10 (2007)
8. I. Solov'yov, A. Yakubovich, A. Solov'yov, W. Greiner, $\alpha$-helix↔random coil phase transition: analysis of ab initio theory predictions. Eur. Phys. J. D. **46**, 227–240 (2008) (arXiv:0704.3085v1 [physics.bio-ph])
9. A. Yakubovich, I. Solov'yov, A. Solov'yov, W. Greiner, Phase transition in polypeptides: a step towards the understanding of protein folding. Eur. Phys. J. D **40**, 363–367 (2006)
10. A. Yakubovich, I. Solov'yov, A. Solov'yov, W. Greiner, Conformational changes in glycine tri- and hexapeptide. Eur. Phys. J. D **39**, 23–34 (2006)
11. A. Yakubovich, I. Solov'yov, A. Solov'yov, W. Greiner, Conformations of glycine polypeptides. Khimicheskaya Fizika (Chemical Physics) (in Russian) **25**, 11–23 (2006)
12. I. Solov'yov, A. Yakubovich, A. Solov'yov, W. Greiner, On the fragmentation of biomolecules: fragmentation of alanine dipeptide along the polypeptide chain. J. Exp. Theor. Phys. **103**, 463–471 (2006)
13. I. Solov'yov, A. Yakubovich, A. Solov'yov, W. Greiner, Ab initio study of alanine polypeptide chain twisting. Phys. Rev. E **73**, 021916 (2006)
14. I. Solov'yov, A. Yakubovich, A. Solov'yov, W. Greiner, Potential energy surface for alanine polypeptide chains. J. Exp. Theor. Phys. **102**, 314–326 (2006)

15. B. Noetling, D.A. Agard, How general is the nucleation condensation mechanism? Proteins **73**, 754–764 (2008)
16. S. Kumar, C.-J. Tsai, R. Nussinov, Maximal stabilities of reversible two-state proteins. Biochemistry **41**, 5359–5374 (2002)
17. J. H. Griffith, H. Scheraga, Statistical thrmodynamics of aqueous solutions. i. water structure, solutions with non-polar solutes, and hydrophobic ineractions. J. Mol. Struc. **682**, 97–113 (2004)
18. A. Bakk, J.S. Høye, A. Hansen, Apolar and polar solvation thermodynamics related to the protein unfolding process. Biophys. J. **82**, 713–719 (2002)
19. Y. Griko, P. Privalov, J. Aturtevant, S. Venyaminov, Cold denaturation of staphyloccocal nuclease. Proc. Natl. Acad. Sci. U.S.A **85**, 3343–3347 (1988)
20. P. Privalov, Thermodynamics of protein folding. J. Chem. Thermodyn. **29**, 447–474 (1997)
21. S. He, H.A. Scheraga, Macromolecular conformational dynamics in torsion angle space. J. Chem. Phys. **108**, 271–286 (1998)
22. S. He, H.A. Scheraga, Brownian dynamics simulations of protein folding. J. Chem. Phys. **108**, 287–300 (1998)
23. W. Scott, W. van Gunsteren, The GROMOS Software Package for Biomolecular Simulations, in *Methods and Techniques in Computational Chemistry: METECC-95*, ed. by E. Clementi, G. Corongiu (STEF, Cagliari, 1995), pp. 397–434
24. W. Cornell, P. Cieplak, C. Bayly et al., A second generation force field for the simulation of proteins, nucleic acids, and organic molecules. J. Am. Chem. Soc. **117**, 5179–5197 (1995)
25. A. MacKerell, D. Bashford, R. Bellott et al., All-atom empirical potential for molecular modeling and dynamics studies of proteins. J. Phys. Chem. B **102**, 3586–3616 (1998)
26. S. Krimm, J. Bandekar, Vibrational analysis of peptides, polypeptides, and proteins v. normal vibrations of $\beta$-turns. Biopolymers **19**, 1–29 (1980)
27. M. Cubrovic, O. Obolensky, A. Solov'yov, Semistiff polymer model of unfolded proteins and its application to nmr residual dipolar couplings. Eur. Phys. J. D. **51**, 41–49 (2009)
28. A. Finkelstein, O. Ptitsyn, *Protein Physics: A Course of Lectures* (Elsevier Books, Oxford, 2002)
29. G. Makhatadze, P. Privalov, Contribution to hydration to protein folding thermodynamics. I. The enthalpy of hydration. J. Mol. Biol. **232**, 639–659 (1993)
30. W. Humphrey, A. Dalke, K. Schulten, Vmd - visual molecular dynamics. J. Molec. Graphics **14**, 33–38 (1996)
31. F.A. Cotton, E.E. Hazen Jr, M.J. Legg, Staphylococcal nuclease: Proposed mechanism of action based on structure of enzyme-thymidine 3',5'-bisphosphate-calcium ion complex at 1.5-a resolution. Proc. Natl. Acad. Sci. U.S.A **76**, 2551–2555 (1979)
32. Protein data bank, www.rcsb.org/ (2009)
33. J.C. Phillips, R. Braun, W. Wang et al., Scalable molecular dynamics with NAMD. J. Comp. Chem. **26**, 1781–1802 (2005)
34. R. Wade, M.H. Mazor, J.A. McCammon, F. Quiocho, A molecular dynamics study of thermodynamic and structural aspects of the hydration of cavities in proteins. Biopolymers **31**, 919–931 (1991)
35. J. Mazur, R.L. Jernican, Distance-dependent dielectric constants and their application to double-helical DNA. Biopolymers **31**, 1615–1629 (1991)
36. J.H. Griffith, H. Scheraga, Statistical thermodynamics of aqueous solutions ii.alkali halides at infinite dilution. J. Mol. Struc. **711**, 33–48 (2004)
37. H.-X. Zhou, Residual charge interactions in unfolded staphylococcal nuclease can be explained by the gaussian-chain model. Biophys. J. **83**, 2981–2986 (2002)
38. J.P. Collman, R. Boulatov, C.J. Sunderland, L. Fu, Functional analogues of cytochrome c oxidase, myoglobin, and hemoglobin. Chem. Rev. **104**, 561–588 (2004)
39. D. Shortle, M.S. Ackerman, Persistence of native-like topology in a denatured protein in 8 m urea. Science **293**, 487–489 (2001)

# Photographs

W. Greiner (eds.), *Exciting Interdisciplinary Physics*,
FIAS Interdisciplinary Science Series, DOI: 10.1007/978-3-319-00047-3,

Conference photograph, Makutsi, November 18, 2011

Preparing for the next talk

Physics seems to be fun

Walter Greiner—explaining it all

Kobus Lawrie and Larry McLerran

Zeblon Vilakazi

Richard Newman

Daily commute to the conference location

Watching some rare animals

Horst Stöcker with the Weber family

Cozy and comfortable: A rondavel

Giraffes in symmetric proportion

White rhinos enjoying themselves

Elephants on the road

Perfect patterns

A watchful cheetah

The king

A traditional dance—Farewell to the guests

Sunset behind the Drakensberg range

Printed by Printforce, the Netherlands